# Physiology for Dental Students

BLOOD GLUCOSE
LOCOMOTION/NEURO
CARDIOVASCULAR
RESPIRITORY

# Physiology for Dental Students

**D. B. Ferguson** PhD, BDS, LDSRCS(Eng)
Senior Lecturer in Physiology for Dental Students, Medical School, University of Manchester

**Wright**
London   Boston   Singapore   Sydney   Toronto   Wellington

**Wright**
is an imprint of Butterworth Scientific

All rights reserved. No part of this publication may be reproduced or transmitted in any form or by any means, including photocopying and recording, without the written permission of the copyright holder, application for which should be addressed to the Publishers, or in accordance with the provisions of the Copyright Act 1956 (as amended), or under the terms of any licence permitting limited copying issued by the Copyright Licensing Agency, 7 Ridgmount Street, London WC1E 7AE, England. Such written permission must also be obtained before any part of this publication is stored in a retrieval system of any nature.

Any person who does any unauthorized act in relation to this publication may be liable to criminal prosecution and civil claims for damages.

This book is sold subject to the Standard Conditions of Sale of Net Books and may not be re-sold in the UK below the net price given by the Publishers in their current price list.

First published, 1988

© **Butterworth & Co (Publishers) Ltd, 1988**

---

**British Library Cataloguing in Publication Data**

Ferguson, D. B.
 Physiology for dental students.
 1. Man. Physiology
 I. Title
 612

ISBN 0-7236-0725-7

---

**Library of Congress Cataloging-in-Publication Data**

Ferguson, D. B.
 Physiology for dental students / D.B. Ferguson.
   p.   cm.
 Includes bibliographies and index.
 ISBN 0-7236-0725-7 :
 1. Human physiology.  2. Dental students.   I. Title.
 [DNLM: 1. Physiology.   QT 104 F352p]
 QP 34.5.F46 1988
 612—dc19
 DNLM/DLC
 for Library of Congress

---

Photoset by Butterworths Litho Preparation Department
Printed and bound in England by Hartnolls Ltd, Bodmin, Cornwall

# Preface

Almost all textbooks of physiology available today are written, explicitly or implicitly, for medical students. This is hardly surprising since the largest single group of students of physiology is that studying medicine. It is probably also true that many of the courses in physiology for non-medical students are based upon those provided for medical students and modified mainly by omissions. The traditional 'systems' approach parallels that of the teaching of clinical medicine and the pattern of specialisation within the medical and surgical professions. There are a number of implications from this. 'Basic' physiology, those principles common to several body systems, tends to be taught in relation to a particular body system and common characteristics are not always identified as such. The interactions and inter-relationships between body systems are understressed and may never emerge. The topic of exercise is a good example. Another implication is that illustrations and examples of relevance are almost always drawn from clinical medicine or surgery.

In preparing this textbook I have attempted to move away from the traditional pattern and present a more generalist and more integrated view of physiological mechanisms and physiological systems. Thus the first section looks at body fluids and at tissue cells, examining the bases of operation of these smallest units of body structure. The second section provides an outline of each physiological system, concentrating on the morphological units and the ways in which they act. The third section deals with physiological control and with whole body physiology.

The book is made more specific for dental students in two ways: by the use of examples and illustrations relevant to clinical dentistry and by the emphasis given to a small group of topics of oral significance. These are often taught in courses of Oral Physiology or Oral Biology. They include saliva, speech, mastication, swallowing, taste, and pain in the area served by the trigeminal nerve. The coverage of these topics is excessive in relation to the normal plan of a medical textbook: this distortion is deliberate, and, in my view, appropriate to a textbook for dental students.

The most difficult decisions facing a textbook writer are those over what to include and what to omit. I have tried to approach the problem by asking myself what I see as the role of a textbook. This question is inevitably linked with what one sees as the role of a lecture course. Which should provide the irreducible minimum of knowledge necessary for the student to pass the examinations? In my experience the lecture course has usually been identified as fulfilling this requirement. If this is so, the textbook is used to enable the student to amplify and put into context the notes derived from lectures – perhaps to explain more fully what was not clear, perhaps to bring out points not obvious in the lectures. Further, the textbook is always there: it ought therefore to function as a source of reference material. I do not see the textbook, then, as containing merely the minimum information necessary for examination purposes; equally I do not see it as a form of instruction to be assessed in isolation. In this book, then, I have tried to provide a basis of facts from which physiological ideas can come and I have further tried to link these facts in a meaningful sequence.

Section I deals with a number of physiological principles and with the physiology of body fluids and particular types of cell. In Section II each of the physiological systems is described in turn. The coverage of neuroanatomy varies in different courses: I have therefore dealt with the nervous system with a slight bias towards this. The endocrine system has usually been covered in basics in school biology courses: I have therefore tried to summarise information on the endocrine organs, the endocrines they produce, and the actions of the individual hormones. The third section of the book, control and integration, assumes a knowledge of the various hormones and therefore the end of the chapter on the endocrine system is intended as a section to which reference back may be made by the reader.

In any scientific subject there must be numerical information; in human physiology one has to add the proviso that when a particular value is given it is usually a number which is relatively easy to remember, somewhere towards the middle of a range, but rarely either a mean or a mode. Biological variation is such that one can only hope to give the student some peg on which to hang an assessment of normality or an indication of magnitude of a particular variable. When one goes back to original data one often finds that the investigations themselves were limited in one way or another.

In a new book it seems sensible to use as far as possible the units and terminology which have been internationally standardised and agreed. British schools teach in SI units and use IUPAC terminology in physics and chemistry. I have therefore used both these conventions. The main changes from traditional physiological textbooks in this respect are in the use of newtons for force units, pascals (newtons/m$^2$) for pressure units, hydrogen carbonate instead of bicarbonate, and iron II instead of ferrous iron.

At the end of each chapter I suggest further sources of information. To these should be added a list of more general textbooks which may, in fact, cover material in many or all of the chapters. These include:

Davson, H. and Segal, M.B. *Introduction to Physiology*. Academic Press, London, Vol.1 1975, vol. 2 1975, vol. 3 1976, vol. 4 1978, vol. 5 1980

Ganong, W.F. *Textbook of Medical Physiology*. Appleton-Lange, Norwalk, 13th edn (1987)
Guyton, W.F. *Textbook of Medical Physiology*. Saunders, Philadelphia, 7th edn (1986)
Mountcastle, V.B. *Medical Physiology*. Mosby, St. Louis, 14th edn (1980)
Ruch, T. and Patton, H.D. *Physiology and Biophysics*. Saunders, Philadelphia, 20th edn (1979)

and that encyclopaedic reference source, *The Handbook of Physiology*, published by Williams and Wilkins, Baltimore, for the American Physiology Society. Sections 1-10, (1977-1986).

There are also a number of textbooks of physiology as applied to dentistry:

Jenkins, G.N. *Physiology and Biochemistry of the mouth*. Blackwell, Oxford, 4th edn (1978)
Osborn, J.W., Armstrong, W.G. and Speirs, R.L. *Companion to Dental Studies Volume I, Anatomy, Biochemistry and Physiology*. Blackwell, Oxford (1982)
Shaw, J.H., Sweeney, E.A., Cappuccino, C.C. and Meller, S.M. *Textbook of Oral Biology*. Saunders, Philadelphia (1978)

Finally, I must record in print my thanks to all those who have helped me along the very long road to the appearance of this book in print: Dr H. Richardson and Dr A. Watt, who looked at the earliest written chapters, Dr R. Foster, who taught me to use, successively, two word-processing programmes, Dr J. Waterhouse, who devoted a very long time to reading, criticising, and offering helpful suggestions towards two consecutive versions of the text, and Prof. R. Green, who repeated that exercise on the third version. I must thank also Dr J. Gilman, of John Wright, who supported me through the early stages of writing, and Mr D. Kingham, who completed the process. They had to be, and were, very patient. The debt to my wife is too great to be spelt out except in relation to her professional skill in producing the index. It is perhaps true to say that many others have suffered: to them also my apologies and thanks.

# Contents

Preface     v

### Section I    A Physiological Basis    1

1. Cells and fluids    3
2. The transfer of materials across epithelial cell layers – secretion and absorption    24
3. Blood plasma    42
4. The cells of the blood    54
5. Excitable tissues    nerves, muscle    730
6. Bulk transport    100

### Section II    Physiological Systems    115

7. The cardiovascular system    117
8. The respiratory system    126
9. The nervous system    132
10. The kidneys (renal system)    168
11. The digestive system    saliva    175
12. Endocrine systems    191

### Section III    Physiological Control    213

13. Mechanisms of control    215
14. Posture, locomotion and voluntary movement    222
15. Control of pressure and flow in the cardiovascular system    229
16. Control of the osmotic activity of the blood and of the blood volume    237
17. Normal control of respiration and the control of blood oxygen and carbon dioxide concentrations    242
18. Control of blood pH    250
19. Mastication and deglutition    257
20. Speech or vocalisation    265
21. Control of digestion    271
22. The menstrual cycle, pregnancy and lactation    281
23. Growth and development    285
24. The control of metabolic processes    294
25. The control of body temperature    309
26. Exercise    315
27. Emergency situations    322

Appendix    333

Index    335

# Section I
## A physiological basis

# 1
# Cells and fluids

## The cell and its contents

Physiology is the study of how living creatures live; human physiology the study of the physical and chemical processes that make up the life of man. It is based on anatomy, the study of structure or morphology; on histology, or micro-morphology; on biochemistry, the study of the chemical processes of life; and on biophysics, the study of the physical parameters of living organisms. In considering physiological principles it is helpful to look first at the basic rules governing the function of all body cells, then to apply these to the operation of the organs of the body, and finally to apply them to the functioning of the body as a whole. This book begins, then, by considering the normal structure of cells, their substructures or organelles, and the ways in which these contribute to the functioning of the cells. Fig. 1.1 shows two adjacent half cells containing most of the possible organelles in some of their typical arrangements. These are hypothetical examples since it is very unlikely that any cell would contain all these organelles.

The cytoplasm of a cell surrounds the nucleus and the organelles, and is bounded by a membrane. In histological sections stained with the standard haematoxylin and eosin stains it appears as a faintly purple-staining, slightly granular background; under the electron microscope it appears clear (or electron-lucent) with varying degrees of fine electron-dense granulation and some fine filaments or rods. Tests with intracellular probes show it to be either a gel or a sol, and to be more gel-like near the cell membrane. It contains cell water, proteins, small organic molecules, and some inorganic ions. Most cells contain about 85% of water. The protein constituents are in various degrees of aggregation. Some are highly structured to form filaments, as in muscle cells. The cell pH is usually slightly acid, and under these conditions most cell proteins behave as anions and contribute to the total anionic charge of the cell. The other major cell anions are chloride, phosphate and hydrogen carbonate. Balancing the anionic charge are the inorganic cations, principally potassium ions.

## The cell membrane

The limiting membrane of the cell, often called the plasma membrane, has been much studied because the properties of this membrane dictate what may enter or leave the cell and hence maintain the identity of the cell contents. Viewed with the electron microscope the typical membrane appears as two electron-dense lines separated by an electron-lucent zone. Membranes of this appearance are often termed 'unit membranes'. The membrane has an average width, after fixation, of around 7.5 nm and each of the electron-dense lines is approximately 2.0 nm wide. Chemical analysis of membrane material and experiments with lipids suggested that a flexible membrane-like structure could be formed from a bimolecular layer of lipid molecules orientated with their hydrophobic fatty acid side chains inwards towards each other, and their hydrophilic phosphate or basic groups pointing outwards. Indeed, it was suggested at one time that the cell membrane was not a true morphological structure but simply a lipid layer on the cytoplasmic surface. However, further study of the properties of membranes showed that they were made up of a bimolecular layer of phospholipid covered on each side by a protein layer; this model accords with the appearance seen with the electron microscope (Fig. 1.2).

It is, however, very simple, and explains only the bounding function of the membrane. Cell mem-

4  Cells and fluids

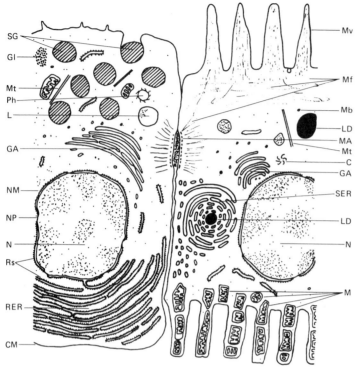

**Figure 1.1** Diagram of two half cells as seen under the electron microscope. These are not actual cells, but have been drawn to demonstrate a number of organelles. Key: C, centriole; CM, cell membrane; GA, Golgi apparatus; Gl, glycogen granules; L, lysosome; LD, lipid droplet; M, mitochondria; MA, macula adhaerens; Mf, microfilaments; Mt, microtubules; Mb, microbody; Mv, microvillus; N, nucleus; NM, nuclear membrane; NP, nuclear pore; Ph, phagosome; RER, rough endoplasmic reticulum; Rs, ribosomes; SER, smooth endoplasmic reticulum; SG, secretory granule.

branes have other properties also. They allow the passage of lipid-soluble substances and, to a lesser extent, small water-soluble substances. The latter property implies that the phospholipid layer is not continuous, but interrupted by 'pores' of small size. Studies of the sizes of water-soluble substances able to cross membranes permit estimation of the effective pore size – around 0.70–0.85 nm in a red blood cell membrane. Comparison of rates of diffusion leads to a calculation that 0.06% of the surface area of these particular membranes is occupied by the pores. No actual pores have been detected by electron microscopy even though pores of this size and density ought to be occasionally visible. Such pores are probably not actual holes in the membrane but areas in which water-soaked hydrophilic proteins penetrate the lipid layer. The membrane is able to permit certain ions and molecules to pass selectively at rates greatly in excess of those predicted from diffusion data. This property is linked with protein, or, more precisely, carrier or enzyme activity. This selective movement of substances in or out of certain cells can be modified precisely and in particular time sequences. Further, certain substances pass through the membrane against concentration and electropotential gradients, implying that the membrane is capable of utilising energy to perform work. All these properties suggest that the membrane contains proteins with specific binding functions. In this respect these proteins resemble enzymes and so they have sometimes been called permeases. Parts of the cell membrane in some instances may enclose external matter and then incorporate it into the cell; other parts of the membrane may apparently be new-formed from fusion of the membranes of organelles. Any model which suggests complete uniformity of the cell membrane is therefore likely to be misleading.

A number of modifications to the original model have been suggested. None is entirely satisfactory but each presents solutions to some of the difficulties with the original model. It is clear that membranes differ from one type of cell to another

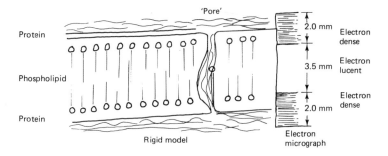

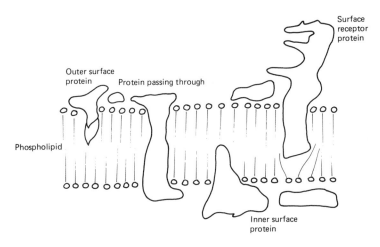

**Figure 1.2** Models of the cell membrane. (*a*) The rigid model with a pore filled with protein; on the right, its appearance under the electron microscope. (*b*) A section through a fluid membrane model at a single instant of time, a protein molecule spanning the membrane again providing a pore for water and small solutes to cross the membrane.

and also from point to point on the surface of one cell. For example, the protein to lipid ratio is different in red blood cell membranes from that in the membranes of nerve cells; indeed, the composition of the lipid material itself is different. Membranes vary in thickness and in biological activity. The current model is of a fluid structure in which the lipid and protein molecules are relatively mobile, so that the appearance of the membrane differs from moment to moment. The structure may become more stable or more rigid when particular protein units are performing particular functions. Chemical substances or drugs may cause changes in the movement of membrane components. Some proteins may extend across the membrane, some may travel from one side to the other, whilst still others may be associated with only the inner or the outer side of the membrane.

## Junctions between cells

Cells are usually separated from each other by a space about 20 nm wide filled with proteoglycans. However, a number of membrane specialisations provide either a means of attaching one cell to another or of barring the space between the cells (Fig. 1.3). The term zonule is used to describe a junction which extends round a cell, and the term macula to describe a junction involving only a small area of cell membrane. Either of these may be 'adhaerens', so that the cells are simply attached to each other, or 'occludens', when the attachment effectively obliterates the intercellular space.

There are three types of junction frequently seen between mammalian cells: desmosomes, tight junctions, and gap junctions. The desmosome (Fig. 1.3a) is a macula adhaerens. It is a small discoid

6  Cells and fluids

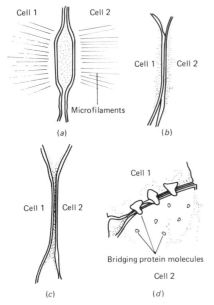

**Figure 1.3** Types of cell junction. (*a*) The macula adhaerens – a desmosome, (*b*) a zonula occludens – a tight junction, (*c*) a close or gap junction, (*d*) a close or gap junction at higher magnification. The protein plugs joining the two cells can be seen on the inner face of the membrane of the cell on the right.

attachment between neighbouring cells, found particularly in epithelial tissues. Dense clusters of filaments are situated in a pair of electron-dense spots on adjacent membranes of two cells and the space between the cells is filled with fibrous material. A tight junction (Fig. 1.3b) is a zonula occludens, usually extending round a cell. In a tight junction the membranes of the two cells are fused together so that their outer layers become one; the membrane thus appears to have three electron-dense lines separated by two electron-lucent lines. The fusion of the membranes means that the intercellular space is obliterated. Tight junctions are seen in cell layers which line body cavities, where they prevent movement of materials into or out of the cavities by intercellular pathways. The gap junction, or nexus (Fig. 1.3c), is a macula occludens, that is, a spot junction closing the intercellular gap. The membranes of neighbouring cells approach closely, to within some 2 nm, and are linked by special membrane proteins which form channels between the cells along which transfer of ions and small molecules and electrical charge can take place. They are found in some excitable tissues such as smooth muscle and cardiac muscle, where they provide pathways for electrical stimuli to pass rapidly across sheets of cells so that cell activities can take place in a highly integrated manner. Gap junctions are labile structures: they close and reform, apparently in response to specific stimuli. They are particularly sensitive to increases in intracellular calcium ion concentration, which cause them to close and disappear. This response may be protective: it could prevent products of damage or malfunction passing from one cell to its neighbours.

## Cell organelles and inclusions

### The nucleus

Within the cytoplasm are varying numbers of organelles, each enclosed by membrane material. The cell nucleus is bounded by a nuclear membrane consisting of two unit membranes 7–8 nm thick separated by a 30–50 nm space. The paired membrane fuses in places to form 'pores' approximately 65 nm in diameter and octagonal in shape. Up to one-third of the nuclear membrane may be occupied by these pores. The nucleus contains the genes which are here associated with strands of chromatin composed of desoxyribonucleic acid (DNA) with some ribonucleic acid (RNA) and nucleoproteins. Within the nucleus there is a large clear area, the nucleolus, containing many granules which are precursors of those on the outer surface of the nuclear membrane. These granules are the ribosomes: they are about 15 nm in diameter and are made up of nucleic acid and nucleoproteins. Similar granules are found on the rough endoplasmic reticulum and are described later. Functionally the nuclear membrane serves to preserve the identities of the nucleoplasm and the cytoplasm. In addition it is the route by which RNA leaves the nucleus after its synthesis against the template of DNA, the nuclear genetic material. There is continuity between the outer layer of the nuclear membrane and the endoplasmic reticulum.

### The endoplasmic reticulum

The endoplasmic reticulum is present in all cells except mature red blood cells and platelets. It is particularly well developed in cells involved in protein synthesis and secretion. There are two types – rough endoplasmic reticulum with ribosomes on the surface, and smooth endoplasmic reticulum which lacks these granules. The reticulum is a system of interconnecting tubes 40–70 nm in diameter whose walls appear as single electron-dense lines about 1.4 nm in thickness (Fig. 1.4). In some areas of the rough endoplasmic reticulum the system of tubules shows close apposition and the tubules are flattened to form cisternae. On the outer surface of the rough endoplasmic reticulum are the granules of RNA and protein which constitute the ribosomes. At these sites amino acids are linked to the transfer RNA to be assembled into new protein molecules under the combined action of the messenger RNA

# Cell organelles and inclusions

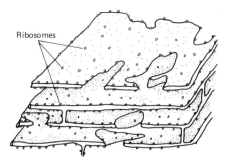

**Figure 1.4** The rough endoplasmic reticulum seen in three dimensions. A section through the reticulum appears as in Fig. 1.1 as a series of parallel tubes. The granules scattered over the surface are ribosomes.

and the ribosomal nucleoproteins. Some of the new protein molecules then pass into the lumen of the tubules. The smooth endoplasmic reticulum transfers some enzymes from the rough endoplasmic reticulum to the Golgi apparatus. It appears to be important in cholesterol metabolism and possibly in general lipid metabolism. Cells involved in the synthesis of steroids have a much more extensive smooth endoplasmic reticulum.

## The Golgi apparatus

The Golgi apparatus is a filamentous or plate-like smooth membrane reticulum (Fig. 1.1). Although present in all cells, it is most highly developed in cells involved in secretion and storage. The secretory cells of the pancreas may have as much as 10% of their volume occupied by the Golgi apparatus. In such cells the apparatus may show cisternae and vacuolar and vesicular regions. The formation of enzyme-containing vesicles, lysosomes and enzyme precursor (zymogen) granules takes place in the Golgi system. There are some enzymes located there, and the linking of sugars and proteins to form glycoproteins takes place in this area.

## Vesicles

A vesicle is a membrane-enclosed sac. Many vesicles have been found to have special functions or to be identifiable with special stains. The larger vesicles are termed vacuoles.

Lysosomes are one group of vesicles with special characteristics and functions. Four types are described:

1. Primary lysosomes containing enzymes
2. Phagosomes containing materials or particles taken up at the cell surface by invagination of the membrane (also called heterophagosomes)
3. Autophagosomes containing cell organelles due for destruction, and
4. Secondary lysosomes formed by the coalescence of primary lysosomes with auto- or heterophagosomes.

The secondary lysosome digests its contents and the products may pass back into the cytoplasm. When digestion is completed the vesicle is termed a post-lysosome; this may remain within the cell or it may release its contents to the environment through the cell membrane (Fig. 2.7). The enzymes found in lysosomes are a group of about 15, which, if free in the cytoplasm, would be capable of digesting the cell itself. They include acid phosphatase, proteolytic enzymes known as cathepsins, α-glucosidase, and enzymes capable of splitting RNA and DNA. The typical enzymes of lysosomes are hydrolases. The relatively high protein content of the lysosomes usually renders them electron-dense. Their bounding membranes are thicker than those of most cell organelles, being around 9 nm in thickness. The lysosomes are thought to develop as vesicles on the Golgi apparatus, although the enzymes are originally synthesised on the rough endoplasmic reticulum. The heterophagosomes are results of pinocytosis (incorporation of the external medium by an invagination of the cell membrane) or phagocytosis (incorporation of solid particles by invagination of the cell membrane). The properties of pinocytosis and phagocytosis are shown by the phagocytic cells of the blood, by their equivalent cells in tissues, and by cells with a 'brush border' – a membrane with many small projections termed microvilli. The autophagosome is produced when smooth membrane surrounds an organelle or cytoplasmic area and then fuses with a primary lysosome. Secondary lysosomes perform their digestive function, small molecules escape across their walls and they become post-lysosomes. Some of these are retained indefinitely, as in the cells of the skin, but others fuse with the cell membrane and release their contents to the extracellular environment. In the liver, for example, waste material is transferred to the bile.

Secretory granules may be regarded as special forms of lysosome.

## Microbodies

Microbodies are membrane-bounded spherical structures containing dense granular or even crystalline material. They are present in large numbers in liver and kidney cells. Their diameters range from 0.3 to 0.6 μm and their unit membranes are about 6.5 nm wide. Microbodies contain enzymes: these include catalase (most consistently, although its function here is unknown), urate oxidase and D-amino-oxidase. The presence of these two enzymes has led to the suggestion that the microbodies are concerned in amino acid and uric acid metabolism. Because of their enzyme content microbodies have also been called peroxisomes or uricosomes. They are probably formed from the endoplasmic reticulum.

Cells of calcifying tissues contain structures similar to microbodies which are able to concentrate calcium ions and form an amorphous calcium phosphate. This later transforms to apatite, the characteristic crystal of calcium phosphate in nearly all mammalian calcified tissues. These structures are referred to as membrane vesicles whilst they are still inside the cells, but when they leave the cell they are known as matrix vesicles.

## Microtubules and microfilaments

Thin cylindrical membranes known as microtubules are found in various forms in different cells. They are usually 20–40 nm in diameter with walls 5–10 nm thick made up apparently of very fine fibrils. One of their functions in cells appears to be that of maintaining rigidity – in blood platelets, for example, they maintain the discoid shape. Microtubules also stabilise the long narrow extensions, or processes, which extend into the tubules of the dentine from the odontoblast cells. Another function appears to be that of defining pathways for movements of cell constituents, as in secretory cells where synthesised material passes from the Golgi apparatus along pathways laid out by the microtubules to the cell surface (see Chapter 2). Microtubules also form the mitotic spindle during cell division. Characteristic patterns of microtubules are found in ciliated cells to form structures capable of producing movement of the cilia. It is probable that the microtubules are assembled by the endoplasmic reticulum.

Microtubules form the walls of the centrioles – a pair of cylindrical bodies about 400 nm long and 150 nm in diameter. The centrioles lie near the nucleus and are an essential part of the mitotic apparatus during cell division.

Specialised microfilaments are found in the cytoplasm of many cells. The most obvious examples are the ordered microfilaments and filaments of the actin and myosin of skeletal and cardiac muscle, but actin filaments are also found in smooth muscle, in platelets, and in the cilia and processes of many other cells.

## Mitochondria

The production of high energy phosphate, in the form of adenosine triphosphate (ATP), takes place in another membranous organelle: the mitochondrion (Fig. 1.5). Each mitochondrion is ellipsoidal in section and consists of an outer double boundary

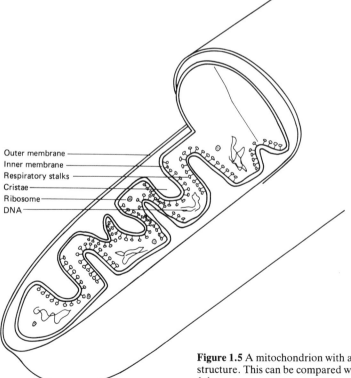

**Figure 1.5** A mitochondrion with a portion removed to show the internal structure. This can be compared with the sections through mitochondria in Fig. 1.1.

membrane, with projections from the inner layer, termed cristae, passing inwards. The number of mitochondria in each cell is related to its energy expenditure. There may be as few as 50 or as many as several thousand. The number of cristae in each mitochondrion is also related to the energy pattern of the cell. The cristae are covered by projections, each having a stalk and a headpiece, arising from an area of the membrane with specific protein composition. Mitochondria, although made up of membranes, contain a large proportion of protein, mainly in the cristae, and much of this protein is associated with enzyme activity. In liver cells, for example, at least 60% of the succinoxidase activity, and at least 80% of the cytochrome oxidase activity, is associated with the mitochondria.

More than 50 different enzymes are now known to be present in the mitochondria. Most are involved in respiratory processes, in the transfer of energy from the substrate glucose to the energy store constituted by adenosine triphosphate (ATP). The breakdown of glucose is divided into three stages: glycolysis, the tricarboxylic (Kreb's) cycle and oxidative phosphorylation. The last two take place in the mitochondrion. Other substrates such as fatty acids or amino acids may also provide energy. The membranes and cristae are so ordered that the transfer of electrons along the dehydrogenase reactions of the Kreb's cycle takes place sequentially along contiguous structures. Such enzymes as $\alpha$-ketoglutarate dehydrogenase, pyruvate dehydrogenase and $\beta$-hydroxybutyrate dehydrogenase, which represent input points of the metabolic chain to the Kreb's cycle, are found in the outer membrane; whilst the enzymes of Kreb's cycle such as fumarase and malate dehydrogenase are either linked to this membrane or lie between it and the inner membrane. The membranes of the cristae carry the enzymes of oxidative phosphorylation such as ATPase (in the headpieces) and the cytochromes of the electron transfer chain (in the bases of the projections). Cytochromes $a$, $a_3$, $b$ and $c$ have all been tentatively located on different sites. Some of the early stages of steroid formation take place in mitochondria.

Mitochondria have been observed to change in shape under physiological conditions, possibly because the energy changes inside them may result in ionic, and hence osmotic, changes. New mitochondria arise by division of existing ones. This, and other characteristics, such as the presence of mitochondrial DNA and the mitochondrion's ability to synthesise its own proteins, suggest that the mitochondrion may have originated as a prokaryotic cell parasite which has attained a symbiotic status within the cell. A particular feature of mitochondrial activity with special importance in the secretory activity of cells, and possibly also in relation to calcification in some cells, is the ability to concentrate calcium ions. Energy is needed for this process and is provided by respiration. Accumulation of calcium takes place by binding to protein or by the formation of electron-dense granules of a calcium phosphate with a Ca:P molar ratio of around 17:10. This is close to the ratio of 3:2 found in apatite crystals and it may be that these granules contain an amorphous precursor of apatite.

## Cilia and flagellae

Cilia are structures found on a number of human cells, particularly in the upper respiratory tract. They are multiple cell projections, 2–10 µm long and 0.5 µm in diameter, made up of a framework of nine pairs of microtubules arranged around two single microtubules. Cilia move by a sliding of the microtubules against each other. Larger versions of these structures, usually found singly on the cells, are called flagellae.

## The movement of materials across cell membranes

Probably the most important of cell structures to the physiologist is the cell membrane. It is the membrane which maintains the identity of the cell and separates it from the surrounding fluids. The cell membrane must preserve the composition of the intracellular fluid as far as possible and yet permit ingress of nutrients and exit of waste and possibly some synthesised products. Since the intracellular fluid differs markedly in composition from the extracellular fluid, the permeability characteristics of the membrane are critically important. The principles which govern the movement of substances across the membranes of cells must now be considered. Movement may occur passively – in response to physical forces, or actively – by biological means involving the expenditure of energy to overcome physical forces, or by some combination of passive and active means.

### Passive movement

Passive movement, or diffusion, occurs when a substance to which the membrane is permeable, crosses the membrane as a result of the kinetic energy of the molecules, but without the expenditure of energy from any other source. Two characteristics are therefore definable – the permeability of the membrane, and the energy gradient. In general cell membranes are freely permeable to lipid-soluble substances. Their permeability to water-soluble substances is governed by apparent pore size, and is therefore dependent upon molecular size. Spherical molecules experience progressively more difficulty in traversing cell membranes as their molecular weight increases; long, thin, molecules may show quite different properties. In the sugars, for example, pentoses usually cross membranes more easily than hexoses, and monosaccharides than disaccharides or polysaccharides. The maximum size of molecule able to penetrate the membrane gives an estimate of the effective diameter of the pores: this is usually around 0.7 nm. Since the forces involved in movement of substances across cell membranes are modified by the relative permeability of the membranes to different molecules, diffusion must be considered in two sets of conditions: those in which the membrane is freely permeable to the molecules on either side of the membrane, and those in which it is differentially permeable to them. The membrane in the latter case is usually termed selectively permeable. The term semi-permeable is reserved for membranes which are permeable only to molecules of the solvent.

### Conditions in which the membrane is freely permeable to the molecules on either side

When the molecules on each side of the membrane are able to cross it freely, three types of diffusion may be distinguished on a basis of the nature of the energy gradient causing diffusion. The gradients are those of concentration, electrical charge and pressure.

### Diffusion down gradients of concentration

When two solutions of a given substance are separated by a membrane permeable to that substance, and the concentration in one solution is greater than that in the other, the solute will pass from the side of high concentration to the side of low concentration until the concentrations are equal. This is a simple consequence of the kinetic movement of the molecules: the number of molecules bombarding the membrane on one side is greater than that on the other, so more molecules cross from the side of greater concentration. This explanation also implies that diffusion occurs in the absence of a concentration difference even though no net change in concentration is observed. If the molecules of solute are in constant movement and the concentrations on the two sides of the membrane are identical, an equal number of molecules cross the membrane in each direction. However, it is the net movement which is usually important.

Water, with its small molecules, also diffuses freely. Its activity is inversely proportional to the activity of the solute. This is because the free energy of the water molecules is reduced by the random collisions with the particles of solute; hence increasing solute concentration reduces the free energy of the solvent. The net effect of this is that there is a movement of water towards the side of the membrane with the more concentrated solution, at the same time as the solute is moving towards the side with the lower concentration.

In living tissues, diffusion down concentration gradients may be maintained by sequential movement across several membranes and compartments, or by active processes which are capable of changing the concentration on one side of the membrane. Thus diffusion from the gut into intestinal cells will continue down a concentration gradient if the diffusing substance can continue out through the membranes on the opposite sides of the cells, through fluid layers and into the blood stream to be removed by bulk flow of the blood: this maintains the gradient across the cell membrane. Again, if glucose passing into a cell is converted therein first to pyruvate and then to lactic acid, the glucose gradient may be maintained at its original level.

The rate of diffusion between two compartments separated by a fully permeable membrane is given by Fick's equation

$$Q_m = \frac{DA(C_a - C_b)}{l}$$

where $Q_m$ is the amount of material passing in unit time, $D$, the diffusion coefficient, $A$, the area of the path available, and $(C_a - C_b)$ the concentration difference between the two compartments and $l$, the path length for diffusion.

This equation applies when the compartments are large in volume and the membrane between them is completely permeable. The biological cell is of limited volume in comparison with the volume of fluids around it and has a membrane of limited permeability. When the diffusion constant is modified to allow for these differences it is termed the permeability constant. Fick's equation may be re-arranged for the calculation of $D$, the diffusion constant, and then modified so that $P$, the permeability constant, can be calculated:

$$P = \frac{V}{At} \log \frac{C_{out} - C_{in}}{C_{out} - C'_{in}}$$

where $P$ is the permeability constant, $V$, the cell volume, $A$, the area of the cell membrane in cm$^2$, $t$, the time in seconds, $C_{out}$ the concentration of penetrating substance outside the cell, $C_{in}$ the concentration of penetrating substance inside cell at zero time, and $C'_{in}$ the concentration of penetrating substance inside the cell after $t$ seconds.

Permeability constants have been calculated for several ions which can pass through cell membranes in the skeletal muscle of the frog and have been shown to be inversely proportional to the size of the hydrated ions (*Table 1.1*). All ions are in practice surrounded by a hydration shell which increases their effective size.

**Table 1.1 Permeability constants of common ions.**

| Ion | Permeability constant (cm/s) | Ionic radius (nm) |
| --- | --- | --- |
| Sodium | $2 \times 10^{-8}$ | 0.18 |
| Potassium | $200 \times 10^{-8}$ | 0.12 |
| Chloride | $250 \times 10^{-8}$ | 0.12 |

The rate of diffusion into or out of a cell, then, depends upon the concentration difference between the sides of the cell membrane and the permeability constant of the membrane for the molecule or ion considered.

### Diffusion down gradients of electrical charge

Where charged particles exist on both sides of a membrane which is permeable to them they will diffuse to balance the charge on the two sides of the membrane. In equilibrium, therefore,

$$A^-_{in} = B^+_{in}$$

and

$$A^-_{out} = B^+_{out}$$

This, the most simple situation, requires substantial modification when the membrane is not fully permeable or when diffusion is uneven. For example, ions of different sizes will diffuse at different rates leading to a temporary imbalance of charge during the period of equilibration of two solutions with differing concentrations of the charged ions.

A concentration difference may be maintained by a potential difference across the membrane if concentration and charge gradients are opposed. The balancing potential is given by the Nernst equation:

$$E = \frac{RT}{zF} \log_e (C_1/C_2)$$

where $E$ is the potential (the Nernst potential), $R$, the gas constant, $F$, Faraday's constant, $T$, the absolute temperature in K, $z$, the valency of the ion, and $C_1/C_2$ the ratio of the concentrations of the ion in the compartments. If the temperature is maintained constant, the term $(RT)/(zF)$ is a constant.

If a concentration difference exists for an ion to which the membrane is permeable, a difference of electrical potential may be shown between the two sides of the membrane, and its magnitude may be calculated from the above expression. Where more than one ion is involved, the equation is modified to the similar but more complex Goldman equation (considered further in Chapter 5).

### Diffusion down gradients of pressure

Pressure gradients are important in passive diffusion mainly in relation to gases. A dissolved gas is said to exert a tension equal to the partial pressure of the gas in equilibrium with the solution. In the equilibrium situation the number of gas molecules escaping from the solution due to molecular movement is equal to the number entering the solution from the gaseous phase. The gas tension in a solution is therefore a measure of concentration and the gradient of pressure is a special case of the more general gradient of concentration. For example, oxygen diffuses from the arterial side of a blood capillary, where its partial pressure, or tension, is about 12.8 kPa to the interstitial fluid and then to the cells of an active tissue where the partial pressure is of the order of 4.0–5.2 kPa. Similarly, carbon dioxide diffuses from venous blood in the capillaries in the lung, where its tension is 6.2 kPa, to the alveolar gas in which its partial pressure is around 5.3 kPa.

### Conditions in which the membrane is selectively permeable (semi-permeable)

The same three types of gradient may be present. Diffusion takes place as already described for the permeant ions or molecules; what must now be considered are the changes imposed on each system by the presence of non-permeant particles.

### Diffusion down gradients of pressure when the membrane is selectively permeable

Where a hydrostatic pressure difference exists across a membrane permeable to water and small solutes, and impermeable to large solutes or particles, filtration takes place. The water and small solutes pass through the membrane leaving behind the solutes unable to penetrate. Such a filtration process takes place in the kidney glomerulus where the blood in the glomerular capillaries has a hydrostatic pressure of around 7 kPa opposed to the hydrostatic pressure in Bowman's capsule of approximately 2 kPa. Water and small solutes pass into the beginning of the kidney tubule, leaving behind the blood cells and the larger molecular weight proteins which are unable to diffuse through the capillary walls and the basement membrane.

### Diffusion down gradients of concentration when the membrane is semi-permeable or selectively permeable

When two solutions of differing concentration are separated by a semi-permeable or selectively permeable membrane, the activity of the solvent molecules becomes important. The activity of solvent molecules is progressively reduced as concentration of the solute increases. This causes a reduction in the number of solvent particles striking the membrane on the side of higher concentration of solute, and therefore a reduction in the number of solvent particles crossing the membrane from that side. The magnitude of the effect is related to the number of particles of solute and thus to molar or ionic concentration – a given weight/volume concentration of a substance of low molecular weight is more effective than the same concentration of a substance of high molecular weight. Thus 1% w/v sodium chloride is much more effective than 1% serum albumin (a protein of 69,000 daltons molecular weight) if the membrane is impermeable to both. If two solutions of differing molarity (concentration in mol/l) are separated by a membrane permeable only to the solvent, the solvent will move to the more concentrated compartment and increase its volume until the molar concentrations are equal. The pressure needed to resist this volume change is termed the osmotic pressure. The solute unable to cross the membrane shows osmotic activity. The magnitude of the pressure needed to balance the activity is given approximately by the van't Hoff equation

$$n = RTc$$

where $n$ is the osmotic pressure, $R$, the gas constant, $T$, the absolute temperature, and $c$, the concentration in mol/l.

More complex equations have been developed to give closer agreement with observed results at higher concentrations when the van't Hoff relationship applies less well because of interactions between molecules.

Since all living cell membranes show selective permeability, osmotic activity and hence osmotic flow are extremely important in the distribution of water throughout the body.

### Diffusion down gradients of potential across selectively permeable membranes

Charged particles which are unable to pass through a membrane down concentration or charge gradients cause a redistribution of other particles in compensation.

This is best illustrated by an example. If two compartments are separated by a membrane impermeable to large negatively charged particles R⁻, and potassium and chloride ions are added to each side to give electroneutrality, the situation may be represented thus:

R⁻    50           | K⁺    100
K+    100          | Cl⁻   100
Cl⁻   50           |

There is a concentration gradient for R⁻ but this ion cannot pass through the membrane. There is no concentration gradient for K⁺ but there is for Cl⁻, so Cl⁻ ions diffuse to the left. This increases the negative charge on the left so that a gradient of potential now exists. K⁺ therefore diffuses to the left down a gradient of charge but this is counterbalanced by the increasing K⁺ concentration gradient from left to right that it produces. It can be shown from thermodynamic considerations that the opposing forces are balanced when

$$\frac{K^+_{left}}{K^+_{right}} = \frac{Cl^-_{right}}{Cl^-_{left}}$$

This balance is known as the Gibbs–Donnan equilibrium. The rule of electroneutrality in solution demands that the sum of the electrical charges in each solution should be zero, that is, that the sum of the anions on one side of the membrane should be equal to the sum of the cations on that side. The final concentrations resulting in this example would therefore be

R⁻    50           | K⁺    86
K+    114          | Cl⁻   86
Cl⁻   64           |

with electroneutrality on both sides of the membrane but a difference in concentration and charge across the membrane. The membrane behaves as if it is charged negatively on its left side and positively on its right side. The magnitude of this charge is given by the Nernst formula

$$E = \frac{RT}{zF}\log_e\frac{Cl^-}{Cl^-} = \frac{RT}{zF}\log_e\frac{K^+}{K^+}$$

that is, the potential across the membrane is directly proportional to the ratio of the concentrations on either side of the membrane of an ion able to cross the membrane freely.

The presence of an ion unable to penetrate the membrane separating two compartments, then, results in the establishment of a Gibbs–Donnan equilibrium and of a Nernst potential across the membrane. In addition, because in the equilibrium position the total concentration of particles on each side of the membrane is different, a secondary osmotic gradient is set up.

### Active transport

Active transport is most simply defined as a movement of substances across the cell membrane as a result of cell activity involving expenditure of energy. Further, this expenditure of energy is usually necessary to overcome a gradient of concentration, charge or pressure. However, active processes are also involved in some bulk movements across cell membranes.

## Bulk movements

**Pinocytosis** (literally 'cell-drinking') is the name given to the process whereby a volume of extracellular fluid is surrounded by the cell membrane, the membrane seals itself around the drop and the drop finally appears as a phagosome within the cell. The contents of the phagosome are usually eventually incorporated into the cell cytoplasm.

**Phagocytosis** (literally 'cell-eating') is the similar process of taking up particulate matter or other cells into the cell. Some digestion of the contents of the phagosome will be necessary before its contents can be incorporated into the cytoplasm or they may be returned to the extracellular environment by exocytosis.

**Exocytosis** (sometimes called emeiocytosis) is the reverse of the previous two. Membrane-enclosed droplets or particles pass to the cell membrane, fuse with it, and lose their contents to the extracellular environment.

## Molecular movements

Molecules or ions are said to be actively transported when they cross a membrane in the opposite direction to a gradient of concentration, charge or pressure, and energy is expended to achieve their passage. Most systems of active transport involve carrier molecules within the cell membrane which are able selectively to bind and release particular substrates.

The most important transport system is that which transports sodium ions out of cells against a concentration gradient. Almost all animal cells exhibit this property of actively transporting sodium outwards across their walls. Indeed the transport of sodium is often used as a means of indirectly causing the movement of other ions or water. Several different transport systems are known for sodium movement but the best understood is that which can be selectively inhibited by the drug ouabain.

The process, by definition, requires energy. This is obtained by the hydrolysis of ATP, each high-energy phosphate bond providing energy for the linked transport of two potassium ions into the cell and three sodium ions outwards. As ATP is hydrolysed, the carrier can be described as a $(Na^+-K^+)$-ATPase. Hydrolysis of ATP will not occur in the absence of $Mg^{2+}$ ions. The carrier has two sites: a $Na^+$ site competitively inhibited by $K^+$, and a $K^+$ site competitively inhibited by $Na^+$. In essence, the system consists of a carrier molecule X, protein or lipoprotein in nature, with specific binding sites for the two ions, a translocation system to move one ion across the membrane, and a means of changing the carrier molecule's properties in order to release the transported ion on the other side of the membrane and pick up the other ion. Fig.

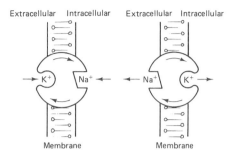

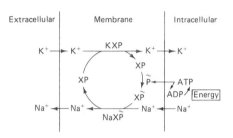

**Figure 1.6** Representations of the sodium/potassium pump mechanism. (*a*) A model with a protein with specific binding sites which can move the ions across the membrane. (*b*) The process summarised as a series of chemical reactions and movements. Phosphate is indicated as P; a wavy line above the P indicates a high energy bond.

1.6 shows how this might operate for a sodium–potassium linked transport. On the outside of the cell $K^+$ links with a phosphorylated protein X–P, which then moves across the membrane and simultaneously releases $K^+$ inside the cell and loses its phosphate. The carrier X must now be re-phosphorylated with phosphate from ATP to form a new active carrier X*P capable of binding $Na^+$. Sodium ions inside the cell are bound and moved to the outside of the membrane. There they are released and the carrier protein undergoes a change, possibly to a lower energy form, becoming once again a carrier of $K^+$ ions.

All the while, diffusion back down concentration gradients continues. This is more significant in relation to potassium ions, mainly because the membrane is not very permeable to sodium ions, but also because sodium ions are larger than potassium ions and therefore move more slowly. The importance of countering the outward leak of potassium ions is shown in human red blood cells, where the normal oxygen consumption is equivalent to an energy production of 168 J/kg/h, a figure of the same order of magnitude as the 54.6 J/kg/h required to maintain the potassium concentration difference across the cell membrane. As red cells age they lose their ability to maintain this gradient and potassium ions leak out. The same effect is observed progressively in stored blood.

Other active transport systems in the body show similar properties to those already described but appear to have a limitation on the maximum load they can transport. This probably means that the number of carrier sites is limited, thus limiting the rate at which material can be transported.

Most of the systems of limited capacity transport organic molecules, usually small organic molecules, although even these are large in comparison with inorganic ions. Amino acids and metabolically useful sugars, such as glucose and galactose, are carried by systems of this type. The ability to transport specific substances, presumably by utilising specific carrier sites on specific membrane proteins, renders the cell selective in its accumulation of ions and organic molecules. Thus glucose and galactose, with six carbon atoms, pass into cells six times as fast as xylose and ten times as fast as arabinose, despite the smaller molecular size of these five-carbon sugars.

Several properties have been described for sugar-transporting systems – they show competitive inhibition, require sodium ions and will carry only sugar molecules of a particular shape.

### Facilitated diffusion

This term is used to describe a transport system which is operating in the same direction as a gradient of concentration or charge. The effective permeability of the membrane is increased by the presence of carrier molecules and so diffusion occurs more readily than would otherwise be the case. For this reason facilitated diffusion can be inhibited by blocking the carrier route and competition between transported substances may occur if the same carrier protein is used to facilitate their diffusion. Simple diffusion is not subject to these limitations. Sugar absorption in red blood cells, muscle cells and liver cells appears to take place by facilitated diffusion when external sugar concentrations are in excess of those inside the cells. In the intestine facilitated diffusion of glucose occurs in the early stages of absorption but as glucose concentration in the lumen of the intestine falls the process becomes one of active transport. It is possible that calcium absorption in the gut similarly occurs in two phases.

# The fluid compartments of the body

### The evolutionary hypothesis

The chemical composition of all animal cells is broadly similar, a fact which suggests that only certain closely defined chemical environments are compatible with life. This common intracellular environment may replicate the environment in which life first developed: its composition may be similar to that of an original primeval sea, with a relatively high concentration of potassium ions (perhaps about 150 mmol/l), a relatively low concentration of sodium ions (perhaps about 20 mmol/l) and chloride as the main anion. It is postulated that at some point a change occurred in the composition of the sea, bringing it to concentrations more similar to those of today, with high concentrations of sodium ions and low concentrations of potassium ions. The primitive cells ceased to be in equilibrium with their environment, and in order to survive they had to develop a cell membrane as a limiting membrane capable of maintaining an intracellular environment different from that outside. The membrane controlled diffusion in and out of the cell, and developed systems capable of transporting substances against concentration gradients. The major transport system was that correcting the sodium/potassium imbalance. Sodium ions which were constantly diffusing into the cell were actively transported out again and the intracellular potassium concentration was maintained by actively transporting potassium ions back into the cell to replace those lost by diffusion. Movement of nutrients into the cell and of waste products out could occur down concentration gradients. However, the development of carriers for facilitated diffusion would make this process more efficient.

Later, when cells linked up to form organisms and specialisation of organs took place, each cell could no longer be surrounded by the original marine environment and so a kind of subsidiary sea, or internal environment, would be necessary within the more complex multicellular organism. This would be a fluid of composition similar to that which had surrounded every cell before aggregation took place. Exchange of nutrients and waste products might occur between cells and this extracellular fluid, and then between the extracellular fluid and the external environment, but this process would be slow over longer distances. As the organism increased in complexity, some of the extracellular fluid became separated off into a system of connecting tubes, or vessels, in which bulk flow was possible – an intravascular fluid in a circulation. Bulk flow would speed up the movement of materials between the external environment and the innermost cells, leaving the slower process of diffusion to take place over only relatively short distances.

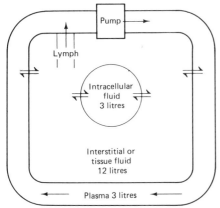

**Figure 1.7** Diagram to illustrate the relationships between the main compartments of body fluids. Arrows pointing in one direction indicate bulk flow in that direction; bidirectional arrows are used to show that exchange occurs across the intercompartmental barriers.

## The major compartments of body water

The water present in an animal's body may be considered as being in two major compartments – the intracellular and the extracellular. The extracellular compartment is further subdivided into the intravascular fluid component of the blood (the plasma), and the intercellular fluid, or interstitial fluid, in the interstices between cells. In a localised area this may also be called tissue fluid (see Fig. 1.7).

In man the total body water amounts to about 63% of body weight – or about 45 litres in an average man weighing 70 kg. There is considerable variation between subjects with a range from 45% to 75% of body weight. This is largely because of the variability in the amount and distribution of fat, or adipose tissue, since fat has a water content of only about 10% in contrast with values of around 80% for most other tissues. Muscle contains almost as high a proportion of water as does blood – 76% against 83%. Bone has a low water content at only 22%. In general, then, lean subjects have higher water contents than obese subjects. The average woman, with her different physical build, and contours smoothed by adipose tissue, has a lower water content than the average man, with only some 52% of her body weight as water. Children usually have a high water content: an average six-month-old infant can be expected to have a total body water content around 73% of body weight.

In any one individual the total body water content is surprisingly constant. The original distribution is restored after an intake of a litre of water within 2–3 hours. Normally there exists a rough balance of water intake and loss over 24 hours (*Table 1.2*).

Figures for the average water intake can be used to estimate the average intake of fluoride from

**Table 1.2 The daily balance of water intake and loss**

|  | Water gain (l) |  | Water loss (l) |
| --- | --- | --- | --- |
| Drinking | 1.5 | Urine loss | 1.5 |
| Food | 1.0 | Evaporation and sweat | 0.7 |
| Metabolic water | 0.3 | Expired air | 0.5 |
|  |  | Faeces | 0.1 |
|  | 2.8 |  | 2.8 |

natural or fluoridated waters. If fluoride supplements are prescribed to help in the prevention of dental disease, the amounts used should be related to the amount already being absorbed by the patient from the drinking water.

Total body water may be measured by a dilution technique. If a substance is introduced into the body fluids – most conveniently by injection into the blood stream – in a known quantity, its concentration in the body fluids after equilibration will enable the calculation of the total volume of the body fluids. The final concentration is also most conveniently measured in a sample of blood. The properties required of the marker substance are that it should be non-toxic, it should distribute evenly throughout the total body fluid, it should undergo no change in the body by binding or metabolic breakdown, it should preferably be rapidly distributed, slowly lost from the body, and, for convenience, it should be easy to estimate. The second and third criteria limit the possibilities to small molecules or ions which are able to cross all cell membranes easily. The fourth and fifth criteria are almost mutually exclusive, since freely diffusible substances also diffuse easily through the filters of the kidneys into the urine. However, if urine as well as blood is analysed, the loss by this route can be allowed for in the calculation

$$V = \frac{I - (v \times c)}{C}$$

where $V$ is the total body fluid volume, $I$, the weight of marker substance injected, $v$, the volume of urine produced over the time of estimation, $c$, the concentration of the marker substance in the urine, and $C$, the concentration of the marker substance in the blood plasma.

Usually the marker is injected and the concentration in the plasma is estimated over several hours so that a curve can be plotted to show how it changes. The loss by all routes is similarly plotted over a period of time. When the plasma concentration plotted on a logarithmic scale ceases to fall in a steep curve and instead begins to follow a straight line of similar slope to that of excretion, equilibration between plasma and total body fluid is assumed to have occurred, and the plasma concentration at a convenient time thereafter read off and used in the calculation. Alternatively, the line of plasma concentrations after equilibrium is extrapolated back to zero time, the concentration at that point ($C_1$) read off, and the calculation simplified to

$$V = \frac{I}{C_1}$$

The substances used for this estimation are the two isotopes of water – deuterium oxide ($D_2O$) and tritiated water ($^3H_2O$). Both of these are easily assayed. The isotopes of water have two disadvantages: water is lost from the body surfaces and the lungs, as well as in the urine, so that an estimate of all losses is required, and, secondly, the marker atoms may be incorporated into other molecules. Fortunately, the rapid diffusion and equilibration of these water isotopes enables the estimation to be carried out in about two hours, during which time little chemical modification takes place.

## Intracellular fluid

The intracellular fluid compartment comprises the fluid present in every cell in the body. As such it is non-homogeneous in composition. Every specialised cell contains characteristic concentrations of particular proteins and ions. However, there are some broad similarities in composition.

The volume of intracellular fluid represents 30–40% of body weight. Mean values are towards the upper end of this range and so the 'standard man' of 70 kg weight has 25–30 litres of intracellular fluid. It is this fraction of body water which is most affected by the proportion of adipose tissue in the body, not only because adipose tissue has cells largely filled with fat droplets, but also because the cells tend to be closely packed and there is less intercellular space. Thus women, with 20–25% of their body weight as adipose tissue, have a lower proportion of their body weight as intracellular water than do men, with 15–20% adipose tissue.

The actual volume of water present intracellularly is not easy to measure. Even if there were a substance which would distribute uniformly throughout this compartment, it would be impossible to administer it so that it would pass freely only to cells. The value, then, must be obtained by a calculation of difference. The total volume of the extracellular fluid compartment can be measured (except for the small volume in the so-called transcellular fluids) and subtracted from the total body water volume measured as above. Any uncertainties in these measurements will be compounded in the calculation of the intracellular fluid volume.

Few values are available for the composition of the fluid part of individual cells, although values have been published for some whole tissues. These are given in *Table 1.3*. There is obviously quite large variation, but these values give some indication of the general pattern in human cells. All cells so far studied in man contain relatively low concentrations of sodium ions but relatively high concentrations of potassium ions. This balance is maintained by active transport systems moving sodium ions out of the cells and carrying in potassium ions. The most constant intracellular anion is protein. Other anions vary from cell to cell. Thus in red blood cells chloride is present in high concentration, but in muscle organic phosphate is much more important. Chloride concentrations are always less than those in extracellular fluids. The ionic balance within cells is difficult to calculate because the magnitude of the charge on the protein molecules is not usually known. However, the pH of cell fluid is usually below that of extracellular fluids. In general, it is assumed that the osmotic activity of the cell contents is equivalent to that of the extracellular fluid, and that exchange of fluids between the compartments will follow any change in the osmotic activity of either compartment.

**Table 1.3 The principal ionic constituents of cells**

| Constituent | Red blood cell | Skeletal muscle cells | Smooth muscle cells | Nerve cells |
| --- | --- | --- | --- | --- |
| Sodium ions (mmol/l) | 12–20 | 10 | 56 | 15 |
| Potassium ions (mmol/l) | 150 | 160 | 119 | 150 |
| Calcium ions (mmol/l) | 0.25 | $10^{-6}$ | $10^{-3}$ | ? |
| Magnesium ions (mmol/l) | 1.4 | 7.5 | ? | 1.5 |
| Chloride ions (mmol/l) | 74 | 2 | 55 | 9 |
| Hydrogen carbonate ions (mmol/l) | 27 | 8 | ? | ? |
| Organic phosphate (mmol/l) | ? | 70 | ? | ? |
| Protein (mmol/l) | 50 | 55 | | |
| Protein (g/l) | 350 | 420 | 400 | 150 |

## Extracellular fluid

The extracellular fluid compartment includes all fluids outside the membranes of the cells. Its major subdivisions are the intravascular blood plasma and the interstitial fluid. Lymph may be regarded as an intravascular fluid in communication with both plasma and interstitial fluid. In addition to these there are a number of other fluids, all of relatively small volume, separated from the plasma by an epithelial cell layer as well as the endothelial lining of the blood capillary walls, and known, therefore, as transcellular fluids. The digestive secretions are sometimes classed with these, but as they pass outside the body into the lumen of the digestive tube, they are strictly extracorporeal rather than extracellular. This is true also of the secretions of the sweat glands. The transcellular fluids include the cerebrospinal fluid, the intra-ocular fluids, the fluids of the ear, and the pleural, peritoneal and synovial fluids. Their total volume is very small: even the largest, the cerebrospinal fluid volume, is only 140 ml in man. The total amount of the extracellular fluid is around 20% of body weight in both men and women. Its volume is therefore about 15 litres in a 'standard man'.

The total extracellular fluid volume may be determined using the same principles as those used in the determination of total body water except, of course, that the marker substance used must not be able to penetrate cells. The properties required of the marker substance, then, are that it must be non-toxic, that it should distribute evenly and, preferably, rapidly throughout the body fluids outside the cells, and that it should remain unchanged in the body before being lost as slowly as possible. The distribution criterion implies that the substance should be small enough to cross the walls of the blood capillaries easily, but too large to cross cell membranes. There is probably no perfect marker which will satisfy all these criteria. Sugar molecules small enough to pass through the pores of a blood capillary but unable to enter cells have been used: these include inulin, raffinose, sucrose and mannitol.

Inulin gives the lowest estimate of the extracellular fluid volume, possibly because its distribution time is so long that it does not attain full equilibration in the time allowed. Sucrose may enter some cells. Some ions have been used: thiosulphate, thiocyanate, radiosulphate, radiochloride and radiosodium. The last two do enter cells to some extent. Thiosulphate, like mannitol, is rapidly excreted. Very small amounts of radiosulphate behave, in many ways, as the ideal marker for extracellular fluid. Although taken up by, and incorporated in, cells, the process is very slow. The relatively small size of the ion – in comparison with, say, inulin – means that distribution is rapid. An intravenous dose of $^{35}SO_4$ is given and its concentration in blood measured over several hours. When the logarithm of the concentration is plotted against time, the concentration is seen to fall rapidly at first as equilibration occurs, and then slowly and linearly as the sulphate is lost from the extracellular fluid to the kidney and to other cells. Extrapolation of the straight line back to zero time gives the concentration of the original dose in the extracellular fluid. The volume, $V_D$, is then

$$V_D = \frac{I}{C_0}$$

where $I$ is the amount injected and $C_0$ the concentration at zero time by extrapolation.

### Interstitial fluid

Like intracellular fluid, this compartment is difficult to quantify. Its volume is calculated by subtracting the plasma volume from the extracellular fluid volume. Errors involved in measuring these will be reflected in the estimate of the interstitial fluid volume. About 75% of the extracellular fluid is extravascular and lies between the cells. This is about 15% of body weight in both sexes and is therefore about 10 litres in the 'standard man'.

The analysis of interstitial fluid is extremely difficult. Although the composition probably varies from site to site it is usually taken to be similar to that of plasma with a much reduced protein content (since the plasma proteins cannot easily cross the endothelial walls of blood capillaries). Sometimes the composition of the interstitial fluid in a particular tissue (the tissue fluid) is assumed to be the same as that of the lymph leaving the tissue. Analysis of fluids obtained from muscle has been attempted, and wicks and porous capsules have been inserted into tissues to collect fluids, but most

**Table 1.4 The ionic composition of interstitial and intravascular fluids**

| Constituent | Interstitial fluid | Plasma water |
|---|---|---|
| Sodium ions (mmol/l) | 144 | 150 |
| Potassium ions (mmol/l) | 4 | 4.3 |
| Calcium (mmol/l) | 1.25 | 2.7 |
| Magnesium ions (mmol/l) | 0.75 | 1.6 |
| Chloride ions (mmol/l) | 114 | 110 |
| Hydrogencarbonate ions (mmol/l) | 30 | 28 |
| Phosphate ions (mmol/l) | 2 | 2.1 |
| Sulphate ions (mmol/l) | 0.5 | 0.55 |
| Organic acids (mEq/l) | 5 | 5.3 |
| Protein (g/l) | <10 | 75 |

of the published figures are based upon the analysis of oedema fluid (which is an excess of tissue fluid forming under pathological conditions) or of lymph from peripheral lymphatic vessels (*Table 1.4*). Most of the differences between the concentrations in plasma and interstitial fluid, apart from the protein difference and the associated difference for calcium (which is bound by plasma proteins) can be attributed to the Gibbs–Donnan equilibrium set up because of the relatively non-diffusible protein inside the vascular compartment.

*Plasma*

Plasma is the principal intravascular fluid. Its composition will be considered in much more detail in Chapter 3 but its volume and the measurement of that volume will be included here.

Plasma comprises some 5% of body weight – about 3.5 litres. It is normally found only in the heart and the blood vessels. It differs from the other extracellular fluids in containing an appreciable concentration of protein. Values for the concentration of protein and other substances in plasma are given in *Tables 1.4* and *1.5*. The data in these tables have been derived from different sources and there are slight differences between them. This is not surprising since there will always be variations in any physiological measurements, due to the differences between subjects, the variations in the state of a subject at the time of measurement, or even differences in the way the measurement was performed. Most single values given as 'normal' are better described as 'typical' (see the comment in the preface) and it is more accurate, if less convenient, to quote ranges.

The volume of plasma is measured by a dilution method like those described earlier. The marker substance employed is usually one which attaches to plasma proteins. This ensures that the labelled combination of protein and marker is one which will not enter cells or leave the blood capillaries. An example would be the dye, Evan's Blue, which binds to plasma albumin. The small loss of plasma albumin due to its escape from the capillaries or its metabolic breakdown can be compensated by plotting the logarithm of concentration against time as described in the estimation of total extracellular fluid volume above. Radio-iodine ($^{131}$I) can also be used as a labelling substance, since it binds to plasma albumin. As the relative volumes of cells and plasma in the blood are easily measured by centrifugation, substances such as radiochromium and radiophosphate, which attach to blood cells, can be used as markers to measure plasma volume indirectly.

*Lymph*

The fluid in the lymphatic circulation amounts to about 2% of body weight – about 1.4 litres. This volume is probably about the same as the amount of lymph formed and returned to the blood stream in 24 hours. Its composition is variable, particularly with regard to its protein content. Lymph from the liver has a high protein content, around 55 g/l, whereas that from, for example, the leg has a lower concentration, around 26 g/l. In other respects the composition of lymph is very similar to that of blood plasma and tissue fluid (*Table 1.5*).

*The relationship between plasma, interstitial fluid and lymph*

The three compartments of the extracellular fluid are in equilibrium across the endothelial walls of the vessels containing either blood or lymph. Before considering the very important relationship between these fluids the nature of the boundaries between the compartments must be understood.

The capillaries of the blood

The blood capillaries are the smallest vessels of the circulatory system. They are the connecting links between the arterioles, the terminal vessels on the arterial side of the circulation, and the venules, the collecting vessels on the venous side. Any capillary has, therefore, an arterial and a venous end.

The blood capillaries are made up of a tube of endothelial cells supported on their outer side by a basement membrane. A cross section through a small capillary may show it to be a tube formed by a single endothelial cell in the plane of the section, but larger ones are made up of a single layer of endothelial cells with more or less complex junctions. In some parts of the circulation additional supporting cells may influence the permeability

Table 1.5 Composition of plasma and lymph

| Constituent | Plasma | Lymph |
|---|---|---|
| Sodium ions (mmol/l) | 140 | 138 |
| Potassium ions (mmol/l) | 4.5 | 3.9 |
| Calcium (mmol/l) | 2.4 | 2.1 |
| Magnesium ions (mmol/l) | 0.9 | 0.85 |
| Chloride ions (mmol/l) | 100 | 100 |
| Hydrogen carbonate ions (mmol/l) | 22 | 23 |
| Phosphate ions (mmol/l) | 3 | 2.9 |
| Protein (g/l) | 70 | 48 |
| Plasma albumin (g/l) | 35 | 26 |
| Plasma globulins (g/l) | 37 | 20 |
| Amino acids (mg/l) | 49 | 48 |
| Urea (mmol/l) | 3.7 | 3.9 |
| Glucose (mmol/l) | 6.8 | 6.7 |

characteristics of the vessel walls. Endothelial cells are usually very thin (100–300 nm), except in the region of the nucleus, which bulges out the cell to some 2–3 µm thickness. In the lungs and in endocrine glands the cell bodies may be even thinner, perhaps a tenth of the thickness.

The outer membranes of the endothelial cells are smooth with very few processes. The only prominent organelles are vesicles some 65–75 nm in diameter, which may make up as much as a third of the cytoplasmic volume. Some vesicles appear to have fused with the outer membrane of the cells: they may be pinocytotic vesicles. These could pick up fluid containing larger molecules such as proteins and carry it across the cell to release it at the other side. Fewer vesicles are observed in the endothelial cells of the capillaries of the brain and in fenestrated capillaries. Where endothelial cells come into contact their cell junctions are long and tortuous. The cell membranes are typical bilayers with outer electron-dense layers about 2.5 nm wide separated by an electron-lucent layer about 3.5 nm wide. The gap between the cell membranes is between 10 and 20 nm on the tissue fluid side but only some 3–4 nm on the luminal side. The intercellular space appears relatively dense and may be almost obliterated to form a zonula occludens, or attachment belt. There may be open channels through these: certainly labelled protein molecules appear to traverse an intercellular pathway. In some capillaries the endothelial cells show specialised pore-like areas which are termed fenestrations. In these areas the cells are attenuated so that the cell membranes from opposite sides of the cell are fused together to form a five-layer structure. These areas are between 20 and 120 nm in diameter, with a typical size of about 70 nm. The fenestration itself forms a diaphragm about 6 nm thick, with a central knob about 15 nm in diameter.

Three types of capillary may be distinguished according to their structure and properties (Fig. 1.8). Continuous capillaries have close or closed junctions between the endothelial cells of their walls, no fenestrations, and a continuous basement membrane outside the endothelial layer. They are found in muscle, skin, adipose tissue, lungs, the central nervous system (particularly in the cerebrum), the retina and the mammary gland. The basement membrane is made up of glycoprotein and is fibrillar in structure. It may contain collagen.

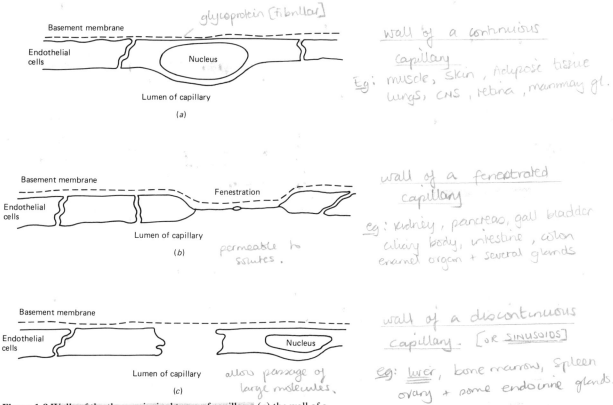

**Figure 1.8** Walls of the three principal types of capillary; (a) the wall of a continuous capillary, (b) the wall of a fenestrated capillary, (c) the wall of a discontinuous capillary. Only one wall of the capillary is shown in each example.

Sometimes cells looking like poorly developed smooth muscle cells, known as pericytes, are found within the basement membrane. No sign of nerve cells has been seen in these or other capillaries.

Fenestrated capillaries contain a high proportion of endothelial cells with fenestrations. These capillaries appear to be more permeable to solutes. They are found in the kidney, the pancreas, the gall bladder, the ciliary body, the intestine and colon, the papillary layer of the enamel organ of the forming tooth, in exocrine glands and in some endocrine glands such as the parathyroid, the posterior pituitary and the adrenal. In all these organs the permeability of the capillaries is functionally advantageous. The endothelial cells of the kidney glomerulus have around a third of their surface occupied by fenestrations. The cells in a fenestrated capillary usually have few vesicles. The basement membrane is continuous outside the capillary and may be relatively thick: 120–150 nm around the glomerular capillaries.

Discontinuous capillaries, or sinusoids, have wide intercellular gaps and show discontinuities in their basement membranes. They are found in organs where the plasma comes into free communication with the tissue fluid and where large molecules, or even cells, pass easily between them. Examples of such organs are the liver, the bone marrow, the spleen, the ovary, and some of the endocrine glands.

### The lymphatic capillaries

Lymphatic capillaries are endothelial tubes, similar to the blood capillaries. Although they are often described as blind-ended, it is probably more precise simply to refer to them as terminal vessels. They collect together to form lymphatic ducts, and these finally join to form the right lymphatic duct and the thoracic duct which drain into the subclavian veins. The function of the lymphatic capillaries is to collect excess tissue fluid and return it to the circulation.

The lumen of a lymphatic capillary is usually wider than that of the corresponding blood capillary. The endothelial cells of the walls are similar in size to those of blood capillaries. There are many vesicles, about 50 nm in diameter with a wide range from 4 to 100 nm, sometimes so large as to almost completely fill the cell. The cells contain cytoplasmic filaments, mitochondria, Golgi complexes, and a scanty endoplasmic reticulum with many ribosomes. Cell processes project into the capillary lumen and into the extravascular space. The junctions between cells may be edge to edge, overlapping, or plicated. In some tissues the lymphatic capillaries have tight junctions between their endothelial cells but more commonly there are gaps or open junctions between the cells. The cells appear to be easily separated and it is possible that the junctions are not stable but vary from time to time. The basement membrane of these capillaries is discontinuous or, more often, completely absent. The capillaries are often surrounded or supported by collagen fibres.

### Movement of fluids between the extracellular compartments

Movement of fluid between the compartments of the extracellular fluid occurs as a result of differences in hydrostatic pressure or osmotic activity. Fluid also moves along tubular structures with valves as a result of the pumping effect of local mechanical activity.

The hydrostatic pressures in the systemic blood capillaries are produced by the pumping action of the heart. Typical figures for a peripheral capillary would be 4.4 kPa at the arterial end and 2.0 kPa at the venous end. The pressure exerted on the interstitial fluid is of the order of 0.4 kPa below atmospheric pressure. Since water passes freely through the capillary walls, a hydrostatic pressure difference of $4.4 - (-0.4) = 4.8$ kPa at the arterial end and $2.0 - (-0.4) = 2.4$ kPa at the venous end, tends to drive water out of the capillary (Fig. 1.9). The process is one of selective filtration since many solutes – in particular the plasma proteins – can diffuse only with difficulty across the endothelial walls.

The presence of these solutes, therefore, establishes an osmotic gradient along which water is forced into the capillary. Since proteins are the main group of substances experiencing difficulty in crossing the capillary wall, the concentration of protein determines the osmotic gradient set up, and the difference in osmotic activity between the interstitial fluid and the plasma is termed the plasma colloid pressure, or the oncotic pressure. With normal plasma concentrations it amounts to some 3.3 kPa. This estimate includes a small osmotic pressure resulting from the imbalance of ions across the capillary wall due to the Gibbs–Donnan equilibrium set up by the presence of the charged non-diffusing protein.

In the capillary, then, a balance of forces is set up: at the arterial end the pressure forcing water out is 4.8 kPa and the pressure opposing this is 3.3 kPa: the net effect is an outward movement of water. At the venous end the pressure forcing water out is only 2.4 kPa but the oncotic pressure is still around 3.3 kPa: water moves into the capillary.

The balance is such that, on the whole, there is a small net loss of water from the blood. About 2% of the fluid passing through a capillary is lost to the interstitial space. This is compensated by lymph formation and flow. Muscular activity around the lymph vessels and the rhythmic contractions of the larger lymph vessels cause movement of fluid along the valved lymph channels, establishing a pressure

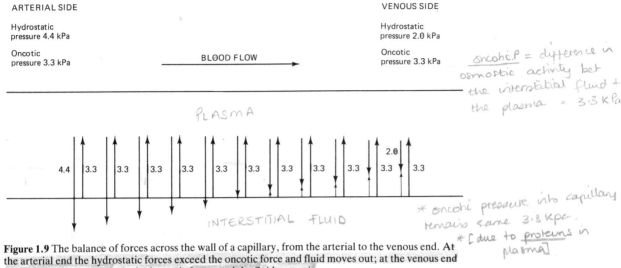

**Figure 1.9** The balance of forces across the wall of a capillary, from the arterial to the venous end. At the arterial end the hydrostatic forces exceed the oncotic force and fluid moves out; at the venous end the oncotic force exceeds the hydrostatic forces and the fluid moves in.

gradient of as much as 3.3 kPa to return fluid from the interstitial space eventually back to the blood.

In the lungs the blood circulates at much lower pressures than in the systemic circuit. The maximum pressures in the pulmonary arteries themselves are only around 3.3 kPa. The pressure in the lung capillary will always be less than this and so the hydrostatic pressure is exceeded by the oncotic pressure along the whole length of the capillaries. This is an important factor in preventing formation of a tissue fluid that would increase the length of the diffusion path for oxygen and carbon dioxide between the air in the lungs and the blood in the capillaries. In heart failure, when the left side of the heart becomes incapable of pumping sufficient fluid out into the systemic circulation, a back pressure builds up in the pulmonary capillaries to such an extent that fluid is forced out of them and pulmonary oedema results – and the uptake of oxygen is reduced.

*Movement of solutes between extracellular fluid compartments*

Small molecules pass easily through cell membranes and also, therefore, through capillary walls. The walls of a typical capillary behave as if they are permeable to molecules up to 7–9 nm diameter, although larger molecules up to 50 nm diameter may pass through in a more restricted fashion. Capillaries appear to have $1-2 \times 10^9$ pores/cm$^2$ in the 7–9 nm range and only a few in the 50 nm range. Most ions, then, such as sodium, potassium, calcium, magnesium, chloride, phosphate, sulphate and hydrogen carbonate, pass readily through the walls, together with organic molecules like glucose, urea and the amino acids. Lipid-soluble molecules diffuse much more rapidly. The gases, oxygen and carbon dioxide, are fat soluble and can diffuse across the entire surface of the endothelial cell. The rate at which molecules reach the tissues from the blood is related to the size of the molecules, since this influences both their ability to cross the vessel walls and their rate of diffusion. Thus in comparison with glucose, urea reaches the tissue fluid from the blood some three times as rapidly, sodium and chloride ions some four times and water some 80 times.

Larger molecules do pass through, though in a more restricted fashion. Experiments with markers of various sizes have shown that molecules between 2 and 7 nm in effective diffusion diameter can pass between the endothelial cells of capillaries in many tissues. Larger molecules cross the endothelial cells in pinocytotic vesicles. Thus ferritin, with an effective diffusion diameter of 11 nm, has been seen in vesicles, but appears to be unable to penetrate between the cells. In fenestrated capillaries molecules between 4 and 10 nm pass more readily through the fenestrations than by vesicular transport. Marker substances around 15 nm pass through the fenestrations but are then unable to pass the basement membrane. In discontinuous capillaries there is easy diffusion through the gaps in the capillary walls.

Plasma proteins and lipoproteins (with diameters ranging from 7 to 60 nm) leave the capillaries mainly by the vesicular route in continuous capillaries, through the fenestrations in fenestrated capillaries, and through the gaps in discontinuous capillaries. The diameters of the vesicles and the fenestrations are approximately equivalent in size to the molecular diameter of the larger lipoproteins. The gaps in discontinuous capillaries may be as large as 1000 nm wide and present no barrier to diffusion. The smaller protein molecules can pass between cells.

There is some evidence that the venous end of a capillary is more permeable than the arterial end.

Because the plasma proteins serve important functions in the transport of other molecules and in the operation of the body's defensive systems, their circulation through extracellular fluid generally may enhance their usefulness. In fact, it has been estimated that 50% of the plasma protein passes through the interstitial fluid and lymph every day. The total amount of the plasma protein in the extravascular fluid at any one time is equal to, or slightly greater than, the total amount present in the intravascular fluid. The actual rate of turnover between plasma and extravascular fluid varies according to the organ studied.

### Movement of large particles and cells between the extracellular fluid compartments

Red blood cells leave the circulation mainly between endothelial cells, usually in the discontinuous capillaries. White blood cells may leave between the endothelial cells if the capillary wall becomes more permeable (*see* p. 66).

### Movement of materials between interstitial fluid and the lymphatic capillaries

In lymphatic capillaries the poorly organised basement membranes and the open junctions between the endothelial cells make ready paths for diffusion. Lymphocytes and red blood cells can pass easily between cells, as do marker substances of all sizes. Neoplastic cells can gain access to the lymphatic vessels through these open capillaries and so the lymphatic system provides an easy channel for the dissemination of tumour cells.

Proteins, lipoproteins, the smaller protein complexes, and even the fat globules sheathed in protein termed chylomicra, enter lymphatic capillaries without difficulty.

### Factors affecting tissue fluid formation

The formation of tissue fluid will be affected by changes in any of the forces or pathways involved.

If capillary permeability is increased, as in inflammation, proteins and large molecules are lost into the interstitial fluid. This decreases the oncotic pressure gradient and so the hydrostatic pressure in the capillaries forces out more water, increasing the production of the tissue fluid. This enhanced rate of tissue fluid formation results in the tissue swelling referred to as oedema. Where capillary permeability is increased locally, either by damage to the vessel wall or by the action of histamine, a local swelling known as a weal is produced.

If the hydrostatic pressure is increased, as in muscle tissue during exercise, the rate of formation of tissue fluid will be increased.

Changes in the plasma content of less permeant solutes will cause changes in the rate of formation of tissue fluid. Haemodilution and a decrease in the concentration of plasma proteins (usually serum albumin) cause increased transfer of water to the extravascular compartment. One effect of continued low protein diets is an increase in the size of the abdomen due to accumulation of extravascular fluid – a condition of oedema (here termed ascites).

An increase in tissue tension will reduce the outward flow of fluid from the capillaries although this is rarely sufficient to stop tissue fluid formation.

Obstruction of lymphatic vessels, as in the parasitic disease of elephantiasis, will prevent the removal of tissue fluid by the lymphatic capillaries and cause oedema.

### Oedema fluid and inflammatory exudates

The formation of oedema fluid has been discussed above. The composition of the fluid is similar to that of tissue fluid, although there is usually a greater protein content.

In inflammation, the reaction to injury or infection, there is an increased permeability of the capillaries due to the presence of peptides from damaged cells and substances like histamine released as a result of tissue damage or immunological

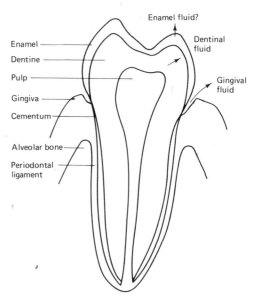

**Figure 1.10** Section through a tooth and the surrounding tissue to show sites where fluid flow is thought to occur.

reactions. The fluids formed are therefore rather like oedema fluid, although they usually contain more of the larger molecular weight plasma proteins involved in defence, and they contain white blood cells that have crossed between the endothelial cells of the capillary walls in response to a number of biochemical attractants. An inflammatory exudate containing dead and dying white blood cells and micro-organisms is usually called pus.

*Specialised fluids related to dental tissues*

Three fluids have been described in relation to dental tissues – the gingival crevice fluid, dentinal fluid, and enamel fluid (Fig. 1.10). The first of these appears to be an inflammatory transudate of plasma; it is considered further in Chapter 11 in relation to saliva, but may be dismissed here as a special example of an inflammatory exudate. Dentinal fluid is the fluid found in the tubules of the dentine, around the processes of the odontoblasts. It is probably best regarded as a special tissue fluid with unusually high concentrations of calcium and phosphate because of its surroundings. Enamel fluid is more of an enigma. It is a fluid which has been observed to collect in droplets when a layer of oil is placed over the enamel surface of the teeth of anaesthetised animals. Its rate of formation is slow, and it is not clear how it is formed. The composition is given in *Table 1.6* but the information is necessarily limited because of the very small amounts which have been analysed.

**Table 1.6 Composition of fluids of dental tissues**

| Constituent | Dentinal fluid (mmol/l) | Enamel fluid (mmol/l) |
| --- | --- | --- |
| Sodium ions | 150 | |
| Potassium ions | 3 | 1.6 |
| Calcium ions | | 2.4 |
| Chloride ions | 100 | |

## Further reading

Coxon, R.V. and Kay, R.H. *A Primer of General Physiology.* Butterworth, London, 1967

Dyson, R.D. *Cell Biology: A Molecular Approach.* 2nd edn, Allyn and Bacon, Boston, 1978

Lockwood, A.P.M. and Lee, A G. *The Membranes of Animal Cells,* 3rd edn, Arnold, London, 1984

Valtin, H. *Renal Function,* 2nd edn, Little Brown, Boston, 1983

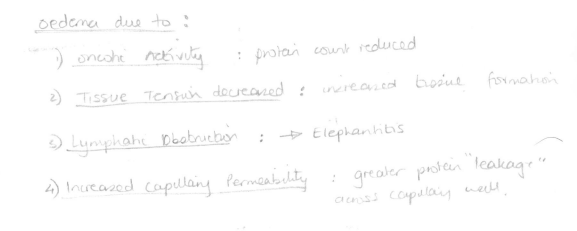

oedema due to:
1) oncotic Activity : protein count reduced
2) Tissue Tension decreased : increased tissue formation
3) Lymphatic Obstruction : → Elephantitis
4) Increased Capillary Permeability : greater protein "leakage" across capillary wall.

# 2

# The transfer of materials across epithelial cell layers – secretion and absorption

Movement of substances across cell membranes has so far been considered in the context of single cells or in relation to exchanges taking place between the major compartments of the body fluids. However, in epithelial cell layers, movements across the cell membrane may be become localised and specific, so that a net flow of particular substances, including water, occurs in a particular direction across a whole cell layer. This polarisation and modification of the activity of the cells can be seen in the formation of the transcellular fluids in their epithelium-lined compartments; in the production of the exocrine secretions (fluids of controlled composition directed by epithelial cells towards the surface of the body), and endocrine secretions (biochemical agents passed into the bloodstream to be carried to distant target organs); in the formation of urine in the kidney, as substances are added or removed; and in the absorption of selected nutrients, salts, and water across the walls of the intestinal tract.

## Transcellular fluids

The greater volume of the extravascular extracellular fluid is separated from the blood plasma by the endothelial walls of the capillaries only; no other cells intervene. Lymph, it is true, is separated from the blood plasma by two endothelial cell layers. There are, however, a number of small fluid compartments which are separated from the plasma by epithelial layers as well as the endothelium: these are known as 'transcellular fluids'. The major ones are the cerebrospinal fluid, the fluids of the eye and ear, and the synovial fluid which lubricates many joints. The spaces between the layers of the sacs which enclose the heart, the lungs, and the intestines (the pericardium, the pleura and the peritoneum) are so small that their contents need not be considered here.

### Cerebrospinal fluid (CSF)

The cerebrospinal fluid fills the ventricles of the brain and the central canal of the spinal cord. The brain has its own tissue fluid in equilibrium with the cerebrospinal fluid, but it has no lymphatic drainage as such (Fig. 2.1). Since the activities of nerve cells are critically dependent upon the composition of the fluid around them, this must be very carefully controlled. The relatively large volume of the cerebrospinal fluid, and the fact that no part of the brain is more than 2 cm away from it, render it a very effective stabiliser of the external environment of the brain cells, in addition to its other important role of providing a protective cushion of fluid between the brain and the skull.

The ventricles communicate with each other and also with the space between the inner two of the membranes covering the brain, the pia mater and the arachnoid mater (Fig. 2.2). This space expands into the so-called cisternae where the pia mater extends into the grooves in the surface of the brain (the sulci) whilst the arachnoid mater bridges over them. The foramen of Magendie in the roof of the fourth ventricle communicates with the cisterna magna, and the two foramina of Luschka in the lateral walls of the fourth ventricle communicate with the cisternae at the base of the brain between the pons and the medulla oblongata. Cerebrospinal fluid is formed continuously from the blood plasma by special structures in the ventricles called the choroid plexuses. It passes through the foramina to circulate through the subarachnoid spaces of the

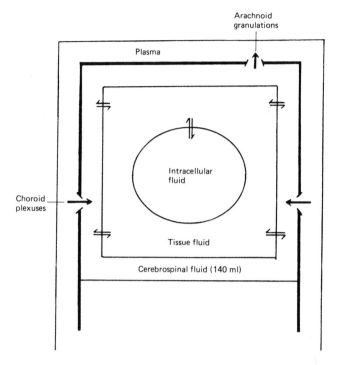

**Figure 2.1** Diagram to show the relationships between the fluid compartments of the brain. The arrows indicate the direction of fluid secretion in the choroid plexuses, and of fluid reabsorption in the arachnoid granulations. The bidirectional arrows indicate exchange between compartments.

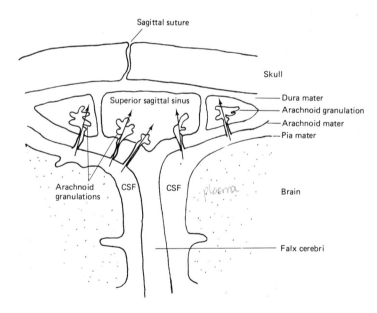

**Figure 2.2** Arachnoid granulations in the venous sinuses of the brain. Arrows indicate the direction of flow of cerebrospinal fluid.

spine, the cerebrum and the cerebellum, and is finally reabsorbed back into the bloodstream through the arachnoid granulations in the large venous sinuses of the dura mater (Fig. 2.3). The total volume of the fluid in man is around 140 ml, the ventricles themselves containing some 28 ml.

The fluid is separated from the blood plasma by the structures of the choroid plexuses and the arachnoid granulations; it is separated from the interstitial fluid of the brain by the pia mater investing the brain or by the epithelium covering of the choroid plexuses, the ependyma.

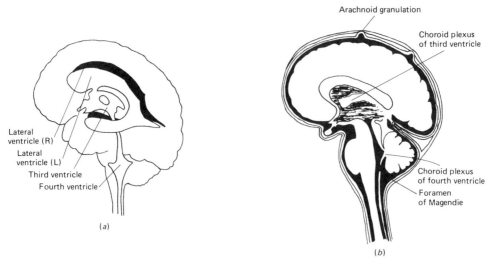

**Figure 2.3** Cerebrospinal fluid: location and circulation: (a) the ventricles of the brain and their connections; (b) the fluid around the brain, in the spinal canal and in the ventricles. The meninges and the arachnoid granulations are exaggerated to make them visible.

## Formation

Some 520 µl/min are formed by the choroid plexuses, highly vascularised outgrowths of the walls of the ventricles, which are covered by a specialised layer of the ependymal lining of the ventricles, the lamina epithelialis. This and the layer of pia mater beneath it are together referred to as the tela choroidea. A rich network of choroidal capillaries is formed from branches of the anterior and posterior choroidal arteries and drains eventually back to the choroidal veins. Frond-like villi extend out into the ventricle from the surface of the choroid plexuses. Each consists of a central stroma of connective tissue and pial cells and contains a single blood capillary. The capillaries are of the continuous type, although some of the endothelial cells may show fenestrations. The villus is covered by a single layer of tall columnar ependymal cells with tight junctions between them, in contrast with the ependyma elsewhere, which has cuboidal cells with few tight junctions. The cells have microvilli projecting into the ventricular fluid and an infolded cell membrane basally. Some cells are ciliated. They contain mitochondria but otherwise no characteristic organelles. The blood supply to the choroid plexuses is very great – some 3 ml/min/g of tissue – being about 5–6 times as great as that to the brain itself and twice as great as that to the kidneys. Cerebrospinal fluid is formed by the movement of fluid from the capillaries of the choroidal plexuses to the interstitial fluid and from there through the tela choroidea into the ventricular fluids.

The composition of cerebrospinal fluid (*Table 2.1*) differs from that of interstitial fluid – sodium and chloride concentrations are higher and potassium and hydrogen carbonate concentrations lower. Equally the fluid is not an ultrafiltrate of plasma – the concentrations do not match with the distribution which might be expected from a Gibbs Donnan equilibrium. Study of the ependymal cells shows that they restrict simple diffusion and they actively transport ions in both directions. Sodium and chloride are actively transported into the cerebrospinal fluid and it is this which generates the osmotic gradient necessary for the formation of the fluid by the movement of water. Calcium concentrations are lower than in plasma because very little protein is able to pass from the plasma into the cerebrospinal fluid and about half of the calcium in plasma is

**Table 2.1 Composition of the cerebrospinal fluid**

| Constituent | Plasma | CSF |
|---|---|---|
| Sodium ions (mmol/l) | 150 | 147 |
| Potassium ions (mmol/l) | 4.6 | 2.9 |
| Magnesium ions (mmol/l) | 0.8 | 1.1 |
| Calcium (mmol/l) | 2.3 | 1.1 |
| Chloride ions (mmol/l) | 99 | 113 |
| Hydrogen carbonate ions (mmol/l) | 27 | 23 |
| Bromide ions (mmol/l) | 1.2 | 0.9 |
| Inorganic phosphate ions (mmol/l) | 1.5 | 1.0 |
| Glucose (mmol/l) | 5.0 | 3.3 |
| Protein (g/l) | 70 | 0.4 |
| pH | 7.4 | 7.3 |
| Partial pressure $CO_2$ (kPa) | 5.5 | 6.7 |

protein-bound. Glucose is known to be transported into the fluid but is used rapidly by the brain cells and so its final concentration is lower than that in plasma. Urea is able to diffuse and balance across the ependymal membrane. The halides, with the exception of chloride, are actively transported out of the fluid. The low concentration of potassium probably reflects its relative exclusion due to the high concentration of positive sodium ions.

Hydrogen carbonate does not pass easily across the membranes of the blood–brain barrier, unlike oxygen and carbon dioxide which, as gases, are lipid-soluble. The dissolved carbon dioxide which reaches the cerebrospinal fluid there hydrates to form carbonic acid. Cerebrospinal fluid is very poorly buffered because of its low protein content, and so changes in the blood carbon dioxide concentration readily produce changes in the pH of the cerebrospinal fluid. This is important in the control of blood carbon dioxide concentrations, since these can be monitored indirectly by cells in the medulla oblongata which are sensitive to hydrogen ion concentration.

## Resorption

Since the cerebrospinal fluid is formed continuously there must be some route by which it can drain back into the bloodstream after its circulation. This is provided by the arachnoid granulations in the venous sinuses of the brain (Fig. 2.2), whose villi enclose tiny canals through which the cerebrospinal fluid flows back to the blood. The cells of the villi are so constituted as to operate in a valve-like manner and allow the cerebrospinal fluid to pass into the sinus when its pressure exceeds the venous pressure there.

## The fluids of the eye

The eye contains two major fluids, the aqueous humour around and in front of the lens, and the vitreous humour behind the lens (Fig. 2.4). The vitreous humour, although gelatinous in nature, contains some 99% water and this solution is in equilibrium with the aqueous humour. The total volume of the vitreous humour is just under 4 ml whilst that of the aqueous humour is about 0.5 ml, of which 0.3 ml are in the anterior chamber. The aqueous humour occupies the space between the vitreous humour and the corneal epithelium; the space is divided into an anterior and a posterior part or chamber by the iris. The function of the aqueous humour is to provide nutrition to the non-vascular lens, which lies in the posterior chamber. The lens is suspended by the ciliary body, made up of the ciliary muscle and connective tissue, with a rich blood supply. The iris also has a rich blood supply. The posterior surface of the iris and the ciliary body are

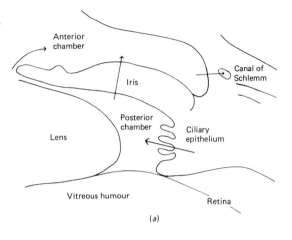

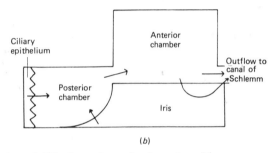

**Figure 2.4** The formation and reabsorption of the aqueous humour of the eye: (a) an anatomical diagram with arrows showing the direction of flow; (b) a schematic diagram of the secretion, flow and reabsorption.

covered by a specialised epithelial layer which actively transports sodium ions, aminoacids and glucose into the posterior chamber. The transport of sodium ions into the posterior chamber produces a gradient of charge along which chloride ions move and a gradient of osmotic activity along which water follows. Like the ependyma, this membrane actively removes other halides from the fluid formed. The vitreous humour is not separately formed but maintains its composition by equilibrating with the aqueous humour. The aqueous humour drains back into the bloodstream through the anterior surface of the iris, which has no covering epithelial layer, and through the canal of Schlemm. Some 2.75 µl/min are formed, and this gives an internal pressure in the eye of around 3.3 kPa. The composition of the aqueous fluid is given in *Table 2.2*; that of the vitreous humour differs slightly because the cells around it secrete hyaluronate. This glycosaminoglycan, with a molecular weight of $2-10 \times 10^6$, exists as long polymeric chains which, because of their extensive overlapping, form very viscous solutions. It is the

**Table 2.2** Composition of the transcellular fluid of the eye

| Constituent | Plasma | Aqueous humour |
|---|---|---|
| Sodium ions (mmol/l) | 150 | 147 |
| Potassium ions (mmol/l) | 5 | 5 |
| Chloride ions (mmol/l) | 107 | 130 |
| Hydrogen carbonate ions (mmol/l) | 27 | 20 |
| Glucose (mmol/l) | 5.7 | 5.25 |
| Protein (g/l) | 70 | 0.1 |

**Table 2.3** Composition of the synovial fluid from the knee joint

| Constituent | Concentration |
|---|---|
| Sodium ions (mmol/l) | 136 |
| Potassium ions (mmol/l) | 4 |
| Calcium (mmol/l) | 2.4 |
| Protein (g/l) | 17.2 |
| Albumin (g/l) | 10.2 |
| Mucoprotein (g/l) | 1.1 |
| Hyaluronate (g/l) | 3.0 |
| Sialic acid (g/l) | 0.3 |
| Glucose (mmol/l) | 0.5 |
| pH | 7.39 |
| Cells (per µl) | 63 |
| Neutrophils (per µl) | 6 |
| Lymphocytes (per µl) | 2 |
| Monocytes (per µl) | 48 |
| Synocytes (per µl) | 4 |

hyaluronate content which gives vitreous humour its viscosity.

## Synovial fluid

Many joints, including the temporomandibular joint, are enclosed in a fibrous capsule lined by an inner membrane termed the synovium, or synovial membrane (Fig. 2.5). The space within the capsule, around the joint surfaces, is filled by the synovial fluid, whose main function is that of a lubricant for the articular surfaces. The outer capsule of the joint is made up of collagen fibres with some blood vessels. There may be some fat cells in the deeper layers. The lining cells are somewhat variable both in number and morphology, looking rather like fibroblasts, but they are larger, stain more readily with basophil dyes and have many vacuoles in the cytoplasm. There are two main types, one with an abundant Golgi system and many vacuoles, the other with an abundant rough endoplasmic reticulum, fewer vacuoles and a much less extensive Golgi system. The first type contains pinocytotic vesicles; they have fine branched filaments in the cytoplasm; and cell processes extend from them into the cavity.

They appear to function, at least to some extent, as macrophages. The second type resembles those mature cells of cartilage and bone which are actively engaged in protein secretion. There is no evidence of tight junctions or of a basement membrane; so it would appear that the synovial fluid constituents reach the fluid by intercellular routes and not by active transport.

The lining cells synthesise hyaluronate and secrete it into the synovial fluid, rendering it highly viscous. The concentrations of ions are similar to those in plasma (*Table 2.3*). The protein content of synovial fluid is less than that of plasma and all the plasma proteins present are small, suggesting that the tissue fluid has been further filtered. In addition to the

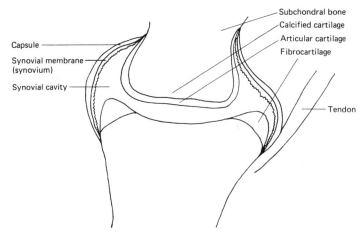

**Figure 2.5** A section through a synovial joint.

plasma proteins there are some proteins probably derived from bone or cartilage. Lymphatic vessels are present in the capsule and it has been shown that there is reabsorption of protein from the synovial cavity, with a flux of about 40% of the protein in 24 h. The fluid also contains a few cells, mainly monocytes, although in the capsule itself the predominant extravascular white cell is the mast cell.

## Exocrine secretions

Single epithelial cells or aggregations, often into distinctive structures, are involved in the production of specialised fluids, termed secretions, which are transported to the body surface. These exocrine secretions have three general functions: the maintenance of the integrity of the body surface by moistening and lubrication, the modification of foodstuffs to enable the nutrients to be absorbed across the walls of the digestive tract, and the cooling of the body surface by providing water for evaporation. In biological terms, the digestive tract from mouth to anus is an enclosed part of the external world: the lining of the digestive tract is an external surface of the body covered by an epithelial layer. Similarly, invaginations of the body, such as the respiratory tract and the urogenital canals, are still parts of the body surface despite their being, apparently, inside it.

The mechanisms by which the exocrine secretions are produced are of two kinds: one primarily involved in the export of large specialised organic molecules, the other in the export of salts and water. In addition to these there may be further mechanisms for the modification of the secretion before it reaches its site of activity.

### Secretion of organic components

The organic components of most secretions are either mucins or enzymes. The lubricant secretion referred to as mucus is actually a mixture of glycoproteins, with a small fluid component. The secretory products of a secretory cell are synthesised within the cell and enclosed in vesicles (Fig. 2.6). A distinction is made between eccrine (or merocrine) secretion, in which the contents of the vesicle are lost when it fuses with the surface membrane and opens to the exterior; and apocrine secretion, in which the whole vesicle is lost, including its bounding membrane. This distinction may not be as absolute as was once thought. Sometimes whole

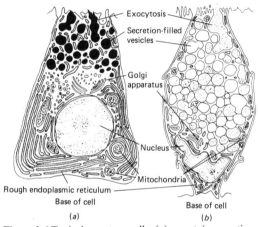

**Figure 2.6** Typical secretory cells: (*a*) a protein-secreting cell; (*b*) a cell secreting glycoproteins. Note the extensive basal endoplasmic reticulum in the protein-secreting cell, and the secretion-filled vesicles in both.

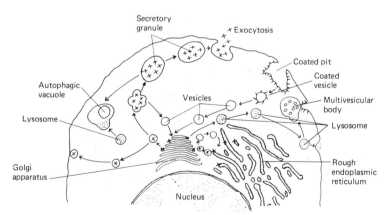

**Figure 2.7** The process of secretion of large molecules. After synthesis by the rough endoplasmic reticulum, the secretory product (x) is packaged by the Golgi apparatus into vesicles and lost from the cell by exocytosis. The membrane material of the vesicles is recycled.

cells appear to disintegrate and lose their contents: this is termed holocrine secretion.

Enzymes and the protein part of mucoproteins are synthesised by the ribosomes on the rough endoplasmic reticulum; a characteristic of enzyme-secreting cells is the extensive endoplasmic reticulum at the base of the cell (Fig. 2.6). The proteins then enter the cisternae of the rough endoplasmic reticulum and from there pass to the Golgi saccules proximal to the nucleus (Fig. 2.7). The Golgi apparatus adds any necessary carbohydrate to the protein molecules, and sends them on in vesicles from the distal saccules to pass to the cell surface. It is thought that microtubules define a path from the Golgi apparatus to the particular face of the cell membrane from which exocytosis will occur: however, in most secretory cells there is an accumulation of secretory vacuoles or vesicles beneath the membrane. This often forces the nucleus to the basal side of the cell. The processes of synthesis and storage usually take place continuously and are divorced from the process of secretion, which occurs only in response to a secretory stimulus. The stimulus to secretion is either a neurotransmitter substance, released as a result of a nerve impulse, or a hormone. Such stimuli can cause the intracellular production of a second messenger substance, although the most characteristic and important effect may be the release of calcium ions in the cell. The calcium ions render the vesicles able to fuse with the plasma membrane, possibly by binding to the vesicle membrane and changing its charge. The vesicles then rupture, and expel their protein and fluid contents to the external environment. The vesicle may reform and pass back into the cell; more usually it simply extends the membrane temporarily, and then, at a later time, pinocytotic vesicles are formed and the membrane reverts to its former size. One unusual gland in this respect is the mammary gland: its secreting cells maintain a fairly constant perimeter because the increase due to the eccrine secretion of casein is balanced by the loss of membrane due to the apocrine secretion of milk fat globules.

The secretion of organic components is seen in the isolated mucous cells of the upper respiratory tract, in the mucous cell aggregations of the digestive tract, and in the individual cells of the tubular glands of the stomach (Fig. 2.8). The enzyme-secreting cells of the salivary and pancreatic glands behave similarly, despite being units of much more complex structures.

## Secretion of salts and water

Many glands produce characteristic mixtures of ions in their secretions; almost all exocrine glands secrete water. In producing these secretions the cells make use of generally available pumping mechanisms, and in most instances the active transport of sodium ions is employed to assist in the movement of other ions or of water. Although water may be exported incidentally in the process of exocytosis, there is no active molecular transport for it. Ion transport systems are usually found in relatively complex glands, either tubular glands (as the gastric pits of the stomach, Fig. 2.8) or acinar glands (as the salivary glands and the pancreas, Fig. 2.9). Acini are the spherical expanded blind ends of the final branches of a tree-like system of ducts. In the bases of the coiled tubes of the eccrine sweat glands and in the acini of the salivary glands a primary secretion is

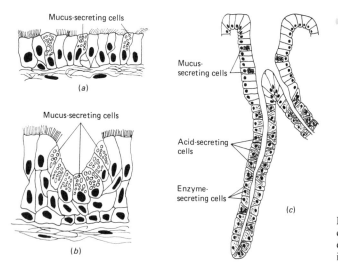

Figure 2.8 Simple secretory structures: (a) single cells in the respiratory tract; (b) aggregations of cells in the respiratory tract; (c) a tubular gland in the stomach (a gastric pit).

Exocrine secretions 31

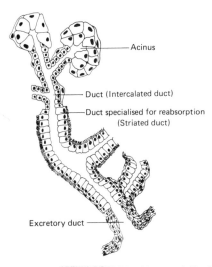

Figure 2.9 An acinar gland, with a specialised group of cells producing the secretion, and a tubular duct which carries the secretion and may modify it in the process. This example is from a 'mixed' salivary gland, and shows specialisation of the cells in the duct as well as in the acinus.

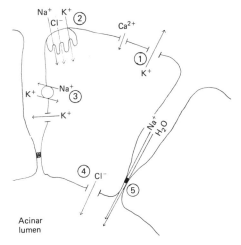

Figure 2.10 The secretion of ions in an acinar gland (a salivary gland). The numbers indicate the sequence of steps from (1) opening of potassium channels to (2) entry of sodium, potassium and chloride, (3) the efflux of sodium by the sodium/potassium ATPase pumps, and then (4) the diffusion of chloride through chloride channels in the luminal membranes followed by (5) diffusion of sodium and water intercellularly down gradients of charge and osmotic activity. The neurotransmitter and second messenger mechanisms which result in these ion movements are shown in *Fig. 21.3*.

Table 2.4 Composition of the primary fluid in sweat and salivary glands (all constituents in mmol/l)

| Constituent | Plasma | Extracellular fluid | Sweat | Saliva |
|---|---|---|---|---|
| Sodium ions | 152 | 144 | 147 | 136 |
| Potassium ions | 4.3 | 4 | 4 | 8 |
| Chloride ions | 110 | 114 | 122 | 112 |
| Hydrogen carbonate ions | 28 | 30 | 28 | |
| Calcium | 2.7 | 1.25 | | 1.25 |
| Magnesium | 1.6 | 0.75 | | 0.75 |
| Lactate | 1.2 | 1.2 | 14 | |
| Urea | | | | 1.5 |

formed with a composition very similar to that of interstitial fluid (*Table 2.4*). When the cells are stimulated by neurotransmitters, the binding of the transmitter to receptor proteins on the cell membrane activates enzyme reactions which result in the basal and basolateral walls of the cells becoming more permeable to potassium ions (Fig. 2.10). This leads to an initial efflux of potassium ions, but is followed by activation of a co-transporter system in the membranes which carries potassium, sodium and chloride ions simultaneously into the cells. The increase in intracellular concentration of sodium ions stimulates the sodium pump mechanisms located laterally in the cells, and sodium is transported by them back out of the cells. The sodium pump mechanisms mobilise energy from ATP to drive the secretion process. The net result of the ion movements is an accumulation of chloride ions in the cells, and these then diffuse across the apical membranes into the acinus or the lumen of the tube. The negative charge of the chloride ions creates a gradient of charge along which sodium ions follow between the cells. Movement of these two ionic species creates an osmotic gradient down which water diffuses. In both salivary secretion and sweat secretion ion transport provides an osmotic pump to carry water to form the secretion and the primary fluid is similar in composition to extracellular fluid.

Some digestive glands modify this basic process in order to produce secretions with a different ionic content. For example, the pancreas is thought to have two different types of acini. In some there is formation of an acinar fluid resembling extracellular fluid in ion content, as above. In other acini the pancreatic acinar fluid is effectively a solution of sodium hydrogen carbonate. Active transport of sodium is still the driving force for secretion but hydrogen carbonate ions replace chloride ions. The cells produce carbon dioxide by metabolism of glucose, and this is hydrated rapidly by the enzyme

**Table 2.5 Composition of the secretion of the gastric acid-producing cells and the mixed secretion found in the stomach (all concentrations in mmol/l)**

| Constituent | Plasma | Extracellular fluid | Canalicular secretion | Gastric secretion |
|---|---|---|---|---|
| Hydrogen ions | $4 \times 10^{-5}$ | $4 \times 10^{-5}$ | 150 | 150 |
| Chloride ions | 110 | 114 | 150 | 150 |
| Sodium ions | 152 | 144 | 4.5 | 40–160 |
| Potassium ions | 4.3 | 4 | 11.2 | 15–25 |
| Magnesium | 1.6 | 0.75 | – | – |
| Calcium | 2.7 | 1.25 | 0.3 | |
| Phosphate ions | 2.1 | 2 | | |
| Hydrogen carbonate ions | 28 | 30 | | |

carbonic anhydrase to form carbonic acid. In the absence of carbonic anhydrase the process would be slow and most of the carbon dioxide would diffuse out of the cells. Carbonic acid dissociates into hydrogen ions and hydrogen carbonate ions. The hydrogen ions exchange for sodium ions by a process of passive diffusion across the basal cell membrane. Outside the cells the hydrogen ions combine with the hydrogen carbonate ions of the interstitial fluid to form carbonic acid, and this breaks down into water and carbon dioxide. The carbon dioxide diffuses back into the cell to be hydrolysed once more and so maintain the gradient. The hydrogen carbonate ions leave the cell either by active transport across the cell membrane into the acinus or by diffusion down gradients of concentration and charge. The gradient of charge is created by active transport of sodium ions across the acinar surface; hydrogen carbonate passes out in preference to chloride.

Yet another variation is seen in the oxyntic cells in the bases of the gastric pits of the stomach. These cells produce a secretion high in chloride ion concentration but with hydrogen ion as the sole cation (*Table 2.5*). It is therefore highly acid with an estimated pH in the region of 0.8. The oxyntic cells have canaliculi extending deep into their cytoplasm and the secretion crosses the membranes of these canaliculi before entering the tube formed by the gastric pit (Fig. 2.11). The canaliculi provide a greatly increased area of membrane surface and allow synthesis of potentially harmful material deep within the cell. The secretion is thought to be produced by a linked active transport system operating in the canalicular membranes (Fig. 2.12). Instead of the sodium pump mechanism providing the secretory drive, hydrogen ion pumping is used. Once again this is the utilisation of an active transport system commonly present in cells – not some unique form of transport. The system which actively transports hydrogen ions out of the cell also pumps potassium ions into the cell. As this pump is activated the canalicular membrane becomes more permeable to chloride and to potassium. These ions

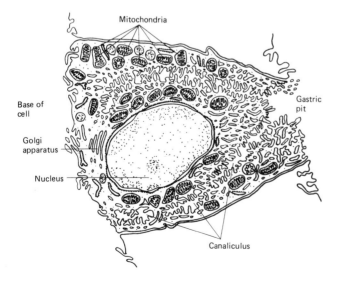

**Figure 2.11** The structure of a gastric acid-producing cell during its active phase. The canaliculi extend deep into the cell, but these would disappear in the resting cell.

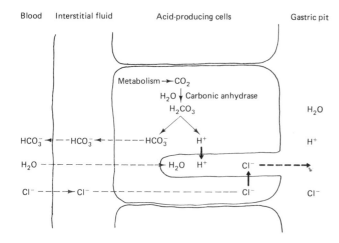

Figure 2.12 A gastric acid-producing cell in schematic form, showing the mechanism by which a specialised ion secretion is formed. The chemical reactions due to metabolism are shown with thin solid arrows, active transport is indicated by heavy solid arrows, diffusion processes by broken lines, and the final flow of secretion by a heavy broken arrow.

diffuse out and the potassium ions are pumped back in exchange for the hydrogen ions. Chloride continues to move out down a gradient of change. The cell chloride concentration is maintained by the inward movement of chloride ions at the base of the cell. The work done in actively transporting the ions requires energy; this is provided by glycolysis, resulting in a build-up of carbon dioxide. The cells contain high concentrations of carbonic anhydrase, so that hydration of the carbon dioxide takes place and the resultant carbonic acid dissociates into hydrogen ions and hydrogen carbonate ions. The cell content of hydrogen ions is thus replenished. Essentially the same mechanism, therefore, provides hydrogen carbonate for the production of an alkaline secretion by pancreatic cells, and hydrogen ions for the production of an acid secretion by the gastric cells. Further, this mechanism is directly dependent upon the normal stimulated metabolic activity of the cells. A concentration gradient is established between the gastric cells and the interstitial fluid with respect to hydrogen carbonate ions and they exchange with chloride ions in the interstitial fluid and are removed in the bloodstream causing the 'alkaline tide' observed when the stomach is secreting acid. There appears to be no outward transport of sodium ions across the canalicular membranes. The active transport of ions provides an osmotic gradient along which diffusion of water to form the volume of the secretion occurs. The hydrochloric acid mixes with the glycoproteins and proteolytic enzymes secreted by other cells in the gastric pit to reach the stomach as a complex mixture with 170 mmol/l of acid and a pH of 0.87.

## The modification of secretions

In the more complex glands, where the secretory cells are situated some distance from the opening of the duct of the gland, passive exchange across the walls of the ducts can modify the secretion. If the acinar secretion is similar to the interstitial fluid, such diffusion produces no net change in composition. When, however, the acinar fluid differs from the interstitial fluid, as in the pancreas, such diffusion will lessen the differences. Thus, fast-flowing pancreatic secretion, spending relatively little time in the ducts, resembles the acinar secretion, but in slow-flowing pancreatic secretion some of the excess hydrogen carbonate is exchanged for chloride ions.

Modification of the secretion in the ducts of a gland can serve a useful purpose. Although active transport of sodium ions is necessary to provide the gradients for diffusion of chloride ions and the osmotic movement of water in the sweat glands and salivary glands, the continuous loss of sodium and chloride ions in the secretions could deplete the body's reserves. The tube of the sweat gland is capable of reabsorbing sodium ions by active transport. Chloride ions follow, but the walls are relatively impermeable to water: the glands can therefore produce a final hypotonic secretion, i.e. a secretion with osmotic activity substantially below that of plasma (Table 2.6). The salivary glands behave similarly. As a result of these ductal modifications to the secretions, slow-flowing saliva or sweat contain little sodium or chloride. At fast flow rates there is insufficient time for enough sodium transport to occur and so the concentrations of these ions rise.

A slightly different process of modification occurs in the formation of bile: sodium ions are actively reabsorbed but water is also removed from the secretion. An initial secretion is continuously produced by the liver, possibly by the active secretion of the bile salts and a consequent osmotic movement of water. The secretion also contains the

Table 2.6 Composition of sweat and saliva secreted at different flow rates (all concentrations in mmol/l). Three values are given for each secretion – the concentrations in the initially formed fluid, in the secretion at slow flow rates, and in the secretion at fast flow rates.

| Constituent | Saliva | | | Sweat | | |
|---|---|---|---|---|---|---|
| | Initial | Slow | Rapid | Initial | Slow | Rapid |
| Sodium ions | 136 | <10 | 80 | 147 | 20 | 100 |
| Potassium ions | 8 | 34 | 36 | 4 | 4 | 4 |
| Calcium ions | | 1.2 | 1.8 | 1.25 | 0.9 | 0.9 |
| Magnesium ions | | 0.75 | | 0.75 | 0.75 | 0.75 |
| Chloride ions | 112 | 10 | 40 | 122 | 20 | 80 |
| Hydrogen carbonate ions | 30 | 4 | 35 | 4 | | |
| Lactate | | | | 14 | 14 | 20 |
| Urea | | | | | 1.5 | 1 |

bile pigments and a number of hormones, hormone metabolites, and detoxified substances, all conjugated with glucuronic acid and secreted by a carrier for organic acids (see p. 39). Bile is produced continuously, so it is stored in the gall bladder until it is needed in the process of food digestion. This organ functions rather like the ducts of the glands described above in modifying the secretion. It actively absorbs sodium ions, and chloride ions follow them across its walls. Unlike the ductal cells, however, those of the gall bladder permit water to pass down the osmotic gradient produced: as a result the bile ejected from the gall bladder is a more concentrated solution (Table 2.7).

Yet another modification is seen in the ducts of the salivary glands and in those ducts of the pancreas with a primary secretion resembling interstitial fluid. In these ducts hydrogen carbonate may be secreted into the ductal lumen and chloride reabsorbed to balance the charge. This results in a retention of chloride and an output of metabolically produced hydrogen carbonate.

Table 2.7 Composition of bile as secreted by the liver and after modification in the gall bladder (all concentrations in mmol/l)

| Constituent | Hepatic bile | Gall bladder bile |
|---|---|---|
| Sodium ions | 174 | 220–340 |
| Potassium ions | 6.6 | 6–10 |
| Chloride ions | 55–107 | 1–10 |
| Hydrogen carbonate ions | 34–65 | 0–17 |
| Calcium ions | 6 | 25–32 |
| Magnesium ions | 3–6 | – |
| Bile acids | 28–42 | 290–340 |

## Secretion in the salivary glands

The general principles already put forward can be expanded to deal in more detail with the secretion of saliva.

### Histological and ultrastructural appearance of the salivary glands

The acini of the salivary glands are spherical groups of cells, polygonal in section, each enclosed by a basement membrane, and enclosing a space, the lumen of the acinus, which is the beginning of the duct system (Fig. 2.9). The acinar cells are classified histologically into two types according to their appearance after staining with haematoxylin and eosin.

In the acini of the submandibular and sublingual glands, but rarely in the parotid glands, there are cells almost filled with large bluish-staining granules or vacuoles, like those in cells elsewhere which produce mucoprotein secretions. Since the secretion of these glands is viscous and rich in protein–carbohydrate complexes, the cells are normally referred to as mucous cells. However, both pairs of glands also secrete other substances. Electron micrographs of the cells (Fig. 2.13) show the greater part of each cell on the luminal side to be filled with large vacuoles, some of which are discharging their contents by exocytosis. The nucleus lies toward the base of the cell, where there is an extensive endoplasmic reticulum and many mitochondria. Smooth parallel membrane structures termed cisternae are also seen. There are tight junctions near the basal ends of the cells and the lateral walls of adjacent cells are tortuously interlinked. Fine intracellular canaliculi are present.

Most of the acini of the parotid glands and the glands of von Ebner are made up of cells whose

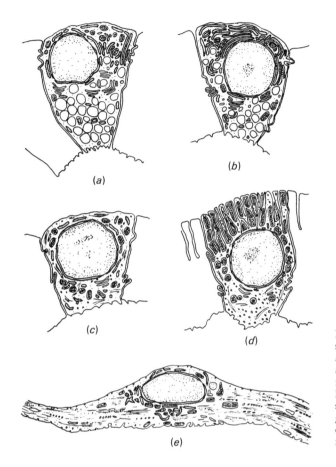

**Figure 2.13** The cells of a salivary gland: (a) a cell secreting glycosamino-glycans; (b) a serous cell; (c) an intercalated duct cell; (d) a striated duct cell; (e) a myoepithelial cell. In (a) and (b) the acinar membrane faces downwards, and in (c) and (d) the luminal membrane faces downwards. Note the secretion-filled vacuoles in (a) and the secretory granules in (b). The striated duct cell (d) has an extensively infolded basal membrane and many mitochondria.

abundant pink-stained granules give them an overall pink appearance. As the secretion of these glands is less viscous (more serous), these cells are termed serous cells. Similar cells in the anterior lingual glands and the submandibular glands are arranged as a layer on the outer surface of acini made up of mucous cells. In sections this layer appears as a crescent shape, or demilune. Although in the parotid glands serous cells can secrete directly into the acinar lumen, in the other glands the secretion passes through fine canaliculi between the mucous cells. The main product of the serous cells is the starch-splitting enzyme α-amylase, but they also produce some mucoproteins. In electron micrographs (Fig. 2.13) the serous cells appear like the mucous cells, but have smaller and darker granules and vacuoles and few cisternae. The vacuoles fill up during resting periods and discharge by exocytosis on stimulation. The granules contain amylase.

The intercalated ducts (Fig. 2.9) consist of small cuboidal cells with large centrally placed nuclei and few organelles (Fig. 2.13). Electron micrographs show little endoplasmic reticulum, and few vacuoles or granules although there are many mitochondria, and tufts of microfilaments are present. The cells interdigitate like the acinar cells. The intercalated ducts are relatively short and their small cells soon give way to the larger cells of the striated ducts. These are almost columnar in section with a centrally placed nucleus. The whole of the basal third of each cell is taken up by deep invaginations of the cell membrane around a large number of mitochondria, giving the cells a striated appearance. The luminal third contains a number of small vacuoles, some mitochondria and tufts of microfilaments (Fig. 2.13). From the striated ducts there is an abrupt transition to the excretory ducts whose epithelial wall has a columnar luminal layer and a flattened basal layer. These ducts join to form the main duct with a multilayered stratified squamous epithelium like that of the oral mucosa (Fig. 2.9).

Another cell type found in the major salivary glands is the myoepithelial cell (Fig. 2.13). Such cells are described as stellate, or basket shaped, since they have long processes wrapped around the acinar or the ductal cells. The thickened part of the cell which contains the nucleus has some mitochondria and strands of endoplasmic reticulum. The remainder of the cell consists of long processes filled with regularly orientated bundles of fibres, termed myofibrils because they resemble the fibres of smooth muscle cells. The myoepithelial cells are contractile cells which may help to expel formed saliva from the lobules.

## The formation and composition of acinar fluid

The acinar fluid consists of water, ions, small molecules and the secretory products synthesised by the cells *(Table 2.4.)* Acinar fluid is produced from the interstitial fluid, which is in equilibrium with the blood plasma across the capillary walls, and therefore contains small molecules and ions in similar concentrations to those in plasma. Molecules such as glucose, urea and the aminoacids can diffuse freely through the capillary wall. The smaller plasma proteins, such as albumin, diffuse more easily than the larger globulins and so are found in greater concentration in the tissue fluid.

The basement membrane which surrounds the acinus defines it as a functional unit; to reach the acinar fluid, ions and molecules must cross the basement membrane and then cross the acinar cells or pass through intercellular spaces. Tight junctions at the bases of the acinar cells restrict the intercellular route. Movement of water and other molecules may occur by diffusion if gradients of concentration or charge exist; as secretion is stimulated the increase in blood flow causes a rise in tissue pressure and this establishes a gradient of pressure towards the acinus.

The acinar cells behave as if freely permeable to lipid-soluble substances and water, but much less permeable to other molecules, even if relatively small. Entry of glucose and aminoacids probably occurs by active transport and since these substances are either used in the acinar cells or have only a limited ability to cross the luminal membrane their final concentration in the acinar fluid is low. Some plasma proteins can pass between the acinar cells even though the tight junctions might be expected to restrict such movement.

Stimulation of the acinar cells causes secretion to begin. All the evidence now available agrees that the process is as described above (and in Fig. 2.10): an opening of basal and basolateral channels in the cell membrane to permit efflux of potassium ions, followed by entry of sodium, potassium and chloride via a co-transporter system, then outward pumping of sodium ions by the lateral sodium/potassium pumps, and finally outward diffusion of chloride ions through channels in the membrane into intercellular clefts or intracellular canaliculi. Sodium ions and water follow the chloride ions into the lumen, probably passing between the cells.

Stimulation also increases the concentration of ionic calcium in the cytosol, both by the release of calcium from mitochondria and microsomes, and by an influx of calcium from outside the cell through calcium 'channels'.

In addition to movements of ions and water, stimulation of the acinar cells initiates the movement of secretory vesicles and granules along pathways delineated by microtubules and microfilaments towards the luminal cell membrane. The increased concentrations of ionic calcium immediately beneath the cell membrane change the relative charges on the plasma membrane and the vesicle membrane to permit their fusion and vesicle contents are lost by exocytosis.

Characteristic potential changes are seen during secretion as the permeability of the cell membranes to different ions changes. The potential of $-20$ to $-30$ mV across the cell membranes of unstimulated acinar cells is probably due to the concentration difference in potassium ions inside and outside the cells as the membranes are only partially permeable to these ions. On stimulation of the cells the potential rises to about $-56$ mV (inside negative) as the permeability to potassium ions increases.

As a result of these processes acinar fluid has an ionic composition broadly similar to that of interstitial fluid *(Table 2.4)* and it contains the characteristic proteins secreted by the acinar cells. Sodium and chloride concentrations resemble those of plasma. The potassium concentrations are greater than those in interstitial fluid, probably because of changes in membrane permeability during secretion, and calcium concentrations are higher.

A fluid of similar ionic composition to the acinar fluid may also be secreted in the intercalated ducts.

## Modification of the acinar secretion in the ducts

### Modification in the striated ducts

The duct system from the beginning of the striated ducts onwards plays an important part in modifying the acinar fluid to become saliva, except at high flow rates. In the striated ducts the acinar fluid is transformed from an isotonic, or slightly hypertonic, fluid into a hypotonic fluid with sodium and chloride concentrations much below those of plasma *(Table 2.6)*. There is a polarisation of sodium pump function, the extensive infolding of the basal cell membranes apparently increasing the pumping capacity on that side (Fig. 2.14). Sodium ions are therefore actively transported out of the cells on the basal side. This increases the concentration gradient between the cell cytoplasm and the fluid in the lumen of the duct and causes sodium ions to diffuse into the cells. The active transport of sodium ions is linked to an active transport of potassium in the opposite direction and associated with a passive

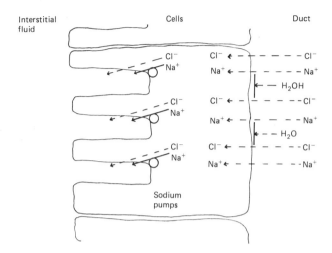

**Figure 2.14** A schematic diagram of a striated duct cell which can reabsorb sodium and chloride ions. The cell is relatively impermeable to water, so that water is not reabsorbed down an osmotic gradient. Diffusion is indicated by broken lines, active transport by arrows on circles.

diffusion of chloride down the gradient of charge it creates. Hydrogen carbonate is actively secreted to the lumen in this part of the ducts. The cells behave as if largely impermeable to water so that, although salts are conserved by this part of the duct, water is not reabsorbed to any significant extent, and a hypotonic secretion results.

The resting membrane potential of the striated duct cells is around −80 mV (inside negative). On stimulation of the glands the transmembrane potential on the luminal side of the cells becomes much less negative (around −20 mV). A similar phenomenon is observed in the cells of the distal tubule of the kidney where there is also active transport of sodium ions across cells.

In the distal parts of the excretory ducts partial re-equilibration of saliva with interstitial fluid may occur and the concentrations of ions may return to values more similar to those of plasma.

## The final secretion

The final product of the salivary glands, resulting from secretion into the acini and subsequent modification of the secretion in the ducts, is described further in Chapter 11.

## Secretion in endocrine glands

The endocrine glands are those which have no functional ductal system, but pass their secretions directly into the blood or interstitial fluid. The hormones secreted pass in the circulation to target organs to produce their effects. All the endocrine glands have a profuse blood supply which brings every cell into a close relationship with the capillary network and the capillaries are often of the discontinuous type. The functions of an endocrine organ can be classified as synthesis, storage and secretion. The greatest variations between glands are found in their storage capacities, ranging from that of the adrenal cortex which stores only sufficient material for about 1.5 minutes secretion to the thyroid gland with enough for 2–3 months secretion.

Glands secreting steroid hormones, such as the adrenal cortex, the ovary and the testis, do not store the secretory product to any extent. Instead they store the precursor molecule, cholesterol, some glands containing as much as 7% cholesterol by weight. The hormone-producing cells contain lipid droplets filled with cholesterol esters, in close proximity to mitochondria and the agranular endoplasmic reticulum (Fig. 2.15). The mitochondrial enzymes break down the lipid esters and synthesise pregnenolone, which is subsequently elaborated into the appropriate steroid hormones by the enzymes of the smooth endoplasmic reticulum and the mitochondria together. It is not known exactly how the final products are released from the cells.

The thyroid gland is exceptional among endocrine glands in storing its secretion extracellularly. The closed follicles of the human thyroid gland probably evolved from the ducts of an exocrine gland found in some invertebrates. This arrangement means that the secretory epithelium must pass the hormone secretion first in one direction to store it in the follicles, and then back in the opposite direction to transfer the hormones from the follicles to the bloodstream (Fig. 2.16). The rough endoplasmic reticulum synthesises a protein known as thyroglobulin, and incorporates mannose into it. This is passed to the Golgi apparatus where galactose is added, and the resultant glycoprotein is packaged into apical vesicles, which migrate towards the luminal side of the cell and are discharged by exocytosis. Iodide is actively pumped into the basal

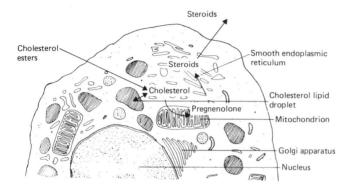

**Figure 2.15** A cell involved in the synthesis and secretion of steroid hormones.

side of the cells from the blood, and passes across the cells to diffuse out down a gradient of charge. Either at the luminal membrane or within the lumen of the follicles the iodide is oxidised by a peroxidase to iodine and bound to tyrosine molecules which themselves have been bound by thyroglobulin. Thyroglobulin-bound thyroxine and tri-iodothyronine are thus stored as a colloidal material within the follicles. The process of storage is probably continuous. When secretion is stimulated the cells take up droplets of colloid by phagocytosis forming vesicles which travel across the cells from the lumen to the basal side. In their passage they fuse with lysosomes and the lysosomal enzymes split the thyroxine and tri-iodothyronine from the carrier thyroglobulin. Exocytosis takes place at the basal sides of the cells, the hormones passing into the bloodstream and the thyroglobulin passsing into the lymphatic system.

Like the enzyme-producing exocrine glands, many endocrine glands maintain a store of secretory products intracellularly. This is often the case when the secretory product is a small molecule which might be easily lost from the cell if it was not bound to create a larger molecule and packaged in vesicular granules. Thus oxytocin and antidiuretic hormone, two of the polypeptides of the hypothalamus, are bound by carrier proteins, the neurophysins, and stored in granules. The catecholamines of the adrenal medulla are thought to form high molecular weight micelles with adenosine triphosphate, the small amount of chromogranin proteins in the granules being insufficient to bind the catecholamines. In all cases the granules contain

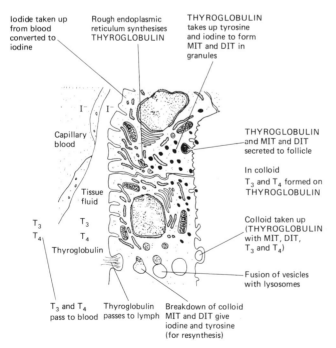

**Figure 2.16** The cycle of events in the synthesis, storage and secretion of thyroid hormones. The sequence begins at the top left of the diagram and reads on clockwise to the lower left

other substances, such as lysosomal enzymes – presumably acquired by fusion with lysosomes – and the phospholipid lysolecithin. This may assist in the fusion of the membrane of the granule with the cell membrane during exocytosis.

## The secretion of substances into the urine in the kidney

The tubules of the kidney, in particular the proximal parts of the tubules, are able to secrete certain substances selectively from the blood plasma into the tubular fluid (the forming urine). These include hydrogen ions and ammonia.

Organic acids and the conjugates of organic acids are secreted into the urine by an active carrier system with transport maximum characteristics. The rate of secretion increases rapidly with increasing concentrations in the plasma up to a maximum level, and then remains at this level regardless of further increases in plasma concentrations. This transport system carries the glucuronic acid conjugates of lipid-soluble drugs and steroid hormone metabolites. It will also carry para-aminohippuric acid. The same carrier appears to operate for all the substances transported: the maximum for the system, therefore, is the maximum for the sum of all the transported molecules.

A similar system secretes some organic bases, such as choline and tetramethylammonium. The neurotransmitters chemically related to these molecules are probably excreted by this route.

## Reabsorption of substances into the plasma from the kidney tubule

The blood plasma filtered through the capillaries of the kidney glomeruli becomes the glomerular filtrate. This is much greater in volume than the final urine excreted and differs from it in composition. The changes which occur in the transformation into urine depend largely upon active transport of materials across the epithelial walls of the kidney tubules back towards the blood plasma.

The small amount of protein which crosses the glomerular filter is reabsorbed by pinocytosis into the cells of the kidney tubules, where it is broken down into aminoacids when the pinocytotic vesicles fuse with lysosomes. The aminoacids then pass out of the basal sides of the cells.

Although the transport of sodium across the cells of the kidney tubules occurs throughout their length, most is reabsorbed in the proximal tubules. Here the cells forming the walls of the tubules are cuboidal in shape with extensively infolded basal cell membranes. They behave as if $(Na^+-K^+)ATPase$ pumps are located in the basal area of the membrane. Whether the pumps are localised in this polarised fashion, or whether they are evenly distributed in the membrane and the membrane area is much greater because of the infolding, is not clear. Whatever the explanation, at the luminal side of the cells there is a net influx of sodium ions diffusing down a concentration gradient from the glomerular filtrate, and at the basal side of the cells the sodium ions are actively pumped out into the interstitial fluid. Chloride ions follow down the gradient of charge thus created, and water follows down the osmotic gradient thus induced. These gradients are not apparent when fluids on either side of the cells are sampled: the absorption appears to be isosmotic. It is thought that the gradients must be produced either between the cells or in the basal clefts.

Nearly all the other reabsorptive transport systems that have been described in the kidneys exhibit a transport maximum. The best known of these is that carrying glucose from the tubular fluid back to the blood plasma. As the concentration of glucose in the blood reaching the kidney is raised, so the concentration in the tubular fluid rises, because glucose passes easily through the glomerular filter. The amount of glucose reabsorbed rises in parallel with this so that, up to a certain concentration in the plasma, no glucose appears in the urine. However, at a plasma concentration of approximately 10 mmol/l, either the capacity of the pump or the availability of pumping sites in some tubules is exceeded. At plasma concentrations above 17 mmol/l a constant amount of the glucose is reabsorbed, and the remainder passes on into the urine. The system is specific: it will carry glucose or galactose and then, to a much lesser extent, other sugars such as mannose or fructose. Even the optical isomer of glucose, L-glucose, has much less affinity for the carrier than has D-glucose itself. The pump is inhibited by drugs like ouabain and phloridzin, which inhibit the sodium-potassium pump: this suggests either that the transport systems are linked, or that the energy generated in pumping sodium ions is necessary for driving the glucose pump.

Aminoacids are also pumped across the walls of the renal tubules. Since some compete with each other for transport, it is thought that three separate transport systems exist. One carries basic aminoacids like arginine and histidine, a second carries acidic aminoacids like aspartic and glutamic, and the third carries neutral aminoacids but possibly some of the basic ones as well. The carrier systems exhibit transport maximum behaviour, although the transport maxima are considerably above the normal concentrations of the appropriate aminoacids in the plasma, so that they do not normally appear in the urine. The transport of aminoacids, like that of glucose, seems to be dependent upon simultaneous transport of sodium ions.

# Absorption of food materials in the intestine

Immediately after a meal, the concentration differences between the contents of the full intestinal lumen and the interstitial fluid and blood on the other side of the epithelial lining are normally sufficiently great to drive rapid diffusion processes provided that the epithelial wall is permeable to the food molecules. The role of the digestive processes in the stomach and intestine is to convert the foodstuffs into their simple nutrient components – aminoacids, monosaccharides, monoglycerides, fatty acids and glycerol – all of which will be small enough to diffuse across the intestinal walls. The other factor in the Fick equation for diffusion, in addition to the concentration difference, is the area available for diffusion: and the surface of the small intestine is modified to provide a maximum area (Fig. 2.17). The intestinal surface is formed into folds, or ridges, which effectively triple its area. On a smaller scale, the actual epithelial wall is formed into a series of finger-like processes, the villi, which give a further ten-fold increase. Finally, each of the columnar surface cells in the single-layered epithelial covering has microvilli over the whole of its luminal surface, increasing the actual cell membrane area by some 20 times. The total absorptive surface of the small intestine has been estimated to be $2\,000\,000\ cm^2 - 200\ m^2$.

The effective pore diameter for diffusion through this epithelial cell layer is between 0.72 and 0.8 nm. Table 2.8 shows how this limits the diffusion of sugar molecules, which are water-soluble but not lipid-soluble. Many lipid-soluble substances cross the cell membrane readily. Indeed, drugs like the barbiturates can diffuse across the intestinal wall in almost any part of the digestive tract: the specialisation seen in the small intestine is not needed for the absorption of such substances.

The rate of diffusion depends upon the concentration difference on the two sides of the membrane; if the absorption of food substances depended solely on diffusion the rate would slow rapidly as the concentration gradient disappeared. Diffusion can continue to operate efficiently, however, if the concentration gradient is maintained. This is achieved by two processes: rapid chemical modification of the diffusing substance within the cells, or removal of the diffused substance from the other

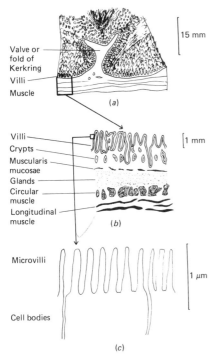

**Figure 2.17** The elaboration of the surface of the small intestine to provide maximum area for absorption: (a) intestinal folds – the valves of Kerkring, which increase the area three-fold; (b) the intestinal villi which increase the area ten-fold; (c) the microvilli on each cell which increase the area twenty-fold.

**Table 2.8 Rates of diffusion in relation to the molecular size of sugars**

| Molecule | MW | Effective diameter (nm) | Rate of diffusion ($\mu$mol/g wet tissue/h) |
|---|---|---|---|
| Polysaccharide: starch | 50 000 | | no penetration |
| inulin | 5000 | 2.96 | no penetration |
| Disaccharide: lactose | 342 | 0.88 | 5 |
| Hexose: mannose | 180 | 0.36 | 19 |
| Pentose: ribose | 150 | 0.30 | 22 |
| Triose: glyceraldehyde | 90 | 0.18 | 45 |

side of the cells by some means of bulk transport. Bulk transport in the body usually means the transport of the substances in blood or lymph. The morphology of the intestinal wall is such that removal of absorbed substances by bulk flow is easy. The villi are formed of a layer of absorbing epithelium around a core of connective tissue which contains an arteriole, a capillary network, a venule, and a small lymphatic vessel. The blood capillaries are of the fenestrated type: they are disposed so that little more than the basement membranes of their endothelial cells and those of the intestinal epithelium separate the capillaries from the absorbing epithelial cells. The lymphatic capillaries begin at the tip of the villus and extend basally. They have no basement membranes. Smaller molecules diffuse into the blood capillaries and are carried away; the larger molecules and the fat particles known as chylomicra pass into the discontinuous lymphatic capillaries. After a fatty meal the lymphatics are milky in appearance and so were given the name of lacteals.

Chemical modification of absorbed substances may permit a diffusion gradient to be maintained. Thus glucose may be converted into glucose-1-phosphate in the cells, or it may be transformed into lactate temporarily. Ions may be complexed into different ionic forms or into non-ionised forms.

The large area available for diffusion and the maintenance of concentration gradients by chemical transformations permit absorption in the small intestine to occur readily in the early stages of digestion, but diffusion becomes less effective as the concentrations in the intestine continue to fall. Absorption of both salts and organic molecules can then remain efficient only if active transport systems are available. This is particularly true of the absorption of salts since the gradient for diffusion of salts is much less than that for organic molecules. In fact, as elsewhere in the body, a polarised transport of sodium ions serves to carry chloride and water across the epithelium of both small and large intestines. The important ions calcium and iron are actively absorbed. Both these ions form many insoluble salts and so the gradients for diffusion of the ions are small despite the low intracellular concentrations. The gradient for calcium ions is maintained by the synthesis of a calcium-binding protein in intestinal cells in response to a circulating metabolite of vitamin D, 1,25-dihydroxycholecalciferol. The binding protein appears to transfer the calcium ions to an active calcium pump on the basal side of the cells, and this pumping maintains the unidirectional flow of the ions. Phosphate absorption seems to be linked to calcium absorption. The mechanism for iron absorption has some similarities: it involves a binding protein called apoferritin which takes up iron II ions to form a complex known as ferritin. Glucose and galactose are carried by a system similar to that in the kidney tubules – the process is, at first, one of facilitated diffusion, but later it becomes active transport against a concentration gradient. Aminoacids are also transported across the wall of the small intestine by pumps like those in the kidney tubules. Both carbohydrate and aminoacid pumps are dependent upon the transport of sodium ions.

The entry of lipids is more complex. Although fats, as lipid-soluble substances, might be expected to cross the intestinal wall easily, access of the lipids to the cell membranes is, in fact, prevented by a water layer on the surface of the cells. Some degree of solubilisation of the fats in water is therefore necessary. The bile salts are detergent molecules, with hydrophilic polar groups on the ends of long hydrophobic chains, and as such they are able to assist the fat droplets in the intestine to form an emulsion: the hydrophobic ends of the molecules are immersed in the fat droplet and the hydrophilic ends form a polar layer over it. This initial solubilisation permits the lipolytic enzymes to reach and break down the neutral fats into 2-monoglycerides, fatty acids and glycerol. Glycerol is readily absorbed, but the other molecules need to be maintained in a water-miscible state. This is achieved by the formation of micelles – groups of molecules aggregated together with polar groups outwards and hydrophobic groups inwards, similar on a smaller scale to the globules in the emulsion. From the micelles absorption into the surface cells of the intestinal epithelium takes place (Fig. 2.18). Inside the cells resynthesis of triglycerides takes place and new droplets of fat with an outer coat of protein are passed into the intercellular clefts and thence to the lacteals. These droplets are the chylomicra. Fatty acids with less than 10 carbon atoms in their chains are sufficiently small and water-compatible to pass directly out of the cells and reach the blood capillaries. The bile salts, having solubilised the lipids in the middle part of the small intestine, the jejunum, are themselves reabsorbed in the final part, the ileum.

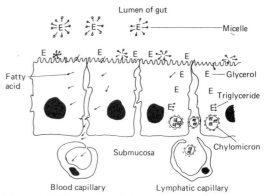

**Figure 2.18** The process of absorption of lipids in the small intestine. On the left, the movement of free fatty acids with less than 10 carbon atoms through the cells, to be taken up by the blood capillaries. On the right, resynthesis of triglycerides, the formation of chylomicra, and the uptake into the lymphatics.

## Further reading

Davenport, H.W. *Physiology of the Digestive Tract,* 5th edn, Year Book Medical Publishers, Chicago, 1982

Pitts, R.F. *Physiology of the Kidney and Body Fluids,* 3rd edn, Year Book Medical Publishers, Chicago, 1974

Sernka, T.J. and Jacobson, E.D. *Gastrointestinal Physiology – The Essentials,* Williams and Wilkins, Baltimore, 1983

Chapters on the secretion of saliva in textbooks of oral biology or oral physiology.

# 3

# Blood plasma

The blood plasma constitutes the intravascular portion of the extracellular fluid of the body. It amounts to some 3.5 litres, and represents, therefore, about one quarter of the total volume of extracellular fluid, and about 5% of total body weight. Plasma is separated by centrifuging whole blood: it is the fluid layer above the packed cells (Fig. 3.1). It normally constitutes around 55% of the blood volume. Plasma differs from serum, the fluid obtained after centrifuging blood that has been allowed to clot, in its greater content of fibrinogen and some of the clotting factors.

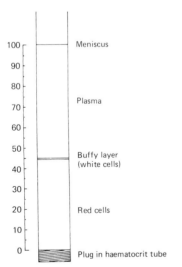

**Figure 3.1** A haematocrit tube after centrifugation. The proportions of cells (denser than plasma, and therefore at the bottom), and plasma (at the top), can be read off as a percentage of the blood volume. The buffy layer includes the white cells and the platelets.

## Composition–inorganic

Although the concentrations of inorganic ions in plasma may vary in different parts of the body, they are in general little different from those in extracellular fluid in the tissues (*Table 1.4*). The blood serves principally as a transport medium, carrying materials around the body, and acts, therefore, as an equilibrating system throughout the body. The values for plasma concentrations given in this chapter are reasonably representative values: they might be different, for example, in blood leaving the absorptive areas of the gut, or in blood entering or leaving the kidneys. It should be remembered here and throughout the book that most values given as normal are figures which are somewhere towards the middle of the normal range but are rounded out or chosen as being fairly easy to remember, rather than precise as in a physical science. It would often be preferable to give normal ranges, but these are less convenient to hold in one's mind, and even then may vary among different populations.

The main cation in the plasma, as in extracellular fluids generally, is sodium. The usually accepted value for its concentration is 152 mmol/l of plasma water, or 140 mmol/l of plasma. This distinction is made because the proteins of plasma contribute to its volume, so that 1 litre of plasma contains only about 930 ml of water. Potassium concentrations in extracellular fluids are low; the values usually given for plasma are around 5 mmol/l of plasma water, and around 4.7 mmol/l of plasma. The other major cations of plasma are calcium, at 2.45 mmol/l of plasma, and magnesium, at 1.5 mmol/l of plasma. The pH of plasma is around 7.40, the hydrogen ions being completely balanced by an excess of hydroxyl ions. Variation in pH occurs as carbon dioxide concentration varies: arterial blood in the lungs has a pH of about 7.40, and mixed venous blood returning to the heart, carrying more carbon dioxide in solution, has a pH of about 7.36. The inorganic ions of plasma are important in maintaining fluid balance in the body since they contribute to the total plasma osmotic pressure exerted across the effectively semi-permeable walls of the cells. Some of the plasma ions help to maintain the concentrations of ions in cells and fluids at optimum levels for the correct functioning of specific body systems – calcium ion concentrations, for example, affect the activity of nerves and muscles.

The anionic components of plasma (excluding the plasma proteins, which are anions at blood pH) are again those found in extracellular fluids generally. Thus the main anion is chloride, present at about 113 mmol/l of plasma water, or 108 mmol/l of plasma. Next in concentration after this is hydrogen carbonate, at up to some 27 mmol/l of plasma water. Here, the proviso about variation in concentration in different parts of the body applies: in a subject in a resting state, breathing quietly, the arterial blood leaving the lungs will contain around 25 mmol/l of hydrogen carbonate in the plasma whilst the mixed venous blood may contain around 27 mmol/l in the plasma. The remainder of the anionic charge is made up of 1 mmol/l of phosphate – mainly monohydrogen phosphate at this pH – 0.5 mmol/l of sulphate ion and some 6 mmol/l of organic acids such as citrate and lactate. The concentrations are per litre of plasma. All these latter components may vary in concentration as a result of dietary intake or tissue metabolism. Clinical chemists often express these concentrations as equivalents rather than moles so that the balance between positively and negatively charged ions will be apparent. The final balancing of charge when expressed in this way is due to some 16 mEq/l of anionic charge contributed by the plasma proteins.

## Composition–organic

The blood plasma contains many small organic molecules, acts as a transport system for many larger ones, and is itself characterised by its content of the plasma proteins. These are described later, but a few of the other organic components may be mentioned here. Their concentrations are given per litre of plasma. The main non-protein nitrogen-containing compound in the plasma is urea, at a concentration of about 4.2 mmol/l. Free aminoacids together make up about 500 mg/l. Glucose is the principal carbohydrate in the plasma, normally at about 5 mmol/l but varying according to the subject's absorptive state. In a fasting subject it may be as low as 4 mmol/l, whilst after a carbohydrate meal it may be as high as 10 mmol/l. Most fats in the plasma are present in a bound form. Hormones, vitamins and enzymes are all found in the plasma in varying amounts.

## The total osmotic activity of plasma

The total osmotic pressure of plasma (that is, the pressure required to prevent water passing through a membrane permeable only to water, from pure water on one side to plasma on the other) is given by an osmolal concentration of 290 mOsmol/l, or about 700 kPa. This value is usually determined by measuring another colligative property (a property dependent upon the number of particles in solution) – the depression of freezing point in comparison with that of pure water. Plasma freezes at -0.54°C, a depression of freezing point less than that predicted from the total concentration of the constituents. This implies that interactions between ions are effectively changing their size and number.

The effective plasma osmotic activity exerted across the walls of the capillaries is much less than 700 kPa since the capillary walls are permeable to small molecules as well as to water (see p. 22, and the section below on functions of plasma proteins).

## The plasma proteins
### Concentrations and classification

Plasma differs from the other extracellular fluids in having an appreciable content of proteins. It is not true to say that plasma alone contains plasma proteins since studies of labelled proteins have shown that there is rapid passage of plasma proteins through the extracellular spaces outside the circulation; nevertheless, the concentrations present outside the circulation at any one time are low because of the limited permeability of most capillaries to molecules of this size and shape.

The protein content of plasma amounts to about 75 g/l of plasma. The plasma proteins contribute little to the total osmotic activity of the blood because their large molecular size means that relatively few molecules, or solute particles, are present in that total – in comparison, for example, with sodium ions whose molecular weight is about 3000 times smaller but whose concentration is only 20 times smaller (see Fig. 3.2).

The original classification of proteins into albumin and globulins was based on fractionation of serum by precipitation, at first by ammonium sulphate, later by various alcohol mixtures. Since serum was used, this classification excluded the fibrinogen present in plasma. The use of increasingly sophisticated electrophoretic techniques, which separate molecules on a basis of charge and size, has led to the separation and identification of more and more

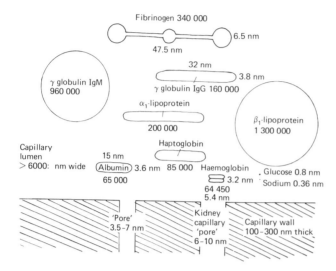

**Figure 3.2** The relative sizes of the plasma constituents. The relation of these sizes to that of the apparent pores in the capillary walls is shown by the size of pore indicated at the bottom of the diagram.

fractions. This in turn has allowed identification of proteins by function as well as by physical properties. As a result, the globulins have been subdivided into $\alpha_1$ and $\alpha_2$, $\beta$ and $\gamma$ globulins, and pre-albumin fractions have been described.

## Albumin

The protein fraction present in the greatest quantity, 35–40 g/l of plasma, is serum albumin. Even this concentration is only about 0.5 mmol/l, since the molecular weight of albumin is 69 000 daltons. Albumin was one of the two original sub-fractions of the plasma proteins and has not been further subdivided by later analytical methods. With the exception of the small amounts of pre-albumins, it is the protein which migrates most rapidly towards the cathode on electrophoresis, since it carries a relatively high charge at pH values above 7.0 and is one of the smallest plasma proteins. Calculation of the theoretical molecular diameter of a protein of molecular weight 69 000 gives a value slightly less than the effective pore size of the capillaries in the kidney glomeruli – 10 nm. The shape of the molecule, however, is narrow and elongated so that only small amounts of the protein normally pass through the glomerular capillary wall. The calculated equivalent pore size in capillaries in general is about 7 nm diameter – approximately the same as the smallest dimension of the albumin molecule and of many other plasma proteins.

Albumin is synthesised in the liver; its turnover rate gives it a biological half-life of about 18 days. Breakdown of the protein is not localised to any specific tissue: albumin functions as part of the mobile aminoacid or nitrogen store of the body. In adults some 10–12 g are produced each day.

## The globulins

The globulins which migrate most rapidly on electrophoresis, the $\alpha_1$ globulins, are a composite group mainly found in association with other smaller molecules. They make up 2–4 g/l of plasma.

The $\alpha_2$ group of plasma globulins is a larger group than the $\alpha_1$, amounting to 4–8 g/l of plasma. A number of identifiable proteins migrate within this group: some mucoproteins, the haptoglobins, prothrombin, caeruloplasmin, complement-inhibiting factor, thyroxine-binding protein, transcortin, and a protein involved in the transport of vitamin D.

The $\beta$ globulins are present in approximately the same concentration as the $\alpha_2$ globulins i.e. 4–8 g/l of plasma. The group includes several proteins of known function: the $\beta$-lipoproteins, transferrin, testosterone-binding protein, some complement factors, blood clotting factors VII, VIII and IX, and plasminogen.

The $\gamma$ globulin fraction is the second largest of the main protein fractions of plasma; the plasma

**Table 3.1 The immunoglobulins**

| Type | Plasma concentration | Polypeptide chains | M.W. (× 000) | Half-life (days) |
|---|---|---|---|---|
| IgA | 1.56–2.94 g/l | K,L, $\alpha$ | 160 and polymers | 5–6.5 |
| IgD | 0.01–0.40 g/l | K,L, $\delta$ | 184 | 2–8 |
| IgE | <10 µg/l* | K,L, $\epsilon$ | 200 | * |
| IgG | 9–18 g/l | K,L, $\gamma$ | 160 | 18–23 |
| IgM | 0.67–1.45 g/l | K,L, $\mu$ | 960 | 5 |

* IgE is rapidly bound by tissues making concentration and half-life difficult to measure

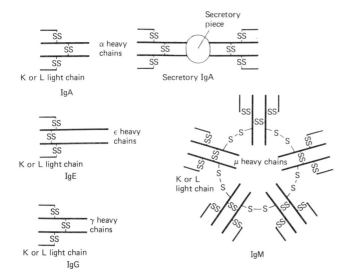

Figure 3.3 The structure of the different classes of immunoglobulin. Sulphur bonds are shown as S. Each protein is made up of either K or L light chains together with α, γ, ε or μ heavy chains. IgD, with δ chains, is not shown, but resembles IgE.

concentration in a normal healthy adult subject is 6–12 g/l. γ globulins are large molecules, with molecular weights between 160 000 and 960 000 daltons. Five main classes of γ, or immunoglobulins are recognised, all based on a structure consisting of two pairs of 'heavy' and 'light' polypeptide chains (*Table 3.1*). The 'light' K and L chains are each built up from two subunits of 110 aminoacids and each of the 'heavy' α, γ, δ, ε and μ chains consist of four of these subunits. The proteins differ in the types of chain they contain and in the structure formed by these chains (Fig. 3.3).

Two types of cell are concerned in the synthesis of immunoglobulins: the plasmacytes and the lymphocytes. Any organ with a large population of these cell types may be a site of immunoglobulin synthesis – lymph nodes and the spleen are the best examples.

## *Fibrinogen*

Fibrinogen is the precursor of fibrin, the protein formed during blood clotting. Its concentration in plasma is between 2 and 4 g/l; it has been removed from serum. The molecule was originally assigned a dumb-bell shape, because of its S–S terminal linkages, but other possibilities have now been suggested. It has a molecular weight of 340 000 daltons and is synthesised in the liver. Two fibrinopeptides, with molecular weights of 3000 daltons and a high content of glutamic acid, are split off from the fibrinogen molecule at arginyl–glycyl bonds to leave the fibrin monomer which can then polymerise into fibrin strands.

## **Functions of the plasma proteins**

Blood circulates throughout the body. Plasma is therefore the only fluid always able to exchange components with every other part of the extracellular fluid compartment. Even lymph does not achieve a similar widespread distribution. The internal environment can be maintained constant in a large multicellular organism only if there exists a transport system capable of smoothing out local concentration changes, carrying nutrients to metabolically active cells, removing the products of metabolism, and providing a means whereby any attack on the cells by external agents can be resisted or neutralised. The blood plasma, and the proteins which it contains, serves these three functions of preserving the constancy of the internal environment, of transport, and of defence.

## **Preservation of the constancy of the internal environment**

### *Osmotic pressure*

The distribution of fluid throughout the body depends largely upon the osmotic pressure differences between fluid compartments. The presence of proteins in plasma, able to penetrate the capillary walls only to a small extent, provides a major difference between plasma and the other extracellular fluids: this results in an osmotic force which retains and reabsorbs fluid into the capillaries (see p. 20). The total osmotic activity of plasma depends upon the inorganic ions present since osmotic activity is a colligative property, proportional to the number of solute particles present, and so lower molecular weight substances contribute a greater osmotic activity at a given concentration in g/l than do higher molecular weight substances. However, within the much smaller activity due to the plasma proteins (the oncotic pressure), albumin, present in the highest concentration and also one of the smallest of the plasma proteins, contributes about 80%. Albumin also has a large anionic charge; as such it makes the greatest contribution to the Donnan effect and the resultant imbalance of cations. Changes in concentration of plasma proteins, particularly of albumin, will cause changes in the osmotic activity of plasma and thus affect movement between the extracellular fluid compartments, and in turn may alter movement between extracellular and intracellular fluid compartments. Low plasma protein concentrations will increase interstitial fluid volume at the expense of plasma volume, whilst high concentrations of plasma protein may cause cell dehydration.

*Control of pH*

Hydrogen ions cross the capillary walls freely, so changes in hydrogen ion concentration of the interstitial fluid are rapidly transferred to the plasma down concentration gradients. Interstitial fluid is buffered only by hydrogen carbonate (a weak buffer) and phosphate (with a buffering range well below the range of pH of normal body fluids); little extracellular buffering, therefore, occurs outside the circulation. In the plasma, however, buffering by protein is added to the buffering by hydrogen carbonate. The proteins of plasma contribute about 42 mEq/l of buffering capacity between pH 7.5 and 6.5. From the plasma, hydrogen ions can diffuse into the red blood cells where haemoglobin contributes some 275 mEq of buffering power/litre of blood in the same pH interval – over six times the buffering capacity of the plasma proteins. Whilst buffering by plasma proteins cannot be ignored, haemoglobin, by virtue of its greater concentration in blood (five times that of the plasma proteins) and its large number of charged sites on such residues as histidine, is the main hydrogen ion buffer of blood.

*Control of the concentrations of other ions*

Plasma serves to equilibrate the concentrations of small ions throughout the extracellular compartments of body water. On the whole, the plasma proteins are little involved in the distribution of the major ions such as sodium, chloride, and potassium; however, many of the ions present in smaller concentrations may be selectively bound by plasma proteins and such binding helps to preserve the normal concentrations of free ions. If ions were not bound by plasma proteins, they could have marked effects on the osmotic activity of plasma and the distribution of body water.

*Maintenance of blood pressure and flow*

The shape of the blood proteins is important in maintaining the smooth flow of blood. As their cigar-like shapes are aligned along the flow axis, they offer little resistance to movement, and the apparent viscosity of blood is low. Proteins of similar molecular size but spherical in shape would offer a greater resistance to flow. However, if the blood flow is slowed or stopped, the molecules become randomly orientated and the apparent viscosity increases. The alignment of the long axes of the molecules with the vessel walls means that relatively large 'pores' may exist in the capillary walls, but the molecules, broadside on, do not escape. When stasis of blood flow occurs, the leakage of protein through the capillary walls is increased because the randomly orientated molecules are much more likely to pass through the narrow 'pores'. Thus in the glomerular capillaries of the kidney, the globular protein haemoglobin, of molecular weight 64 450 daltons, passes readily through the capillary walls but albumin, only a little larger, passes with difficulty, mainly because of its greater charge but also because the molecules must strike the capillary walls almost end on in order to pass through.

Some blood proteins can be split by enzymes, such as kallikrein and renin, to yield peptides which act on the walls of blood vessels, causing them to dilate or constrict. These peptides will change the blood flow locally or may alter the resistance of the vessels and affect the systemic blood pressure (*see* p. 209).

## The bulk transport of materials within the body

Plasma proteins may bind substances and facilitate their transport in the blood. Small molecules which would diffuse rapidly out of capillaries may be carried as larger complexes; active substances may be transported in an inactive form bound to protein; insoluble substances may be solubilised by attachment to soluble, hydrophilic, proteins.

*Transport of inorganic ions*

Sodium ions are to some extent bound by albumin.
About half of the calcium carried in the plasma is bound to albumin; some is also carried by globulins. Since calcium ions are relatively easily precipitated by such ions as phosphate, the concentration of free calcium ions that can be carried is limited; protein binding increases the carrying capacity. As the pH rises the proteins bind more calcium and the concentration of free ions is reduced.

Copper, an essential ion in the formation of red blood cells, is carried partly by albumin, but mainly by the $\alpha_2$ globulin caeruloplasmin. Levels of this protein are low in Wilson's disease, in which there is degeneration of some areas in the basal ganglia of the brain known to contain high concentrations of copper.

Free iron II released from the breakdown of haemoglobin after destruction of red blood cells is carried in the bloodstream bound to the $\beta$ globulin, transferrin. Since 1 g of transferrin can bind 0.02 mmol of iron, the total iron carrying capacity of the blood is about 0.25 mmol. Normally only about 0.06 mmol of iron II is circulating and a considerable reserve capacity exists to prevent its loss from the body.

Zinc, an ion important in the formation of several metalloenzyme complexes (including that of salivary amylase), is carried by albumin.

The harmful effects of the heavy metal ions, including lead and mercury, are somewhat reduced by their binding to albumin.

## Transport of metabolic substrates

### Aminoacids

All the plasma proteins, and particularly albumin, constitute a mobile store of aminoacids and may be considered as possible sources of aminoacids for the tissues. In practice the circulating free aminoacids are used first, and there is little evidence of the breakdown of albumin in any tissue except the liver under normal circumstances.

### Lipids

Lipids are difficult to transport in the bloodstream because of their hydrophobic properties. They are carried, therefore, in association with protein molecules – either in loose combination with albumin, or combined in various proportions with the peptides of the apolipoprotein groups A, B, and C. Three types of lipoprotein are described (Table 3.2). In addition to these three groups of protein and albumin there are small amounts of protein associated with large lipid particles termed chylomicra, which appear in the bloodstream between 2 and 8 h after ingestion of a fatty meal. The chylomicra, like the lipoproteins, contain apolipoprotein moieties. All five lipid–protein complexes may be considered as parts of the same transport system.

When fats are absorbed from the small intestine, the triglycerides are transported in the chylomicra and as the pre-$\beta$-lipoprotein (very low density lipoprotein, VLDL). Free fatty acids released from the triglycerides by adipose tissue and muscle cells may be transported to other tissues by albumin. In the liver the free fatty acids are re-esterified to triglycerides and reincorporated into pre-$\beta$-lipoprotein by activity of the Golgi apparatus and the ribosomes. This is secreted back into the bloodstream. About 50% of dietary cholesterol is absorbed and transported as esters in the chylomicra. These are trapped in the liver and the cholesterol taken up into $\beta$-lipoprotein. Beta-lipoprotein formation may occur in the small intestine and possibly intravascularly; there is evidence that only pre-$\beta$ and $\alpha$-lipoproteins are formed in the liver. Cholesterol is carried by $\alpha$ lipoprotein. Phospholipids are carried by all the groups, but principally by the $\alpha$-lipoproteins. There seems, however, to be an equilibrium resulting from the interchange of the apoliproteins and the lipids themselves between the lipoprotein groups.

## Transport of metabolic products

### Carbon dioxide

The plasma proteins play a minor role in the transport of carbon dioxide in the blood in the form of carbamino-compounds. The reaction

$$-NH_2 + CO_2 \rightarrow NHCOOH \rightarrow -NHCOO^- + H^+$$

occurs between carbon dioxide and all the blood proteins, including haemoglobin. The proportion of carbon dioxide carried in this way by plasma proteins is small, and of little importance; moreover, as a method of carbon dioxide transport it has the disadvantage of generating hydrogen ions and thus maintaining the load upon the blood buffering systems. However, some 20% of the extra carbon dioxide in venous blood is carried as carbamino-haemoglobin.

### Bilirubin

The body tries, as far as possible, to retain its valuable iron supplies; the other breakdown products of haemoglobin are also transported in the blood. Bilirubin, produced from the breakdown of haemoglobin, is carried in the blood in combination with albumin and an $\alpha$ globulin. Both haem and haemoglobin can also be transported by albumin.

Table 3.2 Types of lipoprotein molecule in the plasma

| Type, other names | Plasma globulin | Apolipoprotein type | lipids carried |
| --- | --- | --- | --- |
| Low density lipoprotein (LDL), small | $\beta$ | mainly B | Cholesterol esters |
| Very low density lipoprotein (VLDL), medium | pre-$\beta$ | A, B and C | Triglycerides |
| High density lipoprotein (HDL) | $\alpha$ | mainly A | Cholesterol esters and phospholipids |

## Haemoglobin

Destruction of red blood cells as they age means that there is a turnover of about 25 g of cells per day. If the haemoglobin of these cells (a little over 0.1 mmol) was lost from the body, it would have to be replaced by protein synthesis and iron binding, increasing both the need for raw materials and the energy required for red cell production. As haemoglobin is a relatively small protein (molecular weight 64 450 daltons) of globular shape, it can pass through capillary 'pores', particularly such large ones as those of the capillaries of the kidney glomeruli. Haemoglobin free in the plasma would be lost rapidly into the interstitial spaces and the urine. It is, however, normally bound in the plasma by the haptoglobins, with molecular weights around 85 000 daltons, to form much larger complex molecules. Each litre of blood contains sufficient haptoglobin to bind about 0.03 mmol of haemoglobin – a capacity much in excess of the 0.6 μmol/l normally present. In diseases with excessive destruction of red blood cells, however, the binding capacity may be exceeded; haemoglobin then appears in the urine – the symptom of haematuria.

## Transport of hormones and related substances

The chemical messengers produced in the endocrine glands must reach their target organs to produce their effects. The bloodstream provides the routes by which they do this. Many hormones are small molecules, e.g. polypeptides, and these would readily pass out of the circulation if they were not bound by plasma proteins. Lipophilic molecules such as the steroid hormones are not readily soluble in aqueous solutions: binding to plasma proteins makes them more hydrophilic.

Examples of small molecules bound by plasma proteins include thyroxine, carried by thyroxine-binding protein (an $\alpha_2$ globulin), and calcitonin.

The steroid hormones may be exemplified by cortisol, carried by transcortin, an $\alpha_2$ globulin, and testosterone, carried by a gonadal hormone binding globulin (GBG), a β globulin.

Vitamin D, a steroid which performs hormonal functions in the body, is carried in its various forms around the body attached to an $\alpha_2$ globulin.

## The defence of the body

Since blood is a circulating fluid, reaching all parts of the body, it is an important part of the defence system of the body, carrying protective components in both its cellular and its fluid fractions. Two major defensive functions are performed by the plasma constituents: the initial sealing of damaged area to prevent blood loss, and the transport of proteins active against foreign cells, proteins or toxins.

## Blood clotting

The phenomenon of blood clotting represents a change in the blood proteins which results in the preliminary sealing of damaged blood vessels. It is only a part of the total reaction involving the formed elements of the blood, the blood vessels, and the tissues surrounding the blood vessels, which is termed haemostasis – the stopping of the flow of blood from a wound. The clotting process as a separate entity involving the plasma proteins is described here, but later (Chap. 27) haemostasis will be considered as the co-ordinated response to bleeding.

The substances taking part in the clotting process are given factor numbers agreed by an international committee to prevent confusion over nomenclature (Table 3.3). The process (Fig. 3.4) is described as a cascade because the diagram of the series of reactions looks like a waterfall. This is most obvious on the left of the figure. Sequences of enzyme reactions of this kind are characterised by proceeding slowly in the first instance, but gathering speed and producing more widespread effects as they progress until the final reaction becomes almost explosive in rate and extent. Further amplification of the process may occur, as in this present example of blood clotting, by a kind of feedback when later products potentiate the activity in earlier stages. Thus thrombin, a late product, is capable of activating as many as four of the earlier stages (Table 3.3).

Blood clotting may be described as occurring in three stages. As the mechanisms of the final stage are best known, it is usual to start at the end of the process and work backwards. All the stages are shown together in Fig. 3.4.

The final stage in blood clotting is the formation of insoluble fibrin strands from the soluble precursor protein fibrinogen and their stabilisation by factor XIII. Fibrinogen is a plasma protein which is split by the proteolytic enzyme thrombin to yield fibrin and two small fibrinopeptides. The molecules of fibrin polymerise to form chains with a characteristic striation pattern in electron micrographs. The links between the fibrin molecules are due to interaction of hydrogen bonds and are unstable, so that the fibrin polymers readily break up again. The enzyme thrombin, however, also activates factor XIII, the fibrin stabilising factor, which causes cross-linking to occur between the fibrin chains and thus stabilises the polymers. The main reaction involved is a linkage between terminal lysine and glutamine residues with the elimination of a molecule of ammonia.

The middle stage of blood clotting is the activation of thrombin. Thrombin, an enzyme similar in structure to the proteolytic digestive enzyme trypsin, is not normally present in the

*The plasma proteins* 49

**Table 3.3 Blood clotting factors**

| No. | Name | Molecular size | Activated by | Activated to | Acts on |
|---|---|---|---|---|---|
| I | Fibrinogen | 330 000 | Thrombin | Fibrin + fibrinopeptides | |
| II | Prothrombin | 68 000 | Thromboplastin | Thrombin (MW 32 000) | Fibrinogen, plasminogen, V, VIII, XIII |
| III | Thrombokinase, thromboplastin | Complex of lipoprotein with Ca, $V_a$, and $X_a$ | | | Prothrombin |
| IV | Calcium | | | | Prothrombin, X |
| V | Pro-accelerin | Quadrimer (MW 520 000) | Unknown but thrombin can | Monomer (MW 130 000) | X |
| VI | Not now a recognised factor | | | | |
| VII | Tissue factor | | Contact Product, damaged tissue, tissue extracts, $XII_a$ | $IX_a$ | X |
| VIII | Antihaemophiliac globulin, AHG (complex of VIII and VIII-like antigen or von Willebrand factor) | 250 000 + 880 000 VIIIRAg | Unknown but thrombin can | $VIII_a$ | X |
| IX | Christmas factor | | Contact Product | $IX_a$ | X |
| X | Stuart Prower factor | | $IX_a$ | $X_a$ | Prothrombin with V, Ca and lipoprotein |
| XI | Plasma thromboplastin antecedent | | $XII_a$ | Contact Product | IX, VII |
| XII | Hageman factor, Glass factor | | Contact with tissue | $XII_a$ | XI, plasminogen, kallikrein |
| XIII | Fibrin stabilising factor | | Thrombin | | Fibrin |

blood, but is produced by the activation of a pro-enzyme, prothrombin. This is a plasma protein, which is activated by a complex known variously as thrombokinase, thromboplastin, or prothrombinase. The complex has both proteolytic and esterolytic activity. Calcium ions are necessary for the conversion of prothrombin to thrombin; in the test tube blood may be prevented from clotting by removing calcium ions either by precipitation with oxalate, or by the formation of complexes or chelates of calcium with citrate or ethylene diamine tetra-acetic acid (EDTA). Calcium ions are also necessary for other stages of the clotting process; however, it is extremely unlikely that any patient would have blood so deficient in calcium ions as to prevent or even slow the clotting process. The first stages in the formation of fibrin proceed relatively slowly, but thrombin can also activate factors V and VIII, which will increase the generation of thrombin itself and thus speed up the process. In large amounts thrombin actually inactivates factor VIII: this may be a self-limiting step in the sequence of reactions.

Chronologically, the first stage in blood clotting is the production of the thrombokinase complex. This complex is built up from a phospholipid whose hydrophobic and hydrophilic groups enable it readily to bind the other factors – activated factor X, activated factor V, and calcium ions (Fig. 3.4). The source of the phospholipid may be the platelets or damaged tissue. The mechanism of activation of Factor V is unknown, although in the test tube it can be activated by thrombin. Russell's Viper venom, which was used by dentists many years ago to speed up blood clotting, is an activator of factor V. The key stage, however, in thrombokinase production is the activation of factor X. This may take place by two alternative pathways: the extrinsic pathway, which is set in motion by products of tissue damage, or the intrinsic pathway, so-called because all the factors necessary are present in the blood plasma (Fig. 3.4).

50   Blood plasma

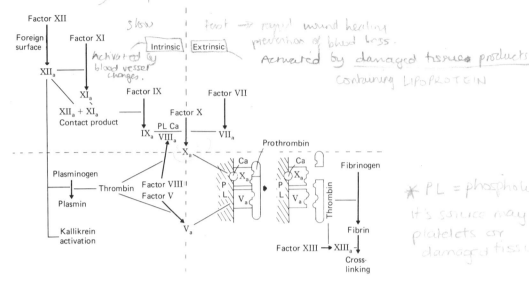

**Figure 3.4** The process of blood clotting. The diagram is divided into four segments by dashed lines. In the top left is the cascade process of activation of Factor X by the intrinsic pathway; in the top right the activation of Factor X by the extrinsic pathway; in the bottom left the subsidiary reactions activated by active Factor X and by thrombin; and in the bottom right the processes which result in the formation of a cross-linked fibrin network from the precursor fibrinogen. The clotting factors are given their numbers in the international system of nomenclature, and the subscript 'a' is used to denote the active form of the factor. PL indicates phospholipid. The course of the clotting process is indicated by the heavy arrows, whilst lighter lines show where the factors act.

The extrinsic pathway of thrombokinase activation requires substances released from damaged tissues, or extracts of damaged tissue containing lipoproteins, which, with factors V and VII are able to activate factor X in the presence of calcium ions. This path way involves few steps and is very rapid in action. Tissue damage, then, is associated with a process capable of rapid sealing of the wound and prevention of blood loss.

The intrinsic pathway is initiated by the activation of factor XII, which occurs when the plasma comes into contact with any surface other than the endothelial wall of a blood vessel or the membrane of a blood cell. Activated factor XII has a number of effects in the body. It acts on a blood protein termed kallikreininogen to produce kallikrein, an enzyme capable of splitting off a vasodilator peptide, bradykinin, from one of the plasma $\alpha_2$ globulins. It acts on plasminogen, another plasma protein, to produce the fibrinolytic enzyme, plasmin. More importantly in the present context, it activates and combines with factor XI to form an active complex known as the Contact Product. The Contact Product in turn activates factor IX and also, incidentally, factor VII. Active factor IX, in the presence of factor VIII, activates factor X. Phospholipid and calcium ions are also necessary in this process of activation of factor X (Fig. 3.4). How factor VIII is activated in the body is unknown, although in the test tube thrombin will activate it. Obviously, in the later stages of clotting this may occur and accelerate the process. The intrinsic pathway is a particularly good example of a cascade reaction and of the advantages of such a reaction. If clotting can occur in a blood vessel from the reactions of proteins already present in the blood, it is essential that this should take place as rarely as possible when it is not needed. The relative slowness of the initial reactions allows time for the body's own anti-clotting mechanisms to stop the cascade if it has begun by chance. On the other hand, the acceleration of the process, if unchecked, still permits clotting to occur rapidly should there be real need to seal a damaged vessel and prevent blood loss. Clotting occurring as a result of activation of the intrinsic pathway is always delayed in comparison with that due to activation of the extrinsic pathway (although the speed of the final reactions is probably greater) but in most situations where clotting is necessary the two pathways complement each other and indeed, stimulate each other.

A deficiency of factor VIII was one of the first conditions to be identified as a cause of defective blood clotting. This deficiency occurs in the

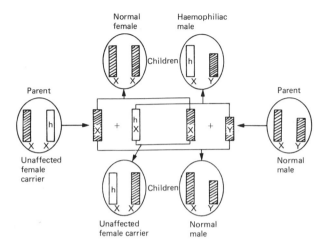

**Figure 3.5** Inheritance of the genes causing haemophilia. The chromosome pairs of the parents are shown to left and right, and the chromosome pairs of their possible offspring above and below. X and Y chromosomes are labelled, and h indicates the gene for haemophilia.

inherited trait of haemophilia caused by an abnormal recessive gene on the X chromosome and therefore transmitted as a sex-linked characteristic. The female usually carries the recessive gene without showing the symptoms of the disease (the chances of having both chromosomes with this gene are extremely remote) but her male offspring suffer excessive haemorrhage from even slight injuries (Fig. 3.5). Such patients present a difficult problem to the dental surgeon.

Factor VIII is a complex of a blood clotting factor and a factor which causes platelet adhesion and aggregation known as factor VIII-like antigen, or von Willebrand factor.

Another cause of defective clotting is a deficiency of vitamin K. This vitamin is essential for the incorporation, by the liver, of a γ carboxyglutamate group into the prothrombin molecule and β-hydroxyaspartate groups into factors VII, IX and X. These groups are necessary for clotting activity since they provide sites for calcium binding.

### The defence of the body against blood clotting

The blood clotting process is an important means of body defence. However, intravascular clotting in the absence of tissue damage is dangerous and potentially fatal. Such clotting is termed thrombosis, or thrombus formation. A number of systems exist in the plasma to prevent or reverse the clotting process. Clearly, they must be capable of dealing only with minor clotting incidents or of removing formed clots after tissue repair has begun.

There are at least six separate plasma fractions which have been observed to inhibit thrombin activity. Of these, the most important appear to be the antithrombins numbered III and IV.

Heparin is a polysaccharide produced in the liver. It contains glucosamine, glucuronic acid, and many charged sulphate groups. By inhibiting factor IX it prevents thrombin formation, but it also prevents formed thrombin from splitting fibrinogen. Clinically, heparin may be used as an anti-coagulant, both in patients and in syringes and vessels used for the collection of blood.

There may be other substances present in plasma which can inhibit thrombin formation.

Some γ globulins appear to inactivate specific clotting factors or prevent fibrin formation. There may in fact be antibodies for all the clotting factors, and some of the deficiency diseases may result from abnormal production of these antibodies.

When fibrin has been formed it may be broken down by the enzyme plasmin, which exists in the blood plasma in an inactive form, plasminogen. In addition to promoting fibrinolysis, plasmin also splits factors V and VII and fibrinogen. The products of this last reaction will inhibit the action of thrombin.

The process of blood clotting is carefully balanced. Not only does it use up the supply of clotting factors, and hence limit its own activity, but at least two of the activated factors also stimulate the transformation of plasminogen to plasmin – factor XII and thrombin itself. Another substance activating plasmin is the enzyme streptokinase, produced by the streptococcal bacteria. Its action enables them to spread through the host tissues even when their passage is blocked by blood clots. The plasminogen–plasmin transformation can be blocked by the drug ε-aminocaproic acid. This has been used in treating haemophiliacs since, by reducing the body's capacity to break up blood clots, it enhances the efficiency of poorly active blood clotting systems.

## Defence of the body against bacteria and foreign proteins

The circulation, reaching every part of the body, is an effective route for transporting defensive materials to the sites where attack is occurring. The main defensive activities of the plasma are found in the plasma protein γ globulin fraction. The immunoglobulins constitute the antibody system which responds to foreign substances or to bacteria – the antigens. This system is part of the so-called acquired immunity system of the body. Man is said to have an innate immunity to certain bacteria and chemical substances which are toxic to other animals; indeed, there is clear evidence of the existence of racial immunity rendering certain races of man less susceptible to certain diseases. Innate immunity is due to a mixture of factors – the nature of the skin barriers, anatomical variation, antibacterial secretions, digestive activity. A group of β globulins, collectively termed complement, forms a further component of the innate immunity system, but also assists in some of the reactions involved in acquired immunity.

The body reacts to bacteria and foreign substances by synthesising the immunoglobulins which act as antibodies. Each antibody has a different structure to its 'heavy' chains – and, to a lesser extent, to its light chains – which enables it to bind specifically to a chemical site on the antigen. This implies that there are separate and distinct antibodies for each antigen unless several antigens share similar chemical structures. These are known as antigenic determinants. In general antigens have molecular weights greater than 5000 daltons and are usually proteins, although some polysaccharides and some lipids can act as antigens. Low molecular weight substances may react with tissue or blood proteins to produce antigenically active complexes. Substances which are not themselves antigenic, but become so after combination with a carrier molecule (often, in the body, albumin) are known as haptens. Aspirin, penicillin and some of the sulphonamides are low molecular weight substances which can become antigenic after combination with tissue proteins: this is why some individuals become allergic to them.

After exposure to an antigen there is usually a period of some 14 days before an antibody can be demonstrated in the blood. This first appearance is the primary response. If a further dose of antigen is given, there is a fall in circulating antibody concentrations as it reacts with the antigen, and then a second and larger rise in antibody concentration which constitutes the secondary response. This immunity reaction is usually referred to as humoral, since it depends upon the production of chemical substances circulating in the bloodstream. The response is mainly from the corticomedullary junctions and medullary cords of the lymph nodes, and from the red pulp of the spleen. In addition to humoral immunity reactions there is also a cellular antibody response which is under the control of thymus-derived lymphocytes (discussed in Chap. 4). Antibodies may inactivate antigens in a number of ways. Some, the antitoxins, usually IgG immunoglobulins, neutralise the activity of toxic substances. Others precipitate antigens out of solution after reacting with them. These precipitins, too, are IgG immunoglobulins. Maximum precipitation occurs when the reacting antigen and antibody are present in specific proportions. Antibodies causing antigenic cells or proteins to stick together are termed agglutinins. Since this process is achieved most readily by antibodies with several antigen-binding sites, they are usually members of the IgM group – indeed, IgM molecules have been calculated to be some 20 times as effective as IgG molecules in causing agglutination. Some IgG immunoglobulins attach to cell walls and cause them to break up, releasing the cell contents. These are termed lysins. Antibodies which attach to cell walls or to foreign particles and change their surface properties so that phagocytic cells can ingest them more easily, are termed opsonins. Once again, IgM immunoglobulins, with their multiple binding sites, are much more effective than the IgG group. Both IgG and IgM antibodies can activate the complement system and so lysis of cells may occur when these complement-fixing antibodies attach to cell walls.

The complement system is a complex system of enzymes, activation of which results in the digestion of small areas of cell membranes or localised alterations in membrane structure. It is usually necessary for an antibody to attach itself to the cell membrane: complement factors then attach to the antibody. There has been much study of the process by which red blood cells with membrane antigenic determinants foreign to a particular individual are lysed in his or her blood. After the antibody has bound to the red cell membrane, a trimolecular complex of molecular weight about 700 000, termed C'1, binds to the antibody–membrane complex. This causes, in the presence of calcium ions, activation of C'1 as an esterase. The enzyme changes the properties of the cell membrane, enabling it to bind a β globulin termed C'4 and this, in turn, reacts with another β globulin, C'2 which has also been activated by the esterase. This new complex triggers a cascade of reactions through six other complement components (C'3, C'5, C'6, C'7, C'8, and C'9) to result in changes in the membrane phospholipids and the development of 8–10 nm diameter holes in the cell membrane. A similar sequence of reactions on the surface of bacterial cells may cause them to be more readily digested by the enzyme lysozyme. The interaction of complement components C'1 esterase and C'2 to C'7 yields a product which attracts white blood cells to the site. Complement activation may also be involved in the production of kinins (active polypeptides increasing the calibre and the permeability of blood vessels) and possibly also in the release of histamine. Components C'1 to C'3 help to promote phagocytosis, acting as opsonins.

Normal human serum contains an $\alpha_2$ globulin which can inhibit activated C'1 esterase. Deficiency of this inhibitor may be a factor in the hypersensivity reaction of angioneurotic oedema.

## Further reading

Roitt, I. *Essential Immmunology*. 6th edn, Blackwell, Oxford 1988

Thompson, R.B. *A Short Textbook of Haematology*. 6th edn, Pitman Medical, Tunbridge Wells, 1984

Turner, M.W. and Hulme, B. *The Plasma Proteins*. Pitman Medical, Tunbridge Wells, 1971

# 4
# The cells of the blood

There are three main types of blood cell: the red blood cells, or erythrocytes, deriving their colour from the respiratory pigment haemoglobin, the white blood cells, or leucocytes, and the platelets. The cells are only slightly more dense than the plasma, with a specific gravity of 1.1 in comparison with 1.03, but this difference is sufficient to cause them to sediment on standing. The rate of sedimentation is slow: at the end of an hour only the top 4 mm of a column of blood 10 cm high will be cell-free. This measurement is used clinically, since any change in the cell membranes or any aggregation of cells, or any changes in blood viscosity, will alter the sedimentation rate. Variations in plasma protein concentrations, often symptomatic of inflammatory disease, usually increase the sedimentation rate. The raised concentrations of fibrinogen when inflammation is present provide a good example. Adhesion of cells causes them to stick together in rods one cell in diameter, termed rouleaux, which sediment more rapidly than single cells. Normal sedimentation rates are around 4 mm/h in males (range 0–6.5 mm/h) and 8 mm/h in females (range 0–12 mm/h).

Centrifugation of blood results in a plug of red cells surmounted by a thin layer of white cells and platelets, the buffy layer, beneath the straw-coloured plasma (Fig. 3.1). The cell layer is usually around 45% of the total volume: this value is termed the haematocrit value (Ht), or the packed cell volume (PCV). It represents a slight overestimate of the total cell volume since it is impossible to obtain a cell layer so tightly packed as to have no interstices of fluid between the cells. The layer of white cells is so thin (there are only about 1/600th as many white cells as red cells) that the haematocrit value is taken to be the volume of red cells present.

## The red blood cells
### Morphology

The erythrocyte is one of the few non-nucleate cells in the body. Despite this it has an active metabolism and survives in the circulation longer than the nucleated white cells. It is a biconcave disc (Fig. 4.1) of mean diameter 8.5 µm in life; histological fixation causes shrinkage and so on sections the mean diameter of the red cell is only 7.5 µm. Because blood vessels and blood cells are found in all tissues the light microscopist has often used the diameter of a red blood cell as a rapid and convenient guide to the size of other structures seen on sections. In cross-section the biconcave disc appears as a dumb-bell with a maximum thickness of 2.4 µm and a minimum thickness of 1.0 µm before fixation. Normal cells can traverse blood vessels with diameters as small as 3 µm.

These dimensions give the red cell a calculated surface area of 163 sq µm and a volume of 87 fl. This latter figure agrees closely with the volume calculated from the red cell count and the haematocrit value – the mean corpuscular volume (MCV) of 90 fl. Cells of normal size (normocytes) should have a mean corpuscular volume around this value. Cells larger than 95 fl are termed macrocytes and cells smaller than 80 fl are termed microcytes.

The shape of the red cell is well adapted to its functions. In order to pick up and release oxygen rapidly it must have a large surface area in relation to volume – that is, the diffusion distance for oxygen must be short. Only a flat disc might have advantages in this respect over the normal shape of the erythrocyte. A sphere of similar volume has a surface area of less than 60% that of a red cell and its radius, or the greatest distance for diffusion, is

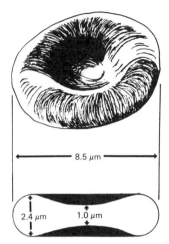

**Figure 4.1** The red blood cell (the erythrocyte) and its dimensions.

more than double. The shape is also well adapted for passing into small blood vessels: its thickness is less than a third of its diameter and the concave surfaces allow it to be readily deformed. In small blood capillaries red cells are often bent double in order to pass through. The easy deformation may assist in the mixing and homogenisation of the red cell contents. Finally, the shape allows the cell to accommodate readily to changes in osmotic activity of the surrounding medium by taking up water and swelling, or by losing water and shrinking. Since the surface area of the membrane is equal to that of a sphere of approximately twice the volume, rupture of the cell membrane does not occur until considerable swelling has taken place. In this respect the shape has advantages over that of a disc. The reverse process of shrinking in fluids of high osmotic activity causes crenation, or wrinkling of the membrane.

The shape of the cell is maintained by the interacting charges of the membrane, the supporting stroma, and the contained haemoglobin. A protein called spectrin may also be important. There is some evidence that maintenance of shape is an active process, since ATP is necessary, and this suggests the presence of a contractile protein. Transformation of shape to a sphere occurs with agents which change the surface charge, or dissolve lipid components of the membrane – such as saponin, lecithin, digitonin, and detergents. Curiously, however, some lipid solvents such as chloroform and ether act as stabilizing, or anti-sphering, agents. There is also an unidentified factor in plasma with anti-sphering activity.

When cells are placed in solutions of osmotic activity below that of the cell contents, water passes in and osmotic swelling results. The ability of the cell to resist damage from this depends in part upon its age. In any normal blood sample the numbers of cells haemolysing in increasing dilutions of sodium chloride are arranged in a Gaussian normal distribution with a median at a concentration of 72 mmol/l, i.e. at half the osmotic activity of the cell contents. A solution of 150 mmol/l sodium chloride is isosmotic and is termed normal saline. The osmotic fragility test is sometimes used to assess the normality of a red cell population in respect of surface structure or age.

In disease states abnormal forms of cell may be present in the blood and these are usually less resistant to mechanical or osmotic stress. Elliptocytes have an elliptical rather than a circular form; meniscocytes or sickle cells are moon-shaped cells observed in the inherited disease of sickle cell anaemia. The disease is a manifestation of the homozygous co-dominant gene, $H^SH^S$, occurring particularly in certain geographical areas and in certain races such as the West Indians and the US Negroes, of whom as many as 8% may be affected. A lesser form is termed sickle cell trait: this occurs with the heterozygote $H^AH^S$ due to inheritance of both the normal and the sickle cell haemoglobin gene, and shows itself as sickling only under extreme conditions. The haemoglobin abnormality due to the gene causes a change in the charge relationships within the cell when it is exposed to lower than usual concentrations of oxygen. The normal shape cannot then be maintained and the cells collapse into the sickle shape. Such cells are unusually sensitive to mechanical damage: haemolysis occurs very readily in the circulation. This greatly reduces the number of cells, producing a state of anaemia. Anaesthetists need to be aware of patients with this condition because during the induction of anaesthesia with gases such as nitrous oxide the levels of oxygen inhaled may be much lower than usual.

## Numbers

The normal red blood cell count in man is around $5 \times 10^{12}$/l (5 million cells/µl) of blood – a total in 5 l of blood of about 25 million million with a weight of about 2.5 kg. The male usually has a higher count at around $5.5 \times 10^{12}$/l, and the female a lower at around $4.8 \times 10^{12}$/l, the difference being due to different rates of production of cells and not solely to the loss of blood during menstruation in the female. The traditional method of counting cells in a blood film of known thickness on a microscope slide with an engraved grid has now given way to electronic particle counting in such counters as the Coulter counter. Although this method eliminates human counting errors there is still the possibility of errors arising during dilution of the blood.

A low red blood cell count is termed anaemia. Strictly this term is defined as a haemoglobin concentration below 1.55 mmol/l (as tetramer, about

6 mmol/l as monomer, or 10 g/dl). However, a red cell count below $4 \times 10^{12}/l$ will normally indicate an anaemic condition. This may arise either from insufficient red cell production or from excessive red cell destruction. A high red cell count is termed a polycythaemia. It is usually observed in response to conditions of low oxygen availability, as at high altitudes, or in chronic lung disease when a layer of fibrous tissue in the alveoli prevents the free diffusion of oxygen. Even at normal red cell counts there is crowding of red cells in the bloodstream: at haematocrit values above 58% close packing occurs and cell distortion is observed. The calculated maximum packed cell volume is 65%: at this value all the cells would be touching each other but not suffering distortion. As the red cell count rises the viscosity of the blood increases markedly and the work necessary to keep it circulating becomes progressively greater.

## Functions

The main functions of the blood cells are those of blood as a whole – transport and defence. The role of the red cells is very largely that of transport, since they are full of the oxygen-transporting protein haemoglobin, but their permeability to ions enables haemoglobin to act also as a buffer and help to maintain the pH of the blood.

There are advantages in having cells whose only function is to contain haemoglobin and carry oxygen, rather than transporting haemoglobin in free solution in the plasma. Haemoglobin has a low molecular weight and is present in high concentration. Free haemoglobin could pass out of the circulation through the capillary walls, both into the tissues and out of the body in the urine. Its concentration would at least double the osmotic activity of the plasma and also render the plasma so viscous that the work necessary to maintain blood flow would be greatly increased. Further, the limited mixing possible in free solution would reduce the uptake of oxygen in the time available in the lungs to about a third. The cells carry on a number of subsidiary functions which provide the right environment for haemoglobin–oxygen association and dissociation. Haemoglobin functions more efficiently in a fluid containing the ions present in the red cells than in any other solutions, so maintenance of these conditions is important. The Embden–Meyerhof pathway and the pentose phosphate shunt pathway of glucose metabolism provide the energy to maintain the sodium pumping mechanism of the red cell membranes and thus preserve the ionic composition.

The maintenance of the haemoglobin molecule itself is important. Throughout the life of the red cell haemoglobin tends to oxidise to methaemoglobin. The regeneration of haemoglobin is another energy-requiring process in the red cell (Fig. 4.2). The energy is provided by the normal glycolytic pathways, the lactate produced diffusing out of the cells. The glycolytic pathways also provide the NAD necessary for the action of methaemoglobin reductase. Unusually, there is during glycolysis a build-up of 2,3-diphosphoglycerate (2,3-DPG), an intermediate which has important effects on the carriage of oxygen by haemoglobin. The –SH groups of the haemoglobin molecule are protected from oxidation by the –SH groups of reduced glutathione, whose resynthesis is activated by the pentose phosphate pathway (accounting for 10–25% of the energy turnover in the red cell). Older cells lose the ability to reduce methaemoglobin. Young cells retain some iron uptake sites from the earlier stages of their development, but older cells have a much reduced ability to take up iron II.

The red cells are also vitally important in the carriage of carbon dioxide, since in contrast with plasma, they contain the enzyme carbonic anhydrase. This enzyme speeds up the conversion of dissolved carbon dioxide into its hydrated form, carbonic acid – or, depending upon the concentrations, the reverse reaction. The transport of carbon dioxide as hydrogen carbonate ion takes place mainly in the plasma but its formation depends upon this enzyme in the red cells. Once formed, the greater proportion of the hydrogen carbonate diffuses out of the cells to be carried in the plasma. In addition to this important function of speeding the uptake of carbon dioxide in the tissues and the loss of carbon dioxide in the lungs, the red cells carry some carbon dioxide bound to the haemoglobin as a carbamino compound.

In the mature red cell there is no evidence of either protein or lipid synthesis.

## Composition

The principal components of the red cells are protein in nature and the major protein is haemoglobin. Electron microscopy of a red cell, either whole or damaged in ways designed to show its structure, reveals an ordered membrane structure some 7.5 nm thick, the membrane not being uniform but covered by discoid plaques about $10 \times 50$ nm in size. Immediately beneath this membrane is a fibrous stroma with which are associated the cell wall antigens of the ABO blood groups, some carbohydrates, and the proteins elenin and stromatin. Another protein, spectrin, forms coiled filaments which may be important in maintaining the shape of the cell. Deeper within the cell the concentration of haemoglobin increases and the structure becomes less ordered. The dimensions of the haemoglobin molecule with its hydration shell are such that the amount of haemoglobin actually found in a red cell can just be contained in the volume available. The

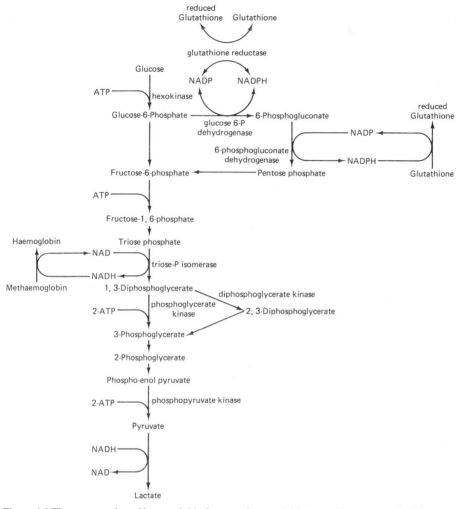

**Figure 4.2** The regeneration of haemoglobin from methaemoglobin takes place as a result of the normal red cell metabolism. Note also the importance of glutathione and 2,3-diphosphoglycerate in the metabolic pathways.

only space for free water and the other components is in the interstices between the packed molecules. This picture is simplified because, in practice, the hydration shell is not rigid but it nevertheless emphasises that the red cell contains the maximum possible amount of haemoglobin. Water, both free and bound as water of hydration, accounts for 70% of the volume of the cells – making the red cell one of the 'driest' cells in the body.

## Haemoglobin

Haemoglobin, the red oxygen-carrying pigment which is the main component of the red cells, is present at a mean concentration of 2.23 mmol/l (14.35 g/dl) of blood, or 5.2 mmol/l (33.5 g/dl) of cells. These values assume that haemoglobin is a tetramer, with four binding sites for oxygen. If values are expressed in terms of monomeric molecules, each with one binding site, then the normal would be about 9 mmol/l of blood. Each monomeric molecule is composed of a protein globin combined with a haem molecule. Haem has a ring structure, with four pyrrole rings linked round an iron II (ferrous) ion – it is therefore a porphyrin, protoporphyrin IX (Fig. 4.3). This molecule is found in many combinations in the body, such as in the myoglobin of muscles and in the cytochromes. Haemoglobin has a molecular weight of 64 450 daltons, and contains 0.338% iron II. It has a characteristic spectral absorption band at 565μm, between red and green. Eighteen variants of

58  *The cells of the blood*

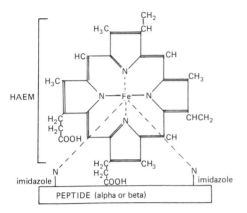

One of the 4 sub-units of deoxygenated Haemoglobin

**Figure 4.3** The structure of haemoglobin – one of the four subunits of the molecule, with its four pyrrole rings around the iron II, making up haem (protoporphyrin IX) attached to the peptide structure.

haemoglobin are known, each differing in the polypeptide chains making up the globin. In normal subjects most of the haemoglobin is $A_1$, but up to 3% $A_2$ is also found. Sickle cell anaemia is characterised by variant S. Fetal haemoglobin, F, also differs from adult haemoglobin in having different polypeptide chains.

The presence of four haem rings containing iron II permits the molecule to take up four molecules of oxygen in a loose bond when the partial pressure of oxygen is high, and to release them when the partial pressure is low. No oxidation occurs and the iron remains in the iron II state. The compound formed is termed oxyhaemoglobin. In a similar way the poisonous gas carbon monoxide may be taken up to give carboxyhaemoglobin. Since the affinity of haemoglobin for carbon monoxide is some 250 times greater than that for oxygen, carbon monoxide binding blocks oxygen uptake. This results in a lack of oxygen in the blood and the tissues however much oxygen is provided to the lungs. Haemoglobin normally appears red when oxygenated, purple when it gives up its oxygen, and cherry pink when it combines with carbon monoxide.

Haemoglobin can carry carbon dioxide only as a carbamino compound (as also can the plasma proteins, p. 47).

The amount of haemoglobin in the blood is measured by destroying the red cell membranes (haemolysis, or laking) so that the pigment is released into the plasma and can be estimated colorimetrically. Usually the haemoglobin is converted to carboxyhaemoglobin by bubbling carbon monoxide through it, or it is oxidised by an appropriate agent to haematin.

When the iron II is converted into iron III by oxidation, a new compound, methaemoglobin, is formed. This is unable to bind and release oxygen and is therefore not a transport protein. Methaemoglobin is formed slowly in the mature red cell throughout its life; the active cell is able to reconvert it to haemoglobin and keep its concentration below 0.5 g/dl. In the laboratory further oxidation converts the haem to haematin whose characteristic absorption can be used in the estimation of haemoglobin.

The estimation of haemoglobin concentrations is a routine clinical laboratory test, and a number of standard values are derived from it. The normal range of haemoglobin concentrations in males is 2.17–2.64 mmol/l (14.0–17.0 g/dl) of blood, with a mean of 2.42 mmol/l (15.6 g/dl), and in females the range is 1.86–2.40 mmol/l (12.0–15.5 g/dl) with a mean of 2.13 mmol/l (13.7 g/dl). All these molar values are expressed in terms of the haemoglobin molecule with four haem subunits. The haemoglobin concentration is sometimes expressed as a percentage of the value suggested by Haldane as normal – 2.30 mmol/l (14.8 g/dl). This value is higher than the mean observed in most populations. Anaemia is defined as that condition when the blood haemoglobin concentration is below 65% of the Haldane standard. The threshold is more conveniently taken as 1.55 mmol/l or 10 g/dl.

Three derived indices use the haemoglobin concentrations to give information about the red blood cells. The Colour Index (CI) is calculated by dividing the haemoglobin concentration, expressed as a percentage of the Haldane value above, by the red cell count expressed as a percentage of a nominal normal value of $5 \times 10^6/\mu l$. This index was derived to give information about the amount of haemoglobin in each red cell, but this is better obtained by calculation of the mean corpuscular haemoglobin content (MCH), obtained by dividing the haemoglobin concentration by the number of cells in the same volume of blood. This gives a value of 0.45 fmol/cell (about 30 pg/cell). The mean corpuscular haemoglobin content is the average amount of haemoglobin in each red cell; sometimes it is more valuable to know the amount of haemoglobin in a given volume of red cells – the mean corpuscular haemoglobin concentration (MCHC). It is calculated by dividing the amount of haemoglobin in the whole blood by the haematocrit value to give a concentration in unit volume of cells only. The normal value would be around 5.43 mmol/l of cells (35 g/dl).

The mean corpuscular haemoglobin content is raised if large cells are present and filled with haemoglobin; it is lowered if each cell contains less than its full complement of haemoglobin, or if the cells are small. The mean corpuscular haemoglobin concentration does not depend directly upon the number of cells present, but only on the volume they constitute.

The red cells are described according to their haemoglobin concentration, or colour, as normochromic when the colour is normal; as hypochromic when the haemoglobin concentration falls below 3.88 mmol/l (25.0 g/dl) of cells; and as hyperchromic when the haemoglobin concentration is higher than normal. These estimations are important in the differential diagnosis of the anaemias.

## Non-haemoglobin proteins

The total non-haemoglobin protein content of the red cells amounts to about 8.3 g/l of cells. The proteins are mainly the structural proteins, stromatin and elenin. Minor constituents include the enzymes involved in the metabolic pathways: notably the Embden–Meyerhof cycle and the pentose phosphate shunt. In addition the cells contain high concentrations of carbonic anhydrase and cholinesterase.

## Blood group antigens

### ABO blood groups

Over half the population in European countries have red blood cells carrying, at or near the cell surface, antigens which can react with appropriate antibodies to cause first agglutination and then lysis of the cells. Two main groups of antigen are involved, A and B. An individual carries on his or her red cells – or, indeed, has present in most of the body tissues outside the central nervous system – either, both or neither of these antigens. Subjects with A or B antigens are said to belong to Group A or B respectively. Subjects with both belong to Group AB, and those with neither to Group O. A person's blood group, then, is conferred by the antigens present on his or her red blood cells. The antigen molecules themselves are complexes made up of carbohydrate (about 75%) and peptides. They differ only in their terminal sugar groups. Individuals with neither antigen (Group O) have instead a glycoprotein known as H substance, which is the basic chain structure lacking the A and B specific terminal sugars.

Blood groups are inherited, with $A_1$, $A_2$, and B behaving as dominant Mendelian characteristics. A subject with B in the genotype is therefore of phenotype B also. The distribution of the blood groups in a Caucasian population is given in *Table 4.1*. Group A consists of five subgroups but only groups $A_1$ and $A_2$ are given in the Table since the other groups occur so infrequently.

The plasma of an individual contains antibodies – agglutinins – to the antigens not present on the individual's own red blood cells. The immune system normally recognises self antigens and does not produce antibodies to them. However, substances with ABO blood group specificity are widespread in nature (*Escherichia coli* organisms, for example, contain B group substances); and it is thought that soon after birth, when the gut flora becomes established, A and B antigens are absorbed through the gut wall. These stimulate rapid production of appropriate antibodies if either or both of the antigens are not present on the individual's own cells. Thus, almost from birth, the blood plasma of a B group subject contains anti-A iso-antibodies, or iso-agglutinins, often referred to as αagglutinins. Group A red blood cells injected into this subject will be first aggregated, and then lysed by complement activity. Group O subjects have both anti-A and anti-B agglutinins. *Table 4.2* shows the implications of this in blood transfusion. Subjects from Group O are universal donors: they can give blood to any other subject since their cells have no AB antigens with which the recipient's antibodies may react. Similarly, subjects of Group AB have no α or β agglutinins in their plasma and can accept blood of any ABO group – they are universal recipients. Normally the volume of blood transfused is less than the recipient's circulating volume, and so it is the transfused cells which may be lysed. However, when

**Table 4.1 Distribution of ABO blood groups in a Caucasian population**

| Group | Occurrence (%) |
|---|---|
| O | 43.5 |
| $A_1$ | 33.8 |
| $A_2$ | 9.6 |
| B | 8.5 |
| $A_1$B | 2.5 |
| $A_2$B | 0.8 |

**Table 4.2 Compatibility of blood transfusions (ABO groups only)**

| Donor's group | Recipient's group | | | |
|---|---|---|---|---|
| | A (β) | B (α) | AB (none) | O (α + β) |
| Cell antigen A | | X | | X |
| Cell antigen B | X | | | X |
| Cell antigen AB | X | X | | X |
| Cell antigen O | | | | |

The cells of the donor's blood carry antigens with the same letter as the blood group; the plasma of the recipient's blood contains antibodies to antigens other than those on the recipient's cells. The β antibody agglutinates B antigens and the α antibody A antigens. The transfusion is incompatible if an X appears above.

very large transfusions are given, the antibodies of the tranfused plasma may become important.

Blood group substances are found in a number of secretions, including submandibular saliva, in about 80% of subjects. Such secretions can therefore be used in forensic tests (*see* Chap. 22).

### Rhesus blood groups

The next most significant blood group system in man is the Rhesus system. Antigens are present on the red blood cells of some 85% of Caucasian subjects, who are termed Rhesus positive, and lacking on the red blood cells of the remainder – the Rhesus negative individuals. The ability to synthesise the Rhesus antigen is inherited through the dominant genes given the letters C, D, and E. Of these the most powerful and important is that for the D antigen. The corresponding recessive genes, c, d, and e, must all be present in the genotype cde/cde for the individual to be Rhesus negative.

The Rhesus antigens are not normally found in non-primate sources, and so anti-Rhesus antibodies are not found in plasma unless the subject has been previously exposed to Rhesus positive blood. A Rhesus negative individual can receive one transfusion or exposure to Rhesus positive blood without danger: but the antibodies then developed will react with a second transfusion. The more important problem, however, arises when a Rhesus negative woman conceives a child by a Rhesus positive father. The child will be Rhesus positive, since this is the dominant characteristic, and at parturition its blood cells may induce the development of antibodies in the mother's plasma. If she then has a second Rhesus positive child, her anti-Rhesus antibodies will reach the child's circulation via the placenta and cause haemolysis of the child's red cells. This condition is termed erythroblastosis fetalis, 'haemolytic disease of the newborn'. Formerly the only treatment available was to exchange the child's blood completely by a massive transfusion, and wash out the antibodies. Now, however, a Rhesus negative mother giving birth to a first baby can be treated with a large dose of anti-Rhesus antibody, which suppresses her normal immune response, possibly by some kind of feedback inhibition.

### Other blood groups

Many other blood group systems have been identified. Examples include the MN, the Lutheran, the Kell, the Duffy, the Lewis, and the Kidd systems. These are all of lesser importance, since no antibodies are produced unless a previous sensitisation has taken place, and they are not usually involved in reactions of the type described above. Nonetheless, it is a general rule that wives should not receive a blood transfusion from their husbands, even though they may be of the same ABO and Rhesus categories, in case any antibodies are produced that might endanger the survival of subsequent children.

### Antigens of other blood cells

White blood cells also share in the ABO antigens, but they have, in addition, unique systems such as the HL-A system. Transfused white cells are often agglutinated and lysed, but this is of little significance. Medicolegally the blood groups of the white cells are of considerable interest, since the combinations present in different individuals are much more characteristic of each individual than the red cell blood groups.

### Other organic substances

Lipids amount to around 4.8 g/l of cells. Although the ratio of lipid to non-haemoglobin protein is approximately the same as the lipid:protein ratio in cell membranes, there is more lipid present than would be needed for a membrane of the thickness known to enclose the red cell: some lipid must be inside the cell.

Glucose is present in red cells at 4.6 mmol/l. It passes into the cells by facilitated diffusion. Another component is glutathione. The amount of 2,3-diphosphoglycerate in the cells is important because this metabolite changes the conformation of haemoglobin and therefore its affinity for oxygen (p. 109).

### Inorganic components

In man the red cell contains potassium, sodium and chloride as its major free ions. The total inorganic content is about 6.7 g/l of cells. The concentrations of ions are given in *Table 1.3*, from which it will be seen that there is some 147 mmol/l of potassium ion, 16 mmol/l of sodium ion, and 80 mmol/l of chloride ion. The sodium and potassium ion concentrations are maintained by an active pumping system which expels sodium ions and accumulates potassium ions against the diffusion gradients. If the ionic concentrations in the cells are changed the ability of the haemoglobin to carry oxygen is impaired. The differences in ionic concentrations inside and outside the red cells result in a charge across the cell membranes of 10–12 mV, inside negative.

The concentrations of hydrogen carbonate and chloride in the cells change with the blood carbon dioxide content (*see* p. 111). The 4.41 mmol/l hydrogen carbonate in venous blood falls to 4.28 mmol/l in arterial blood and as a result of this there is a difference of 0.87 mmol/l of chloride ions between arterial and venous blood. The pH of the cells is about 7.19 in arterial blood, 7.17 in venous; the difference is less marked than in plasma because of the lower arterial pH in the cells, and the greater buffering power of the cell protein which reduces the pH shift as carbon dioxide concentration increases.

## Origin and development of red blood cells

### Sites of red cell production

During intra-uterine life the formation of the vascular system and the blood cells begins. Between conception and two months of intra-uterine age, red cell production takes place in the yolk sac, the body sac, and the placenta. Mesodermal differentiation produces the endothelial lining of the primitive blood vessels. In the next three months red cells are produced in the vessels of the liver, spleen and thymus, and in the bone marrow; but by full term production is limited to the marrow. From birth to about four years, all the bones contain bone marrow actively producing blood cells (the process of haematopoiesis) and, more specifically, red blood cells (erythropoiesis). The red, or erythropoietic, marrow slowly gives way to yellow marrow, which is not erythropoietic, until at age 20 years only the flat bones – the ribs, sternum, pelvis, vertebrae and skull – are actively erythropoietic. If there is a demand for greatly increased red cell production the red marrow spreads outwards again to recolonise the tissues in the reverse order.

Samples of marrow cells may be obtained by inserting a hypodermic needle into the sternum (sternal puncture) and aspirating: this normally yields a sample of marrow cells with a ratio of three or four myeloid (white cell-producing) cells to each one of the erythroid (red cell-producing) cells, reflecting the greater turnover of white cells.

### Course of red cell development

The marrow consists of sinuses lined with endothelium, in which the primitive red blood cells and their precursors are found. The stem cells are the multipotent haemocytoblasts, cells capable of dividing to produce either red or white cell precursors (Fig. 4.4). In red cell development the next recognisable cell is the pro-erythroblast. This marrow cell is capable of taking up circulating iron carried by transferrin; it accumulates iron II in readiness for haemoglobin formation. Subsequent cells in the line of red cell development are also able to take up iron. Another lesser source of iron may be the iron storage protein, ferritin, which can be transferred from reticulo-endothelial cells to the forming red cells. The following stages of cell division through to the normoblast are all nucleated cells, and all take up iron. There are three phases of

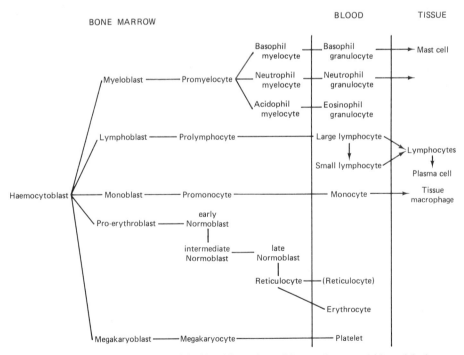

**Figure 4.4** Development of the cells of the blood from the multipotent haemocytoblast of the bone marrow.

normoblast development: early, middle and late. During these phases haemoglobin appears in the cells and increases in concentration, while the cell nucleus becomes vacuolated and dark-staining as it begins to degenerate. The next stage is that of the reticulocyte, a cell with its full complement of haemoglobin and a network, or reticulum, of dark-staining threads which may be chromatin or other nuclear remnants. The reticulocyte is not a spherical cell like its predecessors, but begins to assume the shape of a bi-concave disc. About 1% of the circulating red cells are normally reticulocytes: any increase in this proportion suggests that the marrow is abnormally active in red cell production. This is seen normally as a response to treatment in anaemic patients. The mature erythrocyte is the final stage of development.

## Factors necessary for red blood cell development

### Iron

Iron is essential for the synthesis of haemoglobin, and therefore of red cells. The total body store amounts to about 800 mmol (4.5 g), of which 65% is found in the blood and the remainder in myoglobin (the respiratory pigment of muscle cells), catalase, peroxidase, and the cytochromes. Usually the iron is present as iron II as part of the protoporphyrin IX molecule. In addition to this the total iron pool includes that bound to transferrin in the blood plasma, and that bound to ferritin and haemosiderin in the reticulo-endothelial system (Fig. 4.5).

The average daily intake of iron is 14–15 mg (0.25 mmol), but only some 10–15% is actually absorbed in the gut. The main sources are liver, meat, eggs, legumes, and bread. Loss of iron occurs mainly through the faeces although one third is lost in the urine and a small amount in sweat. The average male subject has a daily loss of 10–20 μmol/day (0.5–1.0 mg) from destruction of red blood cells, and can therefore readily maintain an iron balance. Women during the childbearing years, however, lose blood each month during the menstrual period; and this extra loss of about 600 μmol (36 mg) is equivalent to an additional daily loss of 20 μmol (1.2 mg). The normal dietary intake is therefore barely sufficient to replace the loss, and an iron deficiency anaemia of greater or lesser degree is not uncommon in women in this age group. Pregnancy imposes a different strain on the iron balance, since the fetus requires from the mother some 40 μmol/day (2 mg) of iron for red cell formation, and the needs of the fetus take priority over those of the mother. During pregnancy, therefore, the mother needs to absorb about 60 μmol/day (3 mg); many mothers in the UK receive supplements of iron II during their pregnancies. Even during lactation iron requirements are

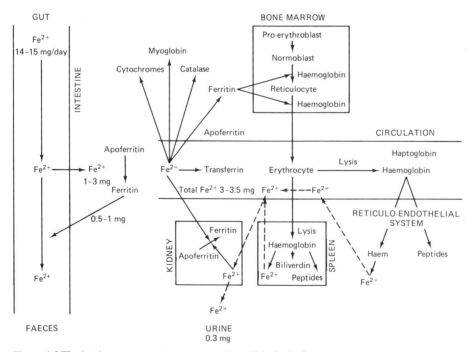

**Figure 4.5** The intake, storage and movement of iron II in the body.

above normal since the litre of human milk produced each day contains 10–30 µmol (0.45–1.5 mg) of iron.

The iron absorbed from the gut, together with that resulting from the breakdown of the red cells, is bound in plasma by transferrin. The normal amount of circulating iron is 60–70 µmol (3.0–3.5 mg), that is, 20 µmol/l of plasma, or about one third of the total binding capacity of transferrin. The circulating iron has to provide for synthesis of haemoglobin and the other haem-containing compounds. Outside the circulation the reticulo-endothelial cells of the liver, spleen and bone marrow store iron as ferritin, a complex of iron with apoferritin. This soluble protein can bind up to 20% of its weight in iron. When large amounts of free iron are present granules of the insoluble storage compound haemosiderin are formed, particularly in the liver and the lungs. Iron can be transferred from ferritin for the synthesis of haemoglobin or other iron-containing molecules.

When the diet contains too little iron to make up body losses a state of anaemia results in which the red cell counts and the total blood haemoglobin concentration are both low. At first the red cells will contain less haemoglobin than usual, but later the numbers decrease and finally the size of the cells becomes less. In iron deficiency anaemia, then, there is a low red blood cell count and the cells are hypochromic and microcytic.

## Vitamin $B_{12}$, Cyanocobalamine

Vitamin $B_{12}$ (also known as cyanocobalamine because the molecule contains the biologically unusual metal, cobalt) is a dietary factor absorbed in the ileum only if it is first combined with a mucoprotein secreted by the neck mucous cells of the gastric pits. Absorption probably occurs by pinocytosis. Once absorbed, the vitamin is stored in the liver. It appears to be essential for the normal process of division or maturation of red cell precursors. Some individuals cannot synthesise the gastric factor (the intrinsic factor) and are therefore unable to absorb the vitamin (the extrinsic factor). In these subjects the pro-erythroblasts do not divide normally but give rise to very large cells, the megaloblasts. These finally divide to produce macrocytic red cells. The development process is shortened, and although the actual amount of haemoglobin in each cell may be normal or even more than usual, few are formed. The resultant anaemia is characterised by a low red cell count (as low as $1 \times 10^{12}$/l of blood) and the cells are macrocytic and normochromic. The mean corpuscular haemoglobin content is usually high but the mean corpuscular haemoglobin concentration is normal or low.

## Folic acid

Maturation is also affected by lack of folic acid so impairment of fat absorption (as in steatorrhea) can cause disturbances of erythrocyte production similar to those in vitamin $B_{12}$ deficiency. Folic acid deficiencies are likely to occur in pregnant women because folic acid is required by the fetus for normal growth.

## Protein

Red blood cells cannot be produced unless protein is available, but anaemia is seen only as a late symptom of protein starvation.

## Copper

Although trace amounts of copper are necessary for haemoglobin synthesis, these are rarely lacking in the diet.

## Stimulation of red blood cell production

Red cell production is stimulated by a hormone, erythropoietin, formed from a plasma α globulin by the action of an enzyme, renal erythropoietic factor. The hormone is produced at a slow constant rate, but when the supply of oxygen to the kidney cells is decreased, more renal erythropoietic factor is secreted.

## The life of a red blood cell

The average life of a red cell is 120 days. This value is estimated by giving a small transfusion of labelled cells and following their rates of disappearance from the circulation. Cells can be labelled by allowing them to take up radioactively labelled glycine or by exposing them to radioactive chromium, which is incorporated into the cells.

As cells age they become more dense and have less resistance to lysis in solutions of low osmotic activity. Enzyme activity is decreased, the sodium pump mechanism is less active, and they are less able to convert methaemoglobin back to haemoglobin. As a result, sodium ion concentrations rise, their potassium ion concentration falls, and they contain more methaemoglobin at the expense of haemoglobin. Most of the old cells are destroyed in the spleen, although some are lysed in the general circulation (Fig. 4.6). About 25 g of cells are lost every day – a daily turnover of 7–8 g of haemoglobin.

In the vessels the haemoglobin released from lysed cells is bound by haptoglobins, and the complex is taken up by reticulo-endothelial cells which can break it down again. Up to 15 mg/h of haemoglobin can be cleared from the blood in this

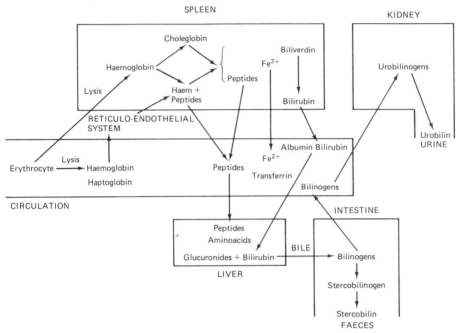

**Figure 4.6** The fate of haemoglobin from the red blood cells.

way. Some of the free haemoglobin is split in the blood and the resulting haem bound by plasma albumin. Further breakdown in the kidney releases free iron, which is then bound by apoferritin. Ferritin, and possibly haemosiderin, is formed.

In the spleen the haemoglobin is transformed either to haem and globin, or to choleglobin. The porphyrin ring can then be opened to give the chain of pyrrole rings in biliverdin, and the iron and globin released. The free iron is taken up by apoferritin, and the globin is returned to the aminoacid pool. The green biliverdin is reduced to reddish bilirubin and carried by albumin to the liver. It is taken up by liver cells, bound by the microsomes, and converted into a water-soluble glucuronide secreted by the liver in the bile. Some escapes from the liver cells back into the plasma, and some reabsorption of the glucuronide in the gut gives a further plasma fraction. In the gut, bacteria convert bilirubin into urobilinogens. Some of these are absorbed in the gut and either excreted via the kidney in the urine – where they are oxidised to the yellow urobilins – or pass again to the liver and the enterohepatic circulation. The remainder stay in the faeces as stercobilinogens and are oxidised to brown stercobilins, giving the faeces their characteristic colour.

Since each gram of haemoglobin yields 35 mg of urobilinogen, the daily excretion is of the order of 200 mg. Normally there is about 5 mg/l in the plasma, bound to an α globulin.

The breakdown of haemoglobin is normally due to aging and destruction of red cells: in abnormal conditions, or in situations outside the blood vessels, even healthy young red cells may be lysed. Bruising is an example of the breakdown of red cells when the blood has been forced out of the vessels and into the tissues. The red cells break down, and the changing colour sequence of the bruise is due to the different colours of the haemoglobin breakdown products. In haemolytic diseases excessive destruction of red cells occurs and iron loss is increased. Unless red cell production can keep pace, anaemia results. Sickle cell anaemia, with its fragile cells in low oxygen tensions, has already been mentioned. Diseases in which red cells are destroyed are characterised by the presence of high concentrations of the yellowish haemoglobin breakdown products in the plasma, giving rise to the symptom of jaundice – the tissues appear yellowish-brown. Jaundice may also arise from diseases affecting the liver or bile duct, when breakdown products of haemoglobin are not removed from the blood or excreted in the bile.

## The white blood cells

The white blood cells, or leucocytes, number between 4 and $9 \times 10^9$/l of blood; the usually accepted mean value is $7 \times 10^9$/l, or $7 \times 10^3$/μl. This is only 1/600th of the red cell count. An increase to over $10 \times 10^9/l$ is termed leucocytosis, and a decrease is termed leucopenia. In neoplasia of the bone marrow, or of white cell-forming organs such as the lymph nodes, the number of white cells can increase enormously to give the condition of leukaemia. Leukaemias are classified according to the dominant cell type. They are often associated with anaemia because the expansion of white cell production in the marrow may restrict red cell production.

Leucocytes are much more typical cells than the red cells – they have nuclei, they lack the concavity of red cells, and their composition is not dominated by one major protein. Whilst the function of the red cells is almost exclusively that of transport, the white cells are associated mainly with protective and defensive functions. As such, their life in the circulation is usually shorter. They appear in the circulation in transit from the forming and storing tissues to the blood or tissue sites where their activity is needed.

The white blood cell count is much more variable than that of the red cells, since it represents only the circulating portion of the white cell population. Thus there is a normal variation throughout the day in white cell numbers, and factors such as high protein meals or exercise can produce transient increases. The turnover of white cells is much more rapid than that of red cells and the number produced is greater. As a result, when the blood volume is reduced after a haemorrhage, white cells are replaced more rapidly than are red cells and a relative leucocytosis may be observed. The increase in body temperature seen as a fever (or pyrexia) is associated with an increased white cell count. In situations requiring defensive action white cells are mobilised; tissue damage or tissue destruction increases the number of circulating white cells. Such increases are often selective, involving one cell type only. A reduction in the number of leucocytes is much less common than an increase: it is usually a sign of excessive destruction of cells or damage to the white cell-producing organs. Any drug which is lethal to bacteria may also kill or damage the cells of the host, and it is only their size, specialisation, and numbers which allow the host cells to survive while the bacteria die. Some drugs, therefore, reduce white cell counts, being toxic to the host cells as well as bacterial cells – for example, chloramphenicol, some sulphonamides, and the antipyretic amidopyrine cause agranulocytosis and hence leucopenia.

There are five major white blood cell types, all related to tissue cells. The first subdivision is into cells with a clear cytoplasm and those with granular inclusions, termed agranulocytes and granulocytes respectively. The granulocytes are further subdivided according to the colour of their granules after staining with haematoxylin and eosin: acidophils or eosinophils have granules staining pink with eosin, basophils have granules staining blue with haematoxylin, and the neutrophils have granules which stain with both dyes and are therefore purple. The agranulocytes include the large and small lymphocytes, with round nuclei filling the cells, and the monocytes with kidney-shaped nuclei.

## Granulocytes

Granulocytes are formed in the bone marrow from tissue similar to that forming the stem cells for erythropoiesis: the stem cells giving rise to white blood cells are termed myeloid cells. The haemocytoblast gives origin by division to the myeloblast (Fig. 4.4). This in turn produces promyelocytes which then acquire their characteristic granules to become granulocytes. One of the functions of the marrow is to serve as a store for the formed granulocytes. The production of granulocytes may be increased by an increase in the number of myelocyte divisions and a shortening of the generation time. In the myeloid series the myelocyte is the most mature cell capable of DNA synthesis and cell division. During maturation RNA and protein synthesis increase and there is progressive condensation of the nucleus and the appearance of granules in the cytoplasm.

The mature granulocyte reserve consists of the $2-3 \times 10^{11}$ cells in the marrow. $10^7-10^8$ cells are released into the bloodstream every minute, the rate of release being proportional to the rate of loss. The rate of production of new cells is so rapid that about one third of the myeloid cells are normally in a state of division and the total daily production of granulocytes is around 1600 million/kg of body weight.

Granulocyte production may be stimulated by a leucopoietin, analogous to erythropoietin, but it is more likely that production is controlled normally by a low molecular weight inhibitor from the granulocytes themselves.

Although granulocytes which are lost because of senescence have usually survived around 30 h in the circulation, the mean half-life of these cells is only 6.5 h. This is because the cells are lost from the bloodstream by many different routes and few reach old age in the circulation. Loss occurs when they are immobilised by the commensal bacteria, or when they are trapped in the liver or spleen. There is also a steady loss from the body into saliva and urine, in the pulmonary system and the gastro-intestinal tract, and in inflammatory exudates.

### Neutrophils (polymorphonuclear leucocytes, 'polymorphs', microphages)

The neutrophils are the most abundant of the white cells (*Table 4.3*). They have multi-lobed nuclei (Fig. 4.7). The number of lobes increases as the cell ages and so a count of the number of cells with two,

**Table 4.3 The cells of the blood**

| Cell type | Numbers (per µl) | White cell count (%) | Diameter (µm) | Shape of nucleus | Total life in body | Life in blood |
|---|---|---|---|---|---|---|
| Erythrocytes | $5 \times 10^6$ | | 7.5 | No nucleus | 120 d | 120 d |
| Leucocytes | $7 \times 10^3$ | | | | | |
| Granulocytes | | | | | | |
|   Neutrophils | $4-5 \times 10^3$ | 60–70% | 10–16 | Multilobed | 13–20 d | |
|   Basophils | $0.1 \times 10^3$ | 0.5–2% | 8–10 | Bilobed | ? | |
|   Eosinophils | $0.2 \times 10^3$ | 2–4% | 10–12 | Round | 3–4 d | hours |
| Agranulocytes | | | | | | |
|   Lymphocytes | $1.5 \times 10^3$ | 20–30% | | | | |
|     (a) Small | | | 7 | Round | 5 d or | hours |
|     (b) Large | | | 12–15 | Round | 2–36 wk months | hours |
|   Monocytes | $0.4 \times 10^3$ | 2–8% | 16–22 | Oval or horse-shoe shaped | | |
| Platelets | $0.25 \times 10^6$ | 2–3 | | No nucleus | 10 d | 10 d |

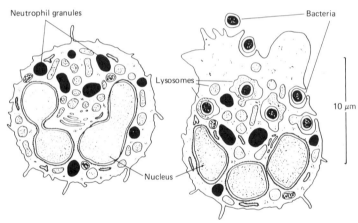

**Figure 4.7** Neutrophil granulocytes, one shown ingesting bacteria by phagocytosis.

three, four, or five lobes to the nucleus (an Arneth count) gives some indication of the proportions of older and younger cells in the neutrophil population. Large numbers of cells with only two- or three-lobed nuclei indicate a high rate of cell production, usually associated with inflammation. The granules of the neutrophils stain with both acidic and basic stains. Their contents have not yet been fully identified although a number of enzymes are known to be present.

Neutrophils are amoeboid and able to squeeze between the endothelial cells of the capillary walls if the blood flow is slow and chemically attractive substances are present extravascularly. Such substances, termed leucotaxins, are usually small polypeptides; they are found particularly at sites of cell damage but are also present normally in a number of gastro-intestinal secretions, notably in saliva. Peptides produced during the breakdown of some plasma proteins are also able to attract leucocytes.

The main function of the neutrophils is to inactivate bacteria and foreign particles by ingesting them (Fig. 4.7). During phagocytosis, cell respiration and glycolysis are increased to provide the necessary energy. As the rate of glycolysis increases, the contribution of the hexose monophosphate shunt rises from about 1% in the resting cell up to about 10%. The charge on the cell membrane is important in phagocytosis; in the neutrophils the main contributor to membrane charge is neuraminic acid, a highly charged molecule present in high concentration and situated near the cell surface. During phagocytosis hydrogen peroxide is produced and this is split by a peroxidase in the presence of halides to yield nascent oxygen – a strong bacteriocide. The most effective of the halides is iodide, which some neutrophils are able to concentrate by taking up iodine.

The cytoplasmic granules and the lysosomes of the neutrophils contain many enzymes and antibacterial substances: acid and alkaline phosphatases; β-glucuronidase and ribonuclease; lysozyme and an acid-soluble bacteriocidal protein termed phagocytin.

## Eosinophils

A diurnal rhythm is observed in eosinophil numbers, the count being high in the early morning and low in the late afternoon.

Eosinophils (Fig. 4.8a) are non-phagocytic and non-amoeboid. They contain histamine, which they take up and neutralise in sites where it is being produced. Studies of the acidophil granule material show it to have powerful anti-histamine activity. The eosinophils, therefore, are important in allergic responses associated with histamine release. Although eosinophils have no direct active role in the complex process of inflammation, they are attracted to the histamine released from the granules of the mast cells (or basophils) and are responsible for its subsequent inactivation. Increased numbers of eosinophils are associated with states of allergy or hypersensitivity. Corticosteroids, which suppress the inflammatory response, reduce the number of circulating eosinophils.

## Basophils

The basophils, with their coarse granules staining with basic dyes (Fig. 4.8b), are the least abundant of the white cells in the circulation (*Table 4.3*). Their lifespan in the circulation is not known. They are the mobile equivalent of the cells in tissues known as mast cells, although they are not motile cells like the neutrophils. Basophils are phagocytic, absorbing dying neutrophils. They are found in the tissues in inflammatory reactions, particularly those associated with hypersensitivity, where their arrival is secondary to that of the neutrophils.

The granules contain bound histamine and heparin, an inhibitor of blood clotting produced in the liver. The cells are also able to release plasmin, an enzyme which breaks down the fibrin of a blood clot.

Basophils or mast cells are stimulated to release their contained histamine as a result of a sequence of reactions involving various complement components, which culminates in the formation of activated products of C'5 and, to a lesser extent, C'3 (anaphylatoxin). The sequence is triggered by the reaction of antigen with antibody in a sensitised subject. The release of histamine, which dilates and increases the permeability of blood vessels to cause swelling (or blistering) and itching, is normally held under control by an inhibitory complement fraction C'1. Some subjects lack this fraction and produce massive histamine reactions almost spontaneously – a condition known as hereditary angioneurotic oedema.

Although most of the phagocytic activity of the basophils is directed towards dead or dying neutrophils, they do also take up iron and calcium.

## Agranulocytes

### Lymphocytes

The lymphocytes are the second most abundant group of white blood cells in the circulation (*Table*

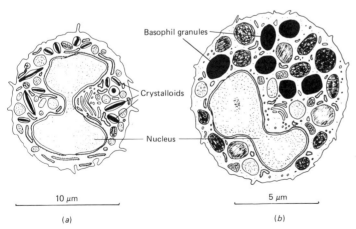

**Figure 4.8** (*a*) An eosinophil with crystalloid material in some vesicles, and (*b*) a basophil with large coarse granules. Note the different scales.

68  The cells of the blood

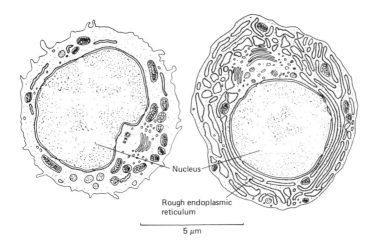

**Figure 4.9** A small lymphocyte and a plasma cell. Note how the amount of rough endoplasmic reticulum is enormously increased in the plasma cell.

4.3). A large and a small type are distinguished histologically. Both types have a large nucleus which fills the cell almost completely (Fig. 4.9). The large lymphocytes are believed to be immature cells or precursors of the small cells. Indeed, the small lymphocytes may not be fully mature cells but may be pluripotent – capable of further development to produce several types of daughter cell. There are four possibilities: the small lymphocyte may be the end of a cell line, it may be capable of further division to produce daughter cells like itself, it may be a parent cell for plasma cells, or it may be a parent cell for some macrophages.

Experiments to determine the lifespan of a small lymphocyte have revealed two populations of cells, one with a life of a few hours in the circulation and about five days *in toto*, and another with a life of at least two weeks, and possibly as long as nine months. These latter cells constitute the mobilisable, or recirculating, lymphocyte pool (Fig. 4.10). The two populations of cells are not of differing origins, cells with both long and short lives being derived from either of the two major cell lines described below.

There are two types of lymphocyte, differing both in origin and function. One of these is the T-cell, derived ultimately from cell lines originating in the thymus gland. The T-cells are also known as thymic-dependent cells, immunologically competent cells, or antigen-reactive cells. Cells from the fetal thymus seed other lymphoid areas, and thymic-dependent strains are subsequently found in the white pulp of the spleen and as cuffs around the germinal centres of the spleen, in the paracortical zones of the lymph nodes, and in the interfollicular

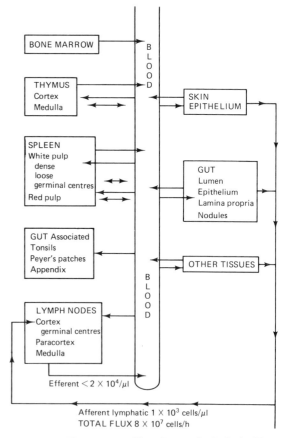

**Figure 4.10** The turnover of lymphocytes in the body. Note the very large number of cells involved, and the traffic between the tissues and the blood.

areas of Peyer's patches in the small intestine. In the adult only a vestigial thymus remains and cells are produced in these other tissues; their development, however, seems to depend upon some chemical signal from the thymic remnants. The second type of lymphocyte is the B-cell. This is histologically indistinguishable from the T-cell but can be identified by electron microscopy, immunological staining and by its activities. These cells are found in the bone marrow, in the germinal centres of the spleen, and in the corticomedullary junctions and medullary cords of lymph nodes. No specific organ controlling their production and development has been identified in man, although it may be the bone marrow itself, or the lymphoid tissue found in close relation to various parts of the gut, the gut-associated lymphoid tissue. Although B-cells are present in both the blood and the lymph, the extent of recirculation of these cells is probably less than that of the T-cells.

In the adult the parent cells of the lymphocytes derive mainly from the bone marrow. The blast cells are similar in appearance to the haemocytoblast cells which give rise to the other blood cells. From the marrow they migrate to other tissues and seed the thymus, the spleen, and lymph nodes to form haemocytoblast colonies. These produce lymphoblasts, prolymphocytes, and then large lymphocytes. In response to immunological challenge the B-lymphocytes may develop in tissues into plasma cells (Fig. 4.9). At first a larger cell, a plasmablast or immunoblast, with more cytoplasm appears in the circulation, but in the tissues those derived from B-cells then form plasma cells, with a smaller nucleus, much more cytoplasm, a highly developed endoplasmic reticulum, and distinctive eosin-staining granules known as Russell bodies. The Russell bodies contain immunoglobulin which the cells secrete into the tissue fluid.

Although lymphocytes are not themselves phagocytic, some of these cells have been observed to develop into macrophages. The change is signalled by an increase in intracellular enzymes – in particular, oxidases, peroxidase, and alkaline phosphatase. Lipid turnover increases and lipid inclusions are seen.

Both T- and B-lymphocytes have on their cell membranes receptor sites for antigens. Exposure to the appropriate antigen causes the cells to enlarge and proliferate to form immunoblast cells. Those derived from B-cells become plasma cells and secrete antibodies: this is the humoral response to antigens. The cells derived from the T-cell line do not secrete antibodies but can produce a cellular reaction to the antigen. Very few antigens can stimulate B-lymphocytes in the absence of T-lymphocytes, although the synthesis of immunoglobulins appears to be the property only of the B-cells and their progeny. It is not clear why the T-cells should be necessary for B-cell activity, although it may be related in some way to the manner in which the T-cells bind antigens. After initial contact with an antigen the sensitised cells retain the ability to synthesise the antibody, and transmit this ability to their offspring. This immunological 'memory' enables a greater and more efficient response on subsequent exposure to the same antigen. The cell line capable of a particular antibody response probably persists indefinitely.

When an antigen enters the body, it is rapidly localised within the macrophages in the marginal zones of the Malpighian bodies of the spleen, or in the medulla of the lymph nodes. This process potentiates the antigenic stimulus, either by trapping the antigen – possibly on the macrophage membranes – or by releasing more active phagocytosed antigen fragments.

Lymphocytes, particularly the blast cells, are extremely sensitive to radiation and may be destroyed when tissues are exposed to it. Large doses of corticosteroids also suppress lymphocyte formation. Both these effects can lead to a reduction or even abolition of the immunological reponse. This can be dangerous, since the reponse to infection is reduced: but radiation and corticosteroids have been used in the preparation of patients for tissue or organ transplantation to reduce the possibilities of transplant rejection.

Several diseases have been characterised over the last 20 years as auto-immune diseases. In such diseases, the patient produces immunological reactions to his or her own body proteins. Lymphocytes normally distinguish between 'self' and 'not-self' – possibly because reactive cells are eliminated early in life. However, cytotoxic, toxic-complex-forming, and cell-sensitive, types of reaction can occur if the normal immune system begins to react to 'self'. IgM and IgG immunoglobulins are involved. The reactions are initiated by T-lymphocytes. Some of the soft tissue diseases of the mouth, such as aphthous ulceration and possibly some forms of gingivitis, may be auto-immune manifestations.

## Monocytes

The numbers and morphology of the monocytes are given in *Table 4.3*. The total lifespan of these cells is unknown, although tissue culture experiments suggest that it may be several months. However, the monocyte is merely one stage in a sequence of cells originating in the bone marrow with the monoblast, and ending with monocytes, which, after spending a few hours in the circulation, reach a tissue where they spend the rest of their lives as tissue macrophages.

The monocyte, like its successor, the fixed macrophage, is phagocytic, amoeboid and responsive to chemotactic stimuli (Fig. 4.11). The cells

70   *The cells of the blood*

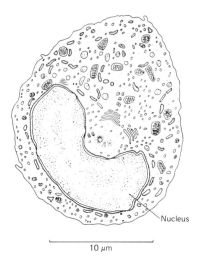

**Figure 4.11** A monocyte.

continue to grow and develop in the tissues, attaining sizes of up to 200 μm, developing mobile cytoplasmic projections, and often becoming multinucleate. Dense granules fill the cytoplasm. The lysosomes contain acid phosphatase, β-glucuronidase, cathepsin, lysozyme, and aryl sulphatases. These cells are the type cells of the reticulo-endothelial system, a defensive system widely scattered throughout the body. Typical cells are given different names in different tissues: the histiocyte, the Kupfer cells of the liver, the alveolar macrophages of the lungs, the macrophages of the spleen, lymph nodes and bone marrow. Osteoclasts, the cells which remove hard tissues, may be derived from the monocytes.

The monocytes have three well established functions. Firstly, they take up and destroy aging or antibody-coated red blood cells, retaining the iron as ferritin or, if there is an excess, as haemosiderin granules. Secondly, they play a number of roles in immune responses. They accumulate antigens, which they may break down, or may maintain in a state and position suitable for producing the maximum response from lymphocytes. Antigens are probably bound to the macrophage membrane, since intimate contact of the macrophage and the lymphocyte membranes has been observed in the initiation of the immunoblast response. The macrophages probably also release antibody fragments which can stimulate the lymphocyte response. Further, the macrophages produce some of the components of complement, the substance interferon, and, possibly, some antibodies. Their ability to phagocytose depends upon complement activation (*see* Chap. 3). The phagocytes are involved, like the T-lymphocytes, in cellular immunity reactions – those in which cell components, rather than circulating antibodies, are the main reactants. The third function of the macrophages is the phagocytosis and killing of bacteria by the lysosomal enzymes. This is extremely important in relation to the acid-fast bacilli and bacteria with lipid cell walls. As in the neutrophils, phagocytosis is associated with an increase in glycolytic activity.

## The platelets (thrombocytes)

Platelets are disc-shaped cell fragments with a smooth outer membrane and no pseudopodia (Fig. 4.12). They have no nuclei but contain a number of subcellular structures. They are 2–3μm in diameter, with a volume of about 5 fl. The cell membranes are trilaminar, the outer layer containing both protein and mucopolysaccharide. Immediately beneath the membrane is a circumferential bundle of microtubules which maintain the discoid shape. The other cell constituents include mitochondria, the electron-dense α granules, the 'very electron-dense' bodies found relatively infrequently in human platelets, some vacuoles, and glycogen granules.

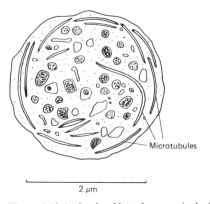

**Figure 4.12** A platelet. Note the many inclusions and the circumferential microtubules.

The number of platelets is $150-500 \times 10^9/l$ (or $\times 10^3/\mu l$) of blood. A reduction in these numbers is termed thrombocytopenia: it may result from lack of production due to bone marrow damage, from an uneven distribution in the circulation (as when there is pooling in the spleen in splenomegaly), from an increased rate of destruction (as in the auto-immune syndrome of idiopathic thrombocytic purpura), or from an excessive consumption of cells in the process of blood clotting. An increase in the platelet count is termed thrombocythaemia: this may be a physiological reaction to trauma or infection, or it may be idiopathic – when it is usually associated with a general polycythaemia.

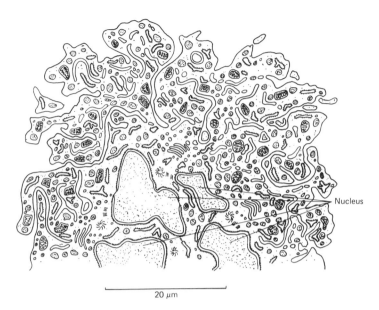

**Figure 4.13** The megakaryocyte – this shows part of a very large cell which will fragment between the deep invaginations to form the non-nucleated platelets.

The precursor cell of the thrombocyte line is the myeloblast or haemocytoblast. The first characteristic cell is the megakaryoblast with two nuclei in a rim of agranular basophilic cytoplasm. The two nuclei fuse to give an irregular single nucleus and mitosis continues without cell division to yield a promegakaryocyte with eight fused nuclei and reddish-purple staining granules in the cytoplasm. The fused nuclear mass is large: the cytoplasm begins to increase in volume also. The next stage, that of the megakaryocyte (Fig. 4.13), is a cell 40–80 μm in diameter, with a 16-fold nucleus, containing in its cytoplasm small aggregations of granules separated by pale-staining agranular zones. These lines represent lines of division between future platelets. There are long pseudopodia containing localised granular areas. At maturation, membranes form and demarcate the platelet subunits, which finally split off from the cytoplasm of the parent cell. The process is thought to be controlled by a hormone, thrombopoietin, analogous to erythropoietin, whose nature and site of formation are unknown. In the bone marrow most megakaryocytes have eight nuclei, but some cells have 16 and, very rarely, some have 32.

Platelets survive about 10 days; the normal turnover is around 35 000/day. Two thirds of the platelets in the body are normally circulating and one third are found in the spleen.

Platelets absorb onto their cell membranes a number of blood clotting factors such as fibrinogen, factor V and factor XI. In addition to these they contain their own blood clotting factors, or factors similar to those in plasma: fibrinogen, some formed fibrin, and a platelet factor XIII. There are other proteins in the cells which are closely similar to plasma albumin, γ globulin, β-lipoprotein, and some complement fractions. Specific platelet factors include an antiplasmin, an antiheparin (platelet factor V), a phospholipid which provides an important link in blood clotting (platelet factor III), a fibrinogen activator, a permeability factor, and some proteolytic enzymes. Both actin and myosin filaments have now been identified in platelets and it seems likely that thrombosthenin, a contractile protein found in platelet plugs and formed blood clots, is actually an actin–myosin complex like the contractile protein of muscle. The 'very electron-dense bodies' and some of the α-granules are storage sites for serotonin (*see Chap. 27*). Some of the α-granules may also store adenine nucleotides.

The functions of the platelets are to maintain the integrity of the vascular walls and to arrest the flow of blood after an injury. The first is achieved by adhesion of the cells to the basement membrane of the blood vessel wall if endothelial cells are lost. The arrest of blood flow is achieved by much more complex means. If the platelet membrane comes into contact with collagen, thrombin, or adenosine diphosphate (ADP) it becomes sticky and the platelet will adhere to the surface or to other platelets. Small amounts of fibrinogen and calcium are necessary for aggregation to take place. Within seconds of adhesion the platelet release reaction occurs as a result of the action of prostaglandins. Platelets undergoing the release reaction change in morphology, losing their discoid shape and becoming spherical with numerous small pseudopodia. The granules gather into the centre of the cell and the microtubules contract round them. In the next stage the platelets become very close-packed and giant pseudopodia are formed; the granules still remain centrally placed. Finally a controlled and selective loss of platelet constituents is observed. Most of the serotonin, some of the adenine nucleotides, histamine, adrenaline, potassium ions, fibrinogen, platelet factor IV, and some enzymes such as acid phosphatase and β-glucuronidase, are lost from the cells. The release reaction may be triggered by adrenaline, bacterial endotoxins, some antigen–antibody complexes, and some viruses. As release of ADP occurs other platelets become adhesive. This sequence of events is part of the process of haemostasis described in Chap. 27.

Thrombosthenin may contract during the process of clot retraction.

A number of platelet defects have been described which affect the size and strength of the blood clot formed even though the biochemical process of blood clotting occurs normally. The usual symptom is a prolongation of the bleeding time. Aspirin, a prostaglandin inhibitor, inhibits the platelet release reaction. In the disease thrombosthenia, the platelets do not aggregate in response to ADP. In von Willbrand's disease, a plasma factor promoting the platelet adhesion reaction is absent. Several drugs such as am idopyrine, oxytetracycline, barbiturates, quinine, streptomycin, and the sulphonamides may cause thrombocytopenia and impair the normal haemostatic process.

## Further reading

Paterson, R.B. *A Short Textbook of Haematology.* 6th edn, Pitman Medical, Tunbridge Wells, 1984

Roitt, I. *Essential Immunology.* 6th edn, Blackwell, Oxford, 1988

# 5
# Excitable tissues

All cells show to some extent the property of irritability – the ability to respond to an external stimulus. Primitive unicellular organisms move away from some stimuli, towards others, the stimulus spreading through the organism probably by chemical means. In the multicellular organism some cells become specialised to perceive stimuli and others to transmit the perception to central cognitive cells. Others transmit impulses from a central control area to peripheral organs to produce specific effects. The cells transmitting impulses are the nerve cells or neurones; their function is to relate the animal's responses to the various stimuli it receives. Other cells – the muscle cells – combine the ability to transmit impulses with the ability to contract: they function as response organs. The more primitive chemical pathways are adapted in the chemical messengers, or hormones, which travel in the bloodstream to target organs.

## Nerve cells
### Structure

The nerve cells, or neurones, are specialised transmitting cells. Each neurone has a body, or soma, with one or more processes (Fig. 5.1). Strictly, a process normally carrying signals away from the soma is referred to as an axon, and the processes carrying signals from the periphery to the soma are termed dendrites. A neurone usually has many dendrites. More usually, the term axon is used of peripheral nerves to denote the longer processes of the cells regardless of the direction of travel of the impulses. Thus the axon of a sensory nerve is strictly a dendrite. The axon of a nerve cell is always surrounded by protective cells known as glial cells in the central nervous system or Schwann cells

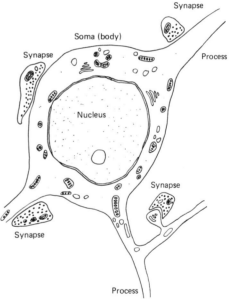

**Figure 5.1** A drawing of an electron micrograph of a neurone.

peripherally (Fig. 5.2). Axons which have only a thin layer of Schwann cell about them – two membranes and a thin band of cytoplasm – are termed unmyelinated; if they have Schwann cells wrapped round and round them so that the axon lies at the centre of concentric layers of Schwann cell membrane making a thick sheath of high electrical resistance lipoproteins, they are called myelinated cells and the multiple layers of Schwann cell membrane are the myelin sheath. Where two Schwann cells meet there is a minute area of bare

74  *Excitable tissues*

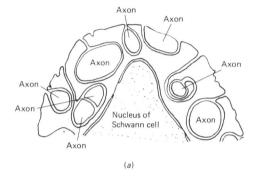

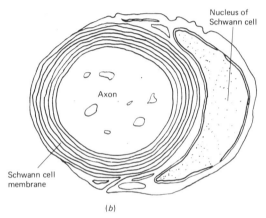

**Figure 5.2** Nerve fibres in section. (*a*) A non-myelinated group of fibres lying in a protective sheath of Schwann cell; (*b*) a myelinated fibre surrounded by concentric layers of Schwann cell membrane, which make up the myelin sheath in the regions between the nodes.

nerve cell – the node; the myelin covered areas are called internodes.

The nerve cell itself contains a nucleus, cytoplasm (or axoplasm), and a membrane, the neurilemma. In the cytoplasm are granules termed Nissl granules and at the termination of true axons are a number of vesicles.

## Resting potential

Like all other mammalian cells the neurone lies in a film of the extracellular fluid. The composition of the intracellular fluid (the axoplasm) is contrasted with that of the extracellular fluid in *Tables 1.3 and 1.4*. The important differences between the two are the higher potassium, and lower sodium and chloride ion concentrations in the neuronal fluid. Like all cells the neurone also contains a much higher concentration of protein than the surrounding fluid. The differences in sodium and potassium ion concentrations are maintained by a sodium–potassium pump in the membrane although the membrane is, comparatively, very permeable to potassium ions and only slightly permeable to sodium ions in the resting state. The chloride ions are in lower concentration inside the cell and the cell membrane is highly permeable to them. The distribution of ions across the membrane with its varying permeability to the ions follows from the Gibbs–Donnan relationship (Chap. 1). This maintains low chloride concentrations within the cell. The discrepancy in ion concentrations inside and outside is balanced by an electrical potential difference across the cell membrane. The magnitude of the potential can be calculated from the difference in concentration by the Nernst equation (Chap. 1). The Nernst equation, however, describes the situation when the membrane is permeable only to one ion, and, further, it relates only to the equilibrium situation. The theoretical potential for each of the major ions can be calculated if it is assumed that the membrane is permeable only to one at the time. The term $RT/zF$, for a monovalent ion at body temperature (310°K), is a constant which, after an adjustment to allow the use of logarithms to the base 10, works out at 61.8. The potential which would exist across the membrane as a result of the difference in potassium ion concentrations would therefore be, if the membrane were permeable only to potassium ions

$$E_K = 61.8 \log 150/4 = -97 \, mv$$

Similarly, for chloride ions

$$E_{Cl} = 61.8 \log 105/9 = -66 \, mv$$

and for sodium ions

$$E_{Na} = 61.8 \log 142/15 = +60 \, mv$$

the sign in each case referring to the polarity of the inside of the membrane.

The potential across the membrane when several ions are present on each side in different concentrations is calculated from the Goldman equation. This equation takes into account the relative permeability of the membrane to the different ions and it is applicable even in non-equilibrium conditions. When the membrane is less permeable to a given ion, the contribution of the ratio of concentrations for that ion will be proportionately reduced. The equation is as follows:

$$E = \frac{RT}{F} \log_e \frac{P_a A^+_{in} + P_b B^+_{in} \, P_c C^+_{in} \ldots +}{P_a A^+_{out} + P_b B^+_{out} \, P_c C^+_{out} \ldots +}$$

$$\frac{P_z Z^-_{out} + P_y Y^-_{out} + P_x X^-_{out} + \ldots}{P_z Z^-_{in} + P_y Y^-_{in} + P_x X^-_{in} + \ldots}$$

where A, B and C .... are monovalent cations, Z, Y and X .... are monovalent anions, and $P$ is the permeability of the membrane to the ion indicated by the subscript. When the membrane is permeable only to one ion, this equation reduces to the Nernst equation; similarly, if the relative permeability of the membrane to one particular ion greatly exceeds that to the other ions, the membrane potential approaches the Nernst potential for that ion. It can be shown that potassium ions, to which the resting membrane is highly permeable, contribute the major part of the resting transmembrane potential, and the contribution of the difference in sodium ion concentrations is very small because of the membrane's low permeability to sodium ions. If the permeability of the membrane to a particular ion changes relative to other ions, the membrane potential changes: an increase in permeability to a particular ion causing a shift of the membrane potential towards the calculated Nernst potential for that ion.

The resting potential of a typical nerve is around −70 mV, the inside of the membrane being negative with respect to the outside. This value is close to the Nernst potential predicted for chloride ions.

## The action potential

The charge across a cell membrane may be changed by applying a stimulus to the cell. An electrical stimulus is the most obvious way to do this but if the permeability of the cell membrane is changed by chemical or mechanical stimuli a new local ionic equilibrium will be set up and the potential due to ionic concentration differences will change. A local change in electrical potential causes current flow to or from neighbouring areas of membrane and so a wave of electrical activity will spread out from the site of stimulation. In non-excitable cells, and in nerve and muscle cells if the stimulus results in a change in potential less than some threshold value, the electrical changes will occur only over a limited area of membrane since the charge leaks away through the surrounding extracellular conductive fluid. Where a cell is surrounded by an insulating layer – the myelin sheath of a myelinated nerve – the electrical pulse can travel relatively long distances, up to several millimetres. This phenomenon is termed electrotonic spread. It is analogous to the spread of a current along a wire and takes place correspondingly rapidly.

Most of the cells of the body are not electrically insulated, and they are surrounded by a fluid medium containing ions which is able to conduct away current. Electrotonic spread of an impulse is not, therefore, a practical means of conduction of an impulse in the body. Despite its rapidity of transmission, the signal is rapidly and progressively attenuated. In the excitable tissues the cell membranes are specialised to respond to stimuli in such a way that an electrical signal can be transmitted over longer distances and the final signal can be maintained at the same magnitude as the original signal induced by the stimulus. This is achieved by a sequence of active changes in permeability and potential which is termed an action potential.

The events which constitute an action potential take place as follows. When a stimulus to a nerve (or a muscle fibre) causes the local membrane potential to move sufficiently towards electrical zero – reaching a threshold or liminal value – an action potential is initiated. A typical threshold for a nerve would be a change of potential of around 15 mV. The stimulus would therefore have to change the resting potential of about −70 mV to about −55 mV. This is a depolarisation – the cell is electrically less polarised. In the nerve or muscle cell membrane there are channels through which sodium can readily diffuse but these are normally closed. As the membrane potential becomes less negative, the membrane conformation changes and, at the threshold level, the sodium channels begin to open, resulting in a much increased flow of sodium ions into the cell. This inflow exceeds the immediate capacity of the outward sodium pump mechanism and, more importantly in relation to the membrane potential, the outward flow of potassium ions. With the now increased permeability of the membrane to sodium ions, the membrane potential is more influenced by the sodium ion concentration difference between the inside and the outside of the cell, and moves towards the positive Nernst potential for sodium. As the potential becomes progressively less negative, more and more voltage-dependent sodium channels open. Fig. 5.3 shows how the membrane potential and the membrane conductance for sodium ions change over the first few milliseconds of the action potential. Although conductance is not precisely the same as permeability the two are closely related. In a matter of 0.2 ms the membrane potential of a nerve cell approaches the Nernst potential for sodium ions (+60 mV) and then the sodium channels close, to remain incapable of further opening for a short period of time thereafter. The permeability of the membrane to potassium ions also increases during this first phase of the action potential, but in contrast to the very rapid rise in permeability to sodium ions, that to potassium ions is relatively slow and does not reach its peak until after that to sodium ions has begun to decrease. As a result of these changes the membrane potential begins to fall back first towards electrical zero and then towards the resting potential. This is a phase of repolarisation. If the permeability to potassium ions is greater than that in the resting nerve, the membrane potential may fall below the resting potential for a short time – a period of hyperpolarisation. The whole process takes some 1–2 ms in a nerve fibre.

The absolute amounts of sodium and potassium ions transferred across the membrane during the

76  *Excitable tissues*

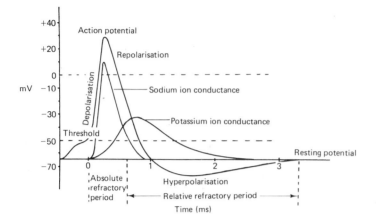

**Figure 5.3** The course of an action potential in a nerve. The upper line shows the changes in electrical potential across the membrane, the two lower lines the changes in membrane conductance to sodium and potassium ions – the basis of the electrical changes. The absolute refractory period ends when the membrane potential reaches -20 to -30 mV again.

action potential are minute: $3–4 \times 10^{-12}$ mol/cm2 of cell surface. In a small axon of 0.5 μm diameter this is less than 0.1% of the total amount of potassium ions in the cell. Even in the absence of an active sodium pump mechanism to restore ion concentrations, the neurone could produce several hundred impulses without using up its store of ions. The action potential is an active process and involves extra work by the energy-consuming sodium pump to restore the resting state: after intense activity a nerve cell can be shown to have produced extra heat and to have used extra oxygen.

## *Liminal stimulation*  THRESHOLD

An action potential is not normally produced unless the membrane is appreciably depolarised – say by about 15 mV. Effective electrical stimulation depends on both the strength and duration of the stimulating current and to some degree upon the rate of change of membrane potential: weak stimuli with a rapid rate of change may initiate action potentials where greater stimuli built up more slowly will have no effect. The weakest stimulus necessary to produce an action potential is termed the threshold, or liminal, stimulus; stimuli below this level cause no propagation of action potential however long they are applied. When a square pulse of current is applied, the current producing the minimal response is termed the rheobase current; the time required for the rheobase current to flow in order to produce the response is called the utilisation time. The time required for activation using a current twice as great as the rheobase current is termed the chronaxie. These measures are used to express quantitatively the excitability of an axon. With very brief current pulses the product of the intensity of stimulation and the stimulus duration required to produce an impulse is a constant in a given axon.

## *Summation*

A stimulus may produce a depolarisation insufficient to trigger an action potential. If another stimulus of subliminal magnitude is given before the effect of the first has died away the two may summate and the resultant depolarisation be sufficient to initiate an action potential. This is temporal summation; a similar phenomenon is observed when two stimuli, each subliminal, are spatially separated on the neurone and summation of the depolarisation occurs at a point between the two stimuli.

## *The all or none rule*

Because an action potential represents a precise series of membrane permeability changes resulting in predictable membrane potential changes, every action potential in any one axon must be identical. If a suprathreshold stimulus is given an action potential is produced. If the stimulus is subthreshold, no action potential is produced. An axon cannot therefore respond to varying stimulus strengths by varying the size of the action potential: it can only code the information by varying the rate of production of action potentials.

## *Refractory period*

Once the axon has received a supraliminal stimulus the sequence of membrane permeability and potential changes is set in motion. During the period in which the changes in permeability to sodium ions are occurring a second stimulus will be ineffective: the rise in potential is self-limiting because a second opening of the sodium channels is impossible during the phase when permeability to sodium ions is being shut off. This period of 1–2 ms is termed the absolute refractory period. When the membrane has repolarised to a level of somewhere between −20 and −30 mV, a second period begins. While the potassium ion conductance remains above the resting level, the outflow of potassium ions is more able to maintain the negative membrane potential and the membrane conformation does not permit the opening of sodium channels to the same extent. During this period an action potential can be initiated but a larger stimulus than usual is required. It is termed, therefore, the relative refractory period. The refractory periods limit the frequency at which the axon may transmit action potentials – the most rapidly transmitting can achieve around 1000/s (i.e. 1 ms between impulses).

## Conduction of the impulse

The action potential is a change in potential across the membrane at the point of stimulation. From this point current flow to neighbouring zones causes a spread of depolarisation. This electrotonic spread produces sufficient depolarisation to initiate a new action potential in the neighbouring area of membrane. Although electrotonic spread can occur in any direction, the next action potential can stimulate a new action potential only in the membrane ahead, since the membrane behind is in a refractory state from the first action potential. Thus the action potential is propagated along the nerve fibre. Normally stimulation occurs at the soma (or cell body) or at the nerve terminal, and so conduction takes place in one direction only. The usual direction of conduction (towards the central nervous system for a sensory nerve, and away from it for a motor nerve) is termed orthodromic. An impulse travelling in the reverse direction is termed antidromic.

This system of conduction results in an impulse of constant size from the beginning to the end of the axon, however long the axon. It is less rapid than electrotonic spread since it involves a time-consuming active series of changes in the axon membrane, and it is energetically more expensive. It is 'all or none' because the nerve responds only to stimuli which are suprathreshold and always with the same magnitude of action potential. The intensity of a stimulus to the nerve can affect only the number and frequency of the action potentials generated – it cannot affect their magnitude.

The rate of conduction of an impulse along a nerve could be increased if it could travel longer distances by electrotonic spread and generate action potentials as signal boosters at intervals. This is what happens in myelinated nerves. A myelinated nerve fibre is surrounded by an insulating layer of myelin which is broken up into segments of a length appropriate for the impulse to remain above the threshold level until the next break in the myelin sheath. Here a new action potential is triggered and can spread along the following segment. The insulation breaks where the action potentials are generated are the nodes, the myelin sheathed sections the internodes. Since the impulse travels very rapidly from node to node and the action potentials appear only at the nodes, it is described as jumping, or leaping, from node to node and this means of conduction is given the name saltatory (Latin *saltare*, to leap). Saltatory conduction is rapid, because electrotonic spread is rapid; it is economical, because action potentials are required only at intervals; and it is a 'no loss' system because the action potential at the end of the fibre is of the same magnitude as at the beginning. The majority of nerves are myelinated and use this rapid means of impulse transmission.

Electrotonic spread can occur past small blocked areas of axon – or even, experimentally, along salt bridges. It can also result in spread of an impulse from one axon to another if they are so close and poorly insulated that the first can sufficiently depolarise the membrane of the second. This occurs only rarely in the peripheral nervous system but happens more frequently in the brain itself.

Anaesthetic substances affect the production and transmission of action potentials in several ways. Most local anaesthetics produce their effects by preventing the opening of sodium channels, possibly by making the nerve membrane locally more rigid and structured. General anaesthetics are usually lipid-soluble substances which dissolve in nerve membranes and change their properties.

## *Velocity of propagation*

The larger a nerve fibre the more readily it will conduct impulses. In part this is a function of the ratio of the volumes, and hence resistances, of the membrane and the axoplasm, and the ratio of ion movement to total ionic concentration; in part it is simply an expression of the core resistance which, like that of a wire, decreases with increase in the cross-sectional area. The velocity of conduction in a nerve fibre, therefore, is usually proportional to the size of the fibre. Nerve fibres have been classified into groups by size, rate of conduction, and function. Two different classifications are used, one for sensory nerves with groups numbered I to IV, and the other for all nerves, with the main groups assigned the letters A to C and the sub-groups given Greek letters (*Table 5.1*).

As a result of the differing nerve conduction velocities, recordings from a mixed nerve with many different fibres show many impulses travelling at different speeds. Recordings from nerves in the cranial area have been used to explore the functions of the nerves supplying the teeth and the gingivae (Fig. 5.4).

The size of an axon is also a factor in the threshold for response to stimuli. Pressure on a mixed nerve

## Table 5.1 Classification of nerve fibres

| Letter | Diameter (μm) | Spike duration (ms) | Conduction velocity (m/s) | Function | Sensory class |
|---|---|---|---|---|---|
| A α | 12–20 | 0.4–0.5 | 70–120 | Proprioceptors, motor to muscle | Ia, Ib |
| β | 2–12 | | 30–70 | Touch, pressure | II |
| γ | 3–6 | | 15–30 | Motor (muscle spindles) | III |
| δ | 2–5 | | 12–30 | Pain, temperature | III |
| B | <3 | 1.2 | 3–15 | Myelinated preganglionic sympathetic | |
| C | 0.3–1.3 | 2.0 | 0.5–2.3 | Unmyelinated postganglionic sympathetic, pain, interoreceptors | IV |

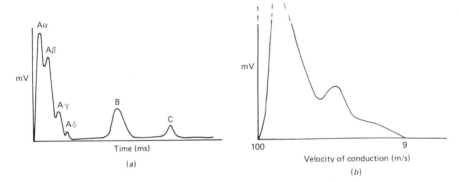

**Figure 5.4** (*a*) The composite action potential recorded from a mixed nerve, showing the sum of activities of fibres of different sizes recorded some distance from the point of stimulation; (*b*) a recording from an inferior dental nerve, showing two major groups of fibres, one with a mean diameter of 10 μm, and the other with a mean diameter of around 5 μm. The smaller fibres are thought to supply the dental pulps; they have conduction velocities below 20 m/s.

causes loss of conduction in large fibres before it affects smaller ones – so the loss of sensation when a tumour is crushing a nerve is selective, the smaller pain fibres remaining unaffected whilst other sensory impulses travelling in larger fibres are lost. The effect of anaesthetics on a nerve is also related to the sizes of its constituent axons. Generally speaking, a local anaesthetic blocks transmission in C fibres first and in A fibres last. A patient may therefore still retain some touch sensation even when pain sensation has been lost: depth of anaesthesia is better tested by asking whether probing hurts, rather than whether it can be felt. A localised lack of oxygen seems to affect B fibres selectively at first, then A fibres, then C fibres, producing typical symptoms when the normal blood supply to an area is reduced.

## Synapses

The nervous system is a transmitting system. The impulses travelling in an axon must pass to another neurone or stimulate an effector organ. The junctional area between two neurones, across which an impulse must be transmitted, is termed a synapse (Fig. 5.1).

Histologically, the synapse consists of the membrane of the pre-synaptic axon, a space usually 15–20 nm wide, and the membrane of the post-synaptic neurone. The membrane is sometimes called the synaptilemma.

Since transmission along an axon is electrical, transmission across a synapse might also be expected to be electrical. However, even though the action potential exceeds the threshold value for stimulation of a nerve by some ten times, the calculated attenuation of a signal across a typical synapse is of the order of 10 000 times. Electrical transmission across a synapse could only occur if the resistance of the terminal membranes was very low, or if the cell membranes approached very closely – even fused –, or if cytoplasmic bridges existed between the cells. Thus electrical transmission does occur between giant nerve fibres in the crayfish, and between muscle cells with tight junctions, and electrical inhibitory impulses cross synapses directly in the Mauthner nerve of the goldfish where a particular area of nerve–nerve contact is very heavily insulated by a high extracellular resistance.

Almost all nerve cells, however, use another mechanism for transmission: the production of chemical substances which diffuse across the synaptic gap to cause depolarisation of the post-synaptic membrane. Several chemical transmitters are now known: acetylcholine, adrenaline, noradrenaline, glutamic acid and γ-amino butyric acid are examples.

The principles of synaptic transmission may be introduced with the example of the nerve–muscle junction where a single axon innervates a small number of muscle fibres. This situation is simpler than that in the true neurone–neurone synapse where many axon terminals may reach a single post-synaptic cell.

## Neuromuscular junction or motor endplate

The anatomy of the area is shown in Fig. 5.5. The structure is a simple relay in which impulses in the axon produce impulses in the muscle fibre. The terminal fibres of the nerve are approximately 1.5 μm diameter. They run in grooves in the muscle surface for about 100 μm, separated from the muscle fibre by a space at least 50 nm wide. An action potential of 150 mV could produce electrically only some 40 μV of depolarisation across a gap of this size.

The nerve fibres and their endings contain granules of acetylcholine together with choline acetyltransferase, the enzyme synthesising acetylcholine. At the neuromuscular junction, particularly on the post-synaptic surface, the enzyme acetylcholinesterase which breaks down acetylcholine is present. When an action potential reaches the axon terminal an increase in permeability to calcium ions occurs. The increase in cell ionic calcium causes the vesicles to move towards the terminal pre-synaptic membrane, fuse with it, and discharge their contents into the synaptic cleft. This process of exocytosis resembles that in secretory cells (Chap. 2). Decreasing calcium levels around the nerve ending or raising magnesium levels greatly reduces the discharge of the vesicles. Although each action potential releases a few million molecules of acetylcholine, the amount of acetylcholinesterase present could split over a thousand times this amount before the end of the refractory period of the muscle. The amount of acetylcholine necessary to cause depolarisation of the muscle is very small: 10–100 million molecules.

When a nerve impulse reaches an endplate, there is a short delay (usually between 0.6 and 0.7 ms) as the vesicles reach and fuse with the membrane, the transmitter substance is released, diffuses across the synaptic cleft and binds to the motor endplate receptors. This is the synaptic delay. Normally, this is followed by a local depolarisation of the muscle termed an endplate potential, or an e.p.p. If the depolarisation is sufficient, a new impulse is generated in the muscle cell.

The e.p.p. rises over about 1 ms and lasts about 20 ms. The maximum depolarisation occurs in the first 2–3 ms. If the action of acetylcholinesterase is blocked the e.p.p. can last as long as 100 ms. The e.p.p. passes by electrotonic spread.

Amplification of the current occurs by changes in the post-synaptic membrane. Acetylcholine reacts with receptor molecules and alters the permeability of the membrane so that it becomes permeable to small cations such as sodium, potassium, calcium and ammonium. The resulting inward flow of

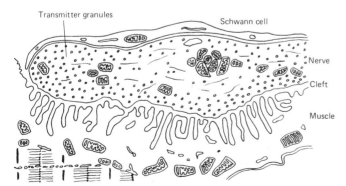

**Figure 5.5** The ultrastructure of a neuromuscular junction.

sodium ions causes a depolarisation (the endplate potential). If no other change occurred the increase in permeability to all the ions would cause the membrane potential to move locally from a resting potential around −80 mV to a value calculated from the Goldman equation of about −15 mV (a change of 65 mV). However, the threshold for stimulation of the muscle cell of about −55 mV is reached with a change of only 35 mV: the current flow from a single endplate potential is therefore theoretically sufficient to trigger an action potential in the adjacent membrane of the muscle fibre. Occasionally, more than one endplate potential has to be summated before the action potential is initiated.

The action of acetylcholine may be altered by a number of drugs. Large amounts of acetylcholine itself prevent the further passage of impulses across the synaptic gap, possibly by producing a continuous depolarisation of the post-synaptic membrane, although exposure of the membrane to acetylcholine also appears to decrease its sensitivity to a subsequent exposure. Molecules structurally similar to acetylcholine, like the relaxant Scoline (suxamethonium, used in anaesthesia) and quaternary ammonium compounds such as decamethonium, block the synapse in the same way as acetylcholine itself. Other drugs compete with acetylcholine for receptor sites, e.g. the curare alkaloids and the synthetic analogue, d-tubocurarine. Atropine also prevents acetylcholine from depolarising the post-synaptic membrane. The drug hemicholinium prevents the re-uptake of choline (see Fig. 5.13 and Chap. 9) and hence causes the synapse to fatigue more readily. Conversely, cholinesterase inhibitors such as physostigmine (eserine) can increase or prolong the post-synaptic depolarisation.

Agents preventing the release of acetylcholine such as botulinus toxin reduce or abolish the e.p.p.

Normally, even in resting muscle there are small changes in endplate potential, each of about 0.4 mV: these are due to the release of quantal amounts of acetylcholine — probably single vesicles. The endplate is subject to fatigue since time is required for the synthesis of transmitter substance. However, repetitive stimulation may at first cause depolarisation more rapidly than usual because the persistence of a high level of ionic calcium in the terminal facilitates further transmitter release, and because the cleft may still contain some acetylcholine from the earlier stimuli.

The system, like the generation of action potentials, is an all or none system producing a post-junctional action potential of fixed size or not at all. The motor endplate provides a junction at which a chemical transmitter bridges the gap between two excitable cells. It does not, however, provide for interaction between more than two cells: each motor endplate receives its stimulus from only one axon. This contrasts with the more complex junctions of nerves with nerves, and nerves with smooth muscle.

## Nerve-nerve synapses

The situation in nerve–nerve synapses resembles that at the motor endplate inasmuch as chemical transmission occurs, and occurs in a quantal fashion.

The substances released at the nerve-nerve junction may be excitatory or inhibitory: the former are said to produce an excitatory post-synaptic potential (e.p.s.p.) which is a depolarisation of the membrane, the latter to produce an inhibitory post-synaptic potential (i.p.s.p.), which is either a stabilising of the resting potential or a hyperpolarisation — a potential even more negative than the resting potential. Both effects result from changes in the permeability of the membrane. Excitatory potentials result from an increase in the permeability of the membrane to sodium ions. As in the production of endplate potentials, this is probably associated with an increase in permeability to all small positively charged ions. Inhibitory post-synaptic potentials produce a hyperpolarisation, probably by opening chloride or potassium channels in the membrane. A transmitter may have different effects on different cells: thus acetylcholine causes depolarisation of skeletal muscle cells by increasing permeability to sodium and potassium but in cardiac (heart) muscle it opens potassium gates only, forcing the potential nearer the Nernst potential of −80 mV for potassium ions and inhibiting action potential production.

Some axons produce an inhibitory effect by altering synaptic transmission between other axons. This is achieved not by a post-synaptic inhibition (i.e. by releasing a chemical to compete for membrane sites with an activating transmitter) but by a pre-synaptic inhibition. A process of the inhibiting axon releases a transmitter which reduces the size of action potentials in the excitatory axon and hence decreases the amount of excitatory transmitter released.

A major difference between the motor endplate and the synapse is in their size. The synaptic area is much less than that of the motor endplate, and the individual pre-synaptic impulses produce depolarisation effects insufficient by themselves to stimulate the generation of an action potential. Transmitters produce, therefore, a local stabilisation of the membrane potential or a local depolarisation. The post-synaptic membranes themselves are relatively insensitive in that a very large depolarisation is necessary to trigger an action potential in these regions; both depolarisation and hyperpolarisation spread outwards from the synaptic area. The cell membrane around the base of the axon, the axon hillock, requires much less depolarisation before it generates an action potential — its threshold is low.

Current flow between the synaptic areas and the axon hillock results in a summation of the inhibitory and excitatory effects at the axon hillock: if this produces a depolarisation of sufficient magnitude at the hillock an action potential is generated. The post-synaptic neurone, therefore, integrates the many excitatory and inhibitory signals arriving at the synapses on the cell body or its dendrites.

In the motor neurone the process of activation of the cell takes place in a series of steps. When depolarisation reaches a sufficient level in the body of the cell an action potential is generated – the soma-dendritic spike of about 60 mV amplitude. This in turn stimulates the initial segment spike in the axon hillock (about 30–40 mV) which spreads down the axon but also renders the soma membrane refractory as it passes back in the opposite direction. This process is another example of integration: it prevents multiple stimulation of the motor nerve.

In a nerve cell, then, local subliminal changes can occur anywhere within a large area of membrane. The cell body with all its synapses, rather than any single synapse by itself, forms the functional unit. This contrasts with the neuromuscular junction which itself functions as an entity. As a consequence it is difficult to predict how many neurones will be stimulated secondarily as a result of a primary stimulus to other neurones. Two variations of this are described: convergence and divergence. These can be illustrated with examples. If three primary neurones are stimulated, and each neurone synapses uniquely with one other secondary neurone, all the secondary neurones may be activated. However, if one of the secondary neurones receives an input from two of the primary neurones, stimulation of the three primary neurones will result in activation of only two secondary neurones. This is termed convergence (or sometimes, occlusion, because one output seems to have been hidden or occluded). A good example of convergence is seen in the innervation of skeletal muscle fibres, where the motor neurone which actually passes to the muscle receives inputs from a number of neurones within the spinal cord. The term 'final common pathway' is used of the $\alpha$ motor neurone in recognition of this. Divergence is a related phenomenon, but here stimulation of the three primary neurones results in the activation of a larger number of secondary neurones, because each primary neurone is able to produce an excitatory post-synaptic potential in more than one secondary neurone. Stimulation of one primary neurone may depolarise only one secondary neurone sufficiently to produce an action potential, but may produce a lesser depolarisation in another secondary neurone. If this is further depolarised by another primary neurone, the two depolarisations together may summate to stimulate an action potential. Thus the stimulation of the original three primary neurones would result in activation of four secondary neurones. These are only examples; the number of neurones involved will differ in different sites.

In the central nervous system passage of an impulse along a particular pathway with its many synaptic junctions often results in subsequent impulses passing more rapidly and producing greater effects. This is termed facilitation; it probably results from summation of subthreshold depolarisations, as well as from increased effectiveness of neurotransmitter release. The sensitivity of post-synaptic cells may also vary. An initial stimulus may 'condition' a pathway to permit easier transmission of a subsequent impulse.

The chemical substances involved in transmission in the central nervous system are still being sought and identified. Glutamate is known to stimulate some receptors, $\gamma$-amino butyrate to inhibit. Acetylcholine stimulates Renshaw cells peripherally and has also been shown to act as a transmitter in the brain. Adrenaline, dopamine, serotonin and a number of peptides have all been localised in the central nervous system and in some areas receptor sites sensitive to these chemicals have been found. In many instances, however, their roles are still uncertain.

It is sometimes possible to stimulate or inhibit neurones by application of substances which produce effects on the cell membrane – thus alcohol can render a membrane more permeable, and potassium chloride can change the local ionic balance. In general the only substances used in dentistry to affect neurotransmission, apart from local anaesthetics, have been those which actually destroy the nerve membrane and completely block transmission: in trigeminal neuralgia surgeons have attempted to reduce pain in the trigeminal area by chemical denervation – injecting alcohol or phenol into the nerve or the ganglion.

## Nerve-muscle junctions in smooth muscle

In smooth muscle, the axons lie in close proximity to the muscle fibres. At intervals they have swellings, or varicosities, containing vesicles which are released to function as chemical transmitters. These junctions, where nerve and muscle fibres lie side by side, are called en passant junctions. Because many axons serve many muscle fibres, depolarisation in any one fibre may be due to release of transmitter from more than one axon. In some smooth muscles there may be changes in the level of the membrane potential, the so-called junction potentials, rather than true action potentials. Junction potentials may be either excitatory or inhibitory.

# Sensory receptors

The nervous system is a transmission system carrying signals to the brain with information about the external and internal environment, and signals from the brain to activate peripheral organs. The form of the signals and the way in which they pass along nerve fibres and from one cell to another has been described above; the way in which nerves stimulate effector organs like muscle and glands is described later. The way in which nerves receive information will now be considered.

## Types of information and sensors

Sensory receptors may be classified as exteroreceptors, which monitor the external environment, and interoreceptors, which monitor the internal environment. Exteroreceptors include receptors for light, taste, smell, touch, temperature and pain or harmful stimuli. These sensations may be subdivided into the special senses, with localised aggregations of receptors into organs – the eye, taste buds, the nasal olfactory mucosa, the ear; and the skin senses, the cutaneous senses of touch, temperature, and pain, distributed over the whole body surface. The interoreceptors include stretch receptors such as those in the lungs or carotid sinus, muscle spindles, chemoreceptors like those monitoring blood oxygen concentrations, osmoreceptors, proprioreceptors, and once again, receptors for painful or harmful stimuli. Receptors in organs within the body are often called visceroreceptors. The receptors monitoring the internal environment are usually linked into nerve pathways which eventually produce a control over the environment, rather than giving rise to consciously perceived sensation. The division of receptors into two major groups reflects both function and, to some extent, anatomical connections, since most exteroreceptors are served by somatic nerves, whilst most interoreceptors are served by autonomic nerves.

At a receptor a process of transduction, or conversion of energy from one form into another, takes place; the stimulus is converted into a series of action potentials along the nerve fibre serving the receptor. Another useful classification of receptors is by the form of energy they transduce. Three main energy forms may be converted into electrical signals – mechanical energy, electromagnetic waves, and chemical energy.

## Mechanoreceptors

These are cells or neurones sensitive to mechanical distortion. Examples are stretch receptors (including muscle spindle receptors), pressure receptors, the hair cells of the cochlea and vestibular apparatus and the skin receptors which detect touch and vibration. They monitor stretch, position, movement, pressure.

## Electromagnetic wave receptors

Some cells and nerve endings respond to electromagnetic waves of light or heat. Usually they show some degree of spectral sensitivity. Thus visual receptors in the eye respond only to certain wavelengths and nerves in the skin respond only to certain ranges of temperature. The ear is able to distinguish sound frequencies over a particular range – although strictly the receptors are mechanoreceptors.

## Chemoreceptors

Chemoreceptors respond to the contact of particular molecules with their membrane receptor sites. The simplest chemoreceptors are, in fact, the postsynaptic membranes. More usually, the senses of taste and smell are linked under this heading. Although the exact mechanism by which pain is sensed is not known, chemoreception may play a part since pain can be evoked by the presence extra-cellularly of normally intracellular molecules or ions – such as high concentrations of potassium ions from damaged cells – and a number of chemical substances, many of them polypeptides, cause a sensation of pain when injected subcutaneously. The monitoring of oxygen and hydrogen ion concentration in the blood is yet another chemoreceptor function.

## The receptor-nerve interaction

Receptors may also be classified according to the number of cells involved in the transduction process. A primary sensory neurone may itself act as the receptor, as in the olfactory epithelium, the carotid body receptors and the annulospiral and flowerspray endings in the muscle spindle. The sensory neurone may be secondary to a primary receptor cell which is the target for the stimulus. Taste bud cells and hair cells are examples of cells which are stimulated by the appropriate stimulus and pass this on as a depolarisation to a sensory neurone. The only example of a tertiary sensory neurone, a sensory neurone separated from the receptor cell by a third cell, is in the nerves supplying the cells of the retina.

## The primary sensory neurone

Sensory nerve terminals may be stimulated in a variety of ways but the net effect is always a depolarisation of the stimulated area of the axon. The depolarisation usually results from a change in permeability of the cell membrane because of mechanical or chemical changes in the conformation of the membrane molecules. Thus stretching of an annulospiral ending in a muscle spindle, deform-

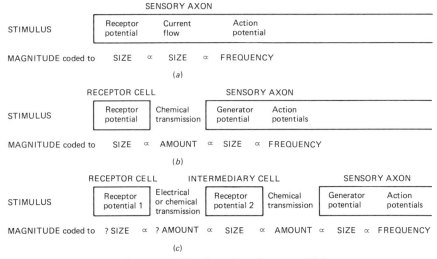

**Figure 5.6** The sequence of events resulting from the action potentials in a sensory nerve which is (*a*) a primary sensory neurone (*b*) a secondary sensory neurone, or (*c*) a tertiary sensory neurone.

ation of the nerve terminal in a Pacinian corpuscle, and the contact of pain-producing peptides with a pain nerve ending, all result in depolarisation of the nerve terminal. This depolarisation is termed a receptor potential. Since the ability to generate an action potential is not uniform along the axon membrane, the receptor potential usually has to travel to a more sensitive area of membrane before an action potential is generated. Current flow from the depolarised area causes depolarisation of the adjacent axon membrane which spreads electrotonically until it reaches a part of the fibre with a lower threshold. If the depolarisation at that point is sufficiently large an action potential is generated. In the primary sensory neurone, therefore, the receptor potential can also be termed the generator potential. It is proportional to the strength of the stimulus (Fig. 5.6). The action potential is 'all or none' in nature – its size and duration are independent of the size of the stimulus. However, another action potential can be generated if sufficient depolarisation remains after the refractory period. The net result is that a chain of action potentials can be induced if the stimulus is great enough: the frequency of action potentials within that chain is proportional to the stimulus strength. The stimulus intensity can therefore be coded despite the constant size and duration of the action potentials.

## The two cell system, the secondary sensory neurone

Specialised receptor cells differ from nerve cells in lacking the property of electrogenesis, the ability to produce action potentials. Instead they respond to stimuli by a depolarisation. This receptor potential travels along the cell membrane and causes the secretion of chemical transmitter substances which diffuse across the intercellular gap to bind to receptors on the secondary sensory neurone. This is equivalent to synaptic transmission between the receptor cell and the neurone. The membrane of the sensory neurone is therefore depolarised. This change in potential, the generator potential, passes along the membrane of the cell to generate an action potential. There is no evidence of electrical conduction between the receptor cell and the secondary sensory neurone. Once again the action potentials evoked are 'all or none' and the coding of the stimulus intensity is by means of the frequency of the action potentials in the chain produced by the generator potential. The receptor potential, the amount of transmitter substance released, and the generator potential, are all proportional to the intensity of the stimulus (Fig. 5.6).

In some examples no actual receptor potential is observed in the receptor cell but a transmitter is released simply as a result of the stimulus.

## The three cell system, the tertiary sensory neurone

In the retina the actual sensory cells are the rods and cones. These respond to light by hyperpolarisation – the membrane potential becomes more negative than the resting potential. Nonetheless, it is probable that these cells secrete a neurotransmitter which passes to the bipolar cells and there causes a depolarisation. Light probably inhibits an otherwise constant release of transmitter. The bipolar cells in turn communicate with ganglion cells to produce a generator potential and stimulate action potentials. There are tight junctions between the bipolar cells and the ganglion cells, so a direct electrical connection is theoretically possible between these two types of cells; however, it is at least as likely that the bipolar cells secrete a neurotransmitter which depolarises the ganglion cells and produces a generator potential.

## Physiological-psychological correlations

The sensation perceived when a stimulus is applied increases with stimulus strength. However, the relationship is not directly linear, since most sensory systems operate in such a way that the brain can detect small increments when the stimulus is small but only large increments when the stimulus is large. As a result the brain can detect a wide spectrum of intensities and distinguish relative changes. The relationship between stimulus intensity and perceived intensity was defined in two slightly different ways by Weber and Fechner. Weber stated that just noticeable differences in intensity result from fixed relative increases in the stimulus, while Fechner said that a relative increment in external energy results in a constant increment in perceived sensation. The two statements can be expressed together as the Weber-Fechner Law

$$\psi \propto \log \phi$$

where $\psi$ is the perceived magnitude and $\phi$ the absolute intensity of the stimulus. This relationship holds for a certain range of certain stimuli. The logarithmic decibel scale of loudness is based on this law – the sound intensity perceived (loudness) is proportional to the logarithm of the sound intensity expressed as energy.

A more general form of the relationship, the Power Law, is based on the observation that the sensation perceived increases in proportion to a power of the stimulus intensity

$$\psi = K\phi^n$$

where K is a constant. The expression gives a straight line when the data is plotted on log-log co-ordinates. The slope of the line is the power to which the stimulus intensity has been raised. Power values have been calculated for several sensory systems: for sound and for smell the exponent is 2/3, for taste sensation and for warmth 3/2, and for cold sensation 1.

In taste nerves the frequency of action potentials is proportional to the stimulus strength as perceived – not the actual stimulus strength. In other words, for this sensation, the coding of stimulus intensity according to a Power Law function takes place in the size of the receptor potential rather than in the generation of action potentials. There is a linear relationship between the amplitude of the receptor potential and the rate of firing of action potentials. This is probably true for other sensations also.

## Sensory pathway

A sensory unit is defined as a single nerve axon and all the sensory receptors which transmit information to it. Each sensory unit, therefore, has a receptive field: in the case of touch or temperature receptors on the body surface, it is the area of skin wherein are situated all the receptors connected to the single sensory axon. For visual sensation the receptive field of a sensory unit is the solid angle monitored by the receptors linked to the nerve axon of the unit. Discrimination is the ability to perceive two or more stimuli as separate. A classical test of touch discrimination is to apply the points of a pair of dividers to the skin surface and determine the least distance between the points at which they are perceived as separate. The lips and the tongue have a high discrimination for touch – the dividers can be detected as two points when they are barely separated – but on the trunk touch discrimination is low. Visually, discrimination represents the ability to see two points as separate. High discrimination implies a low ratio of receptors to nerve fibres: the sensory unit is small and the receptive field is small. High discrimination, however, usually carries the penalty of low sensitivity unless the receptor density is very high. Sensitivity is the ability to measure small changes in stimulus intensity: for this purpose a high ratio of receptors to nerve fibres is preferable. The intensity of stimulus is coded by single receptors but the more receptors there are involved the more effectively changes in intensity can be detected. The bigger the stimulus, the more receptors will be stimulated. Size of receptor field and density of receptor distribution are both important factors in sensitivity of a sensory unit.

The sensory nerve fibre represents the final common pathway for all its receptor areas. Impulses passing to the brain may, however, be modified at any synapse en route, and the final sensation perceived is a result of the original stimulus together with all the interactions that may have occurred at spinal cord, reticular formation, thalamic or cortical levels. It is the product not of discrete sensory impulses but of the pattern of activity in a number of nerves.

## Static and dynamic sensitivity

The immediate response to a stimulus may be different from the long term response to a continued stimulus. Receptors which respond primarily to rate of change of a stimulus (such as touch and pressure receptors) are said to show dynamic sensitivity and are rapidly adapting. They may have a bidirectional response – to both increases and decreases in the stimulus – or they may be 'on' or 'off' receptors which respond to changes in intensity in one direction only. Receptors which respond optimally to a continuous level of stimulation show static sensitivity: they are slow adaptors. Examples include some temperature receptors, taste receptors, stretch receptors in the walls of blood vessels, and the flowerspray endings of muscle spindles.

Some receptors show both a response to rate of change of the stimulus and also to the steady state stimulation itself, e.g. the annulospiral endings of the muscle spindle.

Adaptation, in sensory physiology, is the ability of receptors to adjust to a changed background level of stimulation. A familiar observation of adaptation is the sensation of clothing on the skin, perceived as the clothes are put on, imperceptible as the clothes are worn. Odours, sharply perceived at first, fade slowly as adaptation occurs. The causes of adaptation, and their relative importance, are difficult to assess. In many instances it appears to be a property of the receptor itself: sometimes, perhaps, a phenomenon of fatigue. This might be an explanation in two cell systems with a synapse-like junction. In other cases cells around the receptor may progressively modify the effect of the stimulating factor, as with the Pacinian corpuscle, where pressure stimuli cause deformation of the laminated corpuscle which surrounds the nerve terminal. At first the stimulus is transferred through the laminae to produce deformation of the membrane of the nerve terminal and cause a receptor potential; as the laminae begin to absorb the deforming pressure the stimulus to the nerve terminal is reduced. Further modification of sensory inputs can occur at all levels within the sensory pathway: adaptation may be due also to some kind of filtering of consciousness. The reticular activating system of the brain is one site where this could occur.

# Muscle

The second of the excitable tissues is muscle. Like nerve cells, muscle cells respond to stimulation by producing the sequence of electrical changes termed an action potential. Three types of muscle are distinguished: skeletal, cardiac and smooth. Skeletal and cardiac muscle have fibres which appear striated under the microscope, whilst the fibres of smooth muscle lack striations. The term striated muscle is used to include both skeletal and cardiac. Each type of muscle will be described separately but a comparison of the three types is given in *Table 5.2*.

## Skeletal muscle
### Structure

Skeletal muscle is made up of elongated cells of lengths ranging from a few millimetres to several centimetres. These cells are called fibres. Although the fibre, with its single motor endplate, is the smallest unit of muscle activity, the functional unit of activity is the motor unit, which consists of a single nerve fibre and all the muscle fibres supplied by it.

Each muscle fibre contains numerous contractile elements, termed fibrils, which extend the length of the cell; these in turn are made up of many filaments (Fig. 5.7). The nuclei are pushed to the periphery of the cell by the fibrils. Between the fibrils lie many mitochondria; others are found at the cell periphery. Two sets of tubular vesicles penetrate the cell. Thin transverse tubules, the T-system, arise as invaginations of the cell membrane and are therefore continuous with the outer membrane of the muscle fibre: the space they enclose is extracellular. These invaginations can transmit an action potential rapidly to the interior of the muscle cell. The T-system does not intercommunicate with the second set of tubules, the sarcoplasmic reticulum, which is a specialised modification of the endoplasmic reticulum peculiar to skeletal and cardiac muscle cells. It is made up of longitudinal tubules, which meet the transverse tubules at the junction of the muscle A- and I- bands to give the appearance called the triads. Each triad is made up of three tubules, two from the sarcoplasmic reticulum and the narrower central one from the transverse tubule system. The sarcoplasmic reticulum stores calcium and is important in muscle metabolism.

The muscle fibre has a characteristic striated pattern produced by the striations of the fibrils (Fig. 5.8). From phase contrast microscopy, the striations are identified as isotropic (light) and anisotropic (dark) bands, or I-bands and A-bands. In the resting muscle fibre these are respectively about 1.0 and 1.6 µm long. Midway along each I-band is a dark Z line. The space between two Z-lines is taken as the repeating subunit and termed the sarcomere.

Filaments are made up largely of four proteins, actin, myosin, tropomyosin and troponin. Actin is the main component of the I-bands. It has a molecular weight of 68000 daltons in the monomeric G form, but in the presence of ATP and magnesium ions polymerises into long chains known as F-actin. During polymerisation one phosphate bond of the ATP is split for every molecule of G-actin polymerised. When the chains of F-actin bind to myosin the ADP is released. The F-actin chains are in the form of a two-stranded helix with a regular crossover of the strands at 35 nm intervals. The chains make up the actin filaments which extend from the edge of the H-band to the Z-line, where each joins four

## Table 5.2 Comparison of the three main muscle types

| Characteristic | Skeletal muscle | Cardiac muscle | Smooth muscle |
| --- | --- | --- | --- |
| Cell type and size | Multinuclear cells up to 30 cm long, 10–100 μm thick | Mononuclear cells 100 μm long, branched cells | Mononuclear cells 20–500 μm long, fusiform in shape |
| Cell membrane | Continuous with no special junctions | Membranes form Z-lines; tight and gap junctions | Gap junctions |
| Arrangement of filaments | Ordered: striations | Ordered: striations | Random |
| Sarcotubular system | Well-developed T-tubules | Well-developed T-tubules | No T-tubules |
| Sarcoplasmic reticulum | Terminal cisternae form triads with T-tubules | Flattened saccules form diads with T-tubules | Endoplasmic reticulum only |
| Resting potential | Stable: $-90$ mV | Oscillating: $-80$ mV up to $+10$ mV | Oscillating: about $-45$ mv |
| Action potential | | | |
| peak | $+25$ mV | $+30$ mV | Up to $+50$ mV |
| duration | 1–2 ms | 200+ ms | 50 ms |
| refractory period | 1–2 ms | 250 ms | 50 ms |
| conduction speed | 5 m/s | | 6 cm/sec |
| Ionic basis of action potential | Increased permeability to Na, then K | Increased permeability to Na,Ca,Na then K | Increased permeability to Ca, ?Na, then K |
| Duration of contraction | Latency 10 ms; Duration 10–100 ms | Duration 300 ms | Latency 150 ms; Duration 500+ ms |
| Ca binds to | Troponin | Troponin | Calmodulin |
| Can be tetanised | Yes | No | Yes |
| Exhibits tone | Yes | No | Yes |
| Spontaneous contraction | No | Yes | Yes |
| Length-tension relationship | Tension increases with length up to optimum | Tension increases with length up to optimum | May exhibit plasticity |
| Innervation | Voluntary, motor endplates | None, but activity modified by autonomic | Usually modified by autonomic |
| Transmitter | acetylcholine | acetylcholine | acetylcholine |
| Other agents modifying | | adrenergic transmitters | adrenergic transmitters, some hormones |

rod-like structures connecting across the Z-line with four others in the next sarcomere.

Myosin is the main component of the dark A-band. It has a molecular weight of about 500 000 daltons, and is made up of two identical long chains (M.W. 215 000) forming a helix in the tail of the molecule, and two small chains (M.W. 25 000) which give the molecule a bulbous head. The subcomponents are termed meromyosins. The bulbous heads are the parts of the myosin molecule which can attach to the actin filaments. In free solution actin and myosin aggregate to give a complex known as actomyosin which can split ATP to yield phosphate bond energy. In striated muscle actin and myosin link only in the presence of tropomyosin which is necessary for the formation of ordered helices in the actin filaments.

Tropomyosin is a filamentous protein which lies in the groove between the paired actin filaments. It has a molecular weight of around 74 000 daltons. The fourth protein, troponin, is attached in the resting muscle to the tropomyosin at intervals such that there are two tropomyosin molecules to every one of troponin. The complex of these two molecules has been called relaxing protein because the muscle fibre is relaxed when it is present. Troponin has a molecular weight of about 80 000 daltons but is made up of a number of subunits with different properties. One of these, troponin C, is able to bind calcium ions; this binding causes the troponin–tropomyosin complex to split, and to lose its inhibitory effect on movement of the actin–myosin bonds so that, if ATP is present, shortening or contraction of the muscle fibres can take place.

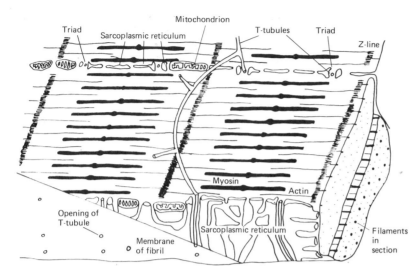

**Figure 5.7** A section through part of a skeletal muscle fibre. The actin and myosin filaments are seen in cross section on the right.

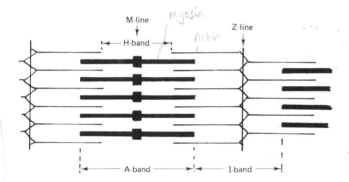

**Figure 5.8** Overlap of the actin and myosin filaments in a relaxed skeletal muscle, showing how these relate to the characteristic pattern of lines and bands seen by light microscopy.

Myosin and actin are found in less ordered forms in many other cells. Actin, in particular, is associated with shortening movements: it is found, for example, in cilia and it forms part of the protein thrombosthenin which assists in clot retraction (Chaps. 3 and 4). Troponin is unique to striated muscle; its role in smooth muscle and in myoepithelial cells is taken by another calcium-binding protein, calmodulin, of similar structure.

In the resting muscle the light I-band is made up entirely of actin filaments together with tropomyosin and troponin, and the actin filaments interconnect with those in the next sarcomere by the rod-like structures crossing the Z-line. Each actin filament shares a group of four rods with three neighbouring actin filaments. Troponin is absent near the Z-line. The myosin filaments extend across the whole width of the A-band. In the H-zone (the lighter zone in the centre of the A-band) only myosin filaments are present, but in the darker bands of the A-band on either side of the H-zone the actin filaments interdigitate with the myosin filaments so that there are two actin filaments between each pair of myosin filaments (see the cross section in Fig. 5.7). The filaments are connected by the crossbridges of the myosin molecule heads which are arranged helically around the myosin filaments with six crossbridges in every turn of the 40 nm helix.

As the fibres contract, the pattern of striation changes. The actin filaments progressively slide between the myosin filaments (Fig. 5.9) until the Z-bands abut onto the myosin molecules and even begin to fold their ends. Such sliding involves the breaking and re-joining of the crossbridge links in a cycle of detachment, movement, re-attachment, conformation to produce a directional force, and then detachment again. The tension produced is proportional to the potential number of crossbridges attached (see later and Fig. 5.9). The number of crossbridges that can occur at any one time will

88  *Excitable tissues*

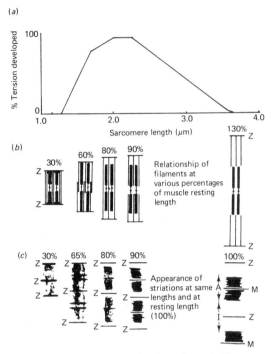

**Figure 5.9** The overlapping of actin and myosin filaments during contraction, and the tension produced as a result. (a) The active tension curve; (b) the pattern of overlap of the filaments corresponding to the tensions in (a). (c) The appearance of the sarcomere at different degrees of contraction.

decrease as the rate of shortening increases so that the force or tension that can be produced will be less during rapid contractions.

Energy is liberated only when a cycle of attachment/detachment occurs – the muscle produces shortening heat. Work is done only when a crossbridge can contribute to tension. There is therefore a relationship between the work done by a muscle and the energy released.

## Mechanical properties of skeletal muscle

Two terms are used in relation to the mechanical effects of muscle contraction – isotonic and isometric. Contraction is isotonic when the tension in the muscle is maintained constant but the length decreases, and isometric when the length is maintained constant and tension increases. Isotonic contraction performs external work, i.e. it moves a load through a distance (work = load x distance moved). Isometric contractions do little external work – the distance the load is moved is either very small or zero – and the work appears as heat. Lifting a weight involves isotonic contraction, holding it in the air, isometric. Most muscular movement in the body is a combination of the two types of contraction. This is apparent in the process of mastication.

### The elastic–contractile model

The properties of a skeletal muscle may be represented by a model with contractile and elastic elements (Fig. 5.10). The contractile element is the myofibrils, a series elastic element is made up of the tendons and connective tissue of the muscle attachment, and a parallel elastic element comes from the connective tissue and the sarcolemma around the cells.

If a force is applied to a structure it deforms. The deformation is given by $dL/dP$ where $dL$ is the change in length of the structure and $dP$ is the change in force producing it. If the structure returns to its original form this ratio is termed the compliance. If the structure does not return to its original form its residual deformation is termed plastic deformation and it is said to show plasticity. The greater the compliance, the greater the deformation produced by a given force. The reciprocal of compliance is stiffness. Rigidity is a very high degree of stiffness. The term compliance will be used later in a slightly different sense in referring to the stretching of lung tissue (Chap. 8). In producing deformation the point of application of the force moves; the energy needed for deformation is given by the work done – the product of the force and the distance moved. As the structure resumes its original form energy is given out – the structure shows elasticity. A perfectly elastic structure gives out as much energy as was put into it.

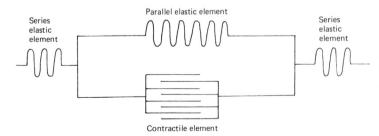

**Figure 5.10** The contractile-elastic model of muscle. The contractile element is represented by the fibres which can slide between each other, the two elastic elements as wavy lines. One elastic element is in parallel with the contractile element, and the other in series.

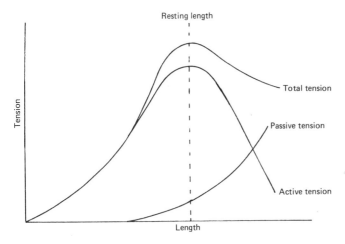

**Figure 5.11** Maximum muscle tension at various lengths. The length/tension curve is the sum of the tension generated by active contraction and the tension resulting from the passive stretching of the muscle at each length. The curve will finally rise as the muscle deforms plastically when it is stretched beyond its elastic limit.

## The length–tension diagram of muscle

The tension in a muscle can be measured in both the active and inactive states (Fig. 5.11). The curve of active tension is the same as that in Fig. 5.9 where it is explained in terms of the overlap of actin and myosin filaments. The tension in the inactive state is due to the parallel elastic component since the series elastic element is not in tension when the contractile element is at rest. The slope of the length–tension curve is the stiffness of the muscle. The curvature shows that stiffness increases with length since the curve becomes steeper as it approaches maximum tension. An ideal spring would obey Hooke's Law

$$T = S(L - L_0)$$

where $T$ = tension, $S$ = stiffness, $L_0$ = the maximum length at which no tension exists (the slack length) and $L$ = the observed length. For an individual spring this is a straight line relationship but in a system made up of a number of springs of differing length the individual straight lines summate to give a composite curve. The length–tension curve of a muscle is a curve of this type.

Tension developed by the contractile element varies during the cycle of activity. In order to transmit the tension when the muscle as a whole is held at a fixed length the series elastic element changes in length in a manner corresponding to the changes in tension shown in the length–tension diagram.

Isometric recording can be carried out on muscles at different lengths under tetanic stimulation. The tension produced is then dependent upon the length at which the muscle is maintained. At stretched lengths there is tension in the parallel elastic element: if this is subtracted from the total tension, a curve of the series elastic tension due to shortening of the contractile elements is obtained (Fig. 5.11). This is parabolic with a maximum value at a length corresponding to the length of the muscle at rest in the body – the so-called resting length. Fig. 5.9 shows that at this length there is the maximum amount of overlap of actin and myosin and therefore the maximum possible number of crossbridges are available.

Isotonic contraction of the muscle can also be studied. If the applied load is less than the maximum isometric tension at the initial muscle length, the muscle will shorten. The shortening is not instantaneous but takes place at a speed which depends on the magnitude of the load. Speed is slow when the applied load is great, but rapid with small loads (Fig. 5.12). The tension velocity curve in isotonic conditions describes the behaviour of the contractile

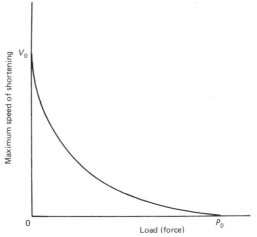

**Figure 5.12** The speed of muscle contraction versus the load applied to the muscle.

90  *Excitable tissues*

element. Load and velocity of contraction are related according to the following equation

$$(P + a)(V + b) = (P_0 + a)b = K$$

where $P_0$ is the maximum isometric tension of the muscle, $P$ is the actual tension, $V$ the velocity of contraction, a and b are constants related to the muscle, and K is another constant. Rearrangement of the equation gives

$$PV + aV = (P_0 - P)b \text{ or } V = (P_0 - P)b/(P + a)$$

showing that the velocity of contraction is inversely related to the ratio of the actual tension to the maximum tension. $PV$ in the second equation is the rate at which the contractile element works upon the load. $aV$ is the rate at which energy is dissipated due to internal damping: the energy appears as heat. $PV$ and $aV$ represent the total energy production by the muscle; this is proportional to the amount by which tension differs from the maximum tension. Rapid movements are inefficient in terms of work since much of the energy is wasted in internal damping – overcoming inertia, taking up the elasticity of the series elastic component, internal friction. Every muscle has its own optimum speed of working.

### The electrical activity of skeletal muscle fibres

Stimulation of the motor nerve to a muscle causes depolarisation and production of an action potential. The sequence of events is summarised in Fig. 5.13.

Concentrations of some of the principal ions in a skeletal muscle fibre are given in Table 1.3 and 5.3. These show that the large amount of protein in the cell contributes most of the anionic charge, and therefore chloride ions are in a much lower

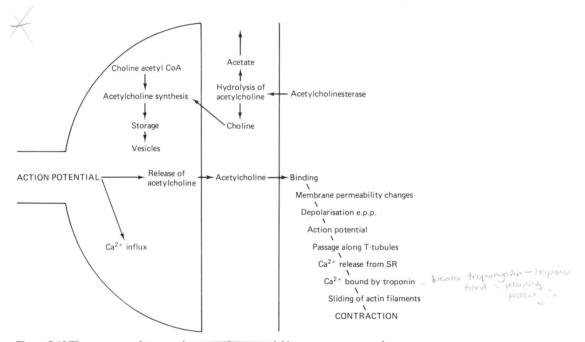

**Figure 5.13** The sequence of events when an action potential in a motor nerve reaches a muscle and causes it to contract.

**Table 5.3** The composition of different types of muscle fibre (concentrations in mmol/l and Nernst potentials in mv)

| Constituent | ECF | Skeletal muscle | Nernst potential | Smooth muscle | Nernst potential | Cardiac muscle | Nernst potential |
|---|---|---|---|---|---|---|---|
| Sodium | 145 | 12 | +65 | 56 | +52 | 10 | +41 |
| Potassium | 4 | 155 | −95 | 119 | −89 | 140 | −94 |
| Chloride | 120 | 4 | −90 | 55 | −24 | 30 | −41 |

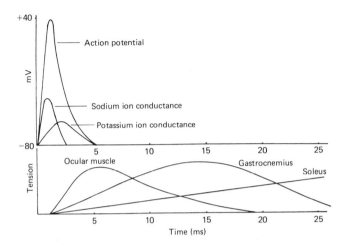

**Figure 5.14** The action potential, the underlying changes in the membrane conductance, and the resulting changes in tension in skeletal muscle. Three muscles with differing speeds of contraction are used as examples in the lower part of the diagram.

concentration than in many cells. The measured resting potential of skeletal muscle is around −90 mV, a value close to the Nernst potential for potassium ions. The action potential (Fig. 5.14), like that in nerve, is generated by a suprathreshold depolarisation resulting in an opening of voltage-dependent sodium channels and a change in permeability to sodium ions. It lasts some 2–5 ms and is conducted at 5 m/s. The absolute refractory period is 1–3 ms and the afterpolarisation is relatively long in extent. Chronaxie is longer than in nerve fibres. The threshold value for stimulation varies between different fibres.

The action potential spreads rapidly through the muscle fibre via the T-tubules. It is possible that there is electrical coupling between the T-tubules and the sarcoplasmic reticulum.

The study of muscle action potentials is termed electromyography.

### The contractile response

The action potential reaches the sarcoplasmic reticulum and causes the release of calcium ions to initiate the molecular sequence of events in the contraction of a muscle fibre. Contraction may be as rapid as 7.5 ms but in a so-called slow muscle may last as long as 200 ms (Fig. 5.14).

The tension in a muscle does not increase immediately after stimulation: there is a latent period of about 3 ms. In the first half of this there is a slight relaxation (latency relaxation). The muscle then shortens rapidly and tension rises to a plateau in as little as 0.5 ms. The contractile response passes over the fibre at the same rate as the action potential.

If the muscle fibre is stimulated again after completion of its refractory period it responds with a second action potential. The time course of the wave of contraction stimulated by an action potential is much longer than that of the action potential and so the second action potential occurs whilst the muscle is still contracted. Part of the tension produced by contraction is lost in the stretching of the series elastic component; since this stretching is achieved by the first contraction the tension produced in subsequent contractions is not damped to the same extent. The tension exerted by the muscle as a whole, therefore, increases to a plateau level as frequency of stimulation increases. Although the contraction of a muscle fibre is described as 'all or none' in response to an action potential, this is only true if the conditions are the same at the time of the action potential. A second action potential in the fibre before the contractile response has faded away results in further sliding of the actin and myosin filaments, if this is possible, and an increase in tension. If further action potentials occur at a sufficient frequency this also causes tension to rise to a plateau level. There is no refractory period for the mechanical phase of the process of contraction. The plateau of tension and contraction reached during repetitive stimulation is known as tetanus, or as tetanic contraction (Fig. 5.15). In complete tetanus there is a plateau of tension up to about four times as high as the tension resulting from a single stimulus (the maximal twitch tension). Incomplete tetanus is marked by some relaxations between contractions. The frequency of stimulation needed to produce tetanus is the rate at which twitch duration is just shorter than the interstimulus interval.

Maximal stimuli below tetanising frequency result in increasing tension – a phenomenon known as treppe. This is due to the inability of the sarcoplasmic reticulum to take up all the ionic calcium that has been released by the first action potential before

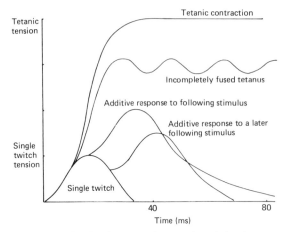

**Figure 5.15** The development of tetanus in a skeletal muscle fibre, as the increasing rate of stimulation causes the contractile responses to fuse together.

the second one reaches it. More ionic calcium is therefore available for the process of contraction.

Each fibre responds electrically in an all or none fashion. The motor unit provides a further means of grading contraction in the muscle as a whole. The motor unit is defined as one nerve fibre and all its associated muscle fibres. A single motor neurone may innervate as few as seven muscle fibres (in the eye muscles) or as many as 2000 (in the calf muscles). A low innervation ratio allows precise movement and a graded response, whereas a high ratio confers strength and is found in large muscles with broad general functions - such as the muscles of the back and the postural muscles of the leg. Grading of force results from the number of motor units active, the asynchronous firing of motor units, the frequency of discharge in the neurone (including tetanising frequencies, although these are rarely observed in physiological conditions) and the length of the muscle.

Two main types of muscle fibre have been described in man: slow twitch fibres and fast twitch fibres. They differ in their speed of contraction, as assessed by the time taken to reach isometric peak tension, and by the latency of contraction, in their content of mitochondria, and in the type of myosin which they contain. Slow twitch fibres (ST, or SO, slow oxidative) have many mitochondria and slow myosin ATPase activity. They use the biochemically more efficient aerobic pathways of oxidation and are fatigue resistant. They are usually found in the muscles of posture and muscles involved in slow repetitive movements. Fast twitch muscle fibres are subdivided into two extreme groups and an intermediate one. The most rapidly contracting and the most easily fatigued are the fast glycolytic group (FG, fast glycolytic, or FF, easily fatigued). They have few mitochondria and a high concentration of rapidly acting myosin ATPase. They use glycogen, ATP and phosphocreatine as anaerobic energy sources to produce contractions of high power and high speed - but these cannot be maintained. Although these fibres contain more glycogen than the slow twitch fibres, they also contain less triglyceride fat. The fast twitch oxidative muscle fibres (FOG, or FR, fatigue resistant) contract less quickly than the FF fibres, but are much more fatigue resistant. They can therefore be used for fast repetitive movements. They contain more mitochondria than the FF fibres because of the need to replace ATP – the FF fibres use energy too rapidly to allow time for ATP replacement. The third type of fast twitch fibre is intermediate between the other two and is called a fast intermediate or F(int) fibre. Motor units are usually built up from fibres of one type only, so that they also can be classified as fast or slow. The fast motor units have fewer fibres per unit. Whole muscles are composed of mixtures of the different types of motor unit, slow muscles having a preponderance of slow units, and fast muscles a preponderance of fast units. The capillary density is greater in slow muscles in order to supply their more constant requirement for oxygen. Slow muscles used to be termed red muscles, because they contain myoglobin, an oxygen-carrying protein (*see* Chap. 4), whilst fast muscles, with little myoglobin, were termed white muscles. Fine, skilled activity, such as eye and hand movements, are performed by muscles with a high proportion of white fibres, whilst the slow muscles are mainly postural.

Normally, many muscles show the phenomenon of tone – a constant degree of muscle contraction maintained by alternating contractions among different motor units. The muscles of the mandible have been quoted as a good example since they maintain the mandible in its rest position with the teeth slightly apart. However, there is no good evidence for or against this being tonic contraction. Postural muscles have also been said to show a constant state of tone but there is in fact no evidence of muscle activity in them during the maintenance of a normal standing posture (*see* Chap. 14). Tonic contraction is associated with a minimum level of fatigue because only a small proportion of the muscle fibres are contracting at any one time and the pattern of contracting fibres is continuously varying.

## Metabolic aspects of muscle contraction

### Heat production in muscles

The basal metabolism of a muscle produces resting heat. During contraction further heat is generated. This is termed the initial heat of contraction and is made up of activation heat, due to the molecular changes in contraction, and shortening heat, due to changes in muscle length. Heat production continues for some 30 minutes after the contraction. This is known as heat of recovery and is approximately equal to the initial heat. In muscle which has performed an isotonic contraction there is an after-release of stored energy from the elastic elements, which appears as relaxation heat. The heat appearing after contraction – the recovery heat and the relaxation heat – is sometimes called waste heat.

The energy supplied to the muscle derives from energy-rich chemical bonds and is equal to the energy released as work and heat. The efficiency of an engine is the percentage of the energy used which is seen as work: the observed heat production from a muscle in isotonic contraction therefore suggests a maximum theoretical efficiency of 50%. In isometric contraction the muscle has 0% efficiency, with all the energy appearing as heat. Measurement of the efficiency of skeletal muscles rarely gives figures as high as 25%, although cardiac muscle can achieve some 30%. This is because work is used in overcoming internal friction and the inertia of the limbs or other body components.

*Metabolism in rest and exercise*

Oxygen and glucose, or glycogen, are used to produce energy in muscle. If the oxygen supply is insufficient, lactic acid is formed from pyruvic acid and this must be further oxidised later after the muscle has relaxed. Muscle also uses free fatty acids – mainly, however, at rest and during recovery from contraction (*see* p. 299).

*Oxygen debt*

At rest some stored energy is transferred from ATP to creatine forming energy-rich phosphocreatine bonds. During exercise these are hydrolysed to reform ATP from ADP and provide an ATP energy source. Normally the energy stored in ATP bonds arises directly from glucose oxidation.

Although during exercise blood flow to a muscle is increased and more oxygen reaches the tissue, the metabolic rate may increase as much as a hundredfold and the aerobic resynthesis of ATP is insufficient to cope with the demand for energy. Phosphocreatine bonds can be resynthesised anaerobically by oxidising glucose to lactic acid. After exertion the lactic acid and any other metabolites must be oxidised and so a further supply of oxygen becomes necessary. This oxygen debt may entail consumption of as much as six times as much oxygen as in the resting muscle.

# Smooth muscle
## Structure

Smooth muscle surrounds the blood vessels and the viscera, and forms an integral part of their walls. It is called smooth because it lacks striations: its contractile elements are randomly orientated within the cells, rather than arranged in the ordered structures of striated muscle. The muscle fibres are shorter than in skeletal muscle and the contractions are usually slower and more sustained. Smooth muscle often shows powerful and sustained tone.

Two types are described: multiple unit smooth muscle and visceral or single-unit smooth muscle. Multiple unit muscles behave rather like striated muscles in that they contract only in response to a stimulus, and the separate fibres require individual stimulation. Single-unit smooth muscle is inherently rhythmic and contracts without prior stimulation by nerves: its activity is, however, modified by the nerves which supply it. The cells of single-unit smooth muscle are electrically linked by gap junctions: the cells of multi-unit smooth muscle do not have gap junctions. Both types are served by autonomic nerves and are not under voluntary control. Although some smooth muscle is clearly one or other type, a wide range of intermediate forms exist, even within similar muscles: vascular smooth muscle, for example, may be multiple unit or single-unit or exhibit characteristics of both types in varying degrees.

Visceral muscle is arranged in a number of layers. The outer layer is usually described as longitudinal and the inner as circular although the fibres of both are actually arranged spirally, the outer as a long spiral and the inner as a tight spiral. Blood vessels also have helically arranged muscle fibres. Each cell (Fig. 5.16) is an uneven spindle shape with a centrally placed nucleus. Arterial muscle cells, however, have irregular branched processes. Smooth muscle cells are between 12 and 600 μm long and between 2 and 12 μm in diameter. They do not form a regular array of cells but they interdigitate in their long axes to form bundles approximately 100 μm in diameter. In visceral muscle the cells are so tightly packed that the extracellular space occupies only some 12% of the tissue, in contrast to about 40% in vascular smooth muscle.

The separation between cells is 50–80 nm in visceral muscle, 200–800 nm in vascular, but there are many areas of close contact between cells (tight junctions) where the separation is only some 10–20 nm. These junctional complexes occupy about 5% of the cell surface. The organelles of smooth muscle include many membrane invaginations and vesicles. The endoplasmic reticulum is not specialised, unlike the sarcoplasmic reticulum of striated muscle, and although there is a longitudinal tubular system, it is not the same as the tubular

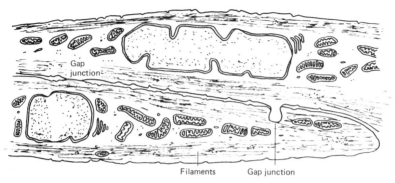

**Figure 5.16** Ultrastructure of the overlapping ends of two smooth muscle fibres.

systems in striated muscle. There are many mitochondria. Granules of lipid and of glycogen are scattered through the cytoplasm.

The myofilaments are found in small bundles 10–14 nm long and about 5 nm thick. They contain actin and myosin. Smooth muscle cells contain much more tropomyosin than striated muscle cells – it makes up between a third and a half of their protein – but they have only about one tenth the amount of actomyosin. Troponin is absent and its place is taken by calmodulin. This protein has some structural affinities with troponin and is similarly able to bind calcium. It is more widely distributed in cells than troponin and it has other functions unrelated to muscle contraction (p. 196). In the resting muscle myosin is probably in a disaggregated, or non-filamentous, form. During contraction the number of filaments appears to increase; contraction probably involves a sliding mechanism as in skeletal muscle. Dark bodies seen in the cells may function as attachment zones during contraction – the equivalent of the Z-lines in striated muscle.

The true neuromuscular junction exists only in the multi-unit type of smooth muscle. In visceral smooth muscle, the nerve fibres lie between the muscle cells and have swellings, or varicosities, along their lengths or at their terminals. In vascular smooth muscle the axons lie on the outside of the muscle. In both types of muscle the space between the varicosities and the muscle cells is only about 80 nm. The varicosities contain many mitochondria and vesicles, within which are granules of transmitter substances.

### Electrical properties of smooth muscle

The ionic concentrations in smooth muscle cells differ from those in skeletal muscle: potassium concentrations are lower and sodium concentrations higher (*Tables* 1.3 and 5.3). The cells are more permeable to sodium ions than those of skeletal muscle so that the resting potential is less electronegative. Single-unit smooth muscle shows slow waves of depolarisation and repolarisation of about 20 mV magnitude, making it difficult to specify a resting potential, but figures of around −55 mV have been given for visceral muscle, and around −40 mV for vascular muscle. The oscillating potential is caused by rhythmic changes in membrane permeability to ions but it is not clear whether depolarisation is due to an increased permeability to sodium or calcium ions, a decreased permeability to potassium ions, an increased permeability to chloride ions, or some combination of these.

In multi-unit smooth muscle, transmitter release causes action potentials to be generated as in striated muscle. This also occurs in some vascular smooth muscle cells although the muscle may resemble and behave otherwise as a single-unit muscle. Some other vascular smooth muscle cells do not produce action potentials as such, but binding of the neurotransmitter noradrenaline results in a graded junction potential, and in these cells the state of contraction is apparently governed by the membrane potential. In most single-unit smooth muscle, however, action potentials are generated spontaneously as a result of the oscillation of the membrane potential. This is termed myogenic activity. Each cell shows a rhythmical variation of resting potential. Where several cells together reach a critical level of depolarisation, synchronisation of their rhythms occurs and eventually the evoked junction potentials produce a threshold response. The cell with the most rapid oscillation of resting potential becomes a pacemaker for cells in its vicinity, triggering action potentials at such a rate that the slower cells entrain into the rhythm of depolarisation and refractory period. Although the initiation of action potentials in most cells is spontaneous, the membrane potential in cells near the varicosities is affected by the local concentrations of neurotransmitter substances, and these may therefore increase or decrease the rate of action potential production. Stretch of smooth muscle can affect the membrane permeability sufficiently to cause depolarisation. This is more likely to produce an effect in thinner muscle fibres. Action potentials can also be triggered by changes in metabolic activity. Sensitivity to stimulation may vary in different parts of the muscle bundle just as in different areas of the membrane of a motor neurone.

In visceral smooth muscle the action potential can reach a peak as high as +50 mV, although there is considerable variability among different muscle sheets in both the resting potential and in the peak polarisation reached during the action potential. The form of the action potential varies in different muscles but an important feature is a plateau which usually follows the initial spike (Fig. 5.17). The action potentials in smooth muscle are due to changes in permeability mainly to calcium but also to sodium ions, the relative importance of the two varying in different muscles. The sodium ion permeability component of the action potential may be relatively small and the major factor may be the increase in permeability to calcium ions – the gradient of concentration of calcium ions from 2.5 mmol/l in the extracellular fluid to between 100 and 1000 nmol/l inside the cell is very great. In most cases a change in permeability to calcium ions causes a spike-like action potential and where a plateau exists this is due to a delayed increase in permeability to sodium ions, but this is by no means universal (Fig. 5.17). Release of membrane-bound calcium ions may also influence the membrane potential. There is no sarcoplasmic reticulum which can store or release calcium ions as in striated muscle: the calcium required for muscle contraction diffuses in during the action potential or is released from the membrane.

In single-unit smooth muscle, both visceral and vascular, impulses pass from one cell to another by electrotonic spread across the gap junctions. The conduction velocity for action potentials is much slower than in striated muscle – only about 6 cm/s, compared with, for example, 160 cm/s in sartorius muscle.

## Effects of neurotransmitters

Acetylcholine usually increases the efflux of potassium ions, the influx of calcium ions and may increase the permeability to sodium ions. The increase in permeability is relatively non-specific: both chloride and bromide ions also enter the cell more readily. When acetylcholine inhibits contraction of the muscle there is a hyperpolarisation mainly due to an increased permeability to potassium ions. This is associated with increased membrane binding of calcium and probably a decreased permeability to calcium ions. Permeability to chloride ions seems to be unaffected. Local release of adrenaline produces excitation or inhibition by similar mechanisms. There is a small but constant release of neurotransmitter from the autonomic nerves in smooth muscle; this can cause small spontaneous junction potentials in the fibres.

Oxytocin and angiotensin, which act on the vascular smooth muscle, seem to produce their excitatory effect entirely by increasing the permeability to sodium ions. In the uterus the main factor affecting the myogenic activity of the smooth muscle is the hormone oxytocin, which again increases permeability to sodium ions.

Serotonin, or 5-hydroxytryptamine, is excitatory. It is continuously released in the intestinal smooth muscle and may be involved in interneural transmission there.

## Mechanical properties of smooth muscle

Smooth muscle usually shows a degree of tone due to myogenic activity. The tension of the muscle changes according to neurotransmitter or drug activity. Relationships between length and tension and between load and velocity of contraction are similar to those already described for skeletal muscle. The maximum isometric tension developed increases up to an optimum length of the fibre. However, the increase in tension due to rapid and pronounced stretch is usually followed by plastic deformation of the muscle and a decrease in tension.

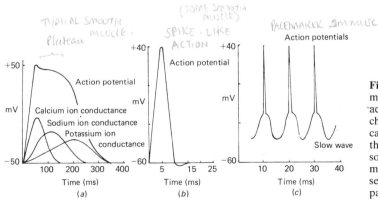

**Figure 5.17** Action potentials in smooth muscles. (*a*) A typical smooth muscle action potential with the associated changes in membrane conductance to calcium, sodium, and potassium ions; (*b*) the spike-like action potentials seen in some smooth muscles; (*c*) the oscillating membrane potential firing a regular sequence of action potentials in a pacemaker area of smooth muscle.

96  *Excitable tissues*

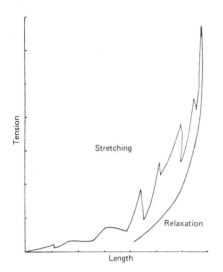

**Figure 5.18** Plastic deformation as a contracting smooth muscle sheet is stretched.

A series of such events is seen in the stretching of muscle from the bladder (Fig. 5.18). The ability to respond to tension by plastic deformation may be due to the random orientation of the actin and myosin fibres and the lack of structures like the Z-lines of striated muscle.

### *Rhythmicity*

The rhythm of contractions in visceral muscle is commonly between 0.5 and 2 cycles/min, whereas that in vascular muscle is around 1.25/min. In the gut the frequency of contractions is characteristic of particular organs or zones.

Although the rhythmic variation of potential may be as much as 50 mV, action potentials are normally triggered with smaller depolarisations.

## Cardiac muscle
### *Structure*

Cardiac muscle makes up the walls of the chambers of the heart. It has some of the properties associated with smooth muscle and others associated with skeletal muscle.

Structurally it has **striations** with ordered fibrils of actin and myosin in I- and A-bands (Fig. 5.19). Z-lines are present. The muscle fibres are branched and interdigitate but the cells are separate rather than forming a syncytium as once described. Each cell has a **peripherally placed nucleus** and many **elongated mitochondria** alongside the fibrils. The spaces between cells are about 30 nm wide but the cell membranes are extensively interdigitated and at times they fuse to form tight junctions situated at the Z-lines – the intercalated discs of the light microscopists. On either side of the tight junctions are **gap junctions**.

The fibrils are surrounded by a sarcoplasmic reticulum similar to that of skeletal muscle and a transverse tubular system crossing the fibres in the region of the Z-lines. The T-tubules are larger than those in skeletal muscle and the sarcoplasmic reticulum less well ordered. Where the sarcoplasmic reticulum in its apparently random course meets a T-tubule there is a small sac-like expansion. This gives the characteristic appearance in cardiac muscle of tubule diads, instead of the triads of skeletal muscle.

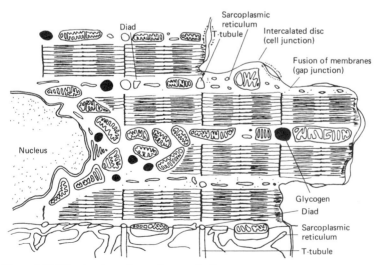

**Figure 5.19** Ultrastructure of a cardiac muscle fibre.

## Electrical activity of cardiac muscle

Specialised versions of the cardiac muscle cells make up the sinoatrial node which functions as a pacemaker area. Other specialised cardiac muscle cells with very weak striations and a much reduced contractile capability have increased ability to conduct impulses. These cells make up the specialised conducting tissue of the bundle of His, which carries the impulse from the atrioventricular node, and also form the Purkinje fibres, which spread the impulse over the ventricular muscle. There are no muscle cells specialised specifically for conduction in the atria: the action potentials pass along cell membranes and electrical transmission occurs between cells across the gap junctions. In general the action potentials of pacemaker cells and cells whose prime function is conduction of the impulse tend to be spike-like, whilst those in ventricular muscle have plateaux resembling those of smooth muscle cells but of longer duration (Figs. 5.17 and 5.20).

All isolated cardiac muscle cells show spontaneous rhythmic contractions, which are associated with a rhythmical pattern of electrical activity. The resting potential of the cells is around $-80\,mV$ and the action potential rises to about $+30\,mV$. Depo-

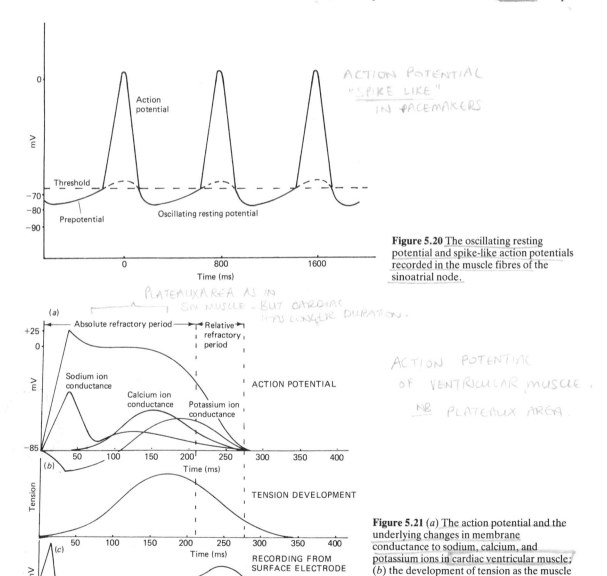

**Figure 5.20** The oscillating resting potential and spike-like action potentials recorded in the muscle fibres of the sinoatrial node.

**Figure 5.21** (*a*) The action potential and the underlying changes in membrane conductance to sodium, calcium, and potassium ions in cardiac ventricular muscle; (*b*) the development of tension as the muscle contracts; (*c*) the form of the electrical changes recorded with electrodes on the surface of the muscle.

larisation is rapid – about 25 ms – but the repolarisation phase lasts about 250 ms. The total time taken for the recovery phase is between 300 and 500 ms. Depolarisation and the first rapid repolarisation are due to changes in the permeability to sodium ions; the slow repolarisation which follows is associated with an increased permeability to calcium ions but also to a second phase of increased permeability to sodium ions (Fig. 5.21). Finally the permeability increase to potassium ions becomes dominant and the cell repolarises more rapidly. The repolarisation time decreases as the rate of contraction increases. At 75 beats/min the duration of the action potential is 250 ms but at 200 beats/min it is only 150 ms.

The cardiac action potential can be recorded only by using intracellular electrodes: if electrodes are placed on the surface of the heart (or even on the body surface) the electrical changes recorded are the sum of the electrical changes occurring in all the cardiac muscle fibres and are known as the electrocardiogram, or ECG.

The long plateau phase of the action potential results in an extended absolute refractory period of 200 ms and a relative refractory period of some 50 ms.

The inherent rhythmicity of the cells is due to slow waves of depolarisation similar to those in smooth muscle cells (Fig. 5.20 ). In the intact mammalian heart two areas, termed pacemakers, are normally capable of initiating impulses – the sinoatrial node and the atrioventricular node. The sinoatrial node has a more rapid rate of production of action potentials and so it imposes its rhythm on the slower atrioventricular node. If impulses from the sinoatrial node fail to reach the atrioventricular node this will then set up a rhythm dominant to the slower random rhythm of the other cardiac muscle cells. The pacemaker cells show a slow rhythmical decrease in their permeability to potassium ions, which causes depolarisation because it reduces the contribution of the potassium Nernst potential to the membrane potential. When depolarisation reaches a threshold level it triggers an action potential. In the human heart the sinoatrial node, the atrioventricular node, and the Purkinje fibres all exhibit inherent rhythmicity; ventricular muscle cells do not.

### Contraction in cardiac muscle

The process of contraction in cardiac muscle is similar to that in skeletal muscle and is similarly linked to the action potential. However, the duration of the action potential is such that the midpoint of the mechanical response at about 175 ms, lies almost at the end of the second phase of increased permeability to sodium ions – that is, towards the end of the absolute refractory period. The relative refractory period ends with the muscle more than half way to relaxation. Normally, therefore, a second stimulus cannot occur early enough to tetanise the muscle. At more rapid rates of contraction the contractility of the muscle may be increased because, as in skeletal muscle, the uptake of ionic calcium by the sarcoplasmic reticulum may not be completed after contraction, and the calcium remaining permits more movement of actin relative to myosin filaments at the next contraction.

The length/tension curve of cardiac muscle is similar to that of skeletal muscle. However, cardiac muscle differs from skeletal muscle in being a hollow sphere and in not being attached to bone. Unlike skeletal muscle, cardiac muscle has no 'resting length'. Up to a certain length cardiac muscle generates more tension as its length increases. In the body, the filling of the heart with blood returning from the circulation causes stretching of the fibres and the force of contraction therefore increases with the degree of filling. The tension/volume curve during diastole (the non-contracting phase of the heart cycle) is equivalent to the passive stretch curve in skeletal muscle, and the tension/volume curve during systole (the phase of contraction) is equivalent to the active contraction curve (Fig. 5.22).

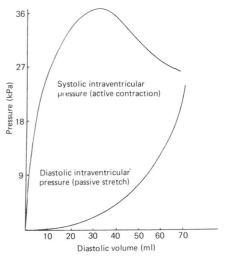

Figure 5.22 The volume/tension curves of the ventricles of the heart during diastole and systole, or passive stretching and active contraction, at varying stretched length. These are equivalent to the components of the length/tension curves of skeletal muscle. These curves were plotted in a dog heart which has smaller volumes and higher pressures than are experienced in the human heart.

Metabolically, mammalian cardiac muscle, with many mitochondria, much myoglobin and a good oxygen supply, behaves like skeletal muscle. Energy is derived from carbohydrate (about 35%), ketones and aminoacids (5%) and lipids (60%) of which about half are free fatty acids. Less than 1% of the energy is due to anaerobic metabolism although in hypoxia this may rise to around 10%.

## Pharmacology of cardiac muscle

Stimulation of the vagus nerve increases the permeability of the cardiac pacemaker cells to potassium ions, rendering them hyperpolarised. This lengthens the repolarisation time and decreases rate of generation of action potentials – and hence the rate at which the heart beats. The ventricular fibres are not affected by the vagus or by acetylcholine.

Adrenaline has the reverse effect, decreasing the permeability of the pacemaker to potassium ions. This increases the rate of production of action potentials and contractions. Substances like adrenaline produce different effects on cells according to the type of membrane receptor with which they interact. When adrenaline, or another adrenergic transmitter, acts on the β receptors of cardiac muscle, it causes an increase in the concentration of cyclic adenosine monophosphate near the cell membrane and this raises the concentration of calcium ions in the sarcoplasm. As a result, the force of contraction of the cells is increased.

The concentrations of ions in the fluids around cardiac muscle cells affect their activity. Calcium ions increase the force of contraction but also increase cell excitability by increasing the permeability to both sodium and potassium ions. An increase in sodium concentration causes an increase in excitability resulting in systolic arrest. It also reduces uptake of calcium ions. A high concentration of potassium ions causes relaxation of the cardiac muscle, and again reduces uptake of calcium ions.

A number of drugs have been used to stimulate the heart, those derived from digitalis being perhaps the best known. The cardiac glycosides, as they are called, inhibit the activity of a sodium/potassium pump in the muscle cell membranes, causing sodium to accumulate in the cells. This, in turn, increases the intracellular calcium ion concentration, which increases the strength of contraction of the muscle fibres.

## Further reading

Bowsher, D. *Introduction to the Anatomy and Physiology of the Nervous System*, 5th edn, Blackwell, Oxford, 1987

Eyzaguirre, C. and Findone, S.J. *Physiology of the Nervous System*, 2nd edn, Year Book Medical Publishers, Chicago, 1975

Katz, B. *Nerve, Muscle and Synapse* McGraw Hill, New York, 1966

Paul, D.H. *The Physiology of Nerve Cells*, Blackwell Oxford, 1975

Roberts, T.D.M. *Basic Ideas in Neurophysiology* Butterworths, London, 1966

Wilkie, D.R. *Muscle. Studies in Biology No. 11*, 2nd edn, Arnold, London, 1976

# 6

# Bulk transport

## Bulk flow of fluids

Transport of materials has so far been considered as local movement of water and solutes. The body can also move gases or fluids, together with suspended particles, along defined internal pathways. The two principal pathways are those of the respiratory system, made up of the nasopharynx, the trachea, the bronchi, the bronchioles and the alveolar sacs; and the circulatory system, made up of the heart, the aorta, the arteries, the arterioles, the metarterioles, the capillaries, the venules, and the veins. Transport in the respiratory system is tidal, with air or gases being moved in and out rhythmically; in the circulatory system a pump drives a fluid round two enclosed circuits in sequence. Other bulk transport systems include the digestive tract, a tube whose lumen is actually external to the body, which carries a semi-fluid mass from mouth to anus, the ureters which transport urine from the kidney to the bladder, and the lymphatic system which carries lymph from the tissue fluid to the thoracic duct and then the subclavian vein.

The physical rules governing flow in any system apply to bulk flow of fluids (liquids or gases) in the body: flow takes place down gradients of pressure, and its rate is affected by the size and resistance of the flow channels.

## The driving force

Fluids flow only when there is a net pressure difference between two points. Flow of blood depends upon contraction of heart muscle to produce pressure within the ventricular chambers. From there to the re-entry of the great veins or the pulmonary veins to the heart there is a steady fall in pressure of some 16 kPa in the systemic circuit and some 3.3 kPa in the lung circuit. The flow is maintained by the heart valves which prevent backflow and maintain a unidirectional pressure gradient.

In the thoracic cavity, expansion of the enclosed volume by depression of the diaphragm (effectively acting as a piston) and by the raising of the ribs to increase the outer radius, reduces the gas pressure in the lungs below atmospheric and allows air to flow in. Recoil of the piston and the walls similarly raises the intrapulmonary pressure and air flows out to the atmosphere down a pressure gradient.

## Laminar and turbulent flow

When a fluid passes along a smooth-walled tube, the layers of fluid in contact with the walls experience drag due to friction. The extent of such frictional drag depends upon the roughness of the wall surface. A fluid moving at low velocity, therefore, travels most rapidly in the centre of the tube, and its velocity decreases smoothly to a minimum along the walls where it is subject to frictional drag (Fig. 6.1). This type of flow is termed streamline flow; it is also termed laminar flow because the fluid may be regarded as made up of an infinite number of cylindrical laminations each with a velocity slightly less than that of its more axial neighbour, and slightly greater than that of its more peripheral neighbour. The smooth profile of flow is maintained until the velocities of the axial and the peripheral layers differ so much that the lines of flow begin to break up. This is more likely in a larger tube because of the greater separation between the axial core and the peripheral laminations. The more viscous the fluid the less likely is the development of discontinuities between laminae.

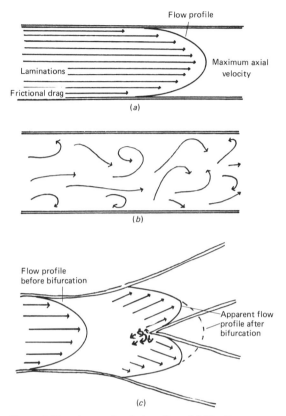

**Figure 6.1** Laminar and turbulent flow. (*a*) The flow profile in laminar flow; (*b*) the random nature of turbulent flow; (*c*) the pattern of flow at a bifurcation with a change in profile in the vessels. Turbulence may occur at the bifurcation itself.

Break-up of streamline flow leads to turbulence in which energy is wasted by the diversion of the fluid into cross currents or eddies (Fig. 6.1). The possibility of turbulent flow may be predicted by calculation of a Reynold's number, a dimensionless number characteristic of the flow of fluid along a uniform cylindrical tube. This is directly dependent upon the velocity of the flow, the diameter of the tube, and the density of the fluid, and inversely related to the viscosity:

$$(Re) = \frac{vdp}{\eta}$$

where $v$ = mean linear velocity of fluid, $p$ = density of fluid, $\eta$ = viscosity, and $d$ = diameter of tube.

An alternative form of this equation allows calculations to be made using the volume flow/s, rather than the linear velocity:

$$(Re) = \frac{2pV}{\pi r \eta}$$

where $V$ = the volume flow/s, $r$ = radius of the tube, and the other symbols are unchanged.

In long smooth-walled tubes Reynold's numbers exceeding 2000 are characteristic of turbulent flow and those below 1000 are always associated with streamline flow.

Low Reynold's numbers are found almost everywhere in the circulatory and respiratory systems. The parts of the systems most likely to show turbulent flow are tubes of relatively large radius with rapid linear flow rates. The maximum rates of flow are in fact observed most frequently in the largest tubes. In the circulation the highest pressures are found near the heart, which is producing the driving force, and the first vessel in the circulatory path is the aorta, which has a large radius. Similarly, in the respiratory system the biggest airway, the trachea, carries air at the greatest velocity.

Figures for flow in the respiratory system at an inspired volume flow of 30 l/min (equivalent to breathing in vigorous exercise) predict that turbulence should occur in the trachea and in the main bronchi (*Table 6.1*). Such turbulence would help in

**Table 6.1 Flow characteristics of the larger lung airways at a ventilatory rate of 30 l/min**

| Tube | Radius (mm) | Velocity (cm/sec) | Reynold's no. |
|---|---|---|---|
| Trachea | 9 | 263 | 3100 |
| Main bronchi | 6 | 286 | 2290 |
| Lobar bronchi | 4 | 314 | 1710 |
| Segmental bronchi | 3 | 334 | 1230 |

the trapping of particles in the airflow. At low air velocities, particles at the periphery are trapped in the mucous layer which covers the ciliated epithelium lining the air passages. Upward movement of the mucus driven by the cilia causes the particles to be washed up to the nasopharynx, where they can be swallowed and then inactivated in the stomach. At greater airflows the number of foreign particles passing is increased and, if laminar flow prevailed, only those in the outer laminations of the airsteam would be trapped. Turbulent flow increases the volume of air coming into contact with the mucous layer.

Even at low flow rates, with Reynold's numbers below 2000, some turbulent flow of air occurs in the upper respiratory tract. The glottis and vocal cords narrow the airway and produce a turbulent cone of air. Discontinuities in the system lower the values for Reynold's numbers at which flow becomes turbulent. Thus, for example, at a bifurcation (Fig. 6.1*c*) the fast-moving axial core stream is split to yield an eccentric laminar flow and turbulence develops as a result of the uneven flow pattern.

In the circulatory system the problem is complicated by the presence of large particles – the blood cells – in the fluid. However, if blood is treated as a uniform fluid with the properties of its components averaged, the peak velocity of 70 cm/s in the thoracic aorta (diameter 2.4 cm) gives a Reynold's number around 4300. For reasons at present unknown, this figure, rather than 2000, appears to be the threshold for turbulent flow in the aorta. Actual velocities are greater on the inner curve of the aortic arch. Turbulent flow has not been observed in any other blood vessels under normal conditions and even in the aorta it is only at peak systolic flow (the maximum output velocity from the heart) that turbulence occurs.

## The effect of suspended particles upon flow

Suspended particles are affected by flow and themselves affect it. In vessels whose diameter is greater than that of the particles the flow velocity profile causes a number of effects. Since the velocity varies from a maximum at the centre of the tube to a minimum at the walls, any particle situated away from the axial core is subjected to uneven forces which cause torsion and hence rotation of the particle. Elastic particles also undergo distortion. The net result of the forces acting on particles such as red blood cells is to cause them to become aligned with their long axes in the line of flow and to rotate around these axes. Columns of aggregated red cells, rouleaux, show a characteristic bending in the flow; this may be an important factor preventing their build-up and survival in the circulation.

The velocity profile results in the more massive white blood cells lying along the axis of flow with the red cells in the surrounding layers (Fig. 6.2a). Further, although at normal haematocrit values the plasma is crowded with cells, the low velocity of the flow around the circumference of the vessels causes a layer of relatively cell-free plasma to appear along the walls. This layer can also be explained in terms of cell size. Since no cell can approach the vessel wall more closely than the point where its centre lies one radius from the wall, the layer one cell-radius wide next to the vessel walls contains fewer cells than any more axial layers. In the blood vessels, therefore, the blood behaves as a system with a core of viscosity roughly that of whole blood and a periphery with a viscosity similar to plasma.

If the forces acting on the suspended particles decrease as a result of a diminution in flow rate, the particles progressively lose their original orientation in the flow stream. Thus in slow flowing blood the long axes of the cells are more randomly orientated and the distribution of red and white cells is altered.

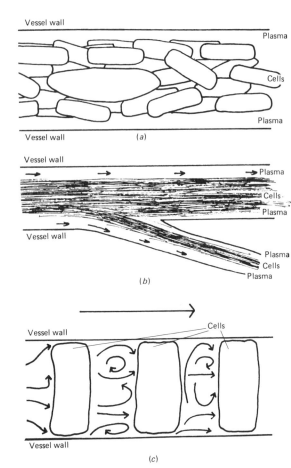

**Figure 6.2** The flow of fluids containing suspended particles. (*a*) Axial flow of cells in the blood; (*b*) plasma skimming occurring in a branch vessel – the branch vessel receives more of the circumferential layer of plasma, and a lower proportion of the axial cells – a situation which increases as the angle at which the branch leaves approaches a right angle; (*c*) bolus flow, with the eddies between the bolus cells, in a small capillary.

If simultaneously the interaction between cell walls and vessel walls is altered to favour adhesion, the increased frequency of collisions between cells and vessel wall results in an apparent migration of cells to the periphery of the vessel. This occurs in sites of inflammation and also when there is stasis due to blood clotting.

The presence of suspended particles itself directly affects the flow of blood. The effective viscosity of a mixture of particles and fluid changes according to the concentration and the orientation of the particles. This causes blood to behave as a non-Newtonian fluid in that the observed viscosity differs at different flow rates. Newton's Law states that the velocity produced in a fluid is proportional to the stress applied, but this is only true if the viscosity is constant. Plasma itself is a non-Newtonian fluid because it contains colloidal particles – the plasma proteins. These are cigar-shaped and, like the cells, have major and minor axes, so they too vary in orientation with the flow rate. At high rates of flow their long axes are aligned with the axis of flow and the viscosity is low; at slow flow rates the long axes are randomly orientated. This is particularly true of fibrinogen, the most elongated of the plasma proteins, and so variations in fibrinogen content markedly affect blood viscosity. Conditions such as inflammatory disease which result in increased plasma fibrinogen concentrations are associated with increased blood viscosity. The erythrocyte sedimentation time (p. 54) in inflammatory diseases is, however, also affected by the tendency of red cells to stick to each other because of the changes in their surface charge in the presence of excess fibrinogen.

Changes in the proportion of blood cells in the plasma also affect blood viscosity. The non-Newtonian behaviour of blood increases rapidly as the haematocrit value exceeds 20%, and is maximal between 40 and 70%.

## Blood clotting

The viscosity of blood increases during the early stages of blood clotting, directly causing some reduction in the velocity of flow. The clot itself affects blood flow: first because its presence on the vessel wall changes the frictional resistance, and then because it increasingly forms an actual obstruction. The adhesion of cells to the vessel walls adds to both these effects.

## Plasma skimming

In theory, branches of vessels leaving the parent vessel at angles approaching 90° may contain blood with a lower haematocrit value because the peripheral layer of flowing blood contains fewer cells (Fig. 6.2b). This phenomenon has been termed plasma skimming. Its importance is difficult to assess. The number of blood vessels with branches leaving at such an acute angle is small and their flow characteristics are such that skimming seems less likely. However, plasma skimming may occur in the kidney and play a part in the auto-regulation of renal blood flow.

## Bolus flow

In very small blood vessels the diameters of the blood cells are of the same order as those of the vessels. Indeed, in the smallest capillaries red blood cells are subject to deformation and folding. This may assist in mixing the cell contents.

Where single cells completely fill the lumen of the vessel, bolus flow takes place. In bolus flow the cells and the fluid between one cell and the next move as units. Within the fluid segments secondary currents are set up. These mix the segment contents more evenly and may therefore contribute to the transfer of dissolved materials across the vessel walls (Fig. 6.2(c))

## The relationship between sound and turbulent flow

The sounds normally associated with flow are always due to turbulence or the breakdown of laminar flow. Perfect laminar flow is silent. When the flow in a normal blood vessel is monitored with a stethoscope no sound can be heard: flow in blood vessels is normally silent. Similarly, a stethoscope will only detect the sound of breathing when there is turbulence in the airways. The sounds due to turbulence caused by pathological changes in blood vessels or the airways may be useful diagnostic aids.

In the heart the turbulence which occurs normally as the valves close generates the heart sounds.

The relationship between turbulent flow and sound is utilised in the measurement of arterial blood pressure (p. 125) when different sounds are heard over the brachial artery as flow through it is obstructed by an inflatable cuff round the upper arm. Obstructing pressures between the systolic and diastolic arterial pressures result in turbulent flow with characteristic sounds.

## Pressure, resistance and flow

The flow of liquids and gases along tubes is described by an equation linking pressure difference and velocity in the same way as Ohm's Law links potential difference and current flow in an electrical circuit. Ohm's Law states:

$$E = IR \text{ or } I = \frac{E}{R}$$

where $E$ = potential difference, $I$ = current, and $R$ = resistance. For the flow of fluids this becomes

$$\Delta P = VR \text{ or } V = \frac{dP}{R}$$

where $\Delta P$ = the pressure difference, $V$ = the volume flow in unit time, and $R$ = the resistance to flow.

For Newtonian fluids in which the velocity attained is proportional to the stress applied, this simple equation may be re-written as the Poiseuille equation:

$$V = \frac{\Delta P}{8\eta L/\mu r^4} = \frac{\Delta P \mu r^4}{8\eta L}$$

where $\Delta P$ = the pressure difference, $V$ = volume flow in unit time, $r$ = radius of the tube, $\eta$ = coefficient of viscosity, and $L$ = tube length.

This equation is valid for liquids and also for air when the pressure difference is small, as in respiration. At higher driving pressures the compressibility of gases affects the relationship. In general, the resistance to flow is given by

$$R = \frac{8\eta L}{\pi r^4}$$

and, if the coefficient of viscosity remains constant,

$$R \propto L/r^4$$

This means that in conditions where Poiseuille's Law is valid, the resistance to flow varies inversely with the fourth power of the radius of the tube. Small changes in the radius of physiological tubes like the arterioles and the bronchioles thus markedly affect their resistance to flow. Although biological flow systems usually conform to Poiseuille relationships only at low flow rates, the relationship between resistance and tube radius is approximately correct in most situations in the respiratory system. In the circulation, changes in the apparent viscosity of blood lead to departures from Newtonian flow. However, in the normal range of pressures in the circulation the behaviour of blood is approximately Newtonian and so flow conforms approximately to Poiseuille's Law. In fact, the effect of decreasing radius in small vessels is not as dramatic as the Poiseuille equation suggests because the apparent viscosity of blood decreases in narrow tubes and at low haematocrit values.

Poiseuille's formula is invalidated by the presence of turbulence. A similar expression has been derived for turbulent flow; in tubes around 1 cm in diameter and with Reynold's numbers of a few thousand, an approximate description of the pressure flow relationship is given by

$$V^2 = \frac{\Delta P}{(8\eta L/\pi R^4)^2}$$

This leads to an empirical and convenient approximation for the pressure flow relationship in the airways in both laminar and turbulent flow:

$$\Delta P = K_1 V + K_2(V)^2$$

where $K_1$ and $K_2$ are constants. This equation can be interpreted in the respiratory system to mean that resistance to flow in the lungs is made up of two independent terms – a fixed resistance to streamline flow and a flow-dependent resistance due to turbulence in the nose, larynx and trachea.

Overcoming the resistance requires work to produce the necessary pressure gradient. However, only about 25% of the total work of breathing is necessary to overcome airway resistance. The remainder is necessary to produce elastic expansion of the lung tissues and walls (70%) and to overcome the drag of the lung tissues (5%).

## Fick's principle and the indirect measurement of flow

It is not often possible to measure flow directly. The flow between two points can be measured, however, if the fluid undergoes a measurable change between the two points and the total effect of that change can be measured.

### Measurement of the cardiac output

The amount of blood pumped by the heart, the cardiac output, is measured by application of Fick's principle. The blood passing through the circulation loses oxygen to the tissues and this is used up in metabolic activity. The total oxygen usage of the body in unit time can be measured from the difference between the oxygen content of inspired and expired air. The oxygen content of both the arterial blood leaving the heart and the venous blood returning to the heart can be measured in mol/l. If the amount of oxygen used (mol/min) is divided by the arterio-venous difference (mol/l), the volume of blood circulating per minute can be calculated.

$$\frac{\text{Cardiac}}{\text{output}} = \frac{\text{oxygen usage}}{\text{arterio-venous difference}} = \frac{0.25\,\text{l/min}}{0.20 - 0.15\,\text{mol/l}}$$

$$= \frac{11\,\text{mmol/min}}{8.9 - 6.7\,\text{mmol/l}} = \frac{11\,\text{mol/min}}{2.2\,\text{mmol/l}} = 5.0\,\text{l/min}$$

### Renal blood flow and clearance measurements

The flow of blood to the kidneys can be measured similarly. Since the oxygen usage of the kidney is difficult to measure, some other marker must be used. The blood reaching the kidneys passes through the glomerular capillaries where a proportion is filtered together with any small molecules or ions. Some of these are reabsorbed in the kidney tubules but others are too large to cross the tubule wall and therefore pass out in the urine. One such substance is the carbohydrate inulin, which can also be used as a marker in measuring extracellular fluid volume because it is small enough to cross capillary walls easily but is too large to enter cells. All the inulin filtered from the blood in the kidney glomeruli passes into the urine. Thus the Fick principle may be applied: if the concentration in the renal artery and in the renal vein are measured, and the difference between these two concentrations divided into the amount of inulin appearing in the urine in unit time, the kidney blood flow can be calculated.

The measurement of kidney blood flow is carried out in practice, however, by an easier technique. A marker substance which can be excreted by the kidney is injected into the bloodstream and the amount excreted in the urine determined. The result is expressed as the volume of blood completely cleared of the marker and is therefore termed the clearance of the marker. When a substance is totally removed from the blood as it passes through the kidney, and no reabsorption occurs, the clearance of that substance is equal to the blood flow to the kidney.

The substance para-aminohippuric acid (PAH) is not normally found in human blood plasma. If it is injected it is then lost: firstly, because as a small molecule it passes through the glomerular filter; and secondly, because it is actively transported by cells of the kidney tubules out of the blood into the forming urine. It is not reabsorbed back into the blood from the kidney. As a result, the blood passing through the kidneys loses all its PAH into the urine. (This is not strictly true, mainly because some of the PAH is bound by plasma protein, but this discrepancy may be corrected for, or even ignored.) If the arterial blood concentration is known, and the concentration in the venous blood leaving the kidney is zero, the arterio-venous difference is equal to the arterial blood concentration. The amount excreted in the urine may be measured over a period of time. This value divided by the concentration/ml in arterial blood gives the volume of arterial blood cleared of PAH in the time and hence the volume of blood flowing through the kidney in that time. Thus, if $U$ is the concentration of PAH in urine in mg/ml and $X$ is the volume of urine in ml/min

$UX$ = excretion of PAH in unit time, in mg/min

and if $A$ is the concentration of PAH in arterial blood plasma in mg/ml, then

$\frac{UX}{A}$ = clearance in ml/min
= plasma flow through the kidney in ml/min or effective renal plasma flow (ERPF).

From this and the haematocrit value the total blood flow may be calculated.

The measurement of blood flow to the kidney is a special example of the use of clearance measurements. It is only because the marker is completely removed from the blood as it traverses the kidney that its clearance can be used in the same way as the Fick principle. More usually, clearance is used to study how the kidney handles particular substances, and to determine other aspects of kidney function. One common usage is to determine the volume of blood actually filtered by the kidney.

This is done by injecting inulin, as before. Inulin is not secreted by the kidney cells into the urine, so it can reach the urine only by the filtration process. All the inulin from the blood plasma which is filtered remains in the urine because the molecule is not reabsorbed. Hence the volume of blood filtered can be calculated if the amount of inulin lost in the urine is measured and the concentration of inulin in the blood is known. The calculation of inulin clearance is exactly the same as that for PAH clearance, but this time the clearance is equivalent to the glomerular filtration rate (GFR). The ratio of filtration rate to effective renal plasma flow is the filtration fraction.

## Countercurrent systems

When energy or chemical exchanges occur across the walls of tubular structures serving as channels for bulk transport, arrangement of the tubes in loops with relatively closely apposed arms can confer special properties on the system. When exchange is passive such loops enable local gradients to be maintained without major effects on the system as a whole. This process is termed countercurrent exchange. When exchange is active and the properties of the arms of the loop differ, the system becomes capable of itself generating local gradients: this is termed countercurrent multiplication. The word countercurrent refers to the opposing directions of flow in the two arms of the loop.

### Countercurrent exchange

There are a number of situations in the circulatory system where blood vessels are arranged in loops such that countercurrent exchange takes place. Examples include the arteriole–capillary–venule loops in the skin which minimise the effect of ambient air temperature on the core temperature of the body (on a larger scale the arteries and veins of the limbs perform a similar function) and the loops of the vasa recta in the kidney medulla whose role is to supply nutrients to the medulla without destroying the gradient of osmolality which extends from the cortex to the pyramids.

The skin circulation provides a good example of the operation of a countercurrent exchange. If blood flow along the surface of the skin took place through widely separated input and output vessels, there would be a marked cooling of the blood from 37°C towards the ambient air temperature of, say, 5°C (Fig. 6.3). The arrangement of the vessels in loops reduces this effect. If the fluid is regarded as a series of separate blocks it can be seen how a gradient of temperature is stabilised between the tip and the base of the loop so that cooling of the returning fluid is kept to a minimum.

106  Bulk transport

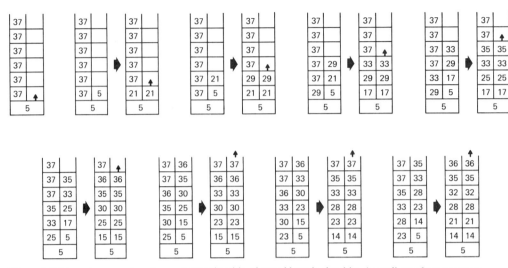

**Figure 6.3** A countercurrent exchange system in a blood vessel loop in the skin. A gradient of temperature develops along the loop. The first diagram shows the warm blood at 37°C reaching the cold surface at 5°C. Each successive pair shows first, a movement of one square round the loop, and then, the equilibration situation that results. This is a convention only: flow is, of course, smooth, and not in steps like this – but the same principle applies.

In the kidney a similar process takes place in the blood vessels which loop down into the medulla. A gradient of salt concentration, and hence of osmotic activity, is present in the medulla from a minimum of 300 mosmol/l at the corticomedullary junction to a maximum of about 1200 mosmol/l at the tips of the pyramids. The mechanism by which this is established is described in the next paragraph. The contents of a blood vessel flowing straight through the renal medulla would equilibrate with the tissue fluid of the medulla and the concentration gradient would disappear as the solute diffused from the tissue fluid into the vessel and was carried away in the bloodstream. The blood supply is, however, arranged in loops (the vasa recta) and exchange occurs between the limbs of the loop and the medullary tissue fluid (Fig. 6.4) so that the gradient of concentration is maintained. The high concentration of solutes at the tips of the pyramids and the relative coolness of the tips of the loops of the skin blood vessels are maintained by similar mechanisms.

## Countercurrent multiplication

If the properties of the two arms of the loop are different, the loop itself may generate a gradient. This occurs in the medulla of the kidney where the nephrons, the tubes carrying the fluid eventually to become the urine, are arranged in loops (the loops of Henle) in order to generate a gradient of osmolality from a low value at the cortical rim to a

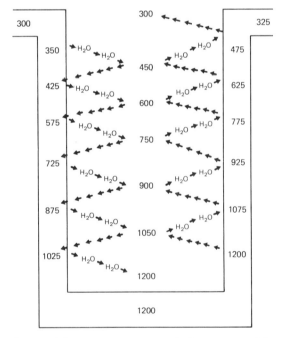

**Figure 6.4** A countercurrent system in the vasa recta of the kidney medulla. The vascular loop permits blood to reach the medulla without destroying the gradient of osmolality through it. The continuous lines of arrows represent movement of salts from the higher osmolality to lower osmolality, and the lines of arrows marked $H_2O$ represent movement of water in the opposite direction.

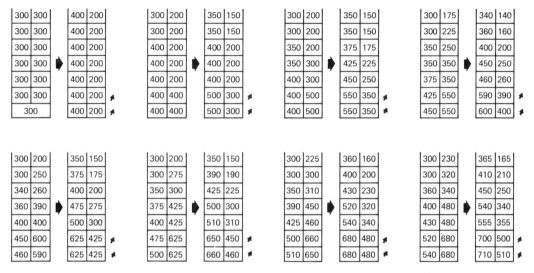

**Figure 6.5** Countercurrent multiplication. The process by which an osmotic gradient is generated in the kidney medulla. At each stage, active transport of salts from the right (or rising) side of the loop – here shown in contact with the descending limb, but actually separated in the kidney – causes a gradient of 200 mosmol to be generated between the two limbs. Again, the development of the vertical gradient is explained by moving the fluid along, two blocks at a time, and then allowing the inter-limb gradient to develop.

high value in the pyramids. This allows the final urine to become hypertonic (with an osmotic activity greater than blood plasma) as exchange of water and salts occurs between the fluid in the collecting ducts and the surrounding kidney medulla. Without this system a concentrated urine could not be produced. The development of the gradient is shown in Fig. 6.5, using a block convention similar to Fig. 6.3. The ascending limb of the loop of Henle can actively transport either sodium or chloride ions from its interior to the exterior. Which of these is not particularly important, since the two ions have opposite charges, and if one ion is actively transported, the other will follow it passively down a gradient of charge. The ascending limb is also impermeable to water, so that water cannot follow the sodium and chloride ions down an osmotic gradient. The descending limb is permeable to water and ions but does not actively transport the ions. The end result of these differences between the arms of the loop is that a high osmolality is established at the tip of the loop and the surrounding medulla, whilst there is relatively little difference in osmotic activity between the fluid in the descending limb and the ascending limb at any one level.

The gradient thus established would rapidly disappear as blood flowed through the medulla if the blood vessels were not arranged so that the gradient could be conserved – and, in fact, the vasa recta are arranged as a countercurrent exchanger as described above (Fig. 6.4).

## The carriage of gases in the blood

Gases may be carried in the blood in simple solution or attached by physical or chemical bonds to other substances soluble in the fluid. The principal atmospheric gases, oxygen and nitrogen, are slightly soluble in blood, and the main product of oxidative metabolism, carbon dioxide, is much more so; but for appreciable amounts of gases to be carried in solution in the blood soluble carrier molecules are necessary. There is no carrier for nitrogen in the blood: only a small amount of nitrogen is carried in simple solution. Oxygen can be physically bound by the haemoglobin of the red blood cells, increasing the oxygen-carrying capacity of the blood by some 60 to 100 times. Carbon dioxide can combine chemically with water itself to form carbonic acid, which is highly soluble both in an ionised and a non-ionised form. Diffusion of solutes down concentration gradients has already been discussed. Similar ideas apply to gases, although concentration as such is difficult to measure. The concentration of a solute is an indication of the number of particles of the solute: diffusion down a concentration gradient is the resultant of all movements of solute molecules across the boundary between two solutions. In a gas the movement of molecules produces the phenomenon known as pressure. The gas laws enable the actual concentration of a gas to be calculated from its pressure. In a mixture of gases the combined movement of molecules produces the total pressure.

Since in a gas mixture all molecules are free to distribute evenly, the proportion of the total pressure due to any one molecular species will be the same as the proportion of that gas species in the mixture. If one fifth of the volume of air is oxygen, then one fifth of the atmospheric pressure will be due to oxygen. This pressure is termed the partial pressure of the gas in a mixture. For oxygen in atmospheric air (made up of 21% oxygen and 79% nitrogen) it is 21% of 101.3 kPa, or 21.3 kPa. The partial pressure of a gas may be defined as the pressure that gas would exert if it were the only gas present, filling the whole volume with no change in the number of molecules. The solubility of a gas in a liquid is directly related to the partial pressure of the gas in contact with the liquid since saturation is reached when there is an equilibrium between the number of molecules entering and leaving the liquid. This is expressed in Henry's Law

$$C = qP,$$

where $C$ is the concentration of the dissolved gas, $P$ the partial pressure of the gas, and $q$ the solubility coefficient. For oxygen in blood plasma, $q$ is 0.23 ml/l/kPa, or approximately 0.01 mmol/l/kPa. A similar figure for carbon dioxide is difficult to derive because of its reaction with water; but if the carbonic acid is ignored and only dissolved carbon dioxide considered, the solubility is around 0.21 mmol/l/kPa. Nitrogen has a solubility of only 0.0044 mmol/l/kPa.

Since the concentration of a gas in solution is related to the partial pressure of the gas in equilibrium with it, concentration can be expressed either in the conventional mol/l, or as a volume/litre, or as the partial pressure of the equilibrant gas. This last is also called the tension of the gas. The tension, or partial pressure, of a dissolved gas is therefore defined as being equal to the partial pressure of the gas in equilibrium with it.

Diffusion of a gas occurs from a higher partial pressure to a lower. Fig. 6.6 summarises the main gradients observed in the body. It is the alveolar partial pressures which are important in driving the whole system of exchange: these are maintained relatively constant by the process of breathing, or ventilation of the lungs.

## Binding of oxygen

In addition to the dissolved oxygen carried in the blood plasma (and in the intracellular water of the red blood cells), oxygen is carried in a physical complex with the haemoglobin molecule. This increases the capacity of the blood for oxygen carriage from 0.13 mmol/l (0.3 ml/dl) of blood to almost 9 mmol/l (20 ml/dl).

The haemoglobin molecule is a tetramer containing four iron II atoms. These can physically bind one molecule of oxygen each. Thus the number of mols of oxygen carried at saturation in a litre of blood is four times the number of mols/l of haemoglobin. The iron II atoms lie within the molecule and the aminoacid chains form pits around them. Access to the iron is limited by the shape of the pit. When oxygen enters the molecule and is bound by the iron atom, the charges within the pit are disturbed, hydrogen ions are released, and the whole molecule changes shape. One effect of this change of shape is that access to a second pit is improved and a second oxygen molecule can enter more easily. This in turn produces further changes, facilitating uptake of a third molecule of oxygen and releasing hydrogen ions. The uptake of the third molecule does not increase access for a fourth to the same extent. In addition to a release of hydrogen ions, there is also a

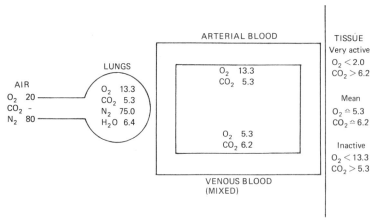

**Figure 6.6** Gradients of oxygen and carbon dioxide partial pressures from the air through the circulation to the tissues (kPa).

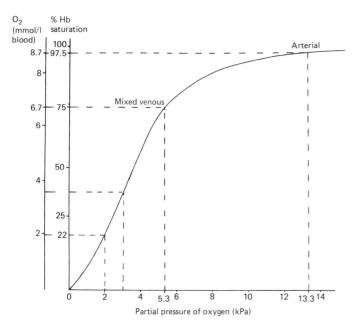

**Figure 6.7** The oxygen-haemoglobin association/dissociation curve, at a pH of 7.4 and a temperature of 37°C (partial pressure of carbon dioxide 5.3 kPa, normal levels of 2,3-DPG). The usual partial pressures of oxygen in the arterial and venous blood are indicated.

release of diphosphoglycerate (DPG), which links together subunits of the non-oxygenated haemoglobin molecule. The haemoglobin molecule carrying oxygen is referred to as oxyhaemoglobin and the non-oxygenated form as deoxyhaemoglobin. The link between iron II and oxygen is a physical one; if actual oxidation to iron III occurs, the molecule loses its capacity to bind and release oxygen. The iron III form is termed methaemoglobin (see p. 58).

The mechanism of binding of oxygen by haemoglobin explains several features of the transport system. The curve of uptake of oxygen by haemoglobin at different partial pressures is not a straight line, but is sigmoid in shape, reflecting the differing affinity of the molecule for oxygen as successive molecules of oxygen are bound (Fig. 6.7). The curve remains the same shape but changes position in relation to partial pressure with differing concentrations of hydrogen ions or diphosphoglycerate. Since the bonding process is exothermic, increase in temperature tends to inhibit attachment of oxygen. The curves can be termed haemoglobin-oxygen association or dissociation curves since they represent both uptake and release of oxygen.

The shape of the curves also has physiological implications. The curve has three parts: a central steeply inclined section and a more gradual curve on either side of this. The shape of the curve is particularly important in the zones corresponding with the range of partial pressures commonly observed in the lung alveoli, in actively metabolising tissues, and in the large veins returning blood to the heart and on to the lungs. Above a partial pressure of 11 kPa the slope of the line is very small. Indeed, the percentage saturation increases from approximately 95% at 11 kPa to 97% at 13 kPa and does not reach 100% until about 250 kPa – two and half times atmospheric pressure. The partial pressure of oxygen in the lungs is around 13.3 kPa. A change of 15% in alveolar oxygen concentration will therefore affect the saturation of haemoglobin with oxygen by about 2%. Thus the uptake of oxygen by the blood in the lung capillaries is only marginally affected by quite large changes in alveolar oxygen concentrations. However, the higher the partial pressure of oxygen in the blood, the greater the gradient to help it diffuse into the tissues.

The partial pressure of oxygen usually observed in the venous blood returning to the lungs is around 5.3 kPa. This value lies on the shoulder of the curve at around 75% saturation of the haemoglobin oxygen-binding capacity. It is known as the mixed venous partial pressure of oxygen; it is a mean value for the oxygen concentration in blood returning from all the tissues. The value indicates that around one quarter of the oxygen transported to the tissues has been used up in metabolism. The oxygen tensions in different tissues will, however, differ enormously: the heart removes almost three quarters of the oxygen from its nutrient blood, and actively exercising skeletal muscle may use even more than this, whilst the kidney uses only about one fifteenth of the oxygen in renal arterial blood. Below partial pressures of 5.3 kPa the curve falls

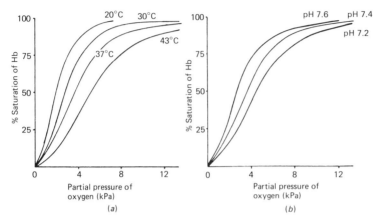

**Figure 6.8** The oxygen-haemoglobin association/dissociation curve as modified by changes in (a) temperature, and (b) pH.

steeply so that very small decreases in partial pressure result in the release of large amounts of oxygen. For example, between 2.0 and 3.0 kPa 16% of the bound oxygen will be released.

Thus the shape of the curve is such that very high levels of saturation are obtained in the alveolar capillaries even if alveolar oxygen concentrations vary widely, and yet oxygen is readily released by very small decreases in oxygen tensions in the tissues.

The modifications to the curve as a result of changes in temperature and hydrogen ion concentration are also helpful (Fig. 6.8). As temperature rises the degree of saturation at a given partial pressure of oxygen is less i.e. less oxygen can be carried. The curve shifts to the right. In metabolically active tissues heat is produced and temperature rises, haemoglobin can bind less oxygen, and so oxygen is released. A fall in pH also decreases the oxygen-carrying capacity and so acid metabolites in an active tissue help to cause haemoglobin to release more oxygen. Carbon dioxide production causes a fall in pH, and the formation of carbamino-compounds with haemoglobin accentuates this effect. Thus, in the presence of carbon dioxide, haemoglobin will release oxygen. Any general factors which increase the concentration of 2,3-DPG should also cause release of oxygen. Exercise, ascent to high altitudes, thyroid hormones, growth hormone, and androgens, have all been said to increase the amount of 2,3-DPG in the red blood cells. Anaemia may also encourage 2,3-DPG production.

The form of the haemoglobin-oxygen dissociation curve is important in relation to the transfer of oxygen from haemoglobin to intermediate carriers or stores such as fetal haemoglobin or muscle myoglobin. Both these molecules have a higher affinity for oxygen at a given partial pressure than does adult haemoglobin (Fig. 6.9). Fetal haemoglo-

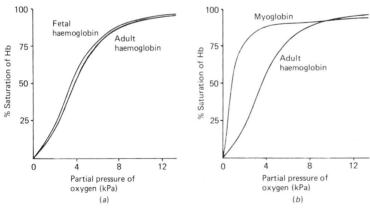

**Figure 6.9** Comparison of the oxygen-haemoglobin association/dissociation curve with those for (a) fetal haemoglobin, and (b) myoglobin.

bin has a different structure in which there are fewer diphosphoglycerate bridges, so that access to the pits in the molecule is easier; myoglobin contains only one haemoglobin-type unit in the molecule instead of four, so that the uptake curve is quite different from that of the adult tetrameric haemoglobin.

## Carbon dioxide transport in the blood

Even though the solubility of carbon dioxide in blood plasma is so much higher than that of oxygen, there are other additional mechanisms of transport.

Carbon dioxide combines chemically with water to form carbonic acid. The reaction is relatively slow but is accelerated by the enzyme carbonic anhydrase. This enzyme, like other catalysts, increases the rate of reaction in either direction. When carbon dioxide concentrations are low carbonic anhydrase accelerates the breakdown of carbonic acid, but when carbon dioxide concentrations are high, the enzyme accelerates its hydration. Although the formation of carbonic acid is slow in the absence of a catalyst, its ionisation to form hydrogen ions and hydrogen carbonate (bicarbonate) ions is rapid:

$$CO_2 + H_2O \rightleftharpoons H_2CO_3 \rightleftharpoons H^+ + HCO_3^-$$

Formation of carbonic acid is slow in the blood plasma, which lacks carbonic anhydrase, but rapid in the red blood cells which are rich in the enzyme. The tissues produce carbon dioxide as a result of metabolism and so the partial pressure of the gas is high, at about 6.0 kPa. It diffuses into the capillaries and reaches the plasma as dissolved gas. The plasma is in equilibrium with the red cell water across the red cell membranes. Carbon dioxide is lipid soluble and diffuses in readily. In the cell it is converted into carbonic acid which ionises to yield hydrogen ions

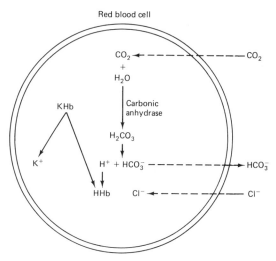

**Figure 6.10** The movement of gases and ions across the membrane of a red blood cell as it takes up carbon dioxide in the tissues. The process proceeds in the reverse direction in the lungs.

and hydrogen carbonate ions. Hydrogen carbonate ions are then in greater concentration inside than outside the cell, giving a diffusion gradient down which hydrogen carbonate diffuses out of the cell. This in turn produces an electrochemical gradient and chloride ions pass into the cell to balance the excess of positively-charged ions. This phenomenon is known as the chloride shift (Fig. 6.10). As haemoglobin gives up its oxygen its ability to buffer hydrogen ions increases. Indeed, at a respiratory quotient of 0.7, i.e. when only 7 mols of carbon dioxide are produced for every 10 mols of oxygen used (as when fats are used as energy substrates),

**Table 6.2 Carriage of carbon dioxide in the blood**

|  | Carbon dioxide (mmol) | | |
| --- | --- | --- | --- |
|  | Arterial blood ($CO_2$ 5.3 kPa; pH 7.40) | Venous blood ($CO_2$ 6.1 kPa; pH 7.36) | V–A difference ($CO_2$ 0.8 kPa; pH 0.04) |
| Blood (1 litre) | 21.53 | 23.21 | 1.68 |
| Plasma (600 ml) | 15.94 | 16.99 | 1.05 |
| Dissolved | 0.71 | 0.80 | 0.09 |
| Hydrogen carbonate | 15.23 | 16.19 | 0.96 |
| Carbamino | <0.10 | <0.10 |  |
| Red cells (400 ml) | 5.59 | 6.22 | 0.63 |
| Dissolved | 0.34 | 0.39 | 0.05 |
| Carbamino | 0.97 | 1.42 | 0.45 |
| Hydrogen carbonate | 4.28 | 4.41 | 0.13 |

the deoxygenation of the haemoglobin provides sufficient buffering capacity to take up all the hydrogen ions generated (*see* p. 252). Because the red cell in venous blood with its increased content of carbon dioxide has more solute particles, its intracellular fluid is slightly more concentrated and a small amount of water enters the cell down an osmotic gradient. The number of particles is increased because although the oxygen molecules have been replaced by an equal number of carbon dioxide molecules, these have been hydrated and ionised to two particles each. The exchange of hydrogen carbonate ions for an equal number of chloride ions does not alter the number of particles further. Red blood cells in venous blood are therefore very slightly larger than those in arterial blood. The whole sequence of events moves in the reverse direction in the lungs where the alveolar gas has a partial pressure of carbon dioxide around 5.3 kPa and carbon dioxide therefore diffuses from the blood plasma into the alveoli.

Carbon dioxide combines chemically with proteins which have free amino groups to form carbamino compounds. This reaction also releases hydrogen ions.

$$-NH_2 + CO_2 \longrightarrow -NHCOO^- + H^+$$

Carbon dioxide, then, is present in the blood in three forms, and unlike oxygen, is carried in relatively large amounts in both cells and plasma. *Table 6.2* summarises the relative importance of the different means of transport, and shows that the principal form in which carbon dioxide is transported is as hydrogen carbonate.

Of the extra 1.68 mmol/l (3.7 ml/dl) carbon dioxide in venous blood, some 70% is carried as hydrogen carbonate, 20% as carbamino-compounds and 10% as dissolved gas.

The hydrogen ions which are produced when carbon dioxide is transported present a problem, since they affect the pH of body fluids, and the maintenance of the pH of body fluids within reasonable limits is essential to the functioning of the body: this is considered further in Chap. 19.

## Uptake and removal of anaesthetic gases

The principles of uptake, distribution, and loss of anaesthetic gases are similar to those of the uptake, distribution and loss of the respiratory gases. The most important factors in uptake are the concentration of the gas in the lung alveoli and the solubility of the gas in blood plasma. The concentration of gas in the alveoli does not approach that in the inspired mixture for some time because of the diluting effect of the dead space volume and the functional residual capacity of the lungs. Of each breath of about 500 ml only some 350 ml enters and mixes with the 2500 ml of alveolar air. There are no specific chemical carriers for the anaesthetic gases in the bloodstream, so that the main factor in uptake is the physical solubility. Since the critical factor in producing anaesthesia is the plasma concentration, induction is less rapid with gases of low water solubility. Thus high concentrations of inspired nitrous oxide (of low water solubility) are necessary to induce anaesthesia at an acceptable rate, while much lower concentrations of water-soluble drugs like ether or halothane are needed to achieve the same result. However, this is not the only consideration. The plasma is not an isolated compartment: gases can diffuse into other tissues. The rate of diffusion depends upon the blood supply to the tissues and the solubility of the gas in water and lipids. All the anaesthetic gases are lipid-soluble – their action on the nervous system depends in part on this property – and they can enter cells. If blood-borne substances are soluble both in blood and in the tissues their rate of uptake is proportional to the blood flow.

Three nominal body compartments may be considered (Fig. 6.11). The brain and the viscera have a large blood flow of around 4 l/min and a relatively small volume of around 6 litres. Muscle and skin have a smaller blood flow of around

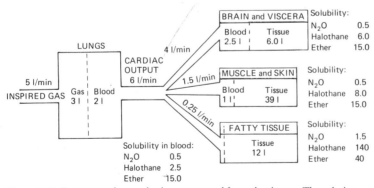

**Figure 6.11** Transport of anaesthetic gases to and from the tissues. The relative volumes of lung gases, blood and tissues, the flow rates of gas and blood, and the relative solubilities of the gases in the different body compartments are shown.

1.5 l/min but a much larger volume of about 39 litres. Fatty tissues have a small blood flow – only about 0.25 litres – with a volume of about 12 litres. The solubility of the gases in the first two compartments is not very different from that in blood, but it is very much higher in the third compartment. Uptake and loss are rapid in the brain and viscera with their rapid blood flow and small volume, slower in muscle and skin with the less rapid blood flow and larger volume of distribution, and much slower in the fatty tissues with their slow blood flow and greater ability to hold the gases in solution. The time necessary for induction depends on how rapidly the anaesthetic level can be achieved in the brain – so this depends upon water solubility. However, the time necessary for complete recovery from the anaesthetic depends upon how much has been stored in the fatty tissues of the body since this can only slowly be eliminated – and this depends upon the lipid solubility of the anaesthetic and the duration of administration.

## Further reading

Coxon, R.V. and Kay, R.H. *A Primer of General Physiology*, Butterworths, London, 1967

Burton, A.C. *Physiology and Biophysics of the Circulation*, 2nd edn, Year Book Medical Publishers, Chicago, 1972

Comroe, J.H. *Physiology of Respiration*, 3rd edn, Year Book Medical Publishers, Chicago, 1987

Pitts, W.E. *Physiology of the Kidney and Body Fluids*, 3rd edn, Year Book Medical Publishers, Chicago, 1974

Rushmer, R.F. *Structure and Function of the Cardiovascular System*, Saunders, Philadelphia, 1972

# Section II
## Physiological systems

# 7

# The cardiovascular system

A transport system capable of stabilising the composition of the body fluids is necessary to maintain the normal function of body cells. The blood, with its cells and plasma, provides the transport medium; the cardiovascular or circulatory system provides a pump driving the blood through a branching system of tubes. The smallest tubes have walls permeable to water and small molecules: materials in the blood may exchange across these walls with the lung gases, the body fluids, and the fluids of the kidney. A consideration of the transport system involves both the cardiac pump and the vascular pipework.

## The cardiac pump
### Anatomy of the heart

The heart is a hollow muscular organ forming two pumps operating simultaneously side by side. It is situated in the thoracic cavity, surrounded by the pericardium, a sac of connective tissue containing a small volume of clear fluid which acts as lubricant to reduce friction during the contraction of the heart. The right pump drives partially deoxygenated blood returning from the tissues back to the lungs for re-oxygenation, and the left drives oxygenated blood through the systemic circulation to reach all the tissues.

The muscle is of a specialised form (*see* pp. 96–99). It has no bony insertions: the fibres are attached to a rigid fibrous tissue septum between the thin-walled atria and the thick-walled ventricles (Fig. 7.1). The septum is made up of four rings closed by valves which permit inflow and outflow of the blood from the ventricles. These are, on the right side, the tricuspid and pulmonary valves, and on the left, the mitral and aortic valves. Above this septum are the two atria, and below are the two ventricles. The atria are separated by the atrial septum, and the ventricles by the interventricular septum, both made up of muscle continuous with that of the walls. Contraction of the muscular walls of these chambers either shortens their circumference and ejects their fluid contents or, if shortening is not possible, increases the muscular tension, and hence the intraluminal pressure. The fibres of the innermost layer of ventricular muscle run transversely, but the fibres of each successive layer make a slight angle with the previous one so that in the outer layers they run from upper right to lower left. This arrangement may enable the muscle to store

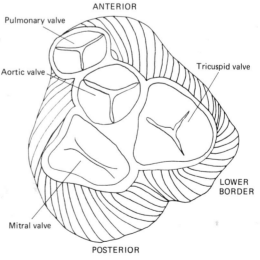

**Figure 7.1** The fibrous tissue septum to which the fibres of the heart muscle are attached.

some elastic energy between layers. The muscle of the left ventricular wall is much thicker than that of the right, enabling it to pump at higher pressures into the systemic circulation, which has a greater resistance to flow than the pulmonary circulation.

The valves are flaps of connective tissue anchored to the walls of the ventricles by the thin chordae tendinae and the papillary muscles. These attachments allow the flaps to act as valves, able to permit flow down pressure gradients in one direction only. There are no valves at the entrance of the great veins or the pulmonary vein.

## Electrical events of the cardiac cycle

Contraction of cardiac muscle, like that of other muscles, is a mechanical process initiated by the spread of an electrical impulse along the muscle. In skeletal muscle the action potential is triggered by an electrical impulse from a motor nerve, but in both cardiac and smooth muscle the impulse is generated by cyclical variations in the permeability of the muscle cell membrane to small ions. The area of muscle fibres with the most rapid oscillation of potential, and therefore of action potential production, superimposes its rhythm on that of all the neighbouring cells since there is electrical continuity between cells via the gap junctions. In the heart, the area of muscle with the most rapid rate of generation of action potentials is a group of modified muscle cells near the point where the superior vena cava enters the right atrium. This small block of tissue, the sinoatrial node, originates action potentials at a rate in excess of 100 per minute. However, a large number of autonomic nerve fibres from the right vagus nerve and from the sympathetic converge on the sinoatrial node. Acetylcholine from the vagus fibre terminals causes the cells of the node to become slightly hyperpolarised and hence less able to reach the threshold level to produce action potentials: this slows the rate of production. Noradrenaline released by the sympathetic nerves has the reverse effect. There is normally an equilibrium between these acceleratory and deceleratory influences which results in a resting heart rate of around 75 beats per minute.

The action potential generated in the sinoatrial node spreads rapidly across the atrial walls from muscle cell to muscle cell (Fig. 7.2), taking only some 0.08 s to traverse every part of the two atria. Near the atrioventricular septum is another area of specialised tissue, the atrioventricular node. It too lies on the right side of the heart and has both sympathetic and parasympathetic nerve terminals around it. The parasympathetic nerves derive from the left vagus nerve. The rate at which the atrioventricular node can generate its own action potentials is greater than that of most cardiac fibres but less than that of the sinoatrial node. It is

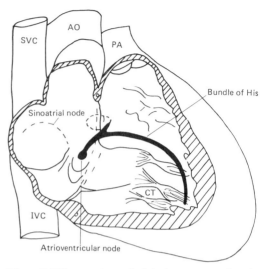

**Figure 7.2** The anterior wall of the heart removed to show the sinoatrial node, and the path of spread of the impulse along the bundle of His from the atrioventricular node. The dashed circle shows a cut into the ventricular septum to demonstrate how the bundle of His divides to pass into the wall of the left ventricle as well as that of the right. IVC inferior vena cava, SVC superior vena cava, CT chordae tendinae of the tricuspid valve, AO the aorta, and PA the pulmonary artery.

normally "driven", therefore, by the sinoatrial node. Excitation of the atrioventricular node does not occur immediately on arrival of impulses from the rest of the atrium but is delayed by 0.08–0.12 s. This is due to the slower transmission of the impulses in the more delicate fibres which reach the atrioventricular node in the fibrous septum between the atria and ventricles. The contraction of the atria which has followed the wave of action potentials is almost completed when the new action potential is stimulated in the atrioventricular node. The delay in ventricular contraction allows the atria to pump as much as possible into the ventricles before an active rise in ventricular pressure. The atrioventricular node may be visualised as a bulbous swelling on the end of a band of specialised cardiac muscle fibres which passes down through the fibrous septum and then splits into two bundles, one on either side of the interventricular septum. The bundle of fibres is called the bundle of His, and the individual muscle fibres are Purkinje fibres. When the bundles reach the apices of the ventricles they fan out as a network of Purkinje fibres on their inner surfaces. Conduction along these fibres is so much in excess of the speed of conduction across the other cardiac muscle cells that the impulse reaches all parts of the ventricles almost simultaneously.

A wave of excitation passing through a mass of muscle of such size as the heart can be detected at the body surface by electrodes connected to a suitable amplifier. As the wave moves towards an electrode it produces a positive potential and as it moves away it produces a negative potential. The potential is conducted by the body fluids. Because of the angulation of the heart a wave of depolarisation moving across the atrial muscle generates a positive potential in the lower left of the body, a negative potential in the upper right, and a sequence of positive and then negative potential in the upper left. The slower wave of repolarisation produces the opposite changes in potential. The changes in potential as a result of the depolarisation of the ventricle are more complex because the wave of depolarisation moves first down the interventricular septum and then back up along the ventricular walls. Electrodes may be placed anywhere on the body surface but are conventionally attached to the arms and legs, or on the surface of the chest in specific positions. The electrical changes they detect are known as the electrocardiogram (ECG, or, in USA, EKG). Such a recording, together with the conventional labelling, is shown in Fig. 7.3. The P wave is identified as representing depolarisation of the atrial muscle and the QRS complex is attributed to ventricular depolarisation. The repolarisation of the atria is lost in the QRS complex but the T wave is associated with the repolarisation of the ventricles. The less-frequently observed U wave has no generally accepted explanation.

Fig. 7.3 also gives the time scale of events at a heart rate of 70 per minute. The first three 0.08 s intervals of the scale are taken up by the P wave, the QRS complex, and the interval between them, respectively. This gives a P–R interval of 0.16 s. The T wave, lasting about 0.16 s follows the QRS complex after about 0.12 s and is followed by a gap of around 0.30 s before the next cycle begins. Important intervals measured in clinical practice are the P–R interval (the time from initiation of the impulse to the stimulation of the ventricles) and the QRS complex (the time required for the impulse to spread across the ventricles). When the heart rate increases, it is mainly the interval between the T wave and the next P wave which shortens, although there is also a lesser reduction in the duration of the electrically active phase.

## Mechanical events of the cardiac cycle

The mechanical events in the cardiac cycle follow the electrical. In an individual fibre the contraction begins as the fibre depolarises but does not reach its peak until 150 ms later – about 50 ms before the longlasting action potential drops back to the resting potential.

The term systole is used to describe a phase of contraction of a chamber of the heart, and diastole to describe a phase when the muscle is relaxed. When the particular heart chambers are not specified, systole and diastole usually refer to the ventricles. The full sequence of events in the cardiac cycle is conventionally divided into five phases. Beginning with the P wave of the electrocardiogram the sequence is as follows (Figs. 7.4 and 7.5).

Blood has been returning to the right side of the heart through the venae cavae and to the left side through the pulmonary vein. The venous orifices are not valved and so blood continues to flow through them as long as a pressure gradient exists. Atrial volumes increase slowly because the pressure gradient into the relaxed ventricles holds open the atrioventricular valves and so the ventricles are filling at the same time. Thus 80% of the ventricular filling takes place before the cardiac contraction begins. This is important because the force of the ventricular contraction is to some extent governed

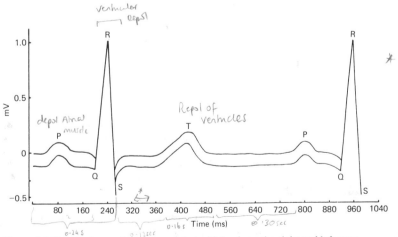

**Figure 7.3** The electrocardiogram recorded with leads on the right and left arms (Lead I) at a heart rate of around 80 beats/min.

120    *The cardiovascular system*

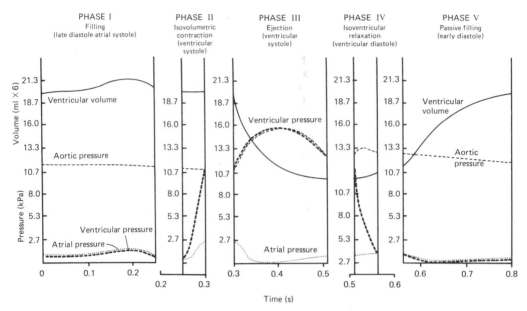

**Figure 7.4** The normal cardiac cycle at 75 beats/min. The pressures in the left atrium, left ventricle, and aorta are shown, together with the volume changes in the left ventricle. The aortic pressure falls steadily through phases V, I and II, but this is less apparent in phase II when drawn to this scale.

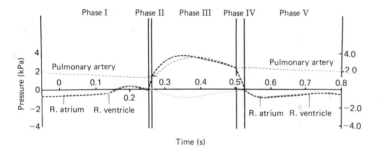

**Figure 7.5** The normal cardiac cycle at 75 beats/min. The pressure changes on the right side of the heart are shown. The volume changes are not shown for the right ventricle as they are broadly similar to those in the left ventricle.

by the volume of blood it contains immediately before contraction begins. After this passive stage of ventricular filling contraction of the atrial muscle then pushes into the ventricles the final 20%, 15–20 ml on each side of the heart. This contraction is seen as a pressure wave back along the great veins – the "a" wave of the venous pulse tracing which is described later. The total pressure achieved by the atria probably does not exceed 0.8 kPa. Some 0.08 s after contraction of the atria the ventricles begin to contract. As ventricular pressure rises above atrial pressure, the atrioventricular valves are forced into the closed position.

In the second phase of the cycle the ventricles are closed chambers, each containing between 120 and 130 ml of blood. The pressures in the aorta and in the pulmonary arteries exceed those in the corresponding ventricles and so the exit valves from the ventricles are closed. The inlet valves closed as pressure began to rise. As contraction continues it is therefore isometric, or isovolumetric, and results in a rapid rise in pressure. This stage ends as pressures in the ventricles reach those in the aorta and the pulmonary arteries of about 10.7 and 1.3 kPa respectively (the diastolic pressures of the systemic and pulmonary circulations), and the valves to these exit vessels open.

The pulmonary valve is forced open about 13 ms after the ventricles have begun to contract, when the pressure in the right ventricle has reached 1.3 kPa; about 25 ms later the pressure in the left ventricle reaches 10.7 kPa and the aortic valve also is forced open. Pressure now rises less steeply and results in ejection of the blood from the ventricles into the arteries. The peak right ventricular pressure reaches 3.3 kPa and the peak left ventricular pressure reaches 16 kPa. These peak pressures in the ventricles and the exit arteries are termed the systolic pressures. As contraction slackens the pressures fall but the ejection of blood continues until the pressures in the ventricles are matched by the pressures in the outflow vessels resulting from the resistance to flow in the vascular networks and the valves close once again. In rest conditions only 70–90 ml of blood is expelled.

In the next 0.1 s the ventricles are again closed chambers, but now they are relaxing and the pressure within them falls rapidly. About 60–80 ml of blood are left in the ventricles – the end-systolic volume. Blood returns from the veins into the atria throughout the whole cycle. As pressure falls the ventricular pressure reaches the same level as that in the atria.

Further relaxation of the ventricles reduces the pressure below that in the atria and the atrioventricular valves open in response to the pressure difference. This final stage of the cycle is the passive filling phase, which lasts for about 0.40 s.

The sequence of events on the right and left sides of the heart is essentially similar although the pressure changes are different. Right ventricular pressure rises more gradually to its peak and falls away more gradually. Systole is thus slightly longer. The acceleration imparted to the blood is only about one half of that on the left side. The acceleration on the left side is in fact so great that the whole body recoils slightly with each beat of the heart. This can be recorded on a device known as a ballistocardiograph, which measures the movement of the body as it rests horizontally on a table and is sometimes used to study cardiac activity.

## Changes in rate and force of contraction

Thus far, only the resting, the base line, cardiac cycle has been described. The heart is capable of more than doubling its resting rate, pulse recordings of 180 being not unusual after exercise. The pattern of the cycle changes at these rates: the major contribution to the speeding being a shortening of the diastolic interval, the fifth phase of the cycle. At 180 beats/min ventricular diastole reduces to about a quarter, about 0.14 s. The systolic part of the cycle also shortens, but only to about 60%, around 0.16 s.

The force of contraction of the heart may also be varied. The term "contractility" refers to the ability of the heart muscle to contract. As the heart rate increases the contraction becomes more powerful: there is an increase in contractility. Contractility is also dependent upon the degree of stretch of the muscle fibres, a measure of which is given by the volume of the heart at diastole (the end-diastolic volume). This property of cardiac muscle is described in the so-called Law of the Heart, or Starling's Law, which says that, in general, the force of expulsion of the ventricular contents depends directly upon the degree of stretch of the ventricular muscle. Both these effects have been mentioned already on p. 98 and Fig. 5.22 showed how the tension generated by cardiac muscle fibres varied with the length of the fibres. However, the effects of these two mechanisms in adjusting the volume of blood expelled from the ventricles is modified in the living animal, firstly because of the arrangement of muscle fibres in the heart, and secondly because of the influence of sympathetic nerves and the hormone adrenaline. Since the cardiac muscle is arranged in the form of four chambers (Fig. 7.1) the distribution of forces in the muscle around each chamber is similar to that around a sphere. These are described by the Law of Laplace: the equilibrium radius of an elastic sphere is equal to the tension in the wall divided by the pressure difference across it:

$$R = T/P$$

Thus the muscle tension necessary to produce a particular pressure in a chamber of the heart increases as the radius increases or, in other words, as end-diastolic volume increases a greater muscle tension is needed to produce the same systolic pressure. This consideration limits the effectiveness of the Starling mechanism. In the diseased or damaged heart it is the operation of the Law of Laplace that leads to heart failure. However, in the healthy individual, sympathetic nerves and circulating adrenaline help to produce the extra tension needed because they stimulate the release of calcium in the muscle fibres and increase their contractility (p. 99). These so-called intrinsic control mechanisms of the heart are considered further on p. 232.

Some special terms are used to describe cardiac activity. A rapid heart rate is known as tachycardia, a slow rate as bradycardia. A rapid beating of the atria without a corresponding rate of beating of the ventricles is called atrial flutter, and an uncoordinated series of rapid random contractions of either the atria or the ventricles, or both, is called fibrillation.

## Cardiac output

The volume of blood pumped by the heart in each minute is the cardiac output, normally around 5 l/min, but capable of being increased to 25 l/min in a normal individual. In trained athletes or sportspersons maximum cardiac outputs of up to 35 l/min have been recorded. The cardiac output is the output per beat, or stroke volume, multiplied by the number of beats per minute, the heart rate. Thus a typical stroke volume of 70 ml and heart rate of 75 l/min give a cardiac output of 5.25 l/min. Since the total blood volume is around 5 litres this means that the total volume is circulated round the body in 1 min. The cardiac output may be measured, as can the blood flow to any organ or vascular bed, by the application of the Fick principle (p. 104). By measuring the concentration of oxygen in the blood entering the right side of the heart and that leaving the left side, the blood flow to the body as a whole (i.e. the cardiac output) can be calculated.

## The vascular tubing

Blood is carried to all parts of the body by a series of tubes made up of smooth muscle and connective tissue. They are lined by a unicellular layer of endothelial cells which provide a smooth inner lining. Blood reaches the tissues via a large artery, smaller arteries, arterioles and metarterioles. In the tissues there are capillaries, vessels composed of only a layer of endothelial calls, and thus specialised – or simplified – for the purpose of permitting the exchange of fluid, salts, and small molecules across their walls. The capillaries link up to form venules, and these in turn join to give first small, and then large, veins. This simple circulation is in parallel with other similar simple circulations. In some parts of the body, however, there are more complex arrangements. The most common of these is a portal circulation, one in which two capillary beds are arranged in series. The best example is that in the gut and liver. The capillaries of the intestinal tract link up to eventually form the portal vein which passes to the liver: there it breaks up into a second set of capillaries in the liver which join to form first venules and then the hepatic veins. Another example of importance in endocrine function is the hypothalamo-hypophyseal system. The capillary bed in the hypothalamus is followed by a second capillary bed in the anterior pituitary gland. The kidney, also, has capillary beds in series, the glomerular capillaries being followed by the capillary networks around the renal tubules.

Structure and function are closely linked in the circulatory system. This relationship is now considered, first on the arterial side of the circulation and then on the venous side (Fig. 7.6).

### Arterial side of the circulation

The aorta is around 25 mm in diameter in man and its wall is about 1 mm in thickness. As the vessels decrease in diameter down the arterial tree, so the ratio of wall thickness to diameter increases from 1:12.5 to 1:8 in the medium-sized arteries, 1:3 in the small arteries and arterioles, and 1:2 in the sphincters of the pre-capillary vessels and the arteriovenous anastomoses. The veins are thinner walled than the arteries, the venae cavae of some 30 mm in diameter having walls less than 1 mm thick. Ratios of wall thickness to diameter are about 1:20 in the venules and do not differ greatly throughout the venous side of the circulation. All the vessels have four layers of tissue in their walls: the endothelial lining, an elastic tissue layer, a muscle layer, and a connective tissue sheath. The relative thicknesses of these layers vary in vessels of different sizes. In the aorta there is some 40% of elastic tissue and more fibrous than muscular tissue. On the arterial side of the circulation the proportion of elastic tissue decreases and the proportion of muscular tissue increases as the vessels become smaller. To a lesser extent this is true in the venous system also despite the thinner vascular walls. The outer connective tissue sheath is relatively thicker in the veins.

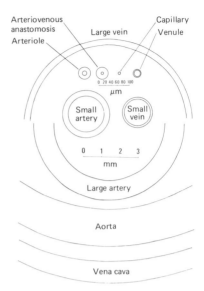

**Figure 7.6** Cross sections of the different vessels in the circulatory path. The upper scale refers to the six small vessels, and the lower scale to the four larger vessels.

These anatomical differences between vessels are associated with differences in function. The large arteries, in particular the aorta itself, act as elastic reservoirs: they absorb the force of the blood expelled from the ventricles during systole and then release the stored energy during diastole to maintain pressure and permit the blood to continue flowing despite the resistance to flow in the smaller vessels. The cycle of pressure from zero to 16 kPa measured in the left ventricle is thus converted into a cycle from 11 to 16 kPa in the large arteries and flow is maintained throughout the cardiac cycle. The elastic recoil of the walls of the aorta after the closure of the aortic valve, together with the recoil of the valve leaflets themselves, produces the characteristic dicrotic wave of the arterial pulse tracing. The pulse wave itself is due to the elastic deformation and recoil of the aorta. Each successive segment is stretched by the pressure and then transmits this to the next, resulting in a pressure wave which travels along the vessels at between 4 m/s in young subjects and 10 m/s in older subjects. Transmission of a wave of pressure is much more rapid than the movement of the fluid which transmits it: the velocity of blood flow averages only about 30 cm/s in the aorta. The velocity of the pressure wave varies with the stiffness of the vessel wall: hence the change with age of the subject.

The smaller arteries and the arterioles are the resistance vessels of the circulation. Because of the high proportion of smooth muscle in their walls their diameters can readily be altered. As the resistance to flow in a vessel varies inversely with the fourth power of the radius (p. 104), the resistance to flow in the arterioles can vary by as much as five-fold even though the actual changes in diameter are relatively small. They constitute the major part of the total peripheral resistance – a concept discussed in more detail in Chap. 15. It is this part of the vascular tree that maintains the blood pressure, particularly the diastolic pressure, at a normal level. A widespread relaxation of arteriolar muscle produces a dramatic drop in blood pressure – as in fainting, when it is also associated with a reduction in heart rate (p. 328).

Arteriovenous anastomoses are short vessels passing directly from an arteriole to a venule, thus lying in parallel with the capillary network linking these two vessels. They have thickened muscular walls and relatively small diameter. When the muscle is relaxed these anastomoses are fully open but as the muscle contracts the tension rapidly exceeds that predicted by the Law of Laplace as necessary to maintain the diameter against the distending pressure of the blood and the anastomoses completely close. This enables them to act as shunts – when closed the blood flows through the capillary network, when open the blood flows through the anastomosis and not through the capillaries. Similarly, constriction in the pre-capillary sphincters which precede certain capillary beds can prevent flow through these beds. These two groups of modified vessels can alter the distribution of blood flow in the body, as the circulation adjusts to the needs of the different tissues or of the body as a whole.

## Capillaries

The different types of capillaries have already been described (pp. 18–20). Since the capillaries generally are made up of only a single layer of endothelial cells, active changes in their diameter are unlikely. The possibility, however, of some capillaries having contractile cells near or around them (pericytes) has never been completely disproved. Observed changes in capillary diameter are almost certainly passive and due to the transmural pressures, which in turn depend in part upon the blood flow.

The post-capillary sphincters act to maintain the capillary and hence the transmural pressures, so adjusting the rate of exchange of fluid between blood and interstitial fluid.

## Venous side of the circulation

The venules, like the arterioles, have a relatively large proportion of muscle in their walls. However, the absolute amount of muscle and the ratio of wall thickness to luminal diameter is so much smaller that the muscle can generate very little resistance to flow. These vessels are distended by the pressure of the blood. Constriction, therefore, has a negligible effect on resistance, but a major effect on the volume of the vessels. This is the case in the small veins also. Thus the venous side of the circulation can act as a blood reservoir and these vessels are therefore referred to as capacitance vessels. They contain some 70% of the blood volume in comparison with around 11% in the large arteries and about 7% in the arterioles and the capillaries.

Because the walls of the veins are relatively so much thinner than those of the arteries the shape of veins is much more dependent upon the transmural pressure. Normally they appear oval but if the pressure inside exceeds that of the tissues outside to a greater degree than usual they adopt a circular shape, and if the external pressure is much greater than that inside the veins they can collapse down to a dumb-bell shape with the walls apposed between two narrow channels which still remain open.

The functional characteristics of the circulation may be described in terms of cross sectional area, volume, flow rates, and pressures. All of these, except pressures, have been described, or are summarised in *Table 7.1*.

Table 7.1 The pressures, velocities and volumes of blood in different parts of the systemic circulation

| | Pressure | | | Velocity (cm/s) | Blood volume (%) |
|---|---|---|---|---|---|
| | Systolic (kPa) | Diastolic (kPa) | Mean (kPa) | | |
| Aorta | 16 | 10 | 12 | 40 | 2 |
| Arteries | 15 | 8 | 10 | 30 | 8 |
| Arterioles | 5.2 | 4.7 | 4.0–5.0 | | 1 |
| Capillaries | | | 4.4–2.0 | 0.07 | 5 |
| Venules | | | 1.6–2.4 | | |
| Veins | | | 0.72 | | 50 |
| Vena cava | | | 0.6 | | |

## Pressures in the cardiovascular system

The mean pressure in the systemic circulation is around 12 kPa (assuming a systolic pressure of 16 kPa and a diastolic pressure of 10 kPa) and that in the pulmonary circulation is about 2.0 kPa (systolic

3.3 kPa, diastolic 1.2 kPa). Although the mean pressure is difficult to calculate because of the shape of the pressure wave (Fig. 7.7) it is conveniently taken as diastolic plus one third of the pulse pressure (the difference between systolic and diastolic pressures).

The pressures observed in the circulation are usually expressed in relation to the horizontal subject, ignoring the effect of gravity at different levels. These pressures, therefore, are those which give a pressure gradient from the left ventricle to the right atrium, or from the right ventricle to the left atrium. In absolute terms the pressure due to a column of blood above or below the effective gravitational zero of the system, the pump itself, must be subtracted or added. Thus the actual transmural pressure in the head is below that at heart level, and the actual transmural pressure in the legs is above that at heart level. However, the transmural pressures are important mainly in relation to two effects: the distension of thin-walled impermeable vessels like the veins, and the transfer of fluids down pressure gradients across permeable vessels such as the capillaries. Flow depends upon the pressure differences within a system: since both arteries and veins at a given level are subject to the same increment due to gravitational forces, this does not affect the flow from artery to vein along the pressure gradient produced by the pumping of the heart. Strictly it is not pressure differences which are the motive forces for flow: some writers prefer to use the phrase "differences in fluid energy". This phrase includes a further mechanism for maintaining flow against the gradient of gravitational energy: the kinetic energy, or momentum, of the flowing blood. The pooling of blood in the dependent parts of the body as a result of the passive dilatation of vessels due to intraluminal pressure – occurring almost solely in the veins, because of their thinner and more distensible walls – is prevented by the action of the skeletal muscles in compressing the vessels and assisting the return of blood to the heart. This pumping action by the muscles, alternately compressing the veins and then allowing them to refill with blood, is an important factor in flow through the veins, particularly in the limbs. The veins have saucer-shaped tissue flaps on the inner aspect of their walls which act as valves and permit flow of blood only towards the heart and prevent any backflow. In the capillaries an increase in transmural pressure causes an increased rate of fluid transfer: tissue fluid accumulates and may cause oedema.

The pressure gradient through the systemic circulation is also assisted by the negative pressures produced in the thorax as a result of breathing. The pressures in the thin-walled great veins, the venae cavae, are about 1 kPa below atmospheric pressure during inspiration. Pressures in the thoracic veins may be influenced by pressures in the thoracic cavity and also by pressures in the abdominal cavity which affect thoracic pressure through the diaphragm. The presence of valves in the veins determines the direction of blood flow in them: thus an increase in abdominal pressure is transmitted onward but no backflow occurs when the intrathoracic pressure exceeds the intra-abdominal pressure.

Abdominal pressure may increase venous return by its effect on the abdominal veins but later, as the pressure is transmitted to the thorax, venous return is inhibited.

The actual pressures in the great veins are important: they maintain the gradient of pressure to allow the return of blood from the periphery, and yet they must exceed the pressure in the right atrium so that blood can pass from the great veins into the heart. The term "central venous pressure" is used in relation to this concept. In a subject in a horizontal position the right ventricular pressure in diastole – the filling phase – is within 0.25 kPa of zero. The venae cavae are distended with blood up to a level just above the right atrium. The pressure in the great veins can be readily estimated since they collapse when the extramural atmospheric pressure exceeds the internal pressure. If, on standing up, the pressure in the great veins fell below atmospheric (i.e. zero blood pressure) at the level of the right atrium there would be no filling pressure for the right side of the heart. The return of blood from the venous reservoirs must be continuously adjusted as

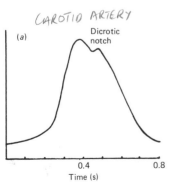
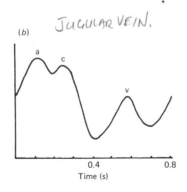

Figure 7.7 The form of the pulse wave in (*a*) the carotid artery, and (*b*) the jugular vein.

the body position changes so that filling of the heart is maintained.

The pressure changes in different parts of the circulation can be seen in the pulse tracings obtained from the arteries and from the great veins (Fig. 7.7). The arterial pulse, with its dicrotic notch, has already been mentioned and the pressure changes shown by it are easy to interpret. The pattern of the venous pulse, however, results from events in the atria rather than in the ventricles. The "a" wave is a back-pressure wave resulting from atrial systole and the "c" wave is a manifestation of a pressure increase in the right atrium caused by the contraction of the ventricles and a bulging of the atrioventricular valve into the atrium. The "v" wave represents the rising pressure as blood returns to the atria and the sudden fall as the atrioventricular valves open.

## Measurement of arterial blood pressure

Arterial blood pressure is customarily measured in the brachial artery. Flow through the vessel is prevented by inflating a rubber cuff wrapped around the upper arm. Pressure in the cuff may be read on a manometer connected to it. As pressure is lowered by allowing air to escape through a screw valve, a point is reached at which the systolic force of the heart is able to force blood past the closed section of the vessel. The resultant turbulent flow with every beat of the heart is heard as a sharp tap through a stethoscope bell placed over the vessel in the antecubital fossa. The sound continues as pressure in the cuff is further reduced until diastolic pressure is reached. The blood pressure is now sufficient to hold the vessel open even during diastole. As a result the tapping sound now becomes blurred. Finally, when the difference between the cuff pressure trying to close the vessel and the blood pressure trying to keep it open is sufficient to permit laminar flow throughout the cycle, the sounds heard through the stethoscope cease. The apparatus is known as a sphygmomanometer; the sounds are named after their discoverer, Korotkov.

## Further reading

Burton, A.C. *Physiology and Biophysics of the Circulation*, Year Book Medical Publishers, Chicago, 1972

Folkow, B. and Neil, E. *Circulation*, Oxford University Press, Oxford, 1975

Rushmer, R.F. *Structure and Function in the Cardiovascular System*. Saunders, Philadelphia, 1972

# 8

# The respiratory system

Although the word respiration is derived from the Latin for breathing, the term has been widened to include those biological oxidations necessary for energy production. Respiration, then, includes inspiration and expiration (the movement of air in and out of the lungs), the exchange of oxygen and carbon dioxide at the lung membranes, the transport of oxygen and carbon dioxide to and from the tissues of the body, and the chemical processes of oxidation of foodstuffs. The last of these topics is usually dealt with by biochemists rather than physiologists.

## The apparatus of breathing

The lungs and respiratory passages, or airways, are an invagination of the body surface specifically modified for the easy and rapid transfer of the respiratory gases, oxygen and carbon dioxide, to and from the blood. Air is moved actively through the nose, passes the oro-nasal cavity, and then reaches the trachea which divides into a right and a left bronchus both of which subdivide into bronchioles. These divide into smaller and smaller branches until they terminate in bubble-like sacs, the alveoli. These are formed of a single layer of epithelial cells and profusely covered with capillaries. Capillary blood is separated from the air in the alveoli by two cell layers only – the alveolar epithelium and the capillary endothelium. Special secretory cells in the lungs, pneumocytes, produce a phospholipid-lecithin material which forms a layer over the inner surface (the air surface) of the alveoli and acts as a surfactant, lowering the surface tension of the fluid lining the walls. This is important because the extremely thin walls of the alveoli are very responsive to changes in pressure. The alveoli behave rather as if they were bubbles of varying sizes on the ends of the bronchiolar tubes. The forces exerted on the walls of bubbles at equilibrium are described by the Law of Laplace

$$P = 2T/r$$

where $P$ is the pressure inside the bubble tending to expand it, $T$ is the tension in the bubble wall tending to contract it, and $r$ is the radius of the bubble. This Law (already mentioned in Chap. 7 to explain some of the characteristics of the heart and blood vessels) predicts that if the surface tension of a bubble remained constant the pressure needed to maintain the bubble would increase as the radius decreased. Thus small bubbles collapse more readily than large bubbles. Collapse of the alveoli is prevented because the surface tension of the surfactant decreases as its film thickness increases. As the alveoli are deflated and decrease in radius, the surfactant lining spread over a smaller surface increases in thickness and lowers the surface tension to equilibrate the forces on the alveolar walls. The total surface area of the alveoli is of the order of $60 m^2$. The blood flow to the lungs is the same as that to the whole of the systemic circulation: 5 l/min at rest, rising to 25 l/min during exercise. Blood reaches the lungs from the right side of the heart via the pulmonary arteries (which carry deoxygenated blood after its return to the heart) and returns to the left atrium in the pulmonary veins. The bronchioles and bronchi have a separate blood supply. The only tissue in the lungs apart from the alveolar membranes and the tissues of blood vessels, bronchi and bronchioles, is the small amount of fibrous and elastic tissue which supports the alveoli.

The lungs are enclosed in a double layer of fibrous and elastic tissue called the pleura. Between the inner layer attached to the lungs themselves and the outer layer attached to the walls of the thorax is a film of fluid. Were the pleura not present to connect the lungs to the thoracic wall, the lungs would be relaxed and much smaller. The chest wall, on the other hand, would expand to give a thoracic cavity with a greater than usual volume. Because chest wall and lungs are held in apposition by the pleura, both are under tension – the chest wall pulled inwards, and the lungs pulled outwards. An equilibrium therefore exists between the opposing pulls of the elastic tensions in these two structures. This is transmitted to the fluid layer between the pleura: the surface tension of this fluid layer prevents the lungs collapsing inwards on themselves. In other words, there is a negative or sub-atmospheric pressure in the potential space between the layers of the pleura. This is the intrapleural pressure. Stretching of the lung tissues helps to maintain the patency of the alveoli because it draws in air to fill them. The lungs themselves have no muscle tissue; changes in their size and shape depend upon the muscular activity of the thorax transmitted to them by the flexible linkage formed by the pleura. This linkage is broken if the layers of the pleura are separated, as for example when air gains admittance to the pleural cavity, either as a result of physical damage to the chest wall or by deliberate injection into the space. When this happens the visceral and parietal layers separate, the nearby lung collapses inward by virtue of its own elasticity and the chest wall recoils outward. This condition, termed pneumothorax, is sometimes induced deliberately to allow a part of the lung tissue to rest. If the pleural space is resealed the air will eventually be absorbed into the bloodstream. The link between lungs and thorax can also be affected by infection or damage to the pleura when this results in formation of fibrous tissue between the two layers – an adhesion – which prevents their movement relative to each other and impairs the normal functioning of the lung locally.

The thorax itself is made up of the ribs and the intercostal muscles, and is separated from the abdominal cavity by the dome-shaped muscle of the diaphragm. There are two layers of intercostal muscles, the external intercostals which slope downward and forward, and the internal intercostals which slope downwards and backwards. Since the ribs themselves slope downwards and forwards from the spine and can rotate at the joints with the vertebrae, the intercostal muscles can change the shape of the thorax when they contract. The external intercostals raise the ribs, rotating the moveable front portion upwards and thus increasing the antero-posterior dimension of the thorax. The internal intercostals have the reverse effect. Thus the external intercostals assist in producing the increase in thoracic volume needed for inspiration. The main muscle of quiet breathing, however, is the diaphragm. This is dome-shaped with its fibres running from their attachment at the lower border of the rib-cage to their insertion into the central tendon. Contraction of the fibres reduces the height of the dome, by pulling the central tendon downwards (Fig. 8.1).

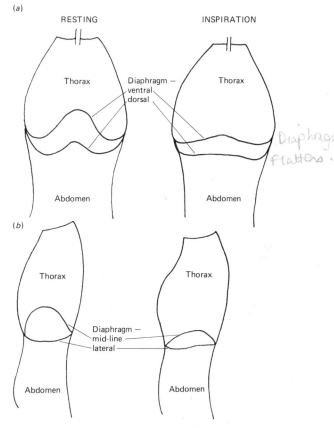

**Figure 8.1** Diagram to show the changes in dimensions of the thorax during inspiration. The changes in horizontal dimensions are exaggerated in order to show them more clearly. (*a*) The thorax before (left) and during (right) inspiration; (*b*) sagittal section through the thorax before (left) and during (right) inspiration.

In normal inspiration the diaphragm contracts, increasing the volume of the thoracic cavity and thus lowering the pressure within so that air flows through the nose down the pressure gradient. When contraction of the diaphragm ceases, the stretched pulmonary tissues recoil inwards and the pressure induced by the decreasing size of the cavity forces the air out again. In quiet breathing the thoracic wall itself will not have been expanded beyond its equilibrium position, so it will still have a tendency to recoil outwards. With greater inflation of the lungs during inspiration it too will have been stretched and will add its inward recoil to that of the lungs. Thus in quiet breathing inspiration is an active process, involving muscular activity, whereas expiration can be achieved by the passive elastic recoil of the lungs. As breathing becomes more vigorous the diaphragm contracts more powerfully; the vertical movement of the central tendon increases as much as six-fold from its usual 1–1.5 cm. Normally, at rest, the lower external intercostal muscles aid in inspiration, and there is also some contraction of the scalene muscles which raise and fix the upper two ribs. In theory, passive relaxation of the lung tissues, and in more vigorous respiration, the thorax, should be sufficient to achieve expiration. In practice, the relaxation is usually accompanied by some contraction of the internal intercostal muscles. As the depth of respiration is increased other muscles are involved. The scalenes increase in activity and the sternocleidomastoid group raise the sternum. Although the flaring of the nose due to the nares muscles is unimportant in man it is often observed with very heavy breathing. In the erect subject some relaxation of contracted abdominal muscles occurs during inspiration. In expiration the internal intercostal muscles become more active as the depth of respiration is increased. Further, the diaphragm is the only barrier between the thoracic and abdominal cavities. If pressure in the abdomen is increased, the diaphragm will be pushed upwards into the thoracic cavity, decreasing its volume and causing expiration. Contraction of the muscles of the abdominal wall, rectus and transversus abdominis and the internal and external oblique muscles, can therefore assist in expiration; as the depth of respiration increases this becomes an important factor. This so-called abdominal breathing may be less important in the female, particularly during pregnancy. Very forceful expirations, as in coughing, are always achieved by contraction of abdominal muscles. The relative mobility of the abdominal viscera has some effect on breathing: in the supine position the viscera press against the diaphragm and decrease the volume of the thoracic cavity. In one method of artificial respiration this effect is utilised by alternately tilting the supine patient from the head high, feet low position to the head low, feet high position. Other methods of artificial respiration involve either compression of the thorax to cause expiration, and then passive inspiration, or movement of the patient's arms so that they pull on the ribs and increase the antero-posterior dimension of the thorax and hence its volume. Nowadays, however, the most common method of artificial respiration is mouth-to-mouth respiration, which does not depend upon the mechanical variation of the volume of the thoracic cavity, but makes use of air pressure and the stimulatory effect of the carbon dioxide in expired air.

When muscle relaxants such as tubocurarine are used in general anaesthesia, the anaesthetist has to ventilate the patient artificially to overcome the paralysis of the repiratory muscles.

## Air flow, lung volumes and ventilation

The normal rate of breathing varies from 5 to 22 breaths/min but in exercise it may rise as high as 40. Children usually breathe more rapidly than adults. With each breath a normal subject inspires and then expires some 400–500 ml of air, but even at rest some subjects may move up to three times this volume (Fig. 8.2 and *Table 8.1*). In exercise the volume can rise to 3000 ml. Because this air moves in and out of the lungs it is termed the tidal volume. The ventilatory flow, or total ventilation, is the

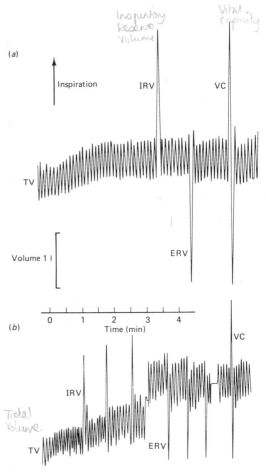

**Figure 8.2** Spirometer tracings from (*a*) a young adult male and (*b*) a young adult female. The arrow indicates the direction of inspiration. TV tidal volume, IRV inspiratory reserve volume, ERV expiratory reserve volume, VC vital capacity (a maximal inspiration followed by a maximal expiration).

**Table 8.1 Some values for respiratory measurements in young adults**

|  | Male | Female |
|---|---|---|
| Total lung capacity (l) | 6.0 | 4.2 |
| Inspiratory capacity (l) | 3.8 | 2.4 |
| Expiratory reserve volume (l) | 1.0 | 0.7 |
| Vital capacity (l) | 4.8 | 3.1 |
| Functional residual capacity (l) | 2.2 | 1.8 |
| Residual volume (l) | 1.2 | 1.1 |
| Respiratory rate (breaths/min) | 12 | 12 |
| Resting tidal volume (l) | 0.75 | 0.34 |
| Minute volume (ventilation) (l/min) | 7.4 | 4.5 |

Timed vital capacity: 83% of vital capacity in 1 s
97% of vital capacity in 3 s
Maximum voluntary ventilation: 125-170 l/min

amount of air passing in and out in unit time – usually around 12 breaths of 500 ml in 1 min, or 6 l/min. In exercise this can rise to 40 breaths each of about 3 litres tidal volume, a ventilation of 120 l/min.

In describing breathing a number of terms are used. An increase in the rate and depth of breathing is termed hyperpnoea if it is appropriate to the subject's metabolic needs, but hyperventilation if it is in excess of the metabolic needs. An inadequate rate and depth of breathing is termed hypoventilation. The term dyspnoea refers to respiration which is difficult for the subject to maintain. Another word used of the pattern of breathing is panting. This describes both rapid deep breathing and also rapid shallow breathing in which the tidal movement of the air does reach beyond the main airways. There are other terms which refer either to the gas being breathed or to the blood gas content. The term hypoxia is used when the subject is breathing a gas mixture with relatively low oxygen content (see also p. 246), anoxia when the gas mixture contains no oxygen; hypercarbia is the corresponding term for the condition in which the subject is breathing gas with a raised partial pressure of carbon dioxide. The condition in which the blood oxygen level is lower than usual is sometimes called hypoxaemia, although clinically it is referred to as hypoxia, and the term hypoxia defined above is then replaced by the phrase hypoxic hypoxia. Raised levels of carbon dioxide in the blood are termed hypercapnia.

The volume of the thorax in an average individual is greater than the 3 litres just quoted as being a maximal breath. The total lung capacity, at full inflation, is around 6 litres in the male and around 4.25 litres in the female. The whole of this lung capacity can never be exchanged with atmospheric air in a single breath. The chest cannot collapse between breaths. There is always a residual volume of between 1 and 1.25 litres of gas in the lungs. The amount of gas remaining in the lungs after a normal expiration is the functional residual capacity. This volume, which at its minimum is the same as the residual volume, is very important since it stabilises the composition of alveolar gases and prevents sudden major changes in the gas at equilibrium with the blood in the lung capillaries. A single 500 ml breath of 100% nitrous oxide would give a concentration in the lungs of less than 20%. In practice it would be much less because not all the gas inspired reaches the alveoli where exchange of gases occurs. The volume of the nose and the airways, about 150 ml, is known as the dead space because exchange of gases cannot take place to any appreciable extent across the walls of the airways. Significant exchange of lung and blood gases takes place only in the alveoli. The term dead space is used in two different ways: the definition above is of the anatomical dead space. The term physiological dead space also includes the volume of any alveoli which are functionally inert, that is, not participating in gas exchange with the blood capillaries. The volume of the dead space is important because it represents a volume of air which is moved by muscular activity but serves no useful purpose in gas exchange. Thus, although the tidal volume may be 500 ml, if the dead space is 150 ml, the effective volume of air inspired is only 350 ml. At a respiratory rate of 12 breaths/min, the alveolar ventilation will be only 4.2 litres instead of the 6 litres of total ventilation. Increasing the depth of respiration is usually therefore more effective in increasing alveolar ventilation than would increasing the rate of respiration. Normally an individual increases alveolar ventilation by increasing both rate and volume. However, in oxygen lack, depth of respiration is often increased before changes in rate are observed. Rapid shallow breathing, as is sometimes seen in hysteria, can even lead to reduced oxygen uptake. Again, in some animals which cool themselves by panting, rapid shallow breathing within the dead space volume aids in heat loss but does not significantly change alveolar gas composition.

In normal breathing the chest is not maximally inflated, nor is its volume reduced to the residual volume between breaths. After a normal expiration a further volume of air may be expired – the expiratory reserve volume of about 1 litres in men and about 0.75 litres in women. Similarly, after a normal inspiration, a further volume of air may be inspired – the inspiratory reserve volume of about 2.5 litres in women and about 3.25 litres in men. The total volume of gas moved by a maximal inspiration followed by a maximal expiration is termed the vital capacity – about 4.75 litres in men and about 3 litres in women. This maximal respiratory gas movement of which the individual is capable is an indication of the efficiency of the pulmonary pump in that individual. As such it has been used as assessment of

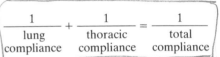

## The work of breathing and lung elasticity

The work of breathing has already been mentioned on p. 104 when it was pointed out that in addition to the work necessary to expand the lungs, work had to be done to overcome the resistance to flow in the pulmonary network, and the frictional resistance of the tissues. In the normal subject the work of breathing amounts to about 0.5 kgm/min, but this increases rapidly as the minute volume increases to reach around 250 kgm/min at 200 breaths/min. The work of breathing is increased if the thorax or the lungs are less elastic and if the airway resistance is increased by narrowing. This may occur in the small thin-walled airways at the end of expiration or during forced expiration when the intrapulmonary pressure is high, in the bronchi when smooth muscle contraction occurs, and in any airways where infection or irritation has caused swelling of the mucosal lining and excess mucus production.

The elasticity of the lungs can be measured by evaluating the increase in volume induced by a given increase in pressure. This factor is termed compliance and is expressed as volume change per unit pressure change (on p. 88 compliance was used in a different context, with length and tension as the variables). Very elastic tissues have a high value for compliance, very stiff tissues a low value. In human subjects it can be measured by placing an inflated balloon, connected to a manometer, in the thoracic part of the oesophagus (the first part of the digestive tract). Pressure is recorded at the end of an expiration, and then again after inspiring a measured volume of air. Repetition of this procedure with a series of different volumes of inspired air gives a series of associated pressure values. When the pressure-volume curve is plotted the slope of the line is the compliance. This method makes use of the fact that the pressure inside the thin-walled tube of the oesophagus is the same as that in the thoracic cavity through which it passes. The value obtained for the compliance is one for the lungs only, the pulmonary compliance, and is around 2 l/kPa. The change in volume is related to the total volume of the lungs, and so, if compliance is to be used as an index of the distensibility of the lungs, volume must be taken into account. The value for compliance divided by the volume of the lungs in litres becomes the specific compliance – normally about 0.67 l/kPa/l; this value is then comparable in lungs of different sizes. The compliance of the thorax–lung complex can be split into the individual compliances of the components – those of the lungs and the thoracic cage separately. All these can be measured; or if any two are measured, the third can be calculated. Since the lung compliances and the compliance of the thoracic cage are effectively in parallel the relationship linking them with the compliance of the whole is

$$\frac{1}{\text{lung compliance}} + \frac{1}{\text{thoracic compliance}} = \frac{1}{\text{total compliance}}$$

The pulmonary compliance and the thoracic compliance are approximately equal.

## Uptake of oxygen

Exchange of gases in the alveoli takes place across the alveolar walls. Diffusion of gases in the alveoli is rapid and although oxygen as a lighter molecule should diffuse faster than carbon dioxide, the difference due to molecular size is in practice negligible. At the gas-tissue interface, and in the fluids of the tissues, the most important considerations are the solubilities of the two gases and the application of Henry's Law which states that the amount of a poorly soluble gas going into solution is proportional to the partial pressure of the gas in equilibrium with the solvent. Because of the greater solubility of carbon dioxide it appears able to diffuse some 20 times as fast as does oxygen in the tissues. This difference is so great that it is necessary to consider only the factors influencing oxygen diffusion as being important in alveolar gas transfer.

The difference between the partial pressures of oxygen in the blood delivered to the pulmonary capillaries and in the alveoli is about 9 kPa; this allows equilibrium to be reached in about 300 ms. The average transit time for blood through the lung capillaries is about 750 ms so that ample time is available for complete diffusion to occur.

The length of the diffusion path is important. Oxygen has to cross the surfactant layer, the alveolar epithelium, the basement membrane, the capillary endothelium, the blood plasma, and the red cell wall. At its thinnest the complete barrier is about 0.2 μm across; this could be increased ten-fold where there is a cell nucleus in either of the cell layers. Disease may lengthen the diffusion path by causing thickening of the alveolar wall or the capillary wall, or by increasing the extracellular fluid layers when oedema fluid or exudate is produced in the alveoli or between the cell layers. Fibrous tissue may be laid down between alveolar and capillary walls. Even dilatation of the capillaries increases the diffusion path because a thicker layer of plasma separates the cells from the capillary wall.

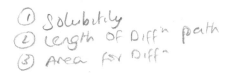
① Solubility
② Length of Diff'n path
③ Area for Diff'n

③ The area of surface available for diffusion is the third important factor. Normally, as has already been stated, this is very large. In disease many of the thin alveolar walls may be destroyed, some parts of the lung may not be adequately ventilated, or the blood flow to some parts may be inadequate. In all these situations there is an actual or effective decrease in the area available for gas exchange.

## Ventilation/perfusion ratios

The uptake of oxygen by, and the release of carbon dioxide from, the blood in the capillaries of the lungs will depend upon the actual ventilation (and therefore gas composition in individual alveoli) and the distribution and volume of blood flow in those alveoli. Since these vary in different parts of the lung, regional variations in the efficiency of gas exchange will occur. In the upright human subject blood flow is greatest at the base of the lungs and least at the apices. This is due to the low pressure in the pulmonary circulation; in the upper parts of the lungs the air pressure in the alveoli can exceed the pressure in the blood vessels, particularly during diastole, and so no flow occurs, whilst in the base of the lungs the arterial pressure exceeds the alveolar pressure. Exercise improves the flow generally but produces the greatest increase at the apices. In the supine subject the flow to the base of the lungs decreases slightly but the flow to the apices is increased to become slightly greater than that to the base.

Ventilation also is greatest at the base of the lungs and least at the apices but the difference is relatively much less than for perfusion. Hence the base of the lungs has a low ventilation/perfusion ratio whilst the apices have a high ventilation/perfusion ratio. The effects of posture and exercise on ventilation are similar to those on perfusion, so that ventilation/perfusion ratios throughout the lungs become more uniform on adopting the supine position or indulging in exercise.

Uneven matching of ventilation and perfusion also occurs in disease - local increases in airway resistance or in compliance will reduce ventilation, and narrowing or compression of blood vessels will reduce perfusion. The results of abnormal ventilation/perfusion ratios can be predicted. If an alveolus is not ventilated the blood passing through the alveolar walls is not oxygenated and joins the rest of the pulmonary circulation as a volume of mixed venous blood in the otherwise arterialised blood. These vessels constitute a veno-arterial blood shunt.

Increased ventilation in the rest of the lung cannot compensate this effect: even breathing pure oxygen cannot saturate the whole of the blood returning to the left atrium. The situation can be corrected only by diverting blood flow away from the poorly ventilated alveoli - as in fact happens in practice (p. 233). If, on the other hand, an alveolus is inadequately perfused with blood but adequately ventilated, the alveolus effectively becomes a part of the dead space volume. Since the unchanged air from the alveolus mixes on expiration with the rest of the alveolar air, the composition of alveolar air moves towards that of the inspired air. In the lungs as a whole mismatching of ventilation and perfusion may vary between the extremes and, if uncompensated, will result in varying degrees of lowered blood oxygen content. This will be particularly noticeable in the dissolved oxygen content (p. 108). The composition of alveolar air will also be affected.

## Lung function tests

Clinically it is sometimes necessary to assess the fitness of a patient to undergo gaseous anaesthesia. A number of lung function tests are available although few of these can be performed without special apparatus. They include measurement of the vital capacity, the timed vital capacity (or forced expiratory volume in one second), and the maximum voluntary ventilation or breathing capacity (the largest volume of gas that can be moved in one minute - 125-170 litres). Vital capacity is the volume of air which can be expelled after a maximal inspiration: it therefore indicates the maximum amount of fresh air which can be taken into the lungs. The timed vital capacity ($FEV_1$), normally expressed as a percentage of vital capacity and around 83%, gives an indication of airway resistance.

## Further reading

Bass, B.H. *Lung Function Tests*. H.K.Lewis, London, 1974

Comroe, J.H. *Physiology of Respiration*, 3rd edn, Year Book Medical Publishers, Chicago 1987

Dejours, P. *Respiration*. Oxford University Press, Oxford, 1966

West, J.B. *Respiratory Physiology*, 3rd edn, Williams and Wilkins, Baltimore, 1985

West, J.B. *Ventilation/Blood Flow and Gas Exchange*, 5th edn, Blackwell, Oxford, 1985

Widdicombe, J. and Davies, A. *Respiratory Physiology*, Arnold, London, 1983

# 9

# The nervous system

## Central nervous system

The central nervous system consists of the brain and spinal cord. From this extend the peripheral nerves carrying sensory and motor information. Nerves carrying impulses to the central nervous system are termed afferent, those carrying impulses from the central nervous system, efferent. The nervous system has two components: the somatic nervous system with its centre in the neocortex of the cerebrum, and the autonomic nervous system with no obvious centre, although associated with the limbic cortex and the hypothalamus. The somatic nervous system is mainly involved in transmitting sensory information to consciousness, and in the innervation of skeletal muscle. Its branches are arranged to provide separate communication to and from each developmental segment or dermatome. The autonomic nervous system, on the other hand, is concerned with the non-conscious regulation of activity in body systems, with sensory nerves from interoceptors carrying impulses which rarely produce conscious sensory perception and with efferent nerves to smooth muscle and glands. Outside the central nervous system the segmental organisation is much less clearly defined and there are groups of nerve cell bodies forming ganglia, or sub-centres of control, more or less distant from the spinal cord.

## The forebrain

The cerebrum is split by the longitudinal fissure into two hemispheres, whose flattened bases are approximated. They are joined by the central commissure. The surface of each hemisphere is made up of grey matter, nervous tissue with a very high proportion of cell bodies, whilst the interior is white matter, consisting almost wholly of nerve fibres, with a number of areas of grey matter of sufficient size and regularity to be known as nuclei or ganglia. The surface layer or cortex is extremely infolded, the major folds being termed sulci. The three largest sulci, the central, lateral and parieto-occipital sulci, divide the cerebrum into lobes, and are sometimes called fissures (Fig. 9.1). An area of cortex delimited by sulci is termed a gyrus. As a result of

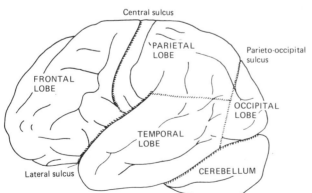

**Figure 9.1** The lobes of the cerebrum in lateral view.

infolding the area of the cortex is greatly increased: it is estimated to contain several thousand million nerve cells. The lobes separated by the fissures are termed frontal, temporal, parietal and occipital. The cortex is associated with a number of functions of which the most clearly defined are motor control (precentral gyrus), conscious somatic sensation (postcentral gyrus), speech and writing, vision, smell and hearing (Figs. 9.2 and 9.3). Learning, memory and association are all thought to be properties of the frontal, temporal and parietal cortex. These functions are mainly related to the neocortex, the ontogenically newer part of the brain found in its most developed form in man. The ontogenically primitive brain, or limbic brain, includes the part of the cortex known as the allocortex, situated mainly in the cingulate and hippocampal gyri lying on the mesial surfaces of the hemispheres. This part of the cortex is sometimes called the rhinencephalon, or 'smell' brain, because smell perception is localised here (Fig. 9.4).

The somatosensory cortex receives inputs from the thalamus (via the corticothalamic tract of nerve fibres) as do all other regions of the cortex except the rhinencephalon. Experimental stimulation of the somatosensory cortex and monitoring of cell activity demonstrate a topographical representation of the body in the sensory cortex with stimuli to the head being perceived in the area near the lateral sulcus and those to the feet in the area nearest to, or within, the longitudinal fissure (Fig. 9.5). The area devoted to each body structure or surface relates to the number of sensory units there and thus to the size of the sensory fields and the number of receptors in that part of the body. Head and hands are therefore represented over a relatively large area of the somatosensory cortex; trunk and legs over a relatively small area. Because the sensory nerve pathways from each side of the body cross en route from the periphery to the cerebral cortex, sensation from the right side is perceived in the cortex on the left, and vice versa.

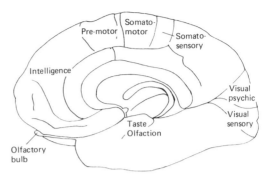

**Figure 9.3** The functions of the cerebral cortex indicated on the medial surface of the cerebral hemisphere.

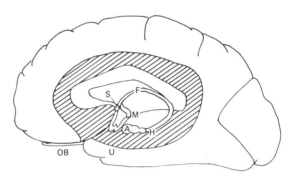

**Figure 9.4** The limbic brain, or rhinencephalon. The shaded area represents the primitive or limbic cortex on the medial surface of the brain. It is also known as the allocortex. A, amygdala; H, hippocampus; OB, olfactory bulb; M, mamillary bodies; U, uncus; F, fornix; S, septum. The septal nuclei are not shown because the septum has been removed.

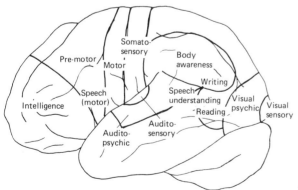

**Figure 9.2** The functions of the cerebral cortex indicated on a lateral view of the cerebrum.

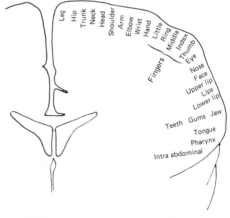

**Figure 9.5** The somatosensory cortex – the postcentral gyrus – showing the topographical representation of the body.

134   *The nervous system*

The occipital cortex is the area in which visual information is perceived. Taste and smell, primitive chemical sensations, are perceived in the infolding of the cortex on the medial surface of the hemisphere by the uncus; neurones involved in the sensation of hearing and the related function of speech lie on either side of the lateral cerebral sulcus in temporal and frontal lobes and the angular gyrus of the occipito-parietal region (Fig. 9.1 and 9.2).

In the precentral gyrus of the frontal lobe is the motor area. Stimulation here causes contraction of skeletal muscle. There is again a topographical representation similar to that in the sensory area although here the area of cortex is related to the number of motor units in a given body zone (Fig. 9.6).

The subcortical areas contain many nerve cells grouped to form the putamen, globus pallidus, and the caudate nucleus: these are termed the basal nuclei and are all concerned in control of posture and/or movement (Fig. 9.7). Their role is described in Chap. 15.

Between these nuclei and on either side of the third ventricle is the thalamus. This functions as a relay station for sensory signals. It can be divided into four zones: the epithalamus, the ventral thalamus, the dorsal thalamus, and the midline and intralaminar nuclei. The epithalamus connects with the olfactory system. The ventral thalamus includes relays to the sensory and motor cortex; since its nuclei receive inputs from specific sensory pathways and transmit them on to the cortex, they are termed specific sensory nuclei. They include the ventrobasal nuclei transmitting general sensation, the medial geniculate bodies transmitting auditory information, and the lateral geniculate bodies receiving optic nerve-optic tract impulses. As in the somatosensory

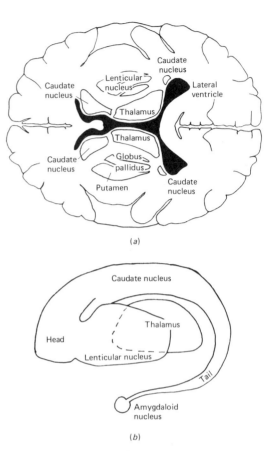

**Figure 9.7** Sections through the brain to show the basal nuclei: (*a*) horizontal, (*b*) sagittal.

cortex the fibres are arranged in positions corresponding to the body segments with which they are associated, forming a somatotopic map. This is less clearly defined than that in the somatosensory cortex. The functions of the dorsal thalamus are still unclear. The midline and intralaminar nuclei are known as the non-specific sensory nuclei of the thalamus. They carry sensory impulses travelling in the reticular activating system and their fibres project to the frontal lobe and the limbic brain as well as, possibly indirectly, all areas of the neocortex. This more primitive sensory pathway does not appear to have somatotopic mapping.

*Electrical activity of the cerebrum*

Although insertion of electrodes into various parts of the brain can reveal the activity of tracts or nuclei, application of surface electrodes to the skull does not give indications of brain activity which can be directly interpreted as related to any special thought process or even physical acts. The 'resting' level of activity in the brain is so high (more than 20% of the

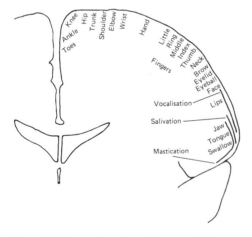

**Figure 9.6** The motor cortex – the precentral gyrus – showing the topographical representation of the body.

resting energy consumption) that even intense mental activity increases it only minutely. Electrodes on the skull usually pick up a non-synchronised pattern of activity in the alert or active subject.

However, electrodes placed on the skull over the somatosensory cortex in the postcentral gyrus can detect electrical activity after stimulation at the periphery. A stimulus to a point on the body surface produces after a 5–12 ms delay, a surface positive electrical wave in the cortex at the appropriate somatotopic locality. This is termed the primary evoked potential. At almost the same time a less specific evoked potential may be observed over the association cortex. Some 20–80 ms after the stimulus there is a second larger and more diffuse wave, known as the diffuse secondary response. Electrodes over the motor cortex are rarely able to detect patterns of activity that can be correlated with specific muscle activity although electrical stimulation of the cortex in this area produces defined and recognisable motor activity.

More generally recognisable patterns of electrical activity are observed in resting subjects with closed eyes and in sleeping subjects (Fig. 9.8). The random signals observed in subjects with open eyes are replaced when the eyes are closed by regular slow waves with a frequency of 8–12 Hz, and an amplitude of around 50 μv, termed α waves. Sleeping subjects show a number of variations in electroencephalogram patterns depending on the depth of sleep. Four stages are described. In Stage I, the α waves disappear and the trace becomes one of low amplitude, high frequency waves. In Stage II bursts of α-like waves appear – the so-called sleep spindles. Stages III and IV show progressive slowing of the waves and an increase in their amplitude. Interspersed through Stages III and IV at about 90 min intervals are periods of rapid irregular electrical activity, which look more like the recordings from an alert subject, and are associated with rapid movements of the eyes beneath the closed lids. These periods are termed paradoxical or rapid eye movement (REM) sleep. They are associated with dreaming and often with the grinding of the teeth termed bruxism. This kind of sleep occurs some four to six times during the night and occupies around 25% of the total sleep time. If subjects are not permitted REM sleep they become anxious and irritable, and the proportion of REM sleep increases in subsequent sleep periods. There is no way of preventing bruxism during sleeping and the subject is usually unaware of the habit. When the practice leads to extensive attrition of the teeth, or when the subject suffers from consequent malfunction or pain of the temporomandibular joints, it may be necessary to fit an overlay or biteguard over the teeth at night. Some grinding of the teeth is often observed in a very early sleep stage even though REM sleep does not usually begin for 60 to 90 min after falling asleep.

## Hypothalamus

Between the thalamus and the pituitary gland are a group of specific nuclei and nuclear areas collectively termed the hypothalamus (Fig. 9.9). The main function of this part of the brain is to act as a connection point between the two communication systems of the body, the nervous system and the endocrine system. It is the principal pathway by which the limbic brain communicates with the body, and hence is the mediator of emotionally charged responses. Hunger and thirst, pleasure and aversions, sexual drives and behaviour all pass through this linkage of primitive brain and body. It is linked to the thalamus and receives sensory inputs from it; additionally it contains cells sensitive to deep body temperature, to blood glucose levels, to the osmotic activity of body fluids and to the concentrations of hormones in the blood. Not only does the hypothalamus receive inputs from sensory pathways, but it is also part of the pathways to and from the limbic brain by which sensory impulses reach the allocortex and by which the limbic brain can modify the transmission of sensory impulses to the neocortex. Thus sensation can be coloured by emotional factors and can be abolished by rage or fear.

The functions of the hypothalamus are summarised in *Table 9.1*. Since many of these relate to the endocrine system and to specific control systems they are described further in Chap. 12 and in Section III. Some of the hypothalamic nuclei and nuclear areas have been identified as having particular functions, although these may vary between species. Those which have been characterised in man are included in *Table 9.1*

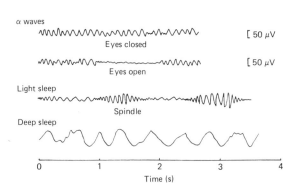

**Figure 9.8** The electroencephalogram (EG). The tracings show, from above downwards, the α waves, the break in the α rhythm observed when the subject's eyes are opened, the burst of α-like waves seen in light sleep and known as sleep spindles, and finally the large slow waves of sleep.

136  *The nervous system*

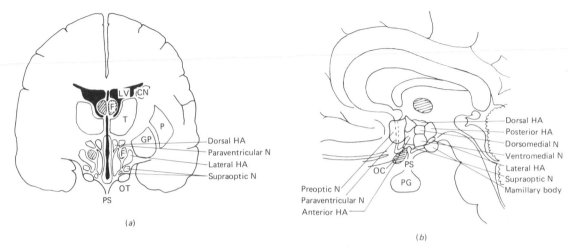

**Figure 9.9** The hypothalamic nuclei seen in (*a*) coronal section of the brain in the plane of the pituitary stalk and (*b*) sagittal section. Nuclei are labelled as N, and hypothalamic areas as HA. Other structures shown are the caudate nucleus, CN; the fornix, F; the lateral ventricle, LV; the optic chiasma, OC; the putamen, P; the pituitary stalk, PS; the pituitary gland, PG; the optic tract, OT; and the globus pallidus, GP.

**Table 9.1 Functions of the hypothalamus**

| Function | Area of hypothalamus involved |
|---|---|
| **Control of endocrine systems** | |
| 1. General control of adrenal medulla and sympathetic nervous system. | Dorsomedial and posterior |
| 2. Control of anterior pituitary function by use of releasing hormones reach pituitary gland via the hypothalamo hypophyseal portal system. | |

| Hormone controlled | Stimulated by | Inhibited by | |
|---|---|---|---|
| Growth hormone | GRH | Somatostatin | Anterior |
| Thyroid stimulating h. | TRH | | Anterior, dorsomedial nuclei |
| Adrenocorticotrophic h. | CRH | | Paraventricular nuclei |
| Follicle stimulating h. | GnRH | | Preoptic and anterior |
| Luteinising hormone | GnRH | | Preoptic and anterior |
| Prolactin | PRH | PIH | Arcuate nucleus |

3. Hormone synthesis and neurosecretion to the posterior pituitary gland
   Oxytocin — Supraoptic and paraventricular n.
   Vasopressin (antidiuretic hormone, ADH) — Supraoptic and paraventricular n.

**Control of body temperature** — Anterior to heat, posterior to cold

Appetitive behaviour
  Hunger — Ventromedial n.(satiety)
    Lateral feeding centre
  Thirst — near paraventricular nuclei
  Sexual drives — Anterior ventral

Sensory receptors
  Thermoreceptors — Anterior
  Osmoreceptors — Anterior (circumventricular organ)
  Glucose receptors — Ventromedial nucleus

## ...rebellum and hindbrain

...rebrum and the spinal cord is the
...either side and behind the upper
...re the cerebellar lobes (Fig. 9.10).
... is connected to the brainstem by
...n each side, the superior, middle
...he pons also connects to the
...erebral peduncle. The cerebellum
...deeply fissured hemispheres and
...an intermediate region between
...about 75% as much surface grey
...erebrum despite being only 10%
...erebellum is responsible for the
...ordination of muscle activity. It
...om the motor cortex, from
...muscle and joints, from the
...receptors and from touch and
...the skin. The vestibular organs
...centre in the cerebellum. As in

systems. It is probably here that the main modulation of sensory inputs occurs (*see* later). General anaesthetics, including the barbiturates and other hypnotics, are particularly effective in this zone because of the many synaptic connections. Ether

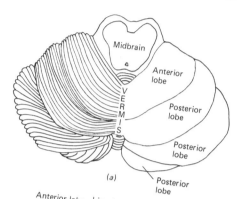

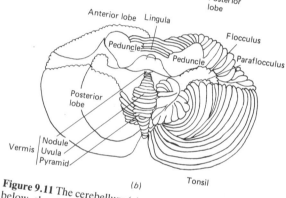

**Figure 9.11** The cerebellum (*a*) from above and (*b*) from below, showing the lobes and their subdivisions.

...ion through the
...obes, separated by heavy

...otor cortex there is
...ugh sensory informa-
...y projects to two areas
....12).

...r of areas involved in
...e medulla upwards
...ne of many small
...nnecting networks.
...cular activating system (Fig. 9.13).
Through the network pass sensory pathways leading
to the non-specific nuclei of the thalamus and
connecting with the hypothalamus and the limbic
system and in it are groups of neurones controlling
vegetative function – the heart rate, blood pressure,
respiratory depth and frequency and the co-
ordinating centres for activities involving many

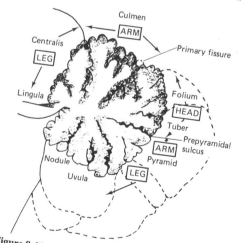

**Figure 9.12** The cerebellum in midline sagittal section, showing the somatotopic representation of the body in the vermis, with the major lobes behind.

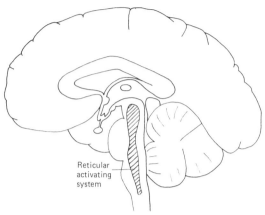

**Figure 9.13** The reticular activating system (shaded) in a section through the brain, midbrain and hindbrain.

also dramatically reduces activity in the reticular activating system. During sleep, activity in the reticular activating system is changed so that the sensory inflow to the cerebral cortex is reduced.

The pons and medulla contain the neurone groups which control the heart rate, peripheral vasoconstriction, and the rate and depth of respiration. In the medulla are also groups of neurones concerned in the control of sneezing, vomiting and coughing, and their co-ordination with other activities. Many of the cranial nerves enter the central nervous system here: these include the trigeminal, facial, glossopharyngeal, hypoglossal and vagus nerves, which are of particular importance to the dentist. Centres for salivation and mastication are found here. The pathways for cranial nerve reflexes are more complex than those for spinal reflexes (Chap. 13) because of the greater complexity of connections within the pons and medulla. Many of the centres in the brain stem are autonomic in function; indeed four of the nerves referred to above carry parasympathetic fibres.

## Spinal cord

In the spinal cord most of the nuclear material is grouped around the central spinal canal although the cell bodies of the sensory neurones are located outside the cord in the dorsal root ganglia. Both the grey and the surrounding white matter can be functionally sub-divided (Fig. 9.14).

The axons carrying sensory inputs travel in specific pathways within the white matter. Impulses carrying information about limb position and movement enter the cord via the dorsal root and then pass in the dorsal columns to the nucleus gracilis and nucleus cuneatus in the medulla, where the nerves synapse and then cross to the opposite side to form the medial lemniscus which ascends to the thalamus (Fig. 9.15). Pain and temperature impulses also enter via the dorsal root, but after ascending a short distance in a spinospinal tract, the fibres end on cell bodies in the dorsal horn in the substantia gelatinosa. After synapsing, the fibres cross over near the central canal to enter the spinothalamic tracts and pass up to the thalamus. Fine touch and pressure signals travel with the proprioceptive fibres; coarse touch and pressure neurones synapse in the dorsal grey matter and then cross the cord to ascend in the spinothalamic tracts. As each new nerve enters the cord it pushes the fibres from lower body regions medially in the dorsal columns, but the cross-over of the pain, touch and temperature fibres causes them to push the fibres from lower segments laterally. In both cases a layering of the representation of the different segments results (Fig. 9.16).

The descending motor fibres are also grouped together in the lateral and anterior corticospinal tracts. (Fig. 9.14). The origins of these tracts and their central connections are shown in Fig. 9.17. Only the direct pathway from the motor cortex will be described here: the central connections and other pathways are treated more fully in Chap. 21. Fibres from the motor cortex pass to the corticospinal tract

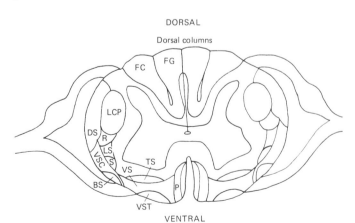

**Figure 9.14** Cross section of the spinal cord to show the principal tracts. These are named as follows (beginning with the dorsal columns and passing round to the ventral aspect): FG funiculus gracilis, FC funiculus cuneatus, LCP lateral corticospinal or lateral pyramidal tract, DS dorsal spinocerebellar tract, R rubrospinal tract, VSC ventral spinocerebellar tract, LS lateral spinothalamic tract, S spinotectal tract, BS bulbospinal tract, TS tectospinal tract, VS vestibulospinal tract, VST ventral spinothalamic tract, P direct pyramidal tract or anterior corticospinal tract.

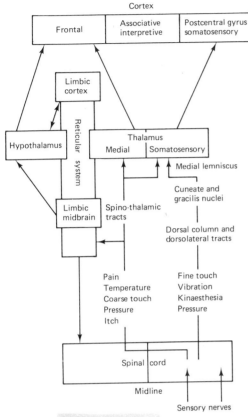

Figure 9.15 The sensory pathways.

in the internal capsule and the pyramids. When they reach the medulla some 80% cross the midline in the decussation of the pyramids and then pass down the cord in the lateral corticospinal tract. The remainder continue on the side of origin in the anterior corticospinal tract and cross the cord only just before their terminations. Fibres of both the lateral and anterior corticospinal tracts (or pyramidal tracts) are classified clinically as upper motor neurones. They terminate at synapses with anterior horn cells, each of which sends its process through the ventral root to pass into the spinal nerve and travel to a muscle. The motor fibres of the trigeminal, hypoglossal and vagus nerves follow corresponding paths in the central nervous system, but the nuclei are in the motor nucleus of the appropriate nerve.

## Synaptic transmission in the central nervous system

Although the majority of central nervous system synapses involve chemical transmission, it now seems probable that electrical transmission occurs across the narrow gaps between many of the densely packed cerebral neurones. Electrotonic spread may contribute to some of the electrical events observed on the cortical surface. A number of possible transmitter substances have been identified apart from the polypeptides of the hypothalamus already mentioned and the encephalins and endorphins to be described later in relation to pain. These include

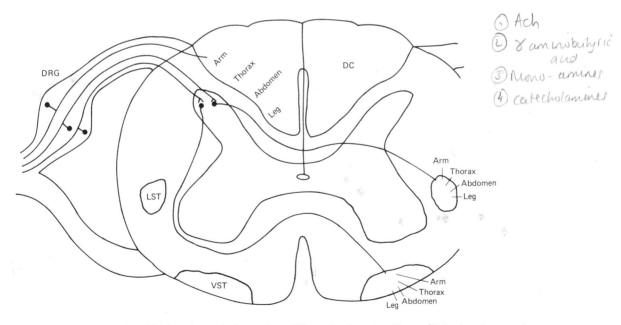

Figure 9.16 The manner in which layering of the inputs from different levels occurs. Shown are the dorsal column, DC and the dorsal root ganglion, DRG

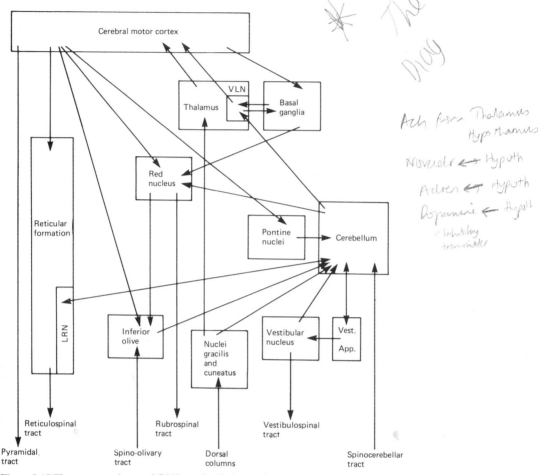

**Figure 9.17** The motor pathways. LRN lateral reticular nucleus, VLN ventrolateral nucleus of the thalamus, Vest App the vestibular apparatus.

acetylcholine, γ aminobutyric acid, and several mono-amines – in particular, the catecholamines.

Acetylcholine is found in the motor cortex, the thalamus, the hypothalamus, and the basal ganglia. In the spinal cord it is found at the synapses on Renshaw cells.

γ aminobutyric acid has been found in many areas of the brain, notably in the basal ganglia, and is normally associated with inhibitory activity. Other inhibitory substances include glutamate, found in the cerebellum, and glycine, in the spinal cord.

Noradrenaline and serotonin are found in the raphe nuclei which lie in the midline of the reticular core of the medulla and upwards to the hindbrain, the hypothalamus, the limbic system, the neocortex and the spinal cord. Increased levels of noradrenaline in the brain are associated with more cheerful moods, decreased levels with depression.

Adrenaline-containing neurones occur in the medulla, the hypothalamus, the thalamus and the spinal cord.

Dopamine is an inhibitory transmitter in the basal ganglia, and is also found in the hypothalamus and the limbic cortex.

## The autonomic nervous system

The autonomic nervous system is a primitive nervous system mainly responsible for the control of the viscera. Although most autonomic pathways are efferent there are also some afferent pathways. Actions mediated by the autonomic nervous system rarely reach consciousness: they are largely automatic and fall into the category of actions termed reflex. They are important in the normal functioning of the body and for the automatic responses to emergency situations. Although the autonomic nervous system has pathways in the brain and spinal cord, the cell bodies of the motor neurones lie outside the central nervous system, either arranged in a chain of ganglia on each side of the vertebral column, the sympathetic chain, or scattered in the organs innervated. The tissues served by the efferent fibres are either smooth muscle or glands, together with the heart, liver and kidneys.

The autonomic nervous system is subdivided into two divisions – the parasympathetic and the sympathetic. Their functions have been traditionally described as vegetative, and 'fright, fight, flight'. This oversimplification tends to exaggerate the contrast in roles. The parasympathetic controls such activities as the heart beat, the secretion of glands, and the passage of material through the digestive tract; the sympathetic is thought of as preparing the body to meet some external challenge. The actions of the two divisions are usually opposite, the sympathetic pattern replacing the parasympathetic when response to a challenge is invoked. Most organs supplied by the autonomic nervous system have a double innervation from sympathetic and from parasympathetic divisions. However, the uterus, the adrenal medulla and the vast majority of the arterioles in the body have a sympathetic innervation only, whilst the stomach and pancreatic glands have a parasympathetic supply only. Collaborative action probably occurs normally, and the minute by minute control of body functions is a balance of activity in the two divisions. The heart rate, for example, depends upon inhibition by the parasympathetic and stimulation by the sympathetic; blood pressure depends upon the cardiac output and the sympathetic control of the calibre of the blood vessels. In sexual intercourse the collaborative action is more obvious: whilst erection of the penis in the male is governed by parasympathetic activity, ejaculation is controlled by the sympathetic.

Although many autonomic activities occur as a result of activity in neuronal circuits involving only a small number of spinal segments, both sympathetic and parasympathetic systems have connections throughout the spinal cord and the brainstem. Over-riding control, at least of sympathetic activity, originates in the limbic brain and is channelled through the hypothalamus. The reticular activating system in the medulla and pons contains many groups of cells belonging to the autonomic nervous system: the centres for vascular and visceral control are obvious examples. The outputs from the central nervous system are grouped as a cranial parasympathetic, a thoracic and lumbar sympathetic, and a sacral parasympathetic.

The main actions of the two divisions of the autonomic nervous system are summarised in Table 9.2.

## The cranial parasympathetic outflow

The cranial parasympathetic outflow consists of fibres originating near the nuclei of the oculomotor, facial, glossopharyngeal and vagus nerves (Fig. 9.18). These fibres pass with the motor roots of these cranial nerves, although some may later leave to become associated with other nerves – some fibres of the facial nerve, for example, leave in the ear as the chorda tympani nerve and then join the lingual branch of the mandibular division of the trigeminal. The fibres synapse in a ganglion near the organ innervated and a short postganglionic fibre carries the final motor stimulus. Such ganglia include the ciliary ganglion (oculomotor fibres), the sphenopalatine ganglion (lacrimal gland fibres), the submandibular ganglion (salivary gland secretomotor fibres) and the otic ganglion (parotid gland secretomotor fibres). Of this group of nerves, the vagus has the widest distribution in the body, serving the respiratory passages and the heart, and also the digestive tract as far down as the caecum.

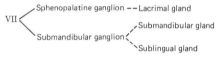
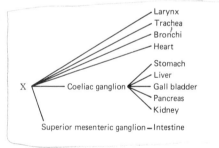

**Figure 9.18** The cranial and sacral parasympathetic outflows, and the main structures innervated. Solid lines indicate preganglionic fibres, dashed lines postganglionic fibres. The large Roman numerals indicate the cranial nerves.

## The sympathetic outflow

The thoracic and lumbar sympathetic outflow is more complex (Fig. 9.19). Fibres pass out of the cord in the anterior or ventral roots of the spinal nerves, from the first thoracic to the second lumbar, and then leave the nerve in the white rami

142   The nervous system

Table 9.2 The principal actions of the autonomic nervous system

| Body system | Parasympathetic nerves | Sympathetic at α receptors | Sympathetic at β receptors | Sympathetic cholinergic |
|---|---|---|---|---|
| **Vision** | | | | |
| Pupil | Constriction | Dilation | | |
| Accommodation | Ciliary contraction | Ciliary relaxation | | |
| **Digestive system** | | | | |
| Mouth | Salivary stimulation (flow and ions) | Stimulation of salivary protein secretion | Stimulation of salivary protein secretion | |
| Stomach | Stimulation of muscle activity<br>Relaxation of sphincters<br>Increased secretion | Decreased muscle activity<br>Contraction of sphincters | Decreased muscle activity | |
| Pancreas | Increased secretion | Decreased secretion | | |
| Intestine | Increased muscle activity<br>Relaxation of sphincters<br>Increased secretion | Decreased muscle activity<br>Contraction of sphincters | Decreased muscle activity<br>Contraction of sphincters | |
| Gall bladder | Contraction | | | |
| Liver | | Glycogen breakdown | Glycogen breakdown | |
| Adipose tissue | | | Lipolysis | |
| **Respiratory system** | Contraction of bronchial muscle | | Relaxation of bronchial muscle | |
| **Cardiovascular system** | | | | |
| Heart | Decreased rate<br>Decrease in atrial contractility | | Increased rate<br>Increased ventricular contractility | |
| Arterioles | | Constriction | Dilatation | Dilatation |
| Veins | | Constriction | Dilatation | |
| Kidney | | Vasoconstriction | Renin release | |
| **Endocrine organs** | | | | |
| Adrenal glands | | | | Release of noradrenaline and adrenaline from adrenal medulla |
| Pancreas | Secretion of insulin and glucagon | Inhibition of insulin and glucagon secretion | Secretion of insulin and glucagon | |
| **Urogenital system** | | | | |
| Bladder | Contraction<br>Sphincter relaxation | Sphincter contraction | Relaxation<br>Sphincter contraction | |
| Male genitalia | Erection (vasodilatation) | Ejaculation | | |
| Sweat glands | | (Adrenaline causes secretion in apocrine sweat glands) | | Stimulation of secretion in eccrine sweat glands |

communicantes to join the sympathetic chain, the chain of ganglia lying on either side of the vertebral column from the region of the third cervical vertebra down to the coccyx (Fig. 9.20). Above the level of the first thoracic vertebra the chain is made up of the inferior, middle and superior cervical ganglia which are not linked directly to the spinal cord at their own levels. After synapses in the ganglia of the chain, the postganglionic fibres pass in the grey rami communicantes to rejoin the spinal nerves. Some preganglionic fibres, however, such as those to the adrenal glands, to the superior and inferior mesenteric ganglia and to the coeliac ganglion, pass to ganglia some distance away from the vertebral column.

The autonomic nervous system    143

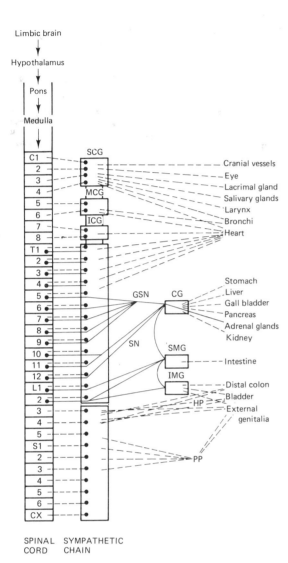

Figure 9.19 The sympathetic ouflows and the structures innervated. C cervical segments of the spinal cord, T thoracic segments, L lumbar segments, S sacral segments, CX the coccyx. SCG superior cervical ganglion, MCG middle cervical ganglion, ICG inferior cervical ganglion, GSN greater splanchnic nerve, SN lesser splanchnic nerve, CG coeliac ganglion, SMG superior mesenteric ganglion, IMG inferior mesenteric ganglion, HP hypogastric plexus, PP pelvic plexus. Solid lines indicate preganglionic fibres, dashed lines postganglionic.

Postganglionic fibres in the sympathetic nervous system are therefore usually relatively long. Sympathetic fibres are widely distributed in the body – every blood vessel, for example, has a sympathetic nerve supply.

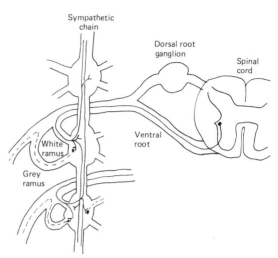

Figure 9.20 The connections between the spinal cord and the chains of sympathetic ganglia lying on either side of the cord.

## The sacral parasympathetic outflow

The sacral parasympathetic outflow passes with the second, third and fourth pelvic nerves. The ganglia in which synaptic connections are made are found near the organs supplied – the fibres pass through the hypogastric and pelvic plexuses to reach the distal colon, the rectum, the kidney, the bladder and the genitalia.

## Autonomic transmitter substances

The transmitter substance at all autonomic synapses between preganglionic and postganglionic fibres is acetylcholine. The receptors on the postganglionic cells are different from those for acetylcholine on effector organs: they are not blocked by atropine although they are affected by ganglion blocking agents such as decamethonium and suxamethonium. Acetylcholine itself in high concentrations will block transmission at the autonomic ganglia by saturating the receptor sites. The drug nicotine produces effects similar to those of acetylcholine at the ganglionic synapse and so these actions of acetylcholine are termed nicotinic.

The transmitter at all neuro-effector junctions in the parasympathetic nervous system is acetylcholine. This is synthesised in the terminal from choline and acetyl CoA (Fig. 9.21). On release it diffuses across the synaptic cleft and binds to receptors on the effector cells. These cells have on their membranes the enzyme acetylcholinesterase which breaks down acetylcholine into choline and acetate. The action of acetylcholine may be potentiated by cholinesterase inhibitors such as physostigmine

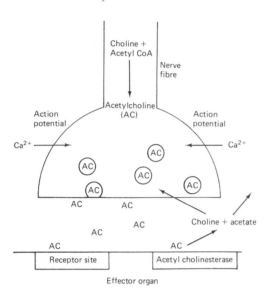

**Figure 9.21** The sequence of events in a parasympathetic nerve terminal (postganglionic) when an action potential reaches it. AC acetylcholine. The circles indicate vesicles.

(eserine). Atropine blocks the action of acetylcholine at effector junctions and can therefore be used to reduce salivation before oral surgery, or to relax the pupillary constrictors during examination of the eye. The peripheral actions of acetylcholine resemble those of the mushroom extract, muscarine, and are therefore termed muscarinic.

The sympathetic system is less simple. The transmitter released at most sympathetic neuro-effector junctions is noradrenaline, but some fibres, notably those to some blood vessels of skeletal muscle, and those to the eccrine sweat glands, are cholinergic – they release acetylcholine. In the majority of sympathetic nerves the aminoacid tyrosine is taken up and converted first to DOPA (dihydroxyphenylalanine) and then to dopamine (dihydroxyphenylethylamine). Dopamine is converted by hydroxylation to noradrenaline and, in the adrenal medulla, noradrenaline can be methylated to adrenaline (Fig. 9.22). In the postganglionic fibre terminal noradrenaline is stored in vesicles. Some of this is constantly being inactivated by the enzyme monoamine oxidase (MAO) and lost from the cell. On stimulation noradrenaline is released into the synaptic cleft. Some binds to post-junctional receptors. The effector cells have the enzyme catecholamine *ortho*-methyl transferase (COMT) on their membranes. This inactivates noradrenaline by methylation to normetanephrine which is lost into the tissue fluid and the circulation. However, a proportion of noradrenaline in the cleft is taken up unchanged by the pre-junctional neurone itself, and returned back to the vesicles.

Dopamine, noradrenaline and adrenaline are collectively called catecholamines. They function as transmitters in the sympathetic nervous system and in several parts of the brain. Adrenaline is produced in the adrenal medulla and circulates in the bloodstream to reach effector organs. All are inactivated by monoamine oxidase or by catecholamine *ortho*-methyl transferase. Psychological depression in many subjects is associated with low concentrations of noradrenaline in the brain, and so

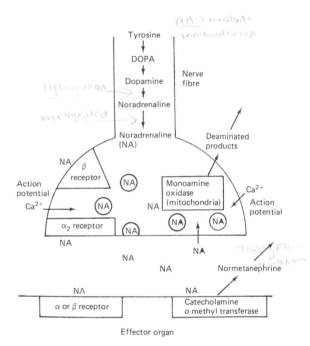

**Figure 9.22** The events which occur at a sympathetic postganglionic nerve terminal in response to an action potential. Note the presynaptic receptors which adjust the release of neurotransmitter, and the two mechanisms for inactivation of the neurotransmitter (catecholamine *ortho*-methyl transferase in the postsynaptic membrane, and monoamine oxidase in the mitochondria of the nerve terminal). NA noradrenaline. The circles indicate vesicles.

monoamine oxidase inhibitors are often used as anti-depressants. Dental surgeons routinely use adrenaline or noradrenaline in local anaesthetic mixtures to produce a local vasoconstriction which reduces the rate at which the local anaesthetic agent is removed by circulating blood. Patients on MAO inhibitor therapy cannot inactivate the adrenaline or noradrenaline and this, at best, causes a severe headache, or, at worst, may result in death. MAO inhibitors are therefore some of the drugs for which a dental surgeon should be alert when asking about a patient's medical history.

## Receptor sites in the sympathetic nervous system

Sympathetic receptor sites appear to be more complex than those of the parasympathetic system. Neuro-effector sites responsive to noradrenaline or adrenaline are of two types, α and β. α receptors are usually associated with contraction of smooth muscle. In the circulation, this means vasoconstriction. β receptors are associated with relaxation of vascular smooth muscle and vasodilatation. Most blood vessels, including arterioles, venules and veins contain α receptors. The blood vessels to skeletal muscle may contain both α and β receptors. The bronchioles, the blood vessels to the heart and the cardiac muscle fibres contain β receptors. Selective inhibition with various drugs shows that the receptors in heart muscle and in adipose tissue differ from the others; they are therefore termed $\beta_1$ receptors whilst the others are $\beta_2$ receptors. The β receptors are involved in the stimulation of secretion of saliva, of insulin and glucagon, and of renin. The drug phenylnephrine stimulates α receptors predominantly whilst isoprenaline stimulates predominantly β receptors. Phentolamine selectively blocks α receptors and propranolol β receptors.

In general the catecholamines stimulate the membrane receptors to cause production of either cyclic adenosine monophosphate (β receptors) or cyclic guanosine monophosphate (α receptors), and these act as second messengers within the effector cells.

Not only are there several types of receptor, but stimulation of a sympathetic nerve may involve several different receptor interactions. Thus, at synapses between preganglionic and postganglionic fibres, acetylcholine is the transmitter. However, a muscarinic receptor on an interneurone in the ganglion may also be stimulated. These interneurones terminate on the postganglionic cell where they release dopamine which produces an inhibitory postsynaptic potential. Postganglionic cell activity depends on the balance between the excitatory and inhibitory potentials generated. At the neuro-effector junction the sympathetic nerve itself has some α receptors ($\alpha_2$ receptors) which are stimulated by the released noradrenaline to inhibit its further release. More recently, β receptors have also been described on the nerve terminal: when they are stimulated more transmitter is released. The amount of noradrenaline released, free in the cleft, bound, destroyed, reabsorbed by the nerve, is presumably monitored and maintained at appropriate levels by these presynaptic receptors.

## The adrenal medulla

The adrenal medulla is a separate part of the sympathetic nervous system which secretes adrenaline and noradrenaline into the circulation. As an endocrine gland, it is discussed in Chap. 12. It may be regarded as a very large peripheral ganglion of the sympathetic nervous system, receiving preganglionic fibres from the splanchnic nerves. These fibres release acetylcholine, rendering the cells more permeable to calcium ions which then enter and trigger secretion of the catecholamine granules.

## Sensation

Sensation includes the transduction of external or internal stimuli into nerve impulses by receptor cells or nerve endings, the transmission of these nerve signals to the thalamus and then the sensory cortex (or the limbic system) and the processing of these signals either in their passage to the brain or in the brain itself so as to cause meaningful perception of the stimulus. The general types of receptor and the manner in which transduction occurs were described in Chap. 5; an outline of the specific sensory systems will now be given. These are the special senses of vision, hearing, taste and smell, position sense (including sensations arising from gravitational, rotational and proprioceptive receptors), the skin sensations of touch, pressure, vibration, and temperature, and the sensation of pain. Special attention will be given to taste and pain.

## Vision

In man the organs which transduce visual signals to yield a pattern of nerve impulses are the eyes. Vision provides a fascinating series of examples of the difference between sensation and perception. The image sensed by the retina is upside down and yet is perceived the right way up. Visual images can be perceived in 'the mind's eye' in the absence of visual stimulation. Optical illusions are perceived, or interpreted, in a manner at variance with reality.

The account which follows of the processes by which light and colour are perceived has deliberately been kept simple: readers interested in more detail should consult the larger textbooks of physiology or more specialised works.

### The visual apparatus

Each eye consists in optic terms of a lens, a clear fluid through which the light rays are diffracted and a mosaic of light sensitive cells on which they are focused (Fig. 9.23). The whole spherical organ is covered by the sclera, and attached to the brain by the optic nerve. The eye is attached to the orbit by the six ocular muscles, the recti superior, lateralis, inferior and medialis, and the superior and inferior oblique. By balanced contraction of these muscles it may be rotated freely in the orbit. The anterior part

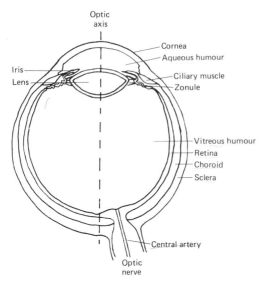

**Figure 9.23** A section through the eye.

of the sclera is clear and is termed the cornea. This forms a tough protective layer, almost perfectly transparent to light. Between it and the lens, in the anterior chamber of the eye, is the clear aqueous fluid. The lens is attached to the lining of the sclera, the choroid, by the lens ligament, or zonule. The choroid is a nutritive layer containing blood vessels, modified near the lens to form the ciliary body, which secretes the aqueous fluid. In front of the lens is the iris, the pigmented portion of the eye. It contains an annulus of muscle whose fibres contract to close the opening of the lens (the pupil), and radial fibres which can contract to open up the pupil. Like the iris diaphragm of a camera, the pupil gives greater depth of field in the constricted state. The lens itself is not rigid but elastic. At rest it is pulled by the zonule into a flattened shape. In order to look at near objects the refractive power of the lens must be increased. This is achieved by contraction of the ciliary muscle which surrounds the zonule and lens. This relaxes the stretched ligament of the zonule and allows the lens to spring back into a less flattened form (Fig. 9.24). The main increase in curvature of the lens as a result of this occurs anteriorly. This change in focal length of the lens from about 1.5 cm to 1.25cm represents an increase in power from 66 dioptres to about 80 dioptres. The nearest point which can be focused by the eye is about 10 cm distance in a young adult, but as the lens becomes less elastic with age, the ability to accommodate (to focus down) is reduced, and by age 60 the near point has retreated to some 80 cm. The mechanism for viewing near objects includes changes other than those in curvature of the lens: there is constriction of the pupil and, because both eyes are involved, a rotation of the optical axes towards the centre line of vision. This is termed convergence.

The light rays pass through the posterior chamber of the eye which contains a jelly-like fluid, the vitreous humour, to reach the light sensitive cells of the retina. The image on the retina is inverted from top to bottom and from side to side, so that the image of the right half of the visual field of each eye is sensed by the left half of each retina and is perceived in the cortex on the left side of the brain.

Impairment of vision may result from anatomical variation in the image-forming mechanism: the eyeball may be shorter or longer than the distance at which the lens brings images into focus, or the cornea may have a non-uniform curvature so that it refracts the light rays striking it from some directions more or less than those from others. A relatively short eyeball causes long-sightedness (hypermetropia) and a relatively long eyeball short-sightedness (myopia). The inability to focus on near objects due to loss of lens elasticity is termed presbyopia. Non-uniform curvature of the cornea causes astigmatism. All these conditions may be corrected by the use of appropriate lenses in front of the eye. A suitable biconcave spectacle lens can correct a condition of myopia by causing divergence of rays before they strike the lens of the eye, thus moving back the focus point. Similarly, a biconvex lens is used to bring forward the focus point to the retina of a short, or hypermetropic, eyeball. Astigmatism is corrected by spectacle lenses of non-uniform curvature.

## The retina

The inner surface of the sclera behind the lens is lined by the nutritive layer of the choroid. Inside this is the retina, histologically described as having ten layers. The surface of the choroid and the first layer of the retina itself contain pigment so that light is not reflected back from them to interfere with the primary image. Next to the pigmented layer of the retina is a layer of receptor cells (Fig. 9.25). These are of two types – the rods, and the cones, with either rod-shaped or cone-shaped pigmented outer segments. The outer segments are made up of parallel saccules containing the photopigments. The next layer of these cells contains the mitochondria. Below this inner layer is the nuclear layer wherein are situated the nuclei of the rods and the cones. Further into the eyeball the receptor cells terminate in the inner synaptic layer where they communicate with the bipolar cells. There are tight junctions between the various receptor cells. The ratio of rods and cones varies in different parts of the retina, but is generally around 20:1. However, both are absent completely from the optic disc (about 3 mm medial to the posterior pole) where the optic nerve and the retinal blood vessels enter the eye, and in the

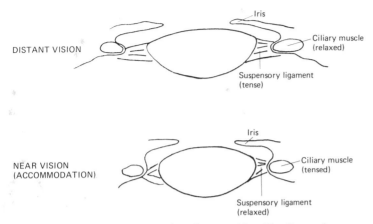

**Figure 9.24** How the eye accommodates for near vision. The diagram is exaggerated to show the change in shape of the lens due to the interaction of the ciliary muscle and the suspensory ligament (the zonule).

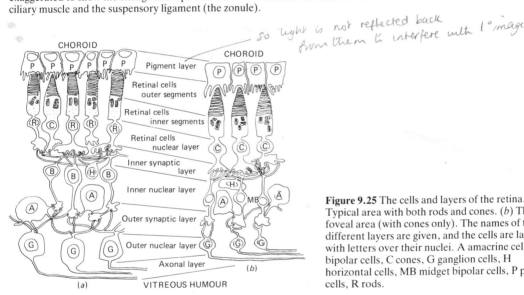

**Figure 9.25** The cells and layers of the retina. (*a*) Typical area with both rods and cones. (*b*) The foveal area (with cones only). The names of the different layers are given, and the cells are labelled with letters over their nuclei. A amacrine cells, B bipolar cells, C cones, G ganglion cells, H horizontal cells, MB midget bipolar cells, P pigment cells, R rods.

macula, a yellowish spot at the posterior pole, there are cones only. In the central part of the macula, the fovea, the inner layers of the retina thin out and are displaced laterally; there is minimum impedance to light falling on that area and it is therefore an especially sensitive area. The bipolar cells usually synapse with several receptor cells, the flat bipolar cells with cones, and the rod bipolar cells with rods. In the fovea, however, each small bipolar cell synapses with a single cone (Fig. 9.25). The nuclei of the bipolar cells form the inner nuclear layer. They then synapse in the second, or outer, synaptic layer with the ganglion cells, whose axons gather together to form the optic nerve itself. Most ganglion cells are synaptically connected with a number of bipolar cells but the midget bipolar cells from the foveal cones link to one ganglion cell only. Since each optic nerve has about 1.2 million fibres and there are some 126 million receptor cells, there is considerable convergence. However, there is wide variation in the size of the receptive field of each sensory unit: in the fovea each cone is connected through to one fibre of the optic nerve, whilst at the periphery of the retina as many as 600 rods may send signals to one fibre. The rods and cones are narrower in the central area: i.e. there are more receptors per unit of area. The effect of these variations in receptor density and convergence is that central, foveal, vision is more acute and more capable of perceiving fine detail, whilst peripheral vision is more sensitive in dim light because as many as 600 input signals may be summated in a single fibre of the optic nerve. The sensitivity of the eye in dim light is similar to that of the photopigment rhodopsin, found only in the rods, with a maximum at a wavelength of 505 nm. Night vision, or scotopic vision, is thought, therefore, to be a function of the rods, whilst vision in bright light, or photopic vision, is associated with the cones. In addition to their differences in acuity and sensitivity, the cones are the sole receptors responding selectively to different colours. Of the three other types of cell in the retina, the Muller cells are glial cells holding together the various elements of the retina and forming with their processes an inner and outer limiting layer; the horizontal cells are connecting cells in the outer synaptic layer, passing between receptor cells; and the amacrine cells pass between ganglion cells and may also link to bipolar cells in the inner synaptic layer.

## Photoreception

Light reaches the retinal receptors by passing through the sclera, the aqueous humour, the lens, the vitreous humour, and the inner neuronal layers of the retina. In the retina itself the receptor cells are the rods and cones which, in the absence of light, are differentially polarised due to the action of sodium-potassium pumps in the inner segments where the mitochondria are localised. This differential polarisation results in a continual flow of current from the inner to the outer segments. Stimulation by light causes changes which inhibit sodium movement, hyperpolarise the cells, and decrease current flow. This in turn decreases the release of a neurotransmitter from the cells. The actual light transducers are photopigments. Each photopigment consists of a protein, opsin, and retinene$_1$, the aldehyde of Vitamin A$_1$. The opsin of the photopigment rhodopsin, visual purple, found in the rods, is called scotopsin. Light breaks the bond between scotopsin and retinene, a change apparent as a bleaching of the visual purple. Rhodopsin is most readily bleached by light with a wavelength of 505 nm. The change in configuration of retinene due to its cleavage from scotopsin probably permits some substance to reach the membrane and decrease its permeability to sodium ions. It is possible that calcium ions are involved in this process. Cones respond abruptly to changes in light intensity but rods, while responding sharply to a light stimulus, are slow to revert to their resting level of activity. The hyperpolarisation of the receptors is a receptor potential (see p. 83): it is transmitted to the bipolar cells as a graded depolarisation in some cells or a graded hyperpolarisation in others. Any horizontal cells synapsing with the bipolar or receptor cells are also hyperpolarised and they cause inhibition of the other bipolar cells with which they connect. This inhibition of response enhances the sharp edge cut-off of light signals: it is called surround inhibition, and is important in increasing acuity. The ganglion cells are the only cells in this neuro-receptor complex in which action potentials are initiated. There are two types: one which continues to generate action potentials in response to the light-induced hyperpolarisation of the receptor cells, and one which gives a burst of action potentials when the illumination is turned either on or off. The amacrine cells also respond to the rate of change of illumination and their depolarisation and action potentials may act as generator potentials in the initiation of ganglion cell action potentials.

The result of stimulation of the photoreceptors is an electrical image in the ganglion cells of the visual image on the retina, modified by convergence and by the interaction of horizontal and amacrine cells. This image reaches the array of ganglion cells through a series of local potential changes and neurotransmitter actions. Although no specific chemical transmitters have been identified in particular synapses, the retina is rich in dopamine, 5-hydroxytryptamine, γ-aminobutyric acid and substance P.

## Optic pathways

The electrical encoding of the image is passed along optic nerve fibres, some of which cross in the optic chiasma (Fig. 9.26). Here the fibres from the nasal

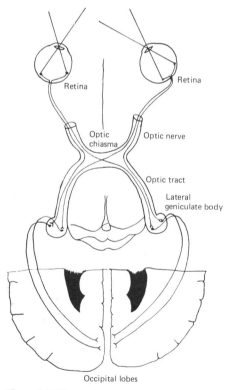

**Figure 9.26** The optic pathways.

side of the retina cross to join the fibres from the temporal side of the other optic nerve to form the optic tracts. These pass to the lateral geniculate body where they synapse. There is topographical localisation here; alternating layers of this structure receive inputs from the ipsilateral temporal retina and from the contralateral nasal retina. Fibres pass directly from the lateral geniculate body in the geniculocalcarine tracts to the primary visual cortex so that the left geniculate body and the left visual cortex receive information from the left side of each retina, and therefore from the right side visual field of each eye. The primary visual cortex is situated in the occipital lobe along the calcarine fissure (Fig. 9.27). It consists of several layers which analyse the input in increasingly complex ways. Fibres pass on from here to other parts of the cortex and also back to the lateral geniculate body.

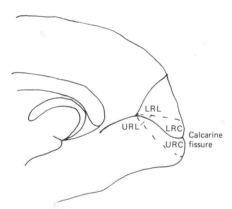

**Figure 9.27** Projection of the fields of view on the occipital cortex (medial view of one side only). LRC lower right centre of retina, URC upper right centre of retina, LRL lower right lateral, URL upper right lateral.

## Perception of visual stimuli

More nerve fibres leave the lateral geniculate body than enter it, and there are more neurones in the visual cortex than fibres in the geniculocalcarine tract. This implies that there is extensive processing of visual information to yield more than the simple light/dark response of the receptors. At the cortical level, contrast seems to be further enhanced. Cells may respond to patterns of light rather than to light or dark. Thus some cells are activated by slits of light, lines, or dark edges; and only when these are in a certain orientation. Others may respond to moving patterns, and still others only to features of a certain size. The cells of the cortex are arranged in columns, each of which responds to a preferred orientation and neighbouring columns have preferred orientations differing only by some 5–10°.

About half of the cortical cells receive inputs from both eyes although the two inputs are not necessarily of equal strength. Columns of cells in the cortex are also arranged so that they alternate between columns influenced predominantly by one eye, and columns influenced predominantly by the other. There is, therefore, localisation in the cortex of inputs from each eye, of inputs from the different parts of the retinal fields, and of more complex information about the orientation of light stimuli.

## Colour vision

In order to simplify this account, vision has so far been described only in terms of light and dark. The perception of colour is even more complex. In the cones there are three pigments, iodopsin (with the opsin, photopsin) and two others as yet unidentified. Names indicating the function of these pigments have been given to them: iodopsin with a maximal sensitivity at 445 nm (blue) is called a cyanolabe, the pigment with a maximal sensitivity at 535 nm (green) is called a chlorolabe, and that with a maximal sensitivity at 570 nm (yellow, but extending into red) is called an erythrolabe.

The retinal cells appear to encode colour information into a form which results in a pattern of on and off responses in the ganglion cells. The ganglion cells are therefore responsible for transmitting information about colour as well as about light intensity. The colour information is encoded as a pattern by the convergence of inputs from the cones onto ganglion cells and this pattern is then modified by inhibitory inputs via the horizontal cells. In the lateral geniculate body, colour information has been transformed into luminance signals and colour contrast signals. In the cortex there appears to be mapping of colours, with cells responding preferentially to certain colours, particularly blue, green, orange, red and purple. In the part of the visual cortex where this occurs the mapping is of colour only, not of position in the visual field, or of light energy.

## Eye reflexes

Three automatic controls of eye function exist in addition to the intentional initiation of conscious movements. The eye makes rapid movements to maintain the desired image on the foveal retina (saccades), movements in response to changing head angle and position, and it rotates to produce the convergence necessary for near vision. Secondly, there is control of lens curvature (accommodation). Accommodation seems to oscillate slightly with a frequency around once per second: this may provide the information necessary for initiating correction signals. Thirdly, the iris of the eye adapts to light intensity: in bright light the pupil may be

Endolymph : high conc K+
low conc Na+

only 1.5 mm in diameter, whereas in dim light or in the dark it is about 8 mm in diameter. This is a parasympathetic nervous system reflex.

## Hearing

The organs of hearing are the ears. Even when sound is transmitted other than by the tympanic membrane – directly through bone, or when the eardrum is perforated – the actual transduction of sound still occurs in the inner ear.

### The auditory apparatus

The auditory apparatus consists of three chambers, two filled with air, and one with fluid (Fig. 9.28). The external auditory meatus at the bottom of the funnel of the external ear leads to the external auditory canal in the temporal bone. The canal is closed by the tympanic membrane, a conical membrane pointing downwards into the middle ear. The middle ear (the second air-filled chamber, Fig. 9.29) has four openings: one closed by the tympanic membrane, two facing into the coiled chamber of the inner ear, closed respectively by the foot of the stapes (the oval window) and by a membrane (the round window), and a fourth in the Eustachian tube which communicates with the nasopharynx. This is normally closed, but when pressure is raised in the middle ear above the external pressure – as during swallowing, yawning, or even during mastication – it opens to permit pressure equalisation. During rapid ascent or descent in an aircraft swallowing may be necessary to permit pressure equalisation and relieve discomfort. The inner ear includes the cochlea as the organ of hearing, and the utricle, saccule and semicircular canals as a composite organ of balance. The cochlea is a complex spiral chamber, 35 mm long, which is effectively three tubes coiled side by side, formed into a spiral of almost three

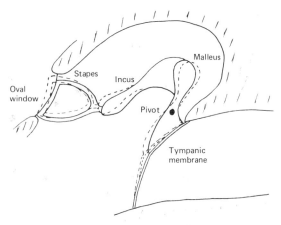

**Figure 9.29** The middle ear and the auditory ossicles. The dotted lines show the positions adopted as the tympanic membrane is forced inward by the pressure of the sound waves.

complete turns. The tubes are the scala vestibuli, the scala tympani, and the scala media (Fig. 9.30). The scala vestibuli opens from the oval window and is connected at the end of the spiral to the scala tympani by a very small opening, the helicotrema. The scala tympani extends back down the spiral to end at the round window. These two tubes contain perilymph, a fluid of similar composition to cerebrospinal fluid. The third tube is the scala media, separated from the scala vestibuli by the thin Reissner's membrane and from the scala tympani by the much thicker basilar membrane which contains blood vessels and supports the organ of Corti. The fluid in the scala media is called endolymph and is remarkable among extracellular body fluids in having a high concentration of potassium ions (around 145 mmol/l) and a low concentration of

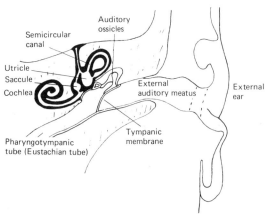

**Figure 9.28** Section through the ear.

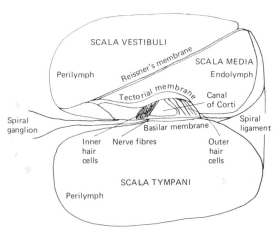

**Figure 9.30** The inner ear, showing the tubes that make up the cochlea: the scala vestibuli, the scala media, and the scala tympani.

sodium ions (16 mmol/l). The membrane of Reissner is thin enough to transmit sound waves to this fluid and yet serves as a barrier to maintain its composition. Endolymph also fills the utricle, saccule and semicircular canals. The organ of Corti extends the length of the basilar membrane up the spiral of the cochlea. The bony core of the spiral contains the spiral ganglion of the auditory division of the acoustic nerve. From there fibres pass into the basilar membrane and terminate in a mesh around the hair cells. The hair cells, one internal row of about 3500, and an external three or four rows of about 20 000 outer hair cells, send up their processes through a stiff reticular lamina to touch or penetrate the surface gel of the thin flexible tectorial membrane (Fig. 9.31).

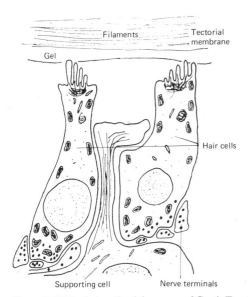

**Figure 9.31** The hair cells of the organ of Corti. Two hair cells, separated by a supporting cell, extend up into the tectorial membrane. Nerve terminals come into close apposition with the bases of the cells.

## Physical characteristics of sound

Sound waves are regular rhythmical movements of the molecules of air set in motion by a mechanical vibration. They are characterised by frequency, or pitch, and amplitude, or loudness. The ear in a young person can detect frequencies between 20 and 20 000 Hz, but with age the high frequency response is progressively lost and the limit may come down as far as 8000 Hz. The ear is most sensitive between 1000 and 4000 Hz, but the sensitivity to a particular frequency varies to some extent with the sound intensity. The scale of sound intensity is the decibel scale, a relative scale with a logarithmic range. The sound intensity in bels is defined as twice the logarithm of the ratio of the sound pressure to that of a standard sound just at the auditory threshold. The human ear can hear sounds from 0 to 140 decibels (db), at which intensity damage to the organ of Corti begins to occur. Since the scale is logarithmic this represents a variation in sound intensity of $10^{14}$. Normal conversation has an average pitch of 120 Hz in the male, 250 Hz in the female, and an intensity of around 60 db.

## Sound transmission in the ear

When a sound wave strikes the tympanum, it causes it to vibrate. The actual movement of the tympanum is less than 100 nm, the extent of the movement varying with the intensity of the sound. This movement is transferred to the foot of the stapes in the oval window by the mechanical linkage of the bony ossicle, malleus, attached to the tympanum rotating and moving the incus which drives the stapes back and forth. The foot of the stapes is attached to the oval window by an annular ligament (Fig. 9.32). The lever system decreases the movement but increases, therefore, the force by about 1.3 times. As the foot of the stapes is only one seventeenth the area of the tympanum, the pressure exerted by the stapes on the cochlear fluid is some 22 times as great as that of the original sound wave. This is necessary to match the differing impedance on the two sides of the tympanum. Transmission of sound in this way is called ossicular conduction. The tympanum is damped so that vibration stops as soon as the actual sound ceases. Sound waves can drive the vibrations of the perilymph directly but in the absence of the tympanum and the ossicles the perceived sound is reduced by about 30 db. The ear can attenuate loud sounds by contraction of the stapedius and tensor tympani muscles to hold the ossicles more rigid and therefore reduce sound transmission. This attenuation is effective up to 1000 Hz and can prevent damage to the cochlea. The vibration induced in the perilymph of the scala vestibuli is transmitted through Reissner's membrane to the endolymph, where it causes distortion of the basilar membrane into the scala tympani. Some sounds can induce vibration in the bones of the skull and these in turn can initiate vibration in the perilymph – a process termed bone conduction. The basilar membrane contains fibres attached to the modiolus which become progressively longer and thinner from the base of the cochlea up to the helicotrema. All the fibres are stiff but because of the changes in shape those at the base are about 100 times stiffer than those at the helicotrema. The fibres at the base of the cochlea therefore vibrate preferentially at high frequencies and those at the apex at low frequencies. The amount of fluid displaced at different levels also favours this

152  *The nervous system*

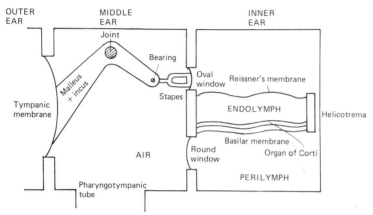

**Figure 9.32** Diagram of the components of the sound transmission system of the ear. The malleus and incus are represented as a single unit pivoting about the joint.

gradient of sensitivity to different frequencies. Thus different sound frequencies cause vibrations at different points on the basilar membrane. Vibration of the basilar membrane causes the cilia of the hair cells to be deflected, and the cilia are therefore bent against the tectorial membrane. This causes alternating changes of potential across them and these receptor potentials stimulate the axons of the cochlear nerve in contact with the hair cells. The receptor potential probably functions as a generator potential to produce action potentials in the nerve (*see* p. 83). Frequency of the waveform is signalled by which cells are stimulated, those close to the base of the cochlea being stimulated by the high frequencies, and those at the apex of the spiral by low frequencies. Intensity of the sound is coded by the rate of action potential generation. The signals in the cochlear nerve are directly related to the sound causing them and the recording of the potentials by an electrode on or near the cochlea yields the cochlea microphonic, an electrical signal which can be amplified to reproduce the original sound. Increasing intensity of a sound stimulates increasing numbers of nerve cells because of the greater displacement of the basilar membrane.

### Auditory pathways

The impulses pass to the cochlear nuclei in the upper part of the medulla oblongata (Fig. 9.33). The pathways from here to the cortex are complex and fibres pass to both the ipsilateral and the contralateral sides. Secondary neurones pass to the superior olives, to the inferior colliculi and to the medial geniculate bodies of the thalamus. In addition to the pathway via the lateral lemniscus there is also a reticular pathway. The tonotopic representation (or frequency pattern) of the organ of Corti is maintained right through to the cortex, where the postero-medial part of the primary auditory cortex in the temporal lobe responds to high frequencies and the antero-lateral part to low frequencies (Fig. 9.34). Cortical cells monitor not only frequency and loudness, but also direction, duration, onset, and repetition of the sound signal. As in the visual cortex there is considerable processing of the original nerve signals.

Sensory systems such as that for hearing are often considered simply as one-way pathways – like one-way telephone lines. It is now clear that there are modifying influences on transmission at many levels. The ossicular attenuating mechanism has already been mentioned. In addition to this, there are outputs in the olivocochlear bundle whose effect is to decrease the effect of auditory stimuli by hyperpolarisation of the terminals of the cochlear nerves. Similarly, there is inhibition and interaction at all levels from the cochlear nuclei upwards. Most sensory inputs pass only to the contralateral side of the brain: the auditory system is unusual in that inputs from both ears reach both halves of the brain; indeed, one cortical neurone may receive signals from both ears. This may be of value in locating the source of a sound, although human beings are less effective in identifying the source of a sound than in identifying position visually.

## Position sense

Kinaesthesia, or the sensation of the spatial position of parts of the body in relation to each other and to the external environment, may be divided into several categories. A simple division is into exteroreceptor sensations and interoreceptor sensations.

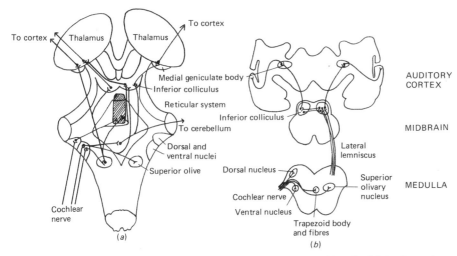

**Figure 9.33** The auditory pathways: auditory input is shown from one side only. (*a*) A schematic view of the dorsal surface of the brainstem; (*b*) sections through the medulla, midbrain, and auditory cortex.

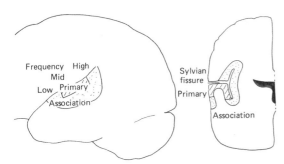

**Figure 9.34** The auditory areas of the cortex. Areas of primary sensation are shaded, the association areas stippled. The localisation of pitch (frequency) is indicated. (*a*) Lateral view of the cortex; (*b*) coronal section through the brain hemisphere in the region of the lateral (Sylvian) fissure to show the cortex in the infolding of the fissure.

## Exteroreceptors

A number of sensory organs contribute information to the cortex which identifies the position of the body in space. Part of the function of the eyes is to relate the head position to other objects. Closing the eyes, or presenting misleading visual information, can disorientate a subject. Touch and pressure sensations (which are considered in detail later) are also important. Touch and pressure are signalled from whatever part of the body is in contact with the ground, or with other objects such as walls which might act as useful references for body position. Suddenly lowering the back of a dental chair removes references of this kind and, together with muscle and joint impulses, can disorient the patient and produce a momentary panic.

## The vestibular apparatus

The ear is important in position sense since as well as the auditory sensors it contains an organ responding to gravitational forces and an organ responding to head movement. The stimulating forces are actually acceleration and change of direction. The part of the inner ear involved in hearing has already been described. The remainder of each inner ear consists of the saccule and the utricle from which open the three semicircular canals, arranged so that in the standing subject one canal is horizontal and the other two vertical but in planes at right angles to each other (Fig. 9.35). The bony canals are membrane-lined and the vestibular apparatus, as the structures are collectively called, is filled with endolymph. There are sensory cells in the expanded ends of the semicircular canals, the ampullae, which open to the utricle and there are sensory cells in the utricle and the saccule. Those in the ampullae are the cristae ampullaris, each a ridge covered with cells sending up tufts of hairs, or cilia, into a gelatinous mass, the cupola (Fig. 9.36). A sudden rotation of the head, or the whole body, in any direction causes a differential force across the hair cell processes as the inertia of the fluid keeps it moving more slowly than the moving environment. The cristae bend because their bases are moving while their tips are held back. This stimulates the haircells which in turn stimulate the fibres of the vestibular nerve. Once the inertia of the fluid in the semicircular canal is overcome the forces on the cristae are equalised and they straighten, halting the stimulation of the hair cells. On ceasing movement of the head or body the reverse process occurs: the endolymph continues in motion whilst the tissues

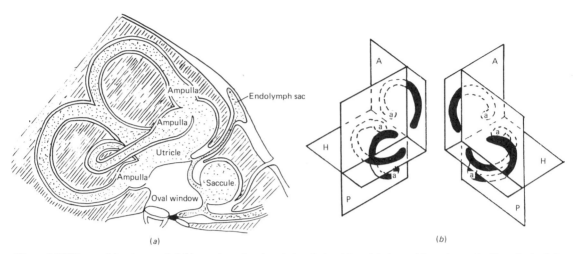

**Figure 9.35** The semicircular canals (*a*) in section, showing their relationships with the utricle and saccule. Bone is shaded, endolymph stippled; (*b*) the planes of the head with the canals superimposed. A anterior plane, H horizontal, and P posterior. The ampulla of each canal is labelled a.

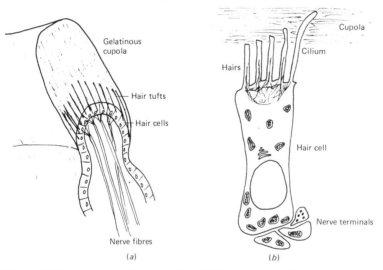

**Figure 9.36** The crista ampullaris in section. (*a*) The structure of the crista ampullaris; (*b*) enlarged view of a single hair cell to show its relation to the nerve fibres and the gelatinous cupola

have stopped, so the cristae are bent the other way. By this means the semicircular canals transmit signals related to change of motion in any one plane. There are structures similar in character and function to the cristae on the floor of the utricle and on the wall of the saccule, where it is tilted 30° away from the vertical. These are the otolithic organ, (or macula of the utricle) and the macula of the saccule (Fig. 9.37). These two organs differ from the cristae in that the gelatinous layer has embedded on its surface crystals of calcium carbonate, or aragonite. These are the otoliths. In the body aragonite crystals are found only in the inner ear; all other calcified tissues are apatite or eventually transform into apatite. The uniqueness of this crystalline form may be due to the unusual composition of the endolymph. The otoliths are the most dense structures in the labyrinth of the inner ear and therefore have greater inertia even than the endolymph: any movement causes their relative displacement. Further, since gravitational forces are sufficient to cause some displacement of the hair cells, the otolith organs signal the direction of gravitational pull as well as the movement of the head or whole body.

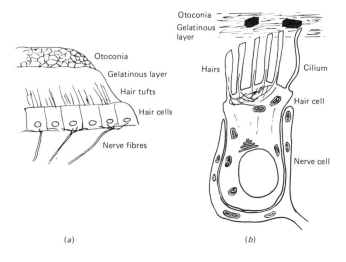

**Figure 9.37** The macula of the utricle. Note the otoliths, crystals of calcium carbonate, in the gelatinous layer. (*a*) The structure of the macula of the utricle, (*b*) enlarged view of a single hair cell to show its relation to a nerve fibre and the gelatinous layer.

The fibres of the vestibular nerve have their cell bodies in the vestibular ganglion and pass from there to the vestibular nuclei. From the vestibular nucleus at the junction of pons and medulla, fibres pass to the cerebellum, to the reticular nuclei in the brainstem, to the cortex and also down to the spinal cord (Fig. 9.38). In addition there are connections with several medullary centres and with the motor nerves to the eye muscles. The functions of these organs and the relevance of their input pathways are described in Chap. 14, in relation to the control of posture and movement.

The stimulus for semicircular canal nerve endings is the relative movement of fluid, so hot or cold water in the external auditory meatus may stimulate the vestibular nerve because convection currents are set up in the endolymph.

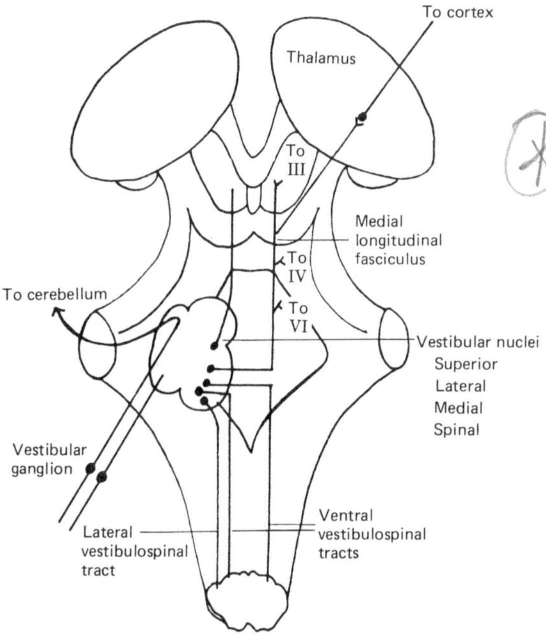

### Interoreceptors

At least three types of receptor give kinaesthetic information from the body interior. They detect the position of body structures by measuring stretch or pressure in muscles, tendons and joints.

### Muscle stretch receptors

Skeletal muscle contains specialised organs, called spindles because of their shape (Fig. 9.39*a*). They are surrounded by a sheath of connective tissue continuous with the connective tissue framework of the muscles. Within the sheath are modified muscle fibres with an innervation different from that of the rest of the muscle. Contraction of these intrafusal muscle fibres alters the length of the spindle relative to the surrounding extrafusal fibres. This allows adjustment of the sensitivity of the spindle to stretching by the surrounding muscle. Around the clear non-striated central portions of the muscle fibres are two types of nerve endings: the coils of the annulospiral endings and multiple terminals of the flowerspray endings. The annulospiral endings respond to stretching of the spindle by increasing their rate of discharge of action potentials – a dynamic response – although they also show a generation of action potentials proportional to their static length – a static response. These action potentials travel to the spinal cord in group Ia fibres.

**Figure 9.38** Pathways carrying position sense from the ear to the brain. The Roman numerals refer to the cranial nerves – III the oculomotor, IV the trochlear, and VI the abducent.

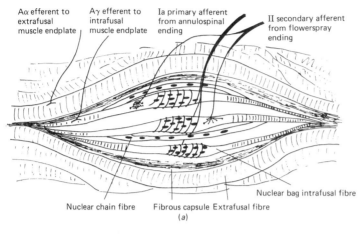

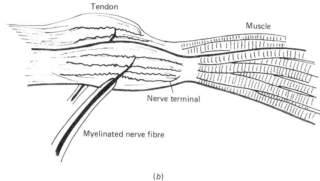

**Figure 9.39** Position sense receptors outside the central nervous system. (*a*) A muscle stretch receptor (the muscle spindle); (*b*) a Golgi tendon organ.

These then synapse with motor neurones to the same muscle and also transmit information onward in the ventral and dorsal spinocerebellar tracts. From the cerebellum information passes to thalamus and cortex. In addition to the annulospiral sensors with their large, rapidly transmitting, nerve fibres, there are slower signals from the flowerspray organs in group II fibres. These endings respond directly to the length of the muscle spindle: they are static receptors. Signals from these are also transmitted in the spinocerebellar tracts.

The muscle spindles are thought to play a role in the maintenance of posture (Chap. 15), and to assist in the control of muscle contraction by matching the degree of the contraction achieved to that intended by the brain in setting up the pattern of muscle activity.

### Golgi tendon organs

Golgi tendon organs are ramifications in the tendons of the nerve endings of large sensory group Ib fibres (Fig. 9.39*b*). The fibrils of the nerve terminal are each in series with some 12–15 muscle fibres inserted into the tendon. Stretch of the tendon (by, for example, contraction of the muscle fibres) causes stimulation of the nerve – so these organs are sensitive to tension in a muscle, in contrast to the muscle spindles which are sensitive to length. The secondary fibres in the spinal cord are mainly the myelinated fibres of the dorsal columns; fibres from the nuclei of the dorsal columns pass both to the cerebellum and the thalamus. Other fibres may travel in spinocerebellar tracts.

### Joint and ligament receptors

Three types of receptors are found in the capsules of joints or in ligaments. Some resemble Golgi tendon organs and are found in both sites. They respond both to steady stretch and to change in tension. Ruffini's end organs, which resemble the flowerspray endings of the muscle spindle but behave like Golgi tendon organs (although less sensitive), occur mainly in the joint capsule. In the tissues around the joint there are not infrequently Pacinian corpuscles responding to the pressures due to joint movement. The axons which pass from all these terminals are group II fibres. They pass in the dorsal columns to the dorsal column nuclei and from there to the cerebellum and the thalamus.

## Sensory pathways

Although most proprioceptive information passes to the cerebellum in the spinocerebellar pathways, some information reaches the cortex and consciousness via the dorsal columns. There is also evidence that information from the muscle spindles in the upper limbs travels in the dorsal columns, and that from the lower limbs in the dorsolateral columns. The cortical cells probably receive integrated signals conveying a pattern of information rather than a point by point consciousness. Neurones may respond to particular movements just as other parts of the sensory cortex respond to particular combinations of stimuli.

## Cutaneous sensations

The skin sensations of touch, pressure and temperature are conventionally grouped as the cutaneous sensations. Pain, which is also normally considered with this group will be considered separately later because of its importance in dentistry. There are a number of other cutaneous sensations, such as tickle and itch, which are less easy to define in terms of actual stimulus, and some, like two-point discrimination and the ability to identify objects by handling them (stereognosis), which are much more complex and probably result from the cortical processing of signals evoked from simpler sensations.

## Sensory receptors

Touch, pressure and vibration are tactile sensations, touch being sensed in the epidermis, pressure below it in the dermis and vibration including a time component. At least four touch receptors have been described – naked nerve endings, Meissner's corpuscles, expanded tip receptors such as Merkel's discs and the nerve terminals around the bases of body hairs (Fig. 9.40). The expanded tip receptors respond strongly at first but then settle to a steady rate of action potential generation as a touch stimulus continues. The encapsulated Meissner's corpuscles and the hair end organs are rapidly adapting. They respond quickly to even very light stimulation but adapt within a second and cease to signal. Free nerve endings are widely distributed

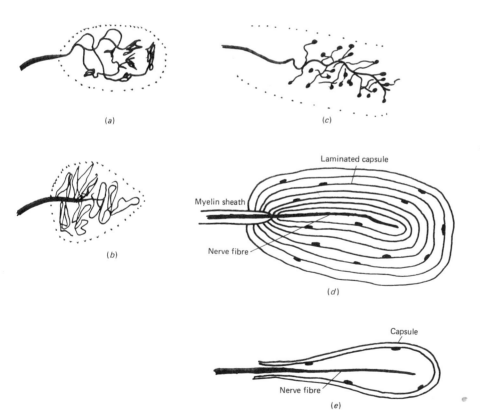

**Figure 9.40** Cutaneous sensory receptors. (*a*) Merkel's disc, (*b*) Ruffini's endorgan, (*c*) Meissner's corpuscle, (*d*) the encapsulated Pacinian corpuscle, (*e*) the encapsulated Krause's corpuscle.

and may act as sensors for any of the skin sensations including tickle and itch, which seem to result from repetitive low frequency stimulation of the free nerve endings of C fibres. A sensation of itching is evoked by release of polypeptide kinins in the skin, or by histamine. Expanded tip receptors are found in the finger tips and sensitive areas such as the lips. There are a few in hairy areas of the body: unlike Meissner's corpuscles which are found in sensitive areas but not in hairy parts of the body. Deeper in the tissues are additional end organs such as Ruffini endings and Pacinian corpuscles responding optimally to high frequency vibratory pressure or touch stimuli. All these produce generator potentials which are converted by the nerve into action potentials when they reach sufficient intensity. The specialised receptors transmit their signals to the spinal cord in A$\beta$ or sensory group II neurones. Free nerve endings and possibly hair end organs transmit in A$\delta$ or group III neurones although some nerves with free endings are C fibres (sensory group IV).

## Sensory pathways

The sensory nerves carrying impulses due to these sensations enter the spinal cord in the dorsal root, pass to the dorsal horn and either synapse there or carry on upwards in the dorsal columns to synapse in the nucleus gracilis and the nucleus cuneatus (Fig. 9.15). After the synapse secondary fibres cross the cord or the medulla. One possible transmitter is substance P; somatostatin has also been found. The interneurones arising in the dorsal horn may synapse with motor nerves at any level or with pain neurones in lamina II of the dorsal horn and modify the sensation perceived. Some degree of integration of sensory inputs occurs at the spinal level. It may be that touch and pressure signals travelling by different routes are perceived differently. The dorsal columns are associated with detailed localisation and also the sensation of vibration, the spinothalamic tracts with more general touch and pressure.

## Oral sensation

There have been few descriptions of touch and pressure endorgans in the mouth although the oral soft tissues are relatively sensitive. The teeth themselves lack these sensations but the periodontal ligaments have mechanoreceptors which respond to the deformation produced by movement of the teeth. This information is valuable in the control of mastication. The sensation is relatively poorly localised: it is not, for example, always easy to detect which teeth are forced apart by impacted particles of food without also using the touch receptors in the tongue.

## Temperature sensation

The other skin sensation is that of temperature. Although termed a skin sensation, it, like the others, is also perceived by some mucous membranes, notably in the mouth and the oesophagus. No specific endorgans for temperature have been described. Temperature sensitive fibres are of two types: those responding maximally to stimuli below normal skin temperature (cold receptors), and those responding maximally to temperatures above that (warm receptors). In the tongue, temperatures high enough to cause tissue damage have been reported to stimulate cold receptors in addition to those for warm; this is not so in the skin. Temperature sensations are probably induced by changes in metabolic activity around the nerve endings. The receptors adapt, but fairly slowly, so that temperature change is perceived more readily than maintained temperature. Specific spots can be identified on the skin which respond to hot or cold with rather more responding to cold. The nerve fibres are A$\delta$ (sensory group III) and C (group IV): they synapse in the dorsal horn and the secondary fibres travel in the spinothalamic tracts. Temperature sensation is not limited to the skin and some mucous membranes: the anterior hypothalamus contains receptors which respond to increases in blood temperature. Changes in temperature may also be sensed indirectly because of their general effects on metabolism and may even influence membrane potentials (the Nernst relationship is temperature-dependent).

## Taste

Taste, like olfaction, is a chemical sensation. It is only one component of what is more properly termed flavour – a combination of taste, smell, texture and other stimuli to oral receptors. These other components are discussed later.

### Taste receptors

The taste receptors are located in specialised groups called taste buds. A typical human subject has about 9000 taste buds, found mainly on the tongue but some are also present on the soft palate, the epiglottis, the pharyngeal wall and the upper third of the oesophagus (Fig. 9.41). The surface of the anterior two thirds of the tongue, in front of the sulcus terminalis, is covered by the filiform and foliate papillae. The total surface of all the filiform papillae together includes very few taste buds but the foliate papillae may have up to 5 per papilla – a total of around 1300. In addition to these there are some 8–12 circumvallate papillae, easily visible on the tongue surface along the V of the sulcus terminalis. In younger subjects (up to 20 years)

Sensation 159

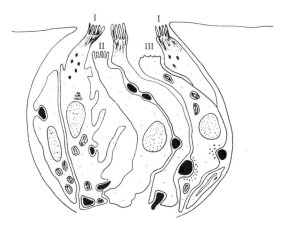

**Figure 9.42** The cells of a taste bud. The cell types indicated by numbers are described in the text. Nerve terminals are shown in black.

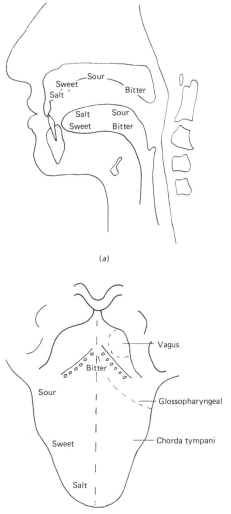

**Figure 9.41** The areas sensitive to taste sensations (*a*) in the oral region as a whole, (*b*) on the tongue itself.

250–300 taste buds are found in the walls of the trough around each circumvallate papilla but the number declines to about 100 over the age of 70. The loss of taste sensitivity with aging does not seem to correlate with the numbers of taste buds in these sites, possibly because these are sensitive only to particular tastes. Although denture wearers may complain of a loss of taste sensation when wearing an upper denture, it may be the sensation of texture rather than actual taste which is lost. Circumvallate papillae and taste buds develop early, reaching their adult form at around 14 weeks intra-uterine life.

Taste buds contain three main types of cell with their tips arranged around a pit or taste pore, filled with dense gel-like material (Fig. 9.42). Type I cells, found peripherally and also between other cells, contain apical granules; the necks of the cells are narrow and dense with fibrillar or tubular elements. Each cell has up to 40 short wide microvilli. A basement membrane lines the floor of the pit. These cells were termed sustentacular cells by the light microscopist. Type II cells have no apical granules, their microvilli are relatively thick and there are vesicles immediately below the microvilli. Type III cells have slender peg-like tips but contain vesicles closely similar to synaptic vesicles. They have synaptic junctions with nerve cells. A fourth type of cell, found in the base of the taste bud appears to be a germinative cell for other types. Nerve fibres enter the bud, lose their myelin sheaths and then appear either to enter Type I cells, coil round Type II cells or synapse with Type III cells. The significance of these different nerve:sensory cell interactions is unknown. All taste buds have the same general appearance. Although they appear as stable structures, in fact there is a fairly rapid cell turnover with cell lives ranging from 2 to 30 days. The taste hairs described in the gustatory pit by the light microscopist have not been identified by the electron microscopist.

### Sensory mechanisms

Sapid (tasty) substances dissolve in the oral fluids and pass into the pore to interact either with the pit substance or with the membranes of the receptor cells themselves. The pit substance stains with reagents specific for ascorbic acid (Vitamin C), acid phosphatase, esterases and ATPase. The sapid molecules may inhibit or potentiate the enzymes or they may bind to the cell membranes. Whatever the mechanism, cellular activity results in generator potentials, which depolarise the nerve terminals, either directly or via the synapse-like areas of Type III cells, and generate action potentials. The frequency of action potentials in gustatory nerve fibres is proportional to the logarithm of the molar concentration of the stimulatory chemical.

### Sensory pathways

Sensory fibres from the anterior two thirds of the tongue travel in the chorda tympani part of the lingual nerve and then in the facial nerve, those from the posterior third in the glossopharyngeal nerve, and those from all other areas in the vagus nerve. The fibres are small, myelinated and slowly conducting. In the medulla oblongata they travel in the tractus solitarius to the nucleus solitarius where they synapse and the secondary fibres cross the midline to travel to the thalamus in the medial lemniscus. From the thalamus neurones pass to the foot of the postcentral gyrus – the taste cortex being a part of the facial cutaneous sensation area. This is only a simple outline of the pathways: there is mixing of inputs at an early stage, there are connections in the pons with salivatory centres, and some part of the input passes to the limbic system probably via the hypothalamus.

### Primary taste stimuli

In 1916 Henning proposed that there were four distinct tastes – sweet, bitter, salt and sour (or acid). He did not define these as primary tastes but rather as extremes of a continuous spectrum, the taste tetrahedron. Subsequent studies have assumed four primary tastes, although sometimes others, such as metallic or alkaline, have been added to the list. Primitive peoples describing taste sensations use terms which are not comparable with these four. Taste is important to the dentist for two reasons. Firstly, taste is a direct stimulus to salivary secretion. Secondly, perhaps more significantly, the taste sensation of sweetness, which most people find pleasing, is usually associated with sucrose, the dietary component most clearly linked with dental decay. Taste specificity in different parts of the mouth has been examined by stimulating as nearly as possible a single taste quality, using sucrose for sweetness, quinine for bitterness, sodium chloride for saltiness and citric or tartaric acid for sourness. Fig. 9.41 shows how sensitivity to different taste qualities varies in the oropharynx generally although individual subjects may differ in their responses. At a neurophysiological level the situation is confusing. Any one taste bud is supplied by some 50 nerve fibres and each fibre may be linked to as many as five taste buds. Cutting a nerve causes degeneration of the taste buds from which it is stimulated. Recording from nerve fibres reveals that taste buds may respond only to one of the taste stimuli, or to any number of the four. It is generally true, however, that each bud is particularly sensitive to one primary taste. In some animals recordings from taste nerves seem to identify distilled water as a specific taste stimulus.

The sensations of taste are markedly affected by temperature: flavourings in ice cream, for example, have to be much stronger than those in warm foods, and very hot foods appear to lose their taste. Electrical stimulation of the tongue gives rise only to a metallic sensation or occasionally to a bitter one.

The threshold at which the four recognised tastes are first perceived varies: sucrose 10 mmol/l (range 5–16 mmol/l), sodium chloride 10 mmol/l (range 1–80 mmol/l), hydrochloric acid 0.9 mmol/l (range 0.05–10 mmol/l) and quinine 8 µmol/l (range 0.4–11 µmol/l).

### The relation of taste to chemical structure

On the assumption that the four recognised tastes are fundamental chemical stimuli, attempts have been made to relate taste sensation to chemical structure and to identify the chemical receptors. With salt and acid stimuli the cation and the hydrogen ion are thought to be the determining factors in taste. However, the anion clearly influences the sensation both in quality and intensity: inorganic acids tend to taste metallic whilst organic acids show varying degrees of fruitiness. In general the perceived acidity relates to hydrogen ion concentration but this is not always true: some organic acids seem more sour than hydrochloric at similar hydrogen ion concentrations. Not all salts taste salty and several in low concentrations taste sweet. The receptor substance for acid and salt sensations is thought to have carboxylic groups; binding of the stimulant to these groups may cause the cell membrane to shrink or swell.

With bitter and sweet tasting substances no common chemical structures have been found. A sweet taste is perceived with most sugars, some aminoacids (glycine, glutamine), halogenated carbon compounds (chloroform), several miscellaneous organic compounds such as saccharin and cyclamates, some salts of beryllium, lead acetate, potassium and lithium chlorides and sulphates in low concentrations. The idea that sweet substances are on the whole nutritionally good and bitter ones poisonous is contradicted by the sweet taste of lead acetate and beryllium compounds. Substances which taste bitter include many alkaloids, glycosides, picric acid, compounds with $NO_2^-$ or S groups, magnesium sulphate and calcium oxide.

The currently fashionable theory of sweet taste sensation suggests that if substances are to taste sweet, they should have a covalent bond to hydrogen ions and that they should be able to bond loosely (either ionically or covalently) to a second grouping. An end group of the form A–H.....B can then bind to a receptor site having a mirror image structure to this: B....H–A. Bitter substances are thought to be similar but to have a shorter bond distance across the end group. Unfortunately this chemical theory will not account for the sweetness of all the substances mentioned above.

When tastes are combined the sensations often change. Addition of sugar as a sweetener to acid foods suppresses the sensation of acid without changing the hydrogen ion concentration.

### Flavour

As mentioned above, taste is only a part of flavour. When a patient complains of a taste, the complaint may relate to texture rather than actual taste. Impression materials are often disliked for this reason. Other sensations also contribute to flavour. Vanillin and citral (a component of oil of lemon) are tasted only if nasal receptors can be stimulated as well as oral. Smelling an onion is said to alter the taste of an apple eaten simultaneously. Wines are tasted by allowing some constituents to volatilise in the mouth. Mustard, pepper, chillis, ginger and such 'hot' flavours arise by stimulation of pain receptors in the mouth. Afferents from these travel in the trigeminal nerve. The flavour sensation from food is a complex of many sensations, integrated in the limbic system and transformed by association and emotional connotations into the final perception.

### Effect of drugs on taste perception

A number of external influences can affect taste sensation. Local anaesthesia suppresses taste in the order bitter, salt, sour, sweet, although the effects on the palate are more marked for sour and bitter, and on the tongue for salt and sweet. There may be differences between anaesthetics. Gymnemic acid selectively depresses the sensation of sweetness whilst miraculin, from the tropical fruit *Synsepalum dulcificiens*, converts the perceived sensation to sweet even when the stimulus is pure acid or salt. Some substances are tasted differently by subjects of different genetic background. Thus inability to taste phenyl thiocarbamide (PTC) is an inherited recessive trait in 30% of the Caucasian population. A similar effect is seen with 6-n-propyl uracil (PROP).

### Other factors affecting taste perception

There is a decline in taste sensitivity with aging but individual variation is so great that this is a trend rather than a statistically valid relationship. The effect of hormones on taste sensation is also debatable: differences between males and females are also within the range of individual variation. Smoking cigarettes is said to affect taste, particularly in relation to bitter, although the data are not very consistent and clear differences have only been seen in subjects smoking over 20 cigarettes a day. Salt deprivation causes a decreased threshold for salt, and water deprivation may increase it. At least part of this effect may be attributable to changes in salivary composition. Bathing the tongue in sodium-free water lowers the salt threshold as much as a hundred-fold.

Finally, although sweet was one of Henning's original four tastes, some authors have suggested that it is an acquired taste which is absent, or at least not present as a preference, in infants. Individuals who reduce their sucrose intake without using other sweeteners develop much lower thresholds for sweetness and find high concentrations unpleasant.

## Olfaction

### Olfactory receptors and sensory pathways

The roof of the nose, near the nasal septum, carries a small patch of yellowish olfactory epithelium, totalling about $5\,cm^2$ of surface (Fig. 9.43). The

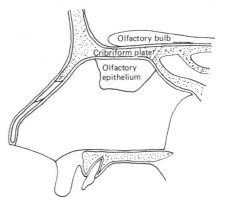

**Figure 9.43** The position of the olfactory epithelium. Bone is shown stippled.

mucosa consists of two types of cell, the sustentacular (supporting) cells which secrete mucus and the olfactory receptors, which are the actual neurones (Fig. 9.44). The nerve cells send out a short thick dendrite, the olfactory rod, with cilia projecting into the layer of mucus on the surface. The supporting cells also have microvilli which project into the mucous film. The axons of the receptor cells pass through the cribriform plate of the ethmoid bone in the roof of the nose to enter the olfactory bulbs (Fig. 9.45). The bulbs contain a complex network of synaptic connections. Near the ethmoid plate are the glomeruli, structures in each of which some 25 000 olfactory cells synapse with some 25 mitral cells from the deeper layers. In the next layer the mitral cells synapse with deeper cells – the granular cells. Two other cell types have been described in

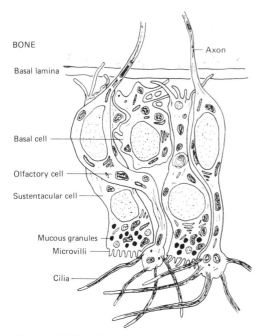

**Figure 9.44** The olfactory epithelial cells.

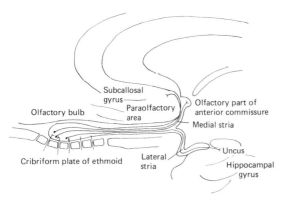

**Figure 9.45** Pathways of olfactory sensation.

the glomeruli: the tufted cells and the periglomerular cells. The axons of the mitral cells pass in the olfactory striae to the olfactory areas. These include the medial olfactory area with nuclei above and anterior to the hypothalamus, and the lateral olfactory area which includes the uncus and amygdala. These last cortical areas are sometimes called the primary olfactory cortex. The olfactory pathway differs from all other sensory systems in having no thalamic relay on the way to the cortex. This part of the brain is the rhinencephalon, in animals the smell brain, but in man is associated with emotions, instincts, and neuroregulation – the limbic brain.

## Transduction of odour sensations and the sensitivity of olfaction

The mechanism by which smell is transduced is unknown. Three theories have been advanced: that odorous molecules affect the activities of enzymes on the surface of the cells or fit into specific receptor sites on the membranes, that they penetrate the receptor cell membranes so as to change the cell membrane permeability, or that their molecular resonances match with those of resonant molecules (such as the yellowish pigment of the olfactory epithelium). The only characteristic common to odorous substances is that they are soluble both in water and in lipids. They are usually organic, with as few as three, or as many as 20, carbon atoms. How many odours can be distinguished is debatable: perhaps several thousand. Very few correlations of smell with chemical structure have been found. One classification of odours was into seven primary odours – camphoraceous, musky, floral, peppermint-like, ethereal, pungent and putrid – although the last two were later eliminated. Other odours are not classifiable in terms of the primary groups. Within this classification there is some correlation of odour with molecular shape – particularly in the musks – and this would support the idea of receptor pits. Proponents of the second theory dismiss this classification and point to relationships between the cell potentials and the lipid solubility and cross sectional area of odorous molecules. There is also evidence that the far infrared spectrum (a measure of molecular vibration) of odorous substances is related to their effect on the receptor cells. The difficulty of classifying odours is similar to the difficulty of classifying tastes: there may be no simple explanation of the mode of action of the primitive chemical sensors.

Smell shows adaptation to specific odours, partly because of receptor adaptation, but probably also by central inhibition. Impulses arising from the sensation are subject to inhibitory inputs at a number of points along the pathway. At least three types of fibre from elsewhere are found in the olfactory bulb bringing signals from the diagonal band, the ipsilateral olfactory nucleus and the contralateral olfactory nucleus. The absolute sensitivity of the olfactory system is remarkably high, the nose being able to detect a few hundred pg of methyl mercaptan in a litre of air. Musk, an odour with possible sexual significance secreted by animal glands similar to the apocrine sweat glands, is detected by the human at around one part in ten thousand million in air. Women are usually more sensitive to odours than are men. The minimum detectable difference with smell is a 30% increase in concentration. This is similar in magnitude to the minimum detectable difference in taste or in auditory stimuli; but very much greater than that in light intensity.

There is probably no such thing as a deodorant, a substance capable of preventing the nasal mucosa responding to other odours, apart from anaesthetic substances, or substances damaging the olfactory mucosa. Commercial deodorants are usually strong odours which mask any weaker ones – there is no evidence that one smell can neutralise another. The olfactory epithelium, despite extensive convergence of nerve fibres, seems to separate smells more readily than the gustatory epithelium can separate tastes.

## Pain

No morphologically distinguishable sensory organs responding to pain have been identified. The naked nerve ending which acts as a pain receptor cannot be distinguished from other naked nerve endings responding to other sensations such as warmth, cold, or touch. Structures such as the dental pulp or the cornea of the eye, which have naked nerve endings only, are now recognised to respond identifiably to stimuli other than pain. It is in fact very difficult to define a pain stimulus. Some have identified pain as excessive stimulation of other sensations but it seems most likely that pain is a recognition of tissue damage. Since pain includes an emotional response which is not quantifiable in terms of stimulus, most experiments on animals give only a limited amount of information on pain in man. The actual stimulus to pain receptors may be the products of cell damage such as hydrogen ions, potassium ions, polypeptides, histamine or 5-hydroxytryptamine. Certainly all these can produce a sensation of pain. One effect of aspirin on pain sensation may be a blocking of the chemical receptors of the nerve endings.

### Pain sensation pathways

The nerves involved in transmitting pain sensation are of two kinds, the myelinated Aδ nerves around 2.5 μm diameter, conducting at 12–30 m/sec and the non-myelinated C fibres, 0.4–1.2 μm in diameter, conducting at only 0.5–2 m/s. Peripheral endings of the Aδ fibres are individually either mechanoreceptors or nociceptors whilst the C fibres in man are apparently purely nociceptive. Both types of neurone may release substance P as a synaptic transmitter. Traditionally it has been held that a painful stimulus causes first a sharp pain, accurately localised, followed by a duller diffuse pain and that these two phases of pain sensation are due to the Aδ and C fibres with their differing speeds of conduction and differing pathways of transmission to the brain.

However, identification of neurones in the dorsal horn of the spinal cord suggests that roles are much less well defined. The dorsal horn can be divided histologically and functionally into eight layers with the most superficial designated I and the deepest VIII (Fig. 9.46). In layer I are the endings of axons from both Aδ and C fibres. These fibres respond to very gross mechanical stimuli but are otherwise almost entirely devoted to pain sensation and have small receptive fields. Large diameter fibres from mechanoreceptors also terminate here. Layer II, the substantia gelatinosa, has Aδ fibre endings but not

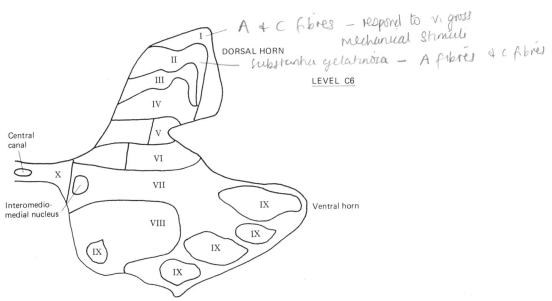

**Figure 9.46** The numbering of the laminae of the grey matter of the spinal cord. Only the grey matter is shown.

C fibre endings, their specificity being similar to that of those in layer I. In layers IV, V, and VI are the terminals of nerve fibres with pain function which may also respond to mechanical or thermal stimuli. These include both Aδ and C fibres but in both groups the receptive fields are relatively large. There is considerable convergence of inputs in this layer, again limiting localisation of sensation. Whilst single pulses of heat in the area served by these neurones are painful, repeated heat pulses may inhibit pain sensations via these Aδ neurones. Interneurones, many of them inhibitory, interconnect between layers. Some of the cells of layer IV may be the transmission cells which carry the signals resulting from integration of three different inputs: pain inputs, inputs from Aβ fibres resulting from other sensations, and inhibitory inputs from nerve fibres which descend from higher regions of the central nervous system. These cells would constitute a dorsal horn gating system by which stimulation of touch or pressure endings, or central inhibitory impulses, might modify transmission of pain sensations, inhibiting them completely, partially, or not at all. Pain fibres do connect with cells in the substantia gelatinosa; and some further inhibition of the transmission of pain sensations may occur at this level by encephalin-containing neurones in the substantia gelatinosa which terminate presynaptically on the primary afferent neurones. Release of encephalin reduces or prevents the release of substance P, a possible transmitter in pain pathways, thus suppressing further transmission of the pain sensation.

From the dorsal horn several pathways carry pain information up the cord. The main and most rapid pathway is in the lateral spinothalamic tract. This pathway is sometimes surgically interrupted to relieve persistent pain. This results in relatively little loss of other sensations apart from those of temperature since even the touch sensations in the anterior spinothalamic tract, which may be inadvertently damaged, are supplemented by the dorsal column pathways. The lateral spinothalamic tract carries pain information to the thalamus from neurones in layer I and also from the less specific neurones of layers IV and V. Perception of pain seems to be largely sub-cortical: stimulation of the sensory cortex rarely gives a sensation of localised pain. There is some somatotopic organisation of neurones associated with pain sensation in the dorsal horn, in the spinothalamic tract and in the thalamus. The second major ascending route is in the spinoreticular tracts, also in the anterolateral quadrant of the cord. Most of the pain inputs to these tracts come from layer V rather than layer I. Information in this pathway passes to the medial thalamus and the hypothalamus. There are probably many interneurones in this route, which has extensive connections with the so-called proprio-spinal system, a mesh of interconnecting neurones within the spinal cord itself. The spinothalamic pathway is concerned more with direct sensation and localisation of pain, the slower spinoreticular pathway with its emotional and motivational aspects. A third spinal tract possibly involved in pain sensation is the dorsal column. Antidromic stimulation of this tract has been used to stimulate cells in the dorsal horn and suppress transmission in pain pathways and it has also been surgically sectioned to relieve some aspects of phantom limb pain (pain experienced apparently from limbs which have previously been amputated).

### Endorphins and encephalins

The identification, first of morphine receptors, and then of endogenous substances capable of binding to them, led to the discovery in the central nervous system of specific transmitter substances whose main effect seems to be to suppress responses to pain inputs. Two classes of substance have been identified: the encephalins, which are pentapeptides, and the longer chain endorphins, – α endorphin with 15 aminoacid residues (including a sequence identical with met-encephalin) and β with the same 15 plus a further 16. All these compounds have a very short life in the body. They have been found in many parts of the brain including the cortex. β-Lipotrophin, a precursor of the endorphins, has been located in the anterior pituitary gland and possibly also in the brain. Increases in encephalin production have been observed in situations where pain is suppressed – in acupuncture, in hypnosis, in the very late stages of pregnancy and early in labour. At least part of this effect is mediated at the primary afferent terminal where cells in the substantia gelatinosa can release encephalins at terminals on the primary afferent pain fibres. This causes presynaptic inhibition of the release of substance P from these synapses, and is thought to prevent or inhibit further transmission of impulses related to pain sensation. Stimulation of cell bodies in the reticular system (the midline raphe nuclei) and in the mid-brain around the aqueduct, causes release of encephalins and inhibits pain sensation. The traditional counter-irritants – touch and pressure stimulation or application of heat to a painful area – may act by inhibiting pain transmission by a gating mechanism, or by stimulating encephalin production and release.

### Modification of pain sensation

Present theories envisage a much more complex idea of a patterned response rather than a simple transmission of impulses along a direct path from receptor to pain fibre to lateral spinothalamic tract to thalamus to cortex. Pain perception may be altered in the dorsal horn where the primary afferent enters, in the ascending pathways, and in the brain itself. The pain pathways to limbic structures are thought to be responsible for the 'affect', or emotional connotations, of painful sensations. Psychological factors are important in the perception and the perceived intensity of pain. Thus worried or frightened patients may have lower pain thresholds than relaxed or happy subjects. Soothing music or 'white sound' can produce a state of 'audioanalgesia'. Electrical stimulation of the skin, or of the dorsal columns, can produce analgesia (absence or reduction of pain sensation). Soldiers wounded in battle have reported feeling little or no pain during the fighting despite the severity of the wounds. All these variations in perceived pain are probably due to variations in the release of endorphins or encephalins within the brain itself.

## Deep pain and visceral pain

Pain is felt in deeper structures as well as in skin. Deep pain is characteristically less well localised than cutaneous pain and may persist longer. There are pain receptors in muscles which are probably stimulated by the release of $K^+$ or kinins from the contracting muscle cells: normally these substances are dispersed by the flow of blood, but if the blood supply is poor or if the muscle goes into spasm, this does not occur. Angina pectoris is a good example: if the blood supply to the heart is insufficient in periods of exercise a substernal pain is experienced. It is relieved by rest, or by the administration of vasodilator drugs. Temporomandibular joint pain is sometimes associated with bruxism, the nocturnal grinding of the teeth often involving spasm of the masseter muscles.

Visceral pain is even less well localised. The few pain receptors in the viscera are stimulated by stretch or by pressure, and in the gut the rhythmic contractions of the organs may give a rhythmic nature to the pain, which is then described as colicky. The afferent fibres from the viscera are sympathetic or parasympathetic.

Pain receptors in the inner ear and in teeth are stimulated by excessive pressure which probably causes tissue damage. Since the pressure is affected by local blood vessels, or is due to the increased blood supply in an inflamed area, the pain experienced is often throbbing or pulsatile.

## Referral of pain

Pain stimuli in deeper organs or viscera are often perceived as pain sensations on the body surface, the pain being 'referred' to a structure in the same embryological segment (dermatome) as the structure stimulated. Classically, pain from the heart is referred to the region of the left arm. For the dental surgeon, referral of pain from a tooth pulp presents considerable problems. Even within the mouth it is difficult to localise pain to a particular tooth; and although the sensation rarely crosses the midline, it may pass from upper to lower jaw and vice versa. Pain from the teeth may be experienced elsewhere in the facial region.

The mechanism by which pain is referred is still not clear. It is possible that somatic and visceral primary afferents may converge on the same neurone in the dorsal horn projecting to lateral spinothalamic tract. There are fewer spinothalamic fibres than there are primary neurones. Since pain is more commonly experienced via the somatic neurone, impulses in the lateral spinothalamic tract are interpreted centrally as somatic rather than visceral. Another possibility is that impulses reaching the dorsal horn in the visceral afferent may lower the threshold of spinothalamic fibres serving somatic sensation, making them responsive to much lower levels of pain stimulation. A third possibility is that the visceral nerve fibres in the dorsal horn may stimulate nearby somatic nerves antidromically, causing the release of amines or peptides at the sensory terminal (as in the 'triple response') and these substances then act as algesics or algogogues (pain producing substances).

## Emotional aspects of pain sensation

Pain, unlike any other sensory modality, has a verbal opposite which cannot be classified as a cutaneous or peripheral sensation. Pleasure is the opposite of pain. There are no receptors for pleasure, no neuronal pathways, no apparent cortical or thalamic representation. Other sensations may cause pleasure: genital stimulation, or taste, or even auditory stimuli. Electrical stimulation of the hypothalamus can localise a pleasure centre – but pleasure is an emotion, a limbic brain response. Pain, too, is an emotion and is affected by the limbic system. Indeed, painful stimuli may be interpreted as pleasurable: even if the extreme of overt masochism is ignored, there are still patients who will suck or wiggle a painful tooth and seem to derive some pleasure from the stabs of pain.

## Pain in the oral region

The trigeminal nerve behaves exactly like a spinal nerve, but the fibres enter the pons rather than the spinal cord. Here they turn downwards in the spinal tract of the trigeminal nerve, which extends down as far as the third or fourth cervical segment of the spinal cord (Fig. 9.47). This tract also receives sensory fibres from the glossopharyngeal, vagus, and facial nerves serving the facial region. The pain fibres pass medially from the trigeminal spinal tract

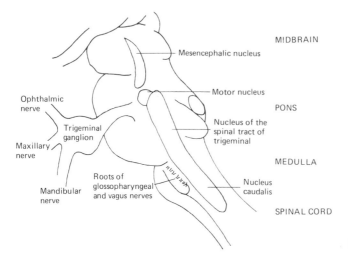

**Figure 9.47** The nuclei of the trigeminal nerve.

to the spinal nucleus with its three subdivisions of oralis, central (interpolaris) and caudalis. In this equivalent of the dorsal horn, the fibres are arranged with the ophthalmic, maxillary and mandibular branches lying progressively more dorsomedial. In the grey matter (Fig. 9.48), layer I contains the endings of mainly small non-myelinated C fibres, but also some Aδ and Aβ fibres. The C fibres are nociceptive, and are arranged somatotopically, particularly in the nucleus caudalis region. They synapse with neurones passing to the spinal tract and to the trigeminothalamic tract as well as with some which pass to the spinal cord. Layers II and III are equivalent to the substantia gelatinosa. Layer II receives inputs from large myelinated fibres and some primary pain fibres, but both layers, together with layer IV, contain many interneurones, both excitatory with connections to layer I and inhibitory with numerous axons to other layers.

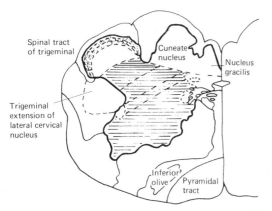

**Figure 9.48** The laminae of the grey matter of the medulla oblongata. The reticular formation is shaded horizontally, and the substantia gelatinosa diagonally.

Layers V to VIII are part of the reticular activating system, and again have many interneurones. Layer V receives fibres of Aβ nerves subserving touch and pressure sensations, Aδ fibres for pain, touch and skin temperature sensations, and nociceptive C fibres. There is considerable convergence, and many of these neurones have large receptive fields. Transmission cells with inputs from many different sensory nerve fibres integrate impulses from small, large and descending fibres to act as gate controls for pain transmission. All the fibres from this area project to the thalamus.

The concentrations of substance P observed in the spinal nucleus of the trigeminal nerve are some of the highest in the nervous system. The nerve cells are stimulated by substance P but inhibited by morphine, endorphins and encephalins. The drug naloxone, a morphine antagonist, allows the nerves to be stimulated even in the presence of these inhibitors. The nucleus caudalis contains substance P, encephalins, and morphine receptors; the application of these substances produces predictable effects. The nucleus oralis receives general sensory neurones and some nociceptive fibres; it is relatively unaffected by morphine. The substantia gelatinosa cells are markedly inhibited by morphine or the encephalins. As in the spinal cord, encephalin seems to produce its effects by presynaptic inhibition, preventing the release of P substance.

Electrical stimulation of the peri-aqueductal grey matter, or the nucleus raphe magnus, or the inferior central raphe nucleus causes widespread trigeminal analgesia without affecting non-noxious sensation. The effect seems to be due to encephalin release. The nucleus raphe magnus has other connections with the trigeminal nerve fibres at which serotonin is thought to be the transmitter.

The response of the caudalis and, to a lesser extent, the oralis nucleus, to painful stimuli is modulated, then, in a manner analogous to that of the spinal neurones, by inputs from other sensory neurones in larger fibres, and by activity in the peri-aqueductal grey matter causing release of encephalins and inhibition of P substance release. Some inhibition of pain sensation also results from impulses passing down from the cortex or thalamus.

*Pain of dental origin – the problem of the transducer*

The distribution of nerves in the individual tooth is limited to the pulp and the innermost layer of dentine. Light microscopic studies show the nerves to lose their myelin sheaths and break up into a plexus of fine fibres and terminations in a layer below the odontoblasts, and separated from them by the so-called cell-free layer. From the sub-odontoblastic plexus, fibres pass to the odontoblast layer and in older teeth some fibres form a second plexus beneath the pre-dentine – the marginal plexus. From here they may extend into the dentinal tubules, spiralling round the odontoblast processes in the innermost layers of dentine. These nerve fibres have been found in the crowns of the teeth only and are most numerous under the cusps.

It is said that the teeth can produce only a sensation of pain whatever the stimulus applied. Although touch and pressure on the teeth are sensed by periodontal membrane receptors, responding to the displacement of the tooth in its socket, there is some evidence that the sensation of warmth or even cold can be appreciated via the dental receptors.

Sensations of pain due to stimulation of pulpal nerves usually arise when there is pressure in the pulp chamber. This is often due to inflammation and an increased blood supply in the confines of a mineralised chamber, and it is the tissue damage due to the pressure which results in stimulation of pain receptors. Toothache may be throbbing as the blood pressure varies with the beat of the heart. Aviators in non-pressurised aircraft may experience toothache from otherwise symptomless, mildly-inflamed, teeth as a result of the pressure differential – a condition known as aero-odontalgia. Inflammation itself also releases kinins and other algogogues.

More physiologically, pain may be experienced from the teeth due to stimuli of hot or cold on enamel or dentine surfaces, or stimuli from solutions of various substances such as salt and sucrose on dentine surfaces where cementum has been worn away from exposed tooth roots. Exposed dentine is also susceptible to touch stimuli. Since neither enamel nor the outer layers of dentine have a nerve supply, or, apparently, any receptor cells, the problem arises of how the stimulus is transmitted. It is possible that nerve endings in dentine are directly stimulated, although these are relatively remote from the site of stimulation. The odontoblasts, with their processes extending up the dentinal tubules, have gap junctions with nerve cells and might therefore be receptor cells, depolarising the nerve cells which they contact. There are two arguments against this: the main role of odontoblasts is that of laying down dentine matrix and assisting in its calcification, and, secondly, odontoblasts have never been found to have demonstrable electrical activity. Neither argument is conclusive. It now seems unlikely that the odontoblast cell with its process is directly stimulated, however, because the most recent evidence suggests that the odontoblast processes themselves extend only a short way up the dentinal tubules. Current theories envisage either the odontoblasts, or the nerve fibres in the odontoblast layer, acting as mechanoreceptors which respond to changes in odontoblast position resulting from movements of, or changes in, the fluid in the dentinal tubules. On this hypothesis, heat and cold cause expansion or contraction of the contents of the dentinal tubules, salt or sugar solutions cause osmotic changes in the dentinal fluid, and the pain experienced when dentine is drilled is due to boiling off of the dentinal fluid and aspiration of the odontoblast cell bodies into the tubules. All these produce movement of the odontoblast cell bodies and initiate a mechanoreceptive impulse. Active research continues on the mechanisms of dental pain and some new explanation may yet emerge.

## Further reading

Bowsher, D. *Introduction to the Anatomy and Physiology of the Nervous System,* the edn, Blackwell, Oxford 1987

Bowsher, D. *Mechanisms of Nervous Disorder: An Introduction,* Blackwell, Oxford, 1978

Carpenter, R.H.S. *Neurophysiology,* Arnold, London, 1984

Eyzaguirre, C. and Findone, S.J. *The Physiology of the Nervous System,* 2nd edn, Year Book Medical Publishers, Chicago, 1975

Mumford, J.M. *Toothache and Related Pain,* Churchill Livingstone, Edinburgh, 1983

# 10

# The kidneys (renal system)

## Functional anatomy and histology of the kidneys

The two kidneys lie on either side of the body against the dorsal body wall in the upper part of the abdomen. Each has a ureter attached to the inner curve and these pass down to the upper corners of the triangular-shaped bladder. The bladder is connected to the external body surface by the urethra. All arteries, veins, nerves and lymphatics enter at the slit-like hilus and pass into the renal sinus, a C-shaped potential space surrounded by the kidney tissue (Fig. 10.1). The space is potential because it is filled by the structures mentioned and by the renal pelvis with its branches, the major and minor calices. These are the tubes connecting the renal papillae with the ureter. Projecting into the sinus are 8–10 papillae, each of which has at its apex some 18–24 openings of the papillary ducts formed by the fusion of many collecting ducts. In section the kidney has a reddish brown granular-looking outer layer, or cortex, and a red- to grey-brown inner portion, the medulla. The outermost part of the cortex is even in appearance and made up mainly of the proximal and distal parts of the nephrons with some glomeruli. The inner part which dips down

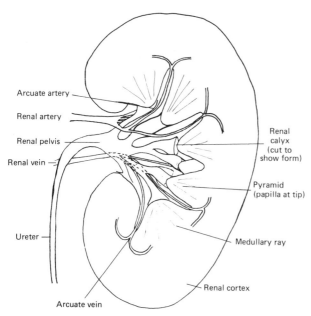

**Figure 10.1** A longitudinal section through a kidney. In the upper part a renal artery and its branches are shown, in the lower part a renal vein and its branches.

into the less even-looking medulla, is marked with lines, or rays, where straight portions of the kidney tubules and blood vessels have been cut longitudinally. The inner cortex contains all parts of the nephron – glomeruli, proximal and distal convoluted tubules, some loops of Henle, and collecting ducts. The medulla presents a rayed appearance because it consists mainly of parallel structures – blood vessels, ascending and descending limbs of loops of Henle, and collecting ducts. Sometimes an inner zone can be distinguished which contains only thin loops of Henle in addition to the collecting ducts and the blood vessels.

## The renal circulation

Branches of the renal artery penetrate between the calices and then pass between the pyramids as interlobar arteries. When these reach the cortex, they curl over the bases of the medullary pyramids as the arciform arteries. These vessels send off radial branches through the cortex in the cortical rays – the interlobular arteries – and these in turn give off short lateral branches, the afferent arterioles, to the glomerular capillary tufts. The glomerular capillaries re-unite to form the efferent arterioles. These are typical arterioles, rather smaller in diameter than the afferent arterioles. They present a high resistance to blood flow and maintain a relatively high pressure in the glomerular capillaries. Both sets of arterioles are supplied by the sympathetic vasoconstrictor nerve fibres. These adjust the resistance of afferent and efferent vessels to maintain an almost constant pressure along the glomerular capillaries and to maintain the filtration pressure despite variations in the general systemic blood pressure. However, in emergency situations the afferent arterioles constrict more than the efferent so that blood flow to the kidney, and the filtration pressure, are reduced. The efferent arterioles rapidly break up into a second set of capillaries, forming a low pressure portal system. The hydrostatic pressure in these vessels is only around 2 kPa. The capillaries form a complex interlinking network around the parts of the kidney tubule in the cortex – mainly the proximal and distal convoluted tubules. Where a glomerulus is close to the medulla (a juxtamedullary glomerulus) and has a long loop of Henle extending deep into the medulla, the vascular arrangement differs. The glomerular capillaries link up to form an efferent arteriole of about the same size as the afferent. This vessel then splits up into a capillary bed supplying the juxtamedullary tubules, but also sends off a major branch which forms a deep straight loop close to the loop of Henle and then drains back into an arcuate vein. These vascular loops are termed the vasa recta. The veins form a system parallel to that of the arteries. The blood supply to the cortex greatly exceeds that to the medulla – all the blood reaching the kidney passes through the glomerular capillaries, but 90% of it then passes on to supply nutrients to the cortex through the peritubular network of vessels leaving only some 9% to supply the medulla and about 1% for the pyramids.

## The nephron

The functional units of the kidneys are the nephrons. Each kidney contains over a million of these units. Since the number actually needed is estimated to be of the order of half a million, a generous reserve is available. The nephron is a blind-ended epithelial tube, continuous with the collecting ducts, the papillary ducts, and, through the calices, the ureter (Fig. 10.2). The blind end of the tube, Bowman's capsule, is wrapped closely around a knot of capillaries, the glomerulus. Each renal corpuscle – the glomerulus together with Bowman's capsule – is about 100 μm in diameter. From the capsule the nephron continues as the proximal convoluted tubule, a tube lined with a

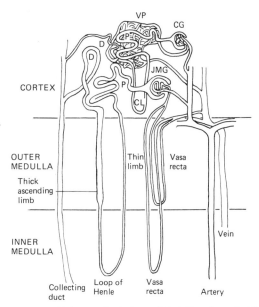

**Figure 10.2** The renal tubules and their blood supply. Two nephrons are shown to the left of the figure, one a cortical nephron, with its glomerulus labelled CG, and the other a juxtamedullary nephron with its glomerulus labelled JMG. On the right are shown an artery and a vein giving a circulation which passes first to the glomeruli, then forms networks (VP) around the proximal (P) and distal (D) convoluted tubules, and then forming loops of the vasa recta in the medulla. Two loops of Henle are shown: one in the cortex (CL) and the other in the medulla. Although for clarity the tubules and the blood vessels are isolated in the diagram, it should be appreciated that, in the kidney itself, all these structures are interwoven in a complex network.

single layer of cuboidal cells with numerous short processes projecting into the lumen. The proximal convoluted tubule coils locally and then extends down a cortical ray to enter the deeper cortex and outer medulla as the relatively thick straight portion which continues as the thin descending limb of the loop of Henle. This second subdivision of the renal tubule is made up of the thin descending limb, the thin segment of the ascending limb, and then the thick segment of the ascending limb which continues on into the distal convoluted tubule. The renal corpuscles in the middle zone of the cortex have loops of varying length which may or may not extend into the medulla and may lack thin segments. The loops entering the medulla from this zone are generally short and thin, in contrast with those originating in corpuscles in the inner zone of the cortex, the juxtamedullary nephrons, which are long and thin. In man these latter constitute only about one eighth of all the nephrons. The thin tubules are made up of only one layer of cells, a thin flattened layer. The ascending limb of the loop of Henle changes abruptly from its thin portion to a thick segment, and this then becomes the beginning of the distal convoluted tubule. From there the distal convoluted tubule passes straight outwards towards its parent corpuscle, and then coils itself in the cortex. There is a transition from cuboidal to columnar cells along its course. The distal convoluted tubule passes close to the afferent arteriole as it enters Bowman's capsule. Here the arteriolar wall forms a thickened cuff in close approximation to a short specialised zone of the distal convoluted tubule with columnar rather than cuboidal cells. This complex is termed the juxtaglomerular apparatus and is an important organ in the control of kidney function. The distal convoluted tubules finally unite to form the collecting ducts, usually via a short transitional segment. Collecting ducts formed in the outer cortex increase in size as they pass towards and into the medulla, picking up more distal convoluted tubules as they go. They fuse to form the papillary ducts which then open into the calices. The total length of a nephron, excluding the collecting duct, is between 20 and 44 mm. and the average length of a collecting duct is around 22 mm. Since the proximal convoluted tubules are almost two thirds of the length of the nephrons they make up the greater part of the cortex.

## Renal function – Production and modification of a tubular fluid and the formation of the urine

### Function of the renal capsules

The relatively short arterial pathway from the heart via wide-bore vessels ensures that the pressure at which the blood reaches the glomerular capillaries is high – around 7.0 kPa (55 mm Hg). This pressure may be controlled directly by the differential constriction of the afferent and efferent arterioles. The structure of the the glomerular capillaries is that of fenestrated capillaries: the endothelium has pores as large as 100 μm in diameter. The epithelium of the Bowman's capsule consists of cells termed podocytes which have foot-like processes attached to the endothelial wall of the capillaries (Fig. 10.3). Between the pseudopodia are slits around 25 μm wide. Separating endothelium and epithelium is a continuous structureless basement membrane. This behaves as if it is permeable to molecules up to 4 nm diameter and partially permeable to those up to 8 nm in diameter. The capillary surface available for the transfer of materials from blood to the kidney tubules totals about 0.8 m². Filtration of fluid and solutes occurs across the triple layer of endothelium, basement membrane and epithelium. About one quarter of the cardiac output, 1.25 litres of blood,

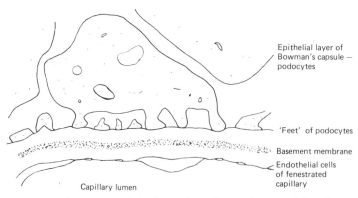

**Figure 10.3** The structure of the glomerular wall – a podocyte and the adjacent glomerular capillary endothelium. Note the 'feet' that give the podocyte its name.

reaches the kidneys every minute; about 1800 litres of blood are passed through every day. Of this about 55% is plasma; and about a fifth of that plasma is actually filtered. The amount of blood plasma filtered in unit time is termed the glomerular filtration rate and the proportion of the total renal plasma flow is termed the filtration fraction. Thus, in the transit through the glomerular capillaries, about 125 ml of plasma leave the bloodstream for the kidney tubules every minute. This is the glomerular filtrate. Since the blood cells and most of the plasma proteins are too large to cross the basement membrane, the fluid reaching the tubules is very similar to plasma without its protein content. In fact a very small amount of protein does get through even in the normal kidney and this may be much increased in inflammation which affects the permeability of the renal corpuscles. There is also a slight difference in ion concentrations between plasma and the tubular fluid because of the Donnan effect due to the non-permeating charged proteins. The blood in the efferent arteriole is richer in cells and in protein than that in the afferent arteriole as a result of the filtration which has occurred: the mean haematocrit value rises from its original 45% up to over 50%. The filtration pressure in the glomerular capillaries is greater than in other capillaries because the hydrostatic pressure is greater for two reasons: the length of the path from the high pressure in the aorta is short, and the presence of the efferent arteriole downstream from the capillaries maintains the pressure. As a result the balance of pressures at the beginning of the glomerular capillaries is much more in favour of egress of fluid than it is in capillaries elsewhere, and the gradient of pressure along the capillaries is almost negligible. The pressures opposing this (*Table 10.1*) are the oncotic

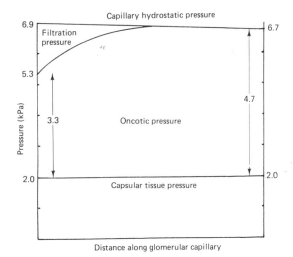

**Figure 10.4** The balancing pressures along a glomerular capillary, and the filtration pressure resulting from them.

Table 10.1 The balance of pressures across the glomerular capillary walls

| Outward pressures (kPa) | | Inward pressures (kPa) | |
| --- | --- | --- | --- |
| Mean hydrostatic pressure | 7.0 | Oncotic pressure | 4.0 |
| | | Tissue pressure | 2.0 |
| | 7.0 | | 6.0 |
| Net pressure 1.0 kPa outwards | | | |

pressure, which at the beginning of the capillary is similar to that in capillaries elsewhere, and the tissue pressure due to the capsule around the kidney and the fluid in the tubules themselves. The oncotic pressure rises along the length of the capillaries as water and small molecules are filtered through the glomerulus, leaving behind the proteins and the cells as described above (Fig. 10.4).

Small changes in systemic blood pressure have little effect on the filtration pressure because the pressure in the glomerular capillaries can be adjusted by constriction of the efferent arterioles. The glomerular filtration rate is maintained fairly constant by adjusting renal blood pressure and flow through the action of sympathetic nerves on the efferent and afferent arterioles. In serious emergencies, blood flow to the kidney is reduced by constriction of the afferent arterioles and glomerular filtration rate becomes progressively reduced.

## Function of the proximal convoluted tubules

The filtered fluid contains all the salts of plasma, glucose, aminoacids, and a small amount of protein (mainly albumin). Since this fluid represents about a quarter of the plasma it is essential that all the components of value to the body be reabsorbed. The proximal convoluted tubule functions as a scavenging device. The kidney operates, not by selectively extracting excess and waste products from the blood, but by taking a filtrate of blood and reclaiming from it those materials the body needs to retain. The work involved in the process can be appreciated by considering the amounts of water and solutes reaching the kidney in the circulation, as compared with the amounts finally lost in the urine (Table 10.2). The reabsorptive mechanisms operating in the kidney have already been discussed on p. 39.

In the course of the proximal convoluted tubule about 70% of water, sodium ions and chloride ions are reabsorbed. No direct mechanisms for the transport of water across cell membranes exist in the body (p. 30): instead, osmotic gradients are created

Table 10.2 The filtered load handled by the kidney in 24 h. The sites of absorption and secretion are abbreviated as follows: PCT proximal convoluted tubule, LH Loop of Henle, DCT distal convoluted tubule, CD collecting duct.

| Substance | Filtered | Reabsorbed | Site | Secreted | Site | Excreted | Reabsorption (net) (%) |
|---|---|---|---|---|---|---|---|
| Water (ml) | 180 000 | 179 000 | PCT 70% LH 10% DCT 15% CD 4.5% | | | | |
| Sodium (mmol) | 26 000 | 25 850 | PCT 70% LH DCT CD | | | 150 | 99.4 |
| Potassium (mmol) | 900 | 900 | PCT | 100 | DCT | 100 | 89 |
| Calcium (mmol) | 225[a] | 225 | PCT 60% LH, DCT | | | 5[b] | 98 |
| Chloride (mmol) | 18 000 | 17 850 | PCT, LH DCT, CD | | | 150 | 99.2 |
| Hydrogen carbonate (mmol) | 4 900 | 4 900 | PCT, DCT | | | 0 | 100 |
| Phosphate (mmol) | 180[a] | 180 | PCT | | | 35[b] | 100 |
| Glucose (mmol) | 800 | 800 | PCT | | | 0 | 100 |
| Aminoacids (mg) | 10 000 | 9 650 | PCT | | | 350 | 96.5 |
| Urea (mmol) | 870 | 460 | PCT, LH | | | 410 | 53 |

[a] indicates that the load may be significantly altered by dietary or hormonal influences
[b] indicates that the urinary output particularly depends upon the dietary input.

and water moves down these. The most important active transport mechanisms in the proximal convoluted tubule, therefore, are the sodium pumps located on the peritubular fluid side of the cells. These actively transport sodium ions out of the cells into the peritubular fluid. Sodium ions diffuse into the cells from the tubular fluid down a concentration gradient, and so are transferred from the tubular fluid to the peritubular fluid and back into the plasma. This active transport of sodium generates a gradient of charge, which causes diffusion of chloride ions across the cells, and an osmotic gradient, which causes the reabsorption of water.

The rich capillary network around the tubules is essential for the uptake of the reabsorbed water, salts and other molecules: the low hydrostatic pressure of around 2 kPa, and the raised oncotic pressure of the blood which has lost a fifth of its smaller constituents by filtration, favour the movement of water and solutes into the capillaries from the interstitial fluid. About 40% of urea is reabsorbed, probably by passive movement down the concentration gradient established by water reabsorption. An active transport mechanism exists for glucose and galactose. This mechanism exhibits a 'transport maximum': it has a limited capacity and is able to reabsorb all the glucose presented to it only up to a concentration of around 17 mmol/l in the tubular fluid (see p. 39). At higher concentrations the excess will appear in the urine. Since different tubules exhibit slightly different maximum levels of reabsorption there is not an abrupt all or none appearance of glucose in the urine as plasma concentrations rise, but a varying degree of efficiency of reabsorption from about 10 mmol/l up to about 21 mmol/l. In the normal subject virtually no glucose appears in the urine unless very high blood glucose concentrations are reached.

Aminoacids are similarly completely reabsorbed. The total plasma concentration is only 2.5–3.5 mmol/l so that the transport maxima of the reabsorptive mechanisms of 0.1 mmol/min and 1.5 mmol/min (for glycine) are adequate to maintain the urine free of aminoacids. At least three transport mechanisms have been described: one reabsorbing lysine, arginine, ornithine, cystine and possibly histidine, one reabsorbing glutamic and aspartic acids, and a third reabsorbing the remainder of the aminoacids. Creatine is also reabsorbed completely, probably by the same pathway as glycine and alanine. In addition to the reabsorption of aminoacids in the proximal convoluted tubule the kidney also participates in their metabolism, removing amides and amino-nitrogen from arterial blood to form ammonia which is excreted in salts of strong acids, and also carrying out transamination reactions with some aminoacids to produce, in particular, alanine and serine.

Hydrogen carbonate ions are effectively reabsorbed in the proximal tubule. If this process of reabsorption is regarded simply as a means of maintaining the plasma bicarbonate concentration at about 24 mmol/l, the proximal tubule can again be seen as performing a scavenging role. However, hydrogen carbonate is one of the factors involved in the control of blood pH and must therefore be considered later in a rather different way (Chap. 18). Hydrogen carbonate ions cannot cross cell membranes easily and are not therefore actually reabsorbed as such. Carbon dioxide produced during the metabolic activity of the proximal tubule cells is hydrated with the aid of carbonic anhydrase and the carbonic acid thus formed ionises to hydrogen and hydrogen carbonate ions. The cells actively transport hydrogen ions out into the lumen. These react with hydrogen carbonate to yield carbon dioxide and water. The carbon dioxide generated is able to diffuse into the cells of the tubule and can be hydrated again inside the cells – thus providing hydrogen ions for secretion to the lumen, and hydrogen carbonate ions which pass out of the other side of the cells to reach the interstitial fluid and then the plasma. Although this mechanism of reabsorption is indirect, it results in an effective renal threshold for hydrogen carbonate above which any hydrogen carbonate present in the glomerular filtrate will pass on into the urine.

Potassium ions are reabsorbed almost completely in the proximal convoluted tubule: the potassium ions in the final urine are those secreted by the distal tubules or collecting ducts.

Approximately 60% of the calcium presented to the tubule is reabsorbed passively.

Phosphate ions are actively reabsorbed by a transport maximum type of mechanism. The maximum is within 85–90% of the normal level of plasma phosphate concentrations so the kidneys play an important role in stabilising plasma phosphate levels at around 1 mmol/l. The transport maximum (Tm) is therefore around 0.125 mmol/l. This transport system is inhibited by parathyroid hormone. Vitamin D does not directly affect the handling of phosphate by the kidney.

Fluoride ions may be passively reabsorbed in the proximal tubule but the major site of reabsorption is thought to be in the distal nephron.

Protein is rapidly reabsorbed, larger molecules probably by pinocytosis in the proximal tubule. These are then metabolised in the cells to aminoacids before passing back into the plasma. The smaller proteins and peptides are broken down to aminoacids by peptidases on the luminal surface of the tubular calls and then absorbed.

In addition to its role in retention of vital solutes the proximal convoluted tubule also plays a part in excretion. Two systems, both of the transport maximum variety, have been described, one secreting organic acid, and the other organic bases, into the tubular fluid. The more important of these from a dental viewpoint is that transporting organic acids. Its importance lies in the way in which many substances, of both intrinsic and extrinsic origin, are metabolised in the liver and conjugated with glucuronide or sulphate. These conjugates are then secreted in the proximal tubule as organic acids and are excreted in the urine. This excretory route is used for breakdown products of steroid hormones, penicillin, barbiturates, and many other drugs. It is also exploited by the physiologist and the physician to measure renal blood flow. Since para-aminohippuric acid (PAH), a substance not naturally found in the human body, reaches the urine by being filtered into the tubular fluid and then secreted by the organic acid pathway into the proximal tubule, it is almost completely removed from the blood flowing through the kidney. If a known blood concentration is established, the amount of PAH excreted in unit time will indicate the volume of blood that has passed through the kidney in that time (p. 105).

## Function of the loops of Henle

The fluid entering the loop of Henle is not very different from that which passed through the glomerular capillary walls in terms of its osmolality and total salt concentration. The loops of Henle are specialised to produce a gradient of osmolality from the cortex to the tips of the medullary pyramids. This gradient is essential for the concentration of the final urine: it causes water to be removed from the tubular fluid in the collecting ducts as they pass down through the medulla. If no osmotic gradient were present, only a dilute urine could be formed and water loss could not be adjusted by varying the permeability of the collecting duct. The gradient is generated by a counter-current multiplier system involving differential transport of sodium and/or chloride ions and differential permeability to water as explained earlier (p. 107). The blood vessels are also arranged in loops to act as a counter-current exchanger (p. 106) and maintain the osmotic gradient. The low blood flow to the medulla also helps to prevent dissipation of the gradient. The changes in tubular fluid which take place in the loop of Henle are a consequence of the mechanism used to produce an osmotic gradient. As a result, the change in composition of the fluid leaving the loop of Henle is relatively small despite its having passed through the highly concentrated environment in the depth of the medulla. It is hypotonic, having lost a further 25% of the filtered sodium load whilst its volume has been reduced by about 5%. Urea diffuses freely through the wall of the loop of Henle as it does elsewhere in the tubule. As water is lost from the collecting duct in the deeper parts of the medulla where the osmotic activity is high, the urea concentration in the tubular fluid rises and urea passes out of the tubule down a concentration gradient. From the interstitium it diffuses back into the loop of Henle and is carried on into the distal tubule again. In the deepest part of the medulla and in the pyramids the urea in the interstitial fluid accumulates, because the vascular counter-current exchanger does not remove it. This helps to maintain the concentration in the collecting tubules.

There is active reabsorption of calcium ions in the ascending limb of the loop of Henle.

## Function of the distal parts of the nephrons

The distal convoluted tubule and the collecting ducts perform similar functions and are sometimes linked in the phrase 'distal nephron'. This part of the nephron is concerned in adjusting the excretion of water and salts to maintain the constancy of the internal environment. It does this mainly by controlled reabsorption, although some secretion occurs here also. As the tubule passes through the increasingly osmotically active interstitium down towards the pyramids, the tubular fluid – and hence, the final urine – may be concentrated if water can diffuse outwards through the tubular walls. Their permeability to water is under the control of antidiuretic hormone (Chap. 16). Although only about 20% of the filtered water load is available for reabsorption in this part of the tubule, this represents about 36 litres of water and adjustment of the reabsorption of this volume allows considerable adjustment of the final urine volume from the normal output of approximately one litre. Sodium reabsorption in this part of the kidney tubule is governed by the hormone aldosterone (Chap. 16). The proportion of filtered sodium entering this section of the tubule is only about 5% but this also represents a considerable amount of sodium. Controlled reabsorption of sodium in the distal nephron allows not only adjustment of body sodium levels but also, because sodium is the major ion in extracellular fluid and therefore the main determinant of extracellular osmotic activity, adjustment of the volumes of the extracellular fluids. Some ionic calcium is actively reabsorbed here under the control of the parathyroid hormones. Secretion of hydrogen ions, production of ammonia, and secretion of potassium ions all occur here but vary in response to changes in plasma pH. A detailed description of these processes is given later (Chap. 18). In the presence of hydrogen ions fluoride forms undissociated HF which is reabsorbed passively down a concentration gradient.

## Function of the ureters and bladder

The fluid reaching the pyramids resembles the final urine in composition; no further changes occur in the ureters or bladder. The ureters are thin tubes consisting of two layers of smooth muscle with a lining of transitional epithelium and a submucous connective tissue layer. Peristaltic waves pass down them at a frequency of between one in three minutes to as many as six in a minute. Each wave sends a jet of urine into the bladder. Pressures are low unless the ureters are obstructed. The ureters enter the muscular wall of the bladder at an angle and the orifices are covered by flaps of mucosa – both arrangements helping to prevent reflux.

The bladder itself is said to be tetrahedral in shape when empty; as it fills it assumes an ovoid shape. A sphincter of smooth muscle surrounds the urethral orifice and there is also an external sphincter of striated muscle.

The normal daily production of urine is around 1.5 litres; the bladder can hold some 600 ml at full capacity.

The process of micturition – or urination – is described later on pp. 240–241.

## Further reading

Valtin, H. *Renal Function*, 2nd edn, Little Brown, Boston, 1983

# 11

# The digestive system

## Anatomy

The digestive system, or the gut, is a muscular tube lined with endoderm which traverses the body from mouth to anus, sealed at various points by muscular sphincters controlled by the autonomic nervous system. The cephalic end of the tube, the mouth cavity, is closed by the lips; entry to the tube is therefore under voluntary control. The caudal end, the anus, is sealed by the anal sphincters, the outermost of which is under voluntary control in the adult. Since the tube may be open to the external environment at both ends, its inside and contents must be regarded as external to the body. Food and drink are taken into this tube, where they are physically and chemically processed. Their components are then absorbed actively or passively across its wall into the body cavity proper, or else remain unabsorbed and are carried down the tube to be excreted as faeces. The walls of the gut are made up of muscle whose contractions knead and mix the food after the initial chopping and grinding in the mouth, and also propel the food mixture onwards. The mural epithelium is specialised to form glands of varying complexity: these produce mucoid substances to lubricate the passage of the food mixture and to protect the walls, and enzymes which break down complex foodstuffs into simple molecules. The glands may be situated in the thickness of the walls or may be so deeply invaginated as to form separate structures – as, for example, the pancreas. Other parts of the tube are specialised for absorption, with infoldings of the surface mucous membrane (the mucosa) which enormously increase the area available for diffusion. The mucosa, like other epithelial surfaces, is constantly losing its outer layer of cells and replacing them. This helps to protect it against the abrasion due to the passage of food. The cells lost in this way in the lower part of the gut are excreted in the faeces, together with some of the gut bacteria and any secretions which have not been reabsorbed.

In life the muscular walls of the gut always maintain some degree of contraction, or tone. Measurements in adults by X-ray and other methods, therefore, suggest that the gut is only about 4.5 m long even though it appears twice this length in cadavers.

The digestive tube is subdivided into a number of organs on a basis of the anatomical and functional breaks due to the sphincters (Fig. 11.1). The mouth, or oral cavity, is separated from the oro-pharynx only by a ring of lymphoid tissue and is continuous through the naso-pharynx with the nasal cavity. The pharynx is not anatomically separated from the oesophagus, although a functional sphincter, the cricopharyngeal sphincter, has been described. The oesophagus extends from there to the cardiac sphincter at the entrance to the stomach. It is situated within the thoracic cavity and at rest its lumen is in equilibrium with thoracic pressures because of the thinness of its walls. A balloon in the oesophagus may therefore be used as a pressure transducer to measure intrathoracic pressures. The next portion is the stomach, separated above from the oesophagus by the cardiac, or gastro-oesophageal sphincter, and below from the small intestine by the gastroduodenal, or pyloric, sphincter. Functionally the stomach has a cardiac portion, a fundus or body, and the tubular pyloric antrum leading to the pylorus. The small intestine is made up of the short C-shaped duodenum holding the pancreas within its loop, and the longer jejunum and ileum. Since there is no obvious division between jejunum and ileum their lengths are arbitrarily estimated as in the ratio of 2:3. The small intestine

176  *The digestive system*

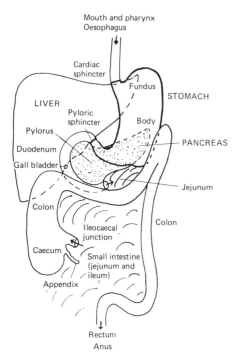

**Figure 11.1** The main organs of the digestive system as seen when the abdomen is opened. The outlines are continued as dotted lines when they pass behind another organ: the complex folds of the small intestine are indicated by curved lines of shading, and only the beginning of the jejunum and the end of the ileum, where it joins the caecum, are shown for localisation.

as a whole is about 270 cm long in the living subject. Its termination is marked by the ileocaecal valve or sphincter. The ileum projects into the colon (the next part of the gut) and since it enters laterally, the opening is closed by colonic contractions. The colon extends a short distance downwards as the caecum, which ends in a small blind-ended extension, the appendix. The colon proper passes up towards the liver, crosses the abdominal cavity from right to left and then continues downwards to join the rectum at the colonic sphincter. The rectum is the final part of the gut and ends in the anus with the internal anal sphincter.

The major part of the tube from oesophagus to anus is made up of layers of muscle separated from each other by connective tissue in which lie blood vessels and nerves. Characteristically the muscle is arranged in three layers: the submucosal, the circular and the longitudinal layers (Fig. 11.2). In fact, both circular and longitudinal layers are spiral muscles, the circular a tightly coiled spiral, and the longitudinal an extended spiral. The lumen of the gut is lined with a mucosa whose surface varies from a stratified squamous epithelium in the mouth and oesophagus to a columnar epithelium in the intestine. The lamina propria is invaginated by glands producing mucus and digestive secretions. The blood vessels and lymphatics are specially arranged in the parts of the intestine whose function is primarily absorptive. Although the gut is composed of smooth muscle with its own intrinsic rhythmicity, it has an extensive innervation to modify and co-ordinate its activity as well as to control its secretions. Two nerve plexuses are described: a submucous plexus (of Meissner) between the muscularis mucosae and the circular muscle layer; and the myenteric plexus (of Au-

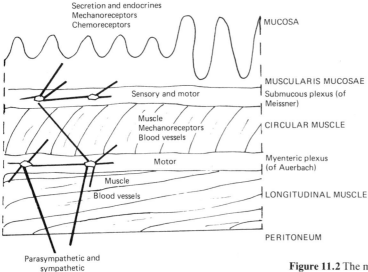

**Figure 11.2** The muscle layers and nerve plexuses of the gastro-intestinal tract.

erbach) between the circular and longitudinal muscle layers. These plexuses are both sensory and motor. The extrinsic nerves connect mainly with the myenteric plexus. Parasympathetic fibres travel in the vagus to the upper part of the gut, and in the pelvic and splanchnic nerves to the colon and rectum (Fig. 9.18). Sympathetic fibres pass from the coeliac plexus to the stomach and small intestine, from the superior mesenteric plexus to the small intestine, caecum, ascending and transverse colon, and from the inferior mesenteric plexus to the descending colon and the rectum (Fig. 9.19).

In different parts of the gut there are variations from the general plan. A marked variation in the muscle layers occurs in the colon where the longitudinal muscle is not uniform over the surface but is arranged in ribbon-like bands, the taenia coli. Their length and state of contraction causes the colon to be pulled together into a series of sacs. The muscle layers of the oesophagus are not so well characterised as those elsewhere and in the stomach there is an intermediate layer of muscle between the circular and longitudinal layers.

## Movements of the digestive tract

The control and co-ordination of the processes of ingestion, digestion and excretion are discussed later (Chaps. 19 and 21) but a general description of the muscular activities seen in the gut is given here.

The characteristic movement of the digestive tract is peristalsis. A peristaltic wave is a wave of contraction originating at the cranial end of a piece of gut and passing caudally. It is usually initiated by stretching of the gut, which probably causes depolarisation and the development of an action potential. The wave of electrical activity passes caudally followed by the contraction. This myogenic effect is supplemented and integrated by nerves in the submucosal plexus and may be modified by the myenteric plexus as a result of sympathetic or parasympathetic stimulation. Generation of the peristaltic wave begins in the longitudinal muscle and, when that is approximately half completed, contraction of the circular muscle follows. The intrinsic plexuses co-ordinate the response of the circular muscle. Distension of the gut not only causes excitation of the longitudinal muscle but also activates the myenteric plexus so that contraction of the circular muscle is briefly inhibited before it is in turn excited. In this way the delay between the two muscles is maintained. In man the peristaltic wave always travels in the same direction. In isolated gut preparations the wave of contraction is preceded by a wave of relaxation. The function of the peristaltic wave is to move the contents of the gut along the tube in the direction of the wave. In the oesophagus peristaltic waves are initiated by swallowing. Some 10 to 20 cm of the oesophagus is seen to be contracting at once and the wave takes about a second to reach its peak, holds it for about half a second, and then dies away over the next second. The wave moves at 2-4 cm/s. Although it generates an internal pressure of up to 1.5 kPa, the actual propulsive force is small – of the order of 0.05–0.10 N. The stomach is traversed by rhythmic peristaltic waves at a frequency of around 3/min generated by a pacemaker area near the cardia. In the duodenum a basic rhythm of 7–8 waves/min is seen. These regular rhythms of muscle contraction all originate in rhythmic action potentials from identifiable pacemaker areas and are known as the basic electrical rhythms (BERs) of the particular sections of the gut.

Peristaltic waves in the small intestine generally are short weak movements travelling at about 1 cm/min which do not pass the length of the small intestine but die away after a short distance. They are only weakly propulsive: as a result a pendular movement of the intestinal contents takes place despite the absence of any actual reversal of peristaltic direction. This slow passage of the intestinal contents provides ample time for absorption to occur across the wall of the gut. A pacemaker area has been described near the entrance of the bile duct but the portion of the duodenum cranial to this has its own independent rhythm of contraction.

The peristaltic waves of the colon have a much slower frequency, between 2 and 10/min, and travel slightly more slowly, at between 1 and 2 cm/min. A massive peristaltic wave is seen infrequently in the colon and extending down into the rectum. It is apparently initiated by the entry of food into the stomach. Such a wave, mass peristalsis, or mass movement, takes about 15 min to traverse the length of the colon. It is not to be confused with defaecation: this also involves a massive peristaltic wave, but one triggered by distension of the rectum which is under over-riding control from conscious centres in the adult.

A number of other movements are described in the large and small intestines. In the small intestine a type of contraction termed segmentation is probably the major form of movement. In this, ring-like contractions appear at intervals, fade away, and then re-appear in new sites between those of the previous contractions. This process is only slowly propulsive and serves to mix the contents of the small intestine. After feeding these contractions have been observed to occur some 10 or 11 times/min. In the colon, similar annular contractions produce what are known as haustral movements. The section between two haustral contractions is called a haustral sac. These haustrations occur between 2 and 10 times/min; like the segmental contractions of the small intestine, they result in kneading of the intestinal contents.

The movements of the muscularis mucosae are independent of those of the longitudinal and circular muscles: they originate as responses to the contact of the mucosal surface with any material within the gut lumen. Extrinsic nerves, however, are able to modify the excitability of the muscularis mucosae via the nerve plexuses in the same way as they modify the excitability of the longitudinal and circular muscles: the parasympathetic generally increasing, and the sympathetic inhibiting, activity.

As a result of the propulsive movements the food ingested is passed through the gastro-intestinal tract. The time taken for this process ranges from a minimum of between 4 and 10 h up to maxima in the range of 68-165 h. A meal rarely disappears from the body in less than 12 h, and it may remain as long as a week. Some mixing of consecutive meals does, in fact, occur in the intestine. If a patient inadvertently swallows a gold inlay it may not appear in the faeces for several days. So-called regular bowel habits do not shorten the period. The time taken to traverse individual segments of the digestive tract has also been measured. Movement is rapid down the oesophagus – partly under the influence of gravity – water taking about 1 s and chewed food about 5 s. If the subject is horizontal the time is extended slightly, but the force generated by swallowing together with the peristaltic force (some 1.5 kPa above resting levels) drives the food on into the stomach. Even a subject in the vertical position with the head downwards can still swallow: the cricopharyngeal sphincter constricts to give a pressure of about 1.2 kpa and this is sufficient to support a column of water as the peristaltic wave carries it upwards. The gastro-oesophageal sphincter can constrict to give a pressure 2.5–4.0 kPa above resting, and this will prevent reflux of material from the stomach.

The stomach in the resting state is of small volume – perhaps 50 ml – but as it fills it stretches, showing the plastic and elastic properties of smooth muscle (see Chap. 5). As food passes into the stomach, layering occurs in the fundus, but by the time the food reaches the pylorus mixing has taken place. The appearance of the stomach in X-ray pictures depends upon the density of its contents – the stomach as a whole is equivalent to a bag of fluid in a fluid-filled cavity and if its density is low it floats. Within the stomach the same principle holds good: solid materials sink within the fundus, liquids 'float', and gases rise to the top – to give the appearance of a cap in some X-ray pictures. The passage of solid material, however, towards the pyloric antrum is relatively slow. Although the rate of emptying varies, when it begins each peristaltic wave transfers about 1 ml to the duodenum, resulting in a transfer rate of about 20 ml/min. About 35% of a 500 ml fluid meal leaves the stomach in one hour and it usually takes around 4 h for a normal meal to leave the stomach. Patients about to undergo general anaesthesia are therefore advised to refrain from meals for some 4 h before the operation, so that the stomach may be empty and the risk of vomiting (with the consequent danger of vomitus passing into the respiratory tract) reduced.

The pressure in the stomach is normally around 2 kPa and hence less than the closing pressure of the gastro-oesophageal sphincter. Increased abdominal pressure, as in pregnancy, may lead to reflux of the stomach contents, giving rise to the sensation of heartburn.

Gases in the stomach may result from the action of the digestive secretions on foodstuffs, or may even reach the stomach from subsequent parts of the digestive tract; most commonly they are simply air which has been swallowed with the food. Such gases may be eliminated if they become uncomfortable by eructation (in U.K. 'belch', in U.S. 'burp'). This swallowing of air (aerophagy) and its expulsion is utilised in patients who have had the larynx removed to eliminate carcinoma: by adjusting the strength and frequency of the eructation they can learn to produce oesophageal speech. The dental surgeon may sometimes assist these patients by fitting an upper denture with a vibrating plate which can be activated by oesophageal airflow.

The minimum transit time through the small intestine is again something of the order of 4 h, with the intestinal contents often delayed at the ileocolic junction for several hours. Movement in the colon is even slower.

Gases in the intestine, often generated by the bacteria normally present in the gut, pass more rapidly than do the solid contents. They contain appreciable quantities of methane. The expulsion of these gases is termed a flatus – strictly, a rectoflatus. Movement of gases around the intestinal coils can produce the sounds known as borborygmi.

## Secretions of the digestive tract – the salivary glands and saliva

The secretions of the oral cavity are collectively known as whole or mixed saliva, although some writers prefer the term 'oral fluid'. All three terms refer to the fluid which is collected by expectoration; secretions of individual glands are named appropriately as parotid saliva, accessory gland saliva, etc. In the resting mouth the lips and cheeks are almost in contact with the teeth anteriorly, and the tongue is in close proximity with the teeth and the palate posteriorly. Only a thin film of whole saliva separates these structures and this constitutes their immediate environment. How far saliva can be considered the environment of the teeth themselves is debatable, since most of the dental enamel surfaces are covered either by a glycoprotein pellicle or the bacterial film known as dental plaque. However, water and small molecules appear to diffuse easily through thin layers of dental plaque.

## Mixed saliva, whole saliva or oral fluid

### Components of the oral fluid

Whole saliva is made up of the secretions of the three pairs of major salivary glands – the parotid, the submandibular and the sublingual – together with the secretions of the minor, or accessory, glands which are distributed in the mucosa of the cheek, lips, hard and soft palates, and tongue. In addition to these secretions, the total oral fluid includes the gingival crevicular fluid, which passes between the teeth and the gingival cuff tissue to reach the oral cavity (p. 23), and, if it is formed by human teeth, enamel fluid (p. 23).

Centrifugation of whole saliva separates a plug of solid material known as salivary sediment. This makes up 5–10% of the volume of whole saliva collected after mechanical stimulation such as chewing wax. Sediment only rarely contains food debris: it is made up of cellular material in varying states of viability, ranging from completely normal cells to unidentifiable fragments. The cells are of three types: epithelial cells which have been desquamated from the oral mucosa, white blood cells derived from the tissues at the gingival margins of the teeth, and the commensal bacteria of the oral and dental surfaces. Whole saliva contains between 6 and $600 \times 10^3$ buccal squames/ml, usually identifiable as intact cells, although many of the unrecognisable fragments may also be derived from this type of cell. In saliva from subjects with natural teeth there are $25-650 \times 10^3$ leucocytes/ml, mainly polymorphonuclear neutrophils, together with a few lymphocytes and basophils. These cells gain access to saliva by diapedesis through the walls of the gingival vessels (which in most subjects show the dilatation and enhanced permeability typical of mildly inflamed tissues) and then through the gingival cuff tissue along the sides of the teeth. The bacterial count is very variable but a representative figure would be between 60 and $70 \times 10^6$/ml. The majority (about 70%) of the bacteria are streptococcal. These include *Streptococcus mutans*, currently identified as the main causative agent of dental caries. Subjects with active dental caries have a higher proportion of lactobacilli in their saliva, but the numbers are very small (1% or less of the total bacterial count).

Estimates based on typical flow rates and durations of flow in response to differing stimuli suggest that the total volume of saliva produced in 24 h is between 0.5 and 0.6 litres. Of this, about half is due to a resting, unstimulated, flow, and the other half is produced in response to various stimuli associated with food intake. Sleeping subjects produce only some 10 ml of saliva from the major glands over 8 h, probably because of the lack of stimulation.

The relative proportions of the secretions in whole saliva vary with the degree of stimulation and with the nature of the stimulus. In sleeping subjects there is no observable flow of saliva from the parotid glands; the submandibular glands contribute 80% and the sublingual 20% of the saliva produced. The contribution of accessory gland saliva during sleep is unknown, but could be substantial at such low total flow rates. Unstimulated, or resting, flow in subjects who are awake amounts to 200–300 ml over 12–14 h: about 20 ml/h. Of this the submandibular glands contribute about 70%, the parotid glands about 20%, the sublingual glands 1–2%, and the accessory glands almost 7%. Gustatory stimulation gives a whole saliva with almost equal contributions from submandibular and parotid glands at fast flow rates, whilst vigorous mechanical stimulation by chewing rubber bands or paraffin wax results in a whole saliva containing twice as much parotid saliva as submandibular saliva. Accessory glands contribute around 7.5% of the volume of stimulated whole saliva, regardless of the nature of the stimulus.

Gingival crevicular fluid may account for 10–100 μl/h of resting flow – less than 0.1%. Despite the small volume, however, it may be very important in the immediate environment near the gingival margin.

## The salivary glands

### Anatomy

Each parotid gland is pyramidal in shape with the base of the pyramid lying directly below the skin (Fig. 11.3). Each is enclosed in a fibrous capsule and weighs about 25 g. The main duct passes forward over the masseter muscle, turns inward to pierce the buccinator muscle and then terminates in a small papilla on the mucosa of the cheek close to the upper first molar tooth. The situation of the duct renders it easy to attach a cup (a Lashley or Carlsen-Crittenden cup) in position over it by suction.

The submandibular glands, irregular in shape, are described as being of the size of walnuts. They lie posteriorly in the floor of the mouth (Fig. 11.3) and the duct of each travels forward to terminate at the summit of the sublingual papilla at the side of the frenulum of the tongue. The devices which have been used to collect submandibular saliva (excluding actual cannulae) are less satisfactory in use than the parotid cups because of the position of the openings of the ducts.

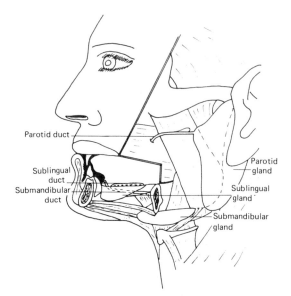

**Figure 11.3** The major salivary glands in man.

The sublingual glands, the smallest of the major glands, are of the size and shape of almonds, weighing 3–4 g each. They lie in the floor of the mouth (Fig. 11.3) and usually have a number of small ducts on their upper surfaces opening on the sublingual folds on either side of the base of the tongue. Suction-attached devices for collection of sublingual saliva are difficult to keep in position on the highly mobile structures of the floor of the mouth.

The minor or accessory glands include the small glands producing mucoprotein secretions scattered over the surface of the tongue, the mucosa of the lips and cheeks, and the mucosa of the soft and hard palates, together with the two anterior lingual glands on either side of the frenulum on the lower surface of the tongue, and the serous glands of von Ebner whose ducts open into the sulci of the vallate papillae of the tongue. The secretions of the glands visible on the inner surface of the lips and on the palate have been collected using filter paper or glass capillaries.

### The nerve supply to the salivary glands

The parasympathetic nerve supply carrying secretomotor fibres to the parotid gland travels in a branch of the glossopharyngeal nerve which synapses in the otic ganglion and passes from there with the auriculotemporal nerve to the gland. Both submandibular and sublingual glands are served by parasympathetic secretomotor fibres originating in the facial nerve, leaving it in the middle ear and then passing as the chorda tympani nerve to join the lingual nerve. These fibres synapse in the submandibular ganglion and the postganglionic fibres pass directly to the glands. In addition to these main secretomotor pathways there are secretomotor fibres to the parotid glands in the chorda tympani nerves, and to the submandibular and sublingual glands in the glossopharyngeal and hypoglossal nerves.

Sympathetic fibres pass from the superior cervical ganglion with the blood vessels to the glands.

Sensory fibres carry pain sensation when the glands are distended, probably in parasympathetic nerves. There may also be a pathway from interoreceptors in the submandibular glands which may influence the rate of secretion.

Within the major glands the distribution of nerve fibres is broadly similar. Main nerve trunks travel adjacent to the main ducts and the blood vessels. Plexuses of small fibres are found around the striated ducts and the fibres then pass along the intercalated ducts to spread out over the surface of the acini. Bundles of nerve fibres frequently include both sympathetic and parasympathetic fibres and both types of nerve have been reported between cells as well as on the surfaces of acini and ducts. Where the nerve axons approach closely (about 200 nm) to acinar, ductal or myoepithelial cells, there are short sections which lack nerve sheaths and contain mitochondria and small granular vesicles; these are probably en passant innervation sites.

### The blood supply to the salivary glands

The blood supply to the parotid gland arises from the facial artery and the external carotid artery. There is a rich vascularisation around the duct system with a much poorer blood supply to the acini. Physiological observations suggest that the blood vessels supplying the ductal systems run in parallel with the ducts and the blood flows in the reverse direction to the saliva flow. The capillaries around the ducts re-join to form portal vessels, which then split up again into capillaries around the acini. The efferent blood passes to the jugular veins. The submandibular glands are supplied with blood by the facial and lingual arteries, and the sublingual gland by the sublingual and submental arteries. The microcirculation in each case resembles that of the parotid glands.

## Composition of saliva

More than 99% of saliva is water: saliva therefore, can make an appreciable contribution to the volume of secretions lost from the body in diarrhoea or vomiting. In whole saliva the total organic content

Table 11.1 Composition of mixed saliva (oral fluid), parotid and submandibular salivas. Concentrations in mmol/l unless otherwise stated. A more detailed version of this table will be found in Tenuovo 'Clinical chemistry of saliva'. Spaces have been left where information was not available.

| Variable | Whole Unstimulated Mean | Whole Unstimulated Range | Whole Stimulated Mean | Whole Stimulated Range | Parotid Unstimulated | Parotid Stimulated | Submandibular Unstimulated | Submandibular Stimulated |
|---|---|---|---|---|---|---|---|---|
| Flow rate (ml/min) |  | 0.33–0.5 | 1.72 | 1.5–2.3 | 0.05 | 1.00 | 0.15 | 1.0 |
| pH | 6.0 | 5.7–6.2 |  | up to 8.0 | 5.5 | 7.4 | 6.4 | 7.4 |
| **Inorganic** | | | | | | | | |
| Sodium | 8 | 6–26 | 32 | 13–80 | 1.3 | 36 | 3 | 45 |
| Potassium | 21 | 13–40 | 22 | 13–38 | 24 | 21 | 14 | 17 |
| Calcium | 1.35 | 0.5–2.8 | 1.7 | 0.2–4.7 | 1.05 | 1.6 | 1.6 | 2.4 |
| Magnesium | 0.3 | 0.1–0.6 | 0.4 | 0.2–0.6 | 0.15 | 0.12 | 0.07 | 0.4 |
| Chloride | 24 | 8–40 | 25 | 10–56 | 22 | 28 | 12 | 25 |
| Hydrogen carbonate | 2.9 | 0.1–8 | 20 | 4–40 | 1 | 30 | 4 | 18 |
| Phosphate | 5.5 | 2–22 | 10 | 2–25 | 9 | 4 | 6 | 5 |
| Thiocyanate | 2.5 | 0.4–5.0 | 1.2 | 0.4–3.0 | | | | |
| Iodide (µmol/l) | | | 14 | 2–30 | 0.5–2.3 | 0.2–1.2 | 1 | 0.5 |
| Fluoride (µmol/l) | 1.5 | 0.2–2.8 | 5 | 0.8–6.3 | 1.5 | 1.0 | | |
| **Organic** | | | | | | | | |
| Protein (g/l) | | | 1.75 | | 2.3 | | 1.1 | |
| Serum albumin (mg/l) | | | 25 | | 10 | | 40 | |
| γ globulins (mg/l) | | | 50 | | 80 | | 60 | |
| Mucoproteins (g/l) | | | 0.45 | | 0.8 | | 0.8 | |
| Amylase (g/l) | | | 0.42 | | 1.0 | | 0.3 | |
| Lysozyme (g/l) | | | 0.14 | | 0.2 | | | |
| Carbohydrate (g/l) | | | | 0.27–0.40 | 0.45 | | 0.3 | |
| Blood group substances (mg/l) | | | | 10–20 | 0 | | 10–20 | |
| Glucose (mmol/l) | | | | 0.02–0.17 | 0.03 | | 0.03 | |
| Lipid (mg/l) | | | 20 | | 20 | | 20 | |
| Cortisol (nmol/l) | | | | 2–20 | | | | |
| Aminoacids (mg/l) | | | 40 | | 10 | | 20 | |
| Urea (mmol/l) | | | | 2.0–4.2 | 2.0–4.2 | | 0.7–1.7 | |
| Ammonia (mmol/l) | | | | 0.6–7.0 | 0.6–7.0 | | 0.2–7.0 | |

amounts to a little over 5 g/l, and the total inorganic content to about 2.5 g/l. The concentrations of the main constituents in whole saliva and in parotid saliva are given in *Table 11.1* and that of gingival fluid in *Table 11.2*.

Table 11.2 Composition of gingival crevicular fluid

| | Concentration (mmol/l) |
|---|---|
| Sodium | 91.6 ± 31.1 |
| Potassium | 17.4 ± 11.7 |
| Calcium | 5.0 ± 1.8 |
| Magnesium | 0.4 ± 0.2 |
| Phosphate (inorganic) | 1.3 ± 1.0 |

Protein (as in plasma) about 75 g/l
   Albumin about 40 g/l
   γ globulin about 7.4 g/l

Flow rate 10–100 µl/h

## Organic components

### Protein

The total protein content of whole saliva is around 2.2 g/l at low flow rates and it increases as the flow rate increases. The mean protein concentration in parotid saliva is nearly twice that of submandibular saliva but the highest concentrations of protein are found in saliva from the sublingual and the accessory glands – unstimulated saliva from the labial accessory glands contains almost 3.0 g/l. Except in the accessory glands the total protein content of gland saliva increases with increasing rate of flow.

The proteins of saliva include proteins synthesised by the gland such as mucoproteins and enzymes, proteins from blood plasma, and the blood group substances, although these are largely carbohydrate in composition. More than 20 individual protein fractions have been separated, but many of these may be present in saliva as a complex mixture linked by calcium bonds.

The total proportion of plasma proteins may be as high as 20% of the salivary protein. They include albumin, γ globulin G (IgG), γ globulin M (IgM), γ globulin A (IgA), and some α and β globulins. Salivary albumin estimations range from 1 to 10% of the total protein. Whole saliva usually contains more albumin than the individual secretions; this may be partly due to the contribution of gingival crevicular fluid. Increased rates of flow are associated with decreased albumin concentrations. IgA is the main γ globulin present and there is some IgM, but very little IgG, the main plasma γ globulin. The total concentration of γ globulins is less than 0.5% of that in plasma; it is not affected by flow rate. The γ globulins do not enter saliva from the blood plasma by diffusion across the acini but are synthesised within the glands, probably in the lymphoid cell clusters around parts of the ducts. They differ from the plasma γ globulins in having a secretory endpiece (p. 45). Although γ globulins in saliva might include antibodies against oral disease, no correlations have been found between salivary γ globulin concentrations and susceptibility to dental caries or gingivitis. Saliva contains some proteins with activities similar to blood clotting factors VII, VIII and IX, and a platelet factor.

The characteristic enzyme of saliva is α-amylase, which splits cooked starch down to maltose. Its optimum pH is at 6.8, and chloride ions are needed for its full activity. It makes up about 30% of the total protein content of parotid saliva. In submandibular saliva the proportion of amylase is much less; sublingual saliva contains very little and saliva from the labial accessory glands appears to have no amylase activity. The concentration of amylase in separate secretions increases as the rate of flow increases; this effect is more marked in whole saliva, since at faster flow rates the proportion of parotid saliva is increased. Most body secretions contain the enzyme lysozyme, or muramidase, which splits the carbohydrate muramic acid found in the cell walls of many bacteria. In parotid saliva it makes up about 10% of the total protein and its concentration is even higher in submandibular saliva. Other enzymes derived from the glands include acid phosphatase, cholinesterase and ribonuclease. Parotid saliva contains a specific lipase, and also a lactoperoxidase which interacts with thiocyanate to give a product with antibacterial activity. The enzyme kallikrein, which splits off the nonapeptide bradykinin from a plasma β globulin, is produced in the ducts of the salivary glands. It may pass back into the gland to reach the blood vessels and cause bradykinin production: if so, this could give a functional vasodilatation to supply the actively secreting gland with a sufficient blood flow.

In addition to the enzymes produced in the salivary glands themselves, whole saliva contains an enormous range of other enzymes, derived from living, dying and dead leucocytes and bacteria. They include proteases, amino-peptidases, carboxypeptidases, lipases, urease, glucuronidase, hyaluronidase, neuraminidase, esterases (cholinesterase, phosphatases, pyrophosphatase, sulphatase), glycolytic enzymes, and acid and alkaline ribonucleases.

The name parotin has been given to a protein with hormonal activity found in parotid saliva. It is said to maintain serum calcium concentrations and to promote calcification. Other biologically active peptides found in saliva include nerve growth factor and an epithelial growth factor.

The majority of the salivary protein molecules contain about 75% protein, with proline, glycine and glutamic acid as the major aminoacids, and about 25% carbohydrate, of which galactose and mannose make up about one half and fucose and the hexosamines (mainly glucosamine) about a quarter each. Sialic acid represents only 1 or 2%. In submandibular saliva glycoproteins are a much higher proportion of the total protein and the carbohydrate:protein ratio is higher. Two major glycoprotein fractions of submandibular saliva contain 80% of the hexosamine and fucose; these glycoproteins differ from those of parotid saliva in having different proportions of galactose, fucose, glucose and mannose. Another glycoprotein in submandibular saliva contains only 44% protein but has an aminoacid composition similar to the parotid glycoproteins.

The blood group substances are carbohydrate-protein complexes present in the cell walls of the red blood cells. About 80% of the population (known as secretors) have present in tissue fluids and secretions either the antigens corresponding to their blood group (A, B or AB) or the related glycoprotein H substance (p. 59). These substances are secreted by all the salivary glands except the parotids. The only other blood group substances in saliva are the Lewis group. Lewis antigen is found in body fluids and adsorbs onto the walls of the red cells; its appearance in saliva may therefore be unrelated to its presence in blood. Non-secretors may have Lewis-A antigen in their saliva, whilst secretors may have either Lewis-A or Lewis-B antigen.

## Non-protein-bound carbohydrates

A small amount of free glucose is found in both parotid and submandibular saliva, which also contains other hexoses and fucose. The concentration of free carbohydrate, and free hexose in particular, is higher in whole saliva which contains a number of enzymes capable of splitting glycoproteins. Glucose concentrations in parotid saliva have been used as a measure of blood glucose concentrations.

## Lipids

The lipid content of saliva is very low although the concentration in whole saliva is raised slightly by cell wall lipids. In parotid saliva β-lipoprotein has been identified and diglycerides, triglycerides, cholesterol and cholesterol esters have been found in whole, parotid and submandibular salivas.

The very small amounts of the steroid hormones in whole and parotid saliva can be assayed by immunological methods. These hormones reach the saliva because they are lipid soluble and their concentrations are directly related to the concentrations of free, or non-protein-bound, hormone in the plasma. Circulating hormone concentrations can therefore be measured without taking blood samples. Hormones assayed in this way now include cortisol, oestrogens, progesterone, testosterone, DHEA, and aldosterone.

## Miscellaneous low molecular weight compounds

Whole saliva contains some 18 aminoacids, the most abundant being glycine. Many of these probably arise from the breakdown of proteins by bacterial enzymes, since the range in saliva from individual glands is much smaller. γ-aminobutyric acid is found only in whole saliva while phosphoethanolamine is present only in separate secretions.

There are also some small peptides, including sialin, a tetrapeptide (glycine-glycine-lysine-arginine) which can act as a substrate for alkali production in dental plaque, and statherin, a large proline-containing peptide, which is an inhibitor of calcium salt precipitation in saliva.

Other nitrogen-containing compounds present in saliva include urea, uric acid, ammonia, and creatinine. The concentration of urea in saliva (1–3.5 mmol/l) is related to plasma levels (usually 1.7–7.0 mmol/l). As a small molecule urea can diffuse freely across the acinar epithelium: the absolute amount diffusing in unit time changes very little as flow rate alters, and so concentration usually falls as flow rate increases. Uric acid, another small molecule, shows similar changes in concentration with flow rate. Ammonia is present in the separate unstimulated secretions but its concentration can be up to ten times higher in whole saliva where urea and aminoacids are broken down by bacterial enzymes.

Citrate and lactate, both probably arising from the metabolism of carbohydrate, have been identified in whole saliva. Lactate, in particular, increases after food intake.

Vitamin C, ascorbic acid, is normally present in unstimulated whole saliva, but its concentration falls as the flow rate increases. The B complex of vitamins and also vitamin K are present at very low levels.

## Inorganic components

Sodium, the most abundant ion in extracellular fluids, appears in the acinar secretion at 140–150 mmol/l, a concentration similar to that in the extracellular fluids. It is reabsorbed in the striated duct systems of the glands and is, therefore, low in concentration in saliva which has passed slowly through these parts of the glands. As flow rate increases, sodium concentrations rise towards the plasma level. Some analyses of fast flowing saliva have given concentrations of over 100 mmol/l.

Extracellular fluids normally contain low concentrations of potassium. However, all parts of the salivary acinus and ductal systems except the final ducts opening to the oral cavity appear to transport potassium into the secretion, and at flow rates above 0.2 ml/min the potassium concentration in saliva stays constant at between 20 and 30 mmol/l. At very low flow rates potassium concentration is high; up to 80 mmol/l in unstimulated saliva. This may be related to the length of time the saliva spends in the ducts. On stimulation there is an immediate loss of potassium from the acinar cells and this leads to a high concentration of potassium in the first few drops of saliva obtained after stimulation – a phenomenon termed a 'rest transient'.

The calcium content of parotid saliva is approximately half that of saliva from the other glands. The concentration is high in unstimulated saliva: levels in submandibular saliva may exceed blood levels, suggesting that calcium is actively secreted into saliva. As flow rates rise above resting rates the concentration at first falls but then increases again until fast flowing saliva from the individual glands contains almost the same concentration as that in the resting saliva. In whole saliva concentrations decrease as flow rate increases because of the greater proportion of parotid saliva. About 50% of salivary calcium is in an ionic form, the remainder being complexed by phosphate, citrate and lactate (40%), or bound by proteins (10%), with amylase being particularly important. The binding by amylase is unaffected by pH whereas the binding by other proteins becomes more effective at higher pH values. As the protein and hydrogen carbonate content of saliva increases with flow rate, the binding capacity for calcium also increases.

The pH of unstimulated saliva is between 6.0 and 6.5 but it can rise as high as 8.0 in fast flowing saliva because of the increased hydrogen carbonate content.

Chloride is the major anion of extracellular fluids: its concentration in the acinar secretion mirrors that in the interstitial fluid at about 100 mmol/l. Reabsorption in the striated ducts reduces its concentration to as little as 10 mmol/l in unstimulated saliva. As the flow rate increases chloride concentrations at first rise, but the increased metabolic activity leads to the formation of more and more hydrogen carbonate, which then becomes available for active transport into the secretion. Chloride ions are reabsorbed in exchange for the hydrogen carbonate and the final concentrations of chloride never rise much above 40 mmol/l.

An active transport system for iodide, thiocyanate and nitrate is present in the ducts of the salivary glands. These ions, therefore, appear in higher concentration in saliva than in plasma, although the saliva and plasma concentrations are related. The salivary glands, like the thyroid gland, concentrate iodine, and so can be used to study iodine turnover in the body. Concentrations of thiocyanate are higher in the saliva of cigarette smokers than in that of non-smokers. Thiocyanate is converted by salivary lactoperoxidase into hypothiocyanate, a bacteriostatic product.

Saliva contains 0.02 ppm, or 1.04 μmol/l, of fluoride, a concentration of the same order as that in plasma. Very small variations in the total fluoride level are observed as the fluoride intake changes or after ingestion of fluoride.

Hydrogen carbonate is in very low concentration in unstimulated saliva. Increased metabolism in the cells of the glands results in more carbon dioxide production: this is rapidly hydrated by carbonic anhydrase to form carbonic acid which ionises. The hydrogen carbonate ion is actively secreted into the saliva so that its concentration in saliva may rise to as high as 60 mmol/l as the flow rate increases. Hydrogen carbonate is the principal buffer in saliva.

**Table 11.3 Sites of secretion and absorption in the gut**

| Site | Enzymes | Substrates | Other secretions | Absorbate |
|---|---|---|---|---|
| Salivary glands | α-amylase | Starch | Mucus | |
| Stomach | Pepsin | Protein | Hydrochloric acid<br>Intrinsic factor<br>Mucus | (Alcohol)<br>(aspirin)<br>Fluoride |
| Pancreas | Trypsin<br>Chymotrypsin<br>Carboxypeptidase<br>Elastase<br>Lipase<br>Esterase<br>α-amylase<br>Ribonuclease<br>Deoxyribonuclease<br>Phospholipase | Proteins<br>Proteins<br>Peptides<br>Elastin<br>Triglycerides<br>Cholesterol esters<br>Starch<br>RNA<br>DNA<br>Lecithin | Hydrogen carbonate | |
| Liver | | | Hydrogen carbonate<br>Biliverdin<br>Bilirubin<br>Glycocholic acid<br>Glycochenodeoxycholic acid<br>Taurocholic acid<br>Taurochenodeoxycholic acid<br>(Deoxycholic acid)<br>(Lithocholic acid) | |
| Intestinal mucosa | Enterokinase<br>Aminopeptidases<br>Dipeptidases<br>Maltase<br>Lactase<br>Sucrase<br>Dextrinase<br>Nuclease | Trypsinogen<br>Peptides<br>Peptides<br>Maltose<br>Lactose<br>Sucrose<br>α limit dextrins<br>Nucleic acids | Mucus | Monosaccharides<br>Aminoacids<br>Lipids<br>Free fatty acids<br>Calcium<br>Fluoride<br>Sodium<br>Chloride<br>Vitamins<br>Iron<br>Water |
| Colon | | | Mucus | Sodium<br>Chloride<br>Water |

When saliva with its high content of both dissolved carbon dioxide and hydrogen carbonate is exposed to air it loses carbon dioxide and becomes more alkaline. The solubility of calcium salts in saliva seems to be related to the carbon dioxide content and the loss of the carbon dioxide can therefore also cause precipitation of calcium phosphates.

Almost all the phosphate in saliva is inorganic. In samples collected during the day the organic phosphate is probably about 10% of the total. Concentrations in parotid and submandibular saliva are not very different and in saliva from both pairs of glands there is less phosphate at higher flow rates. As a result the phosphate concentration in whole saliva decreases with increasing flow rate. The concentration of phosphate is low in saliva from the labial accessory glands. A small amount of pyrophosphate, a salt which inhibits calcification in bone, has been found in saliva in some subjects and may be a factor influencing the formation of dental calculus (tartar).

## The secretions of the gastro-intestinal tract

The total volume of secretions produced by the digestive tract in 24 h amounts to between 7 and 10 litres of fluid. Loss of this fluid by vomiting or diarrhoea (rapid movement of the gut contents which allows insufficient time for absorption of fluid) represents loss of an appreciable proportion of total body water: an even more serious loss if only extracellular water is considered. In the healthy individual all the fluid, together with ingested fluid, is absorbed across the intestinal wall. The secretions also contain salts; and therefore loss of secretions also means loss of salts.

The estimated contributions to the total of fluid secreted are as follows: salivary glands about 0.6 l/day, stomach 2–3 l/day, pancreas 1.2-2 l/day, bile 0.7 l/day, small intestine 2 l/day (the duodenum contributing only about 50 ml), and the colon about 60 ml/day.

The mechanisms of secretion in most of the glands have already been described in Chap. 2; in this chapter the composition of the secretions will be discussed. *Tables 11.3* and *11.4* summarise the sites of secretion and the actions of the main gastro-intestinal secretions.

Although saliva contains many enzymes in trace amounts, usually produced by bacteria or by the disintegration of bacterial, epithelial and white blood cells, the only one of digestive importance is α-amylase. This begins the digestion of cooked starch in the mouth, but the interval between intake and swallowing is so short that little breakdown of the starch occurs. The swallowed mass, or bolus, travels down the oesophagus, again with too short a time interval for appreciable digestion. In the stomach it meets the very acid stomach secretions and these inactivate and denature the enzyme. The optimal pH for the enzyme is about 6.8. Since the acid takes some time to penetrate the bolus, the α-amylase continues to act within the bolus until the acid reaches it. Experiments on the gastric digestion of starch suggest that about half of an intake of 500 g of starch may be digested in the stomach before all the amylase is inactivated.

Table 11.4 The principal enzymes of the digestive tract

| Enzyme | Activator | Optimum pH | Secreted by | Substrate |
|---|---|---|---|---|
| Carbohydrate-splitting enzymes | | | | |
| Amylase | | 6.8 | Saliva, pancreas | Starch |
| Maltase | | 6.6 | Intestinal mucosa | Maltose |
| Lactase | | 6.6 | Intestinal mucosa | Lactose |
| Sucrase | | 6.6 | Intestinal mucosa | Sucrose |
| Dextrinase | | 7.0 | Intestinal mucosa | α-limit dextrins |
| Protein-splitting enzymes | | | | |
| Pepsin | HCl | 1.8–3.8 | Stomach | Proteins |
| Trypsin | Enterokinase | 7.0–8.0 | Pancreas | Proteins polypeptides |
| Chymotrypsin | Trypsin | 8.0 | Pancreas | Proteins polypeptides |
| Carboxypeptidase | Trypsin | 7.5–8.5 | Pancreas | Peptides |
| Aminopeptidase | | 7.5–8.5 | Intestinal mucosa | Peptides |
| Tripeptidases | | 8.0 | Intestinal mucosa | Tripeptides |
| Dipeptidases | | 8.0 | Intestinal mucosa | Dipeptides |
| Fat-splitting enzymes | | | | |
| Lipase | | 7.8 | Pancreas | Triglycerides |

The stomach is primarily concerned with the digestion of proteins. It produces the greater volume of its secretion from the tubular glands of the cardia and the body of the stomach (Chap. 2, Fig. 2.8) which contain a number of types of secretory cell. The oxyntic cells (Figs. 2.11 and 2.12) produce what is effectively a solution of hydrochloric acid of very low pH, whilst the peptic, or chief, cells produce the inactive precursors of the pepsin enzymes, the pepsinogens. These are converted to pepsins by the gastric acid. These cells also produce other enzymes of lesser importance such as a lipase and a gelatinase. The neck mucous cells produce a mucoid secretion and the intrinsic factor which is essential for the absorption of vitamin $B_{12}$ in the ileum. The mucous cells on the surface of the stomach lining contribute a steady flow of mucus. This maintains a protective aqueous layer, probably rich in hydrogen carbonate, over the stomach wall to prevent attack by the acid and the pepsins of the gastric secretion.

Acid denaturation of proteins in the stomach prepares them for the action of pepsins. These break down proteins to polypeptides by cleaving the links between phenylalanine or tyrosine and other aminoacids. A gelatinase and a lipase are also found in small amounts in gastric secretions. Although the action of the stomach acid is mainly digestive, it is also held to have a protective role, since few of the bacteria gaining entrance to the body by this route can survive the pH of 2.0 in the stomach.

In the small intestine the breakdown of foodstuffs into easily assimilable units takes place. The small intestine itself secretes mucus and enzymes but the main secretions in the small intestine are those from the pancreas and the liver. The pancreatic acini have been described in Chap. 2, where their production of an alkaline secretion of sodium hydrogen carbonate was described: in addition to salts the gland also secretes a complex mixture of enzymes – such as a lipase, an amylase like that in saliva, trypsinogen, chymotrypsinogens, carboxypeptidase, ribonucleases (*Table 11.3*). The powerful proteolytic enzyme, trypsin, is activated by enterokinase from the small intestine, which splits off a terminal peptide. Trypsin is similar in many ways to thrombin, the enzyme which splits fibrinogen in the process of blood clotting, and the activation of trypsin and prothrombin take place in similar ways. The chymotrypsinogens are activated by trypsin itself.

The liver produces a continuous secretion of bile which is stored and concentrated in the gall bladder (p. 34). The secretion contains hydrogen carbonate ions, together with the bile pigments, bilirubin and biliverdin, derived from the breakdown of haemoglobin (p. 64), and the bile salts, detergent molecules derived from cholesterol. The liver has many metabolic and other functions (*see* summary in *Table 11.5*); references to some of these occur in the chapters on blood cells and on the control of metabolism (Chaps. 4 and 24).

In the small intestine the hydrogen carbonate of the pancreatic and other secretions remains in the intestine at a concentration around 66 mmol/l, maintaining a pH between 7.6 and 8.2. Carbohydrates are broken down by pancreatic amylase and intestinal disaccharidases. Starch is split at $\alpha$ 1–4 linkages in the straight chains only. The end-products of this digestion are $\alpha$ limit dextrins (short branched chains of glucose molecules), the trisaccharide maltotriose, and the disaccharide maltose. Further breakdown by maltase and isomaltase on the surface of ileal mucosal cells results in all these polymers being broken down to glucose. The other digestible carbohydrates of the diet are the disaccharides lactose (from milk) and sucrose, and the monosaccharides glucose and fructose. The disaccharides are split by other intestinal enzymes present on the cell surfaces – a lactase and a sucrase. Cellulose is not digested in the human gut.

The proteins, now mainly polypeptides, are split by the pancreatic proteases, trypsin and chymotrypsin, down to small polypeptides and dipeptides. The bonds cleaved are those of lysine and arginine at the carboxyl ends of the chains. Pancreatic juice also contains a carboxypeptidase and this, together with intestinal aminopeptidase and dipeptidases, completes the breakdown of most proteins to their aminoacid subunits. There is an elastase and a collagenase in the pancreatic secretion.

Ribonuclease and deoxyribonuclease from the pancreas split nucleic acids into polynucleotides. These in turn are converted into phosphoric acid and nucleosides by intestinal enzymes, and finally to sugars, purines and pyrimidines.

Fats are digested by pancreatic lipase. The gastric lipase has little effect, partly because it is water-soluble and cannot penetrate the fat globules. In the small intestine, bile salts, in the presence of phospholipids and monoglycerides, act as detergents, emulsifying the fats to produce particles of the order of 0.2–5.0 µm in diameter. These can then be digested by pancreatic lipase, which hydrolyses the 1- and 3-bonds of the triglycerides to yield free fatty acids and 2-monoglycerides. Some glycerol is also formed. The bile salts are sodium and potassium salts of the bile acids, cholic and chenodeoxycholic acids, which have been synthesised in the liver from cholesterol, and then conjugated with glycine or taurine to form glyco- and tauro- acids. Bile salts are reabsorbed in the ileum after completing their task, but those not reabsorbed are then converted by the colonic bacteria into deoxycholic and lithocholic acids: of these, the deoxycholic acid is then absorbed, passes back to the liver and is secreted into the bile with the other reabsorbed and newly synthesised bile salts. If sufficient bile salts are present in the small intestine they interact physically with lipids in the form of fatty acids, monoglycerides and cholesterol to form micelles. These are aggregates of some 3–10 nm in diameter, which are made up of a lipid core surrounded by an envelope of bile salts arranged with their steroid lipid-soluble groups inwards to the lipid core, and their polar aminoacid or hydroxyl groups outwards to the aqueous medium.

Some of the enzymes of the pancreas and the small intestine are also necessary to activate proteolytic enzymes by splitting off the terminal groups which render them inactive until after secretion (see Table 11.3). Gastric pepsin is activated from pepsinogen by the gastric acid, but pancreatic trypsinogen needs a duodenal enzyme, enterokinase, to activate it. The chymotrypsinogens are activated by trypsin itself.

Throughout the gut, from mouth to anus, there is a secretion of mucus to lubricate the passage of food or chyme (as food is termed after the action of enzymes upon it) and thus to protect the mucosal surface. Mucus is the only secretion of the colon and rectum. In the stomach it may be important in protecting the wall against excessive acidity, elsewhere its protective function is completely mechanical. Even with this protection, some 300 g of cells are lost daily from the mucosal surface. Cell loss together with the loss of intestinal bacteria results in the production of some faeces even when there is no food intake. An incidental effect of cell loss in this way is the loss of ions from the cells – notably calcium and iron.

## Absorption in the gut

In theory, absorption can occur passively across any mucous membrane. In practice, the amount of absorption occurring across the gut wall is small except in the intestine, and it is only in the small intestine, with its enormous luminal surface area, that absorption of food materials occurs to any significant extent. There is also absorption of water and salts in the colon. It is possible that if foodstuffs were not absorbed in the small intestine there would still be absorption further on in the system: indeed nutrients such as glucose can be administered per rectum if necessary.

Little absorption occurs in the mouth: this route is used clinically for the rapid uptake of glyceryl trinitrite tablets, placed under the tongue to relieve the pain of angina pectoris. Medicaments used on the oral surface to treat, for example, oral ulceration, may also be absorbed. Thus corticosteroid ointments can produce systemic effects when applied only to oral mucous membranes.

The stomach functions as an absorbing organ for a few readily diffusible substances. Alcohol is an example. Aspirin may be absorbed here because in the acid environment it is undissociated and lipid-soluble: the ionised acetyl salicylic acid is polar and will not cross cell walls. Fluoride can be absorbed as undissociated hydrogen fluoride.

In the duodenum, jejunum and ileum, absorption of most necessary nutriments occurs. Active transport systems exist for glucose (linked with sodium transport) and galactose. Transport of these and structurally similar sugars depends upon the shape of the molecule itself and its conformation to the transporting molecule. The carrier does not bind the sugar unless it has first bound sodium; and the energy necessary to transport the sugar against a concentration gradient is derived from the energy due to the gradient of sodium ion concentration across the luminal cell membranes. Modified glucose molecules can be carried if small changes are made which affect the first carbon atom or the hydrogen atom attached to the second carbon atom, but changes affecting the hydroxyl group at the second carbon atom – as in deoxyglucose – prevent transport. The carrier for glucose and galactose is the same, and is shared with some other sugars. The two sugars compete for the carrier. Other sugars using the same carrier will inhibit galactose transport before they begin to affect glucose transport. Fructose and lactose cross the luminal wall passively. The only pentose actively transported is xylose. Most of the absorption of sugars takes place in the jejunum and upper ileum; although active transport of glucose can be observed even down as far as the colon, the process of absorption is usually complete before the chyme has passed very far into the small intestine.

Aminoacids are taken into the body in the same parts of the digestive tube. Three active transport systems have been described: one each for the acidic, basic, and neutral aminoacids. Aminoacids and monosaccharides are both removed rapidly in the bloodstream, thus maintaining a concentration gradient.

Lipids are handled somewhat differently. Although they can cross the lipid layers of the cell membranes without difficulty, they are prevented from reaching the intestinal cells by a relatively fixed layer of water – the unstirred water layer. However, the micelles produced by the bile salts and lipase are able to pass into the unstirred water layer and the monoglycerides and free fatty acids pass through. Once inside the cells two alternative routes present, one for fatty acids of less than 12 carbon atoms, and one for the larger fatty acids. The smaller ones pass directly to the bloodstream and are referred to as the free fatty acids. The larger are re-esterified to triglycerides in the smooth endoplasmic reticulum. Some glycerophosphate from the metabolism of carbohydrate is also incorporated into the triglycerides. The triglycerides, together with cholesterol, are surrounded by a layer of lipoprotein, cholesterol and phospholipid in small spherical particles termed chylomicra. These leave the cells by exocytosis and pass into the lymphatic vessels to be transported by that route. About 95% of ingested fat is absorbed, mainly in the upper small intestine. In the absence of bile salts the proportion falls to about 75%. The sites of absorption of cholesterol are the same as those for other lipids.

Water is absorbed throughout the intestinal tract. The daily intake of 1–2 litres, plus the water in food, together with the digestive secretions, makes a total load for absorption of around 10 litres. Most of this is achieved in the small intestine. Around 500 g of chyme is presented to the colon each day but only some 200 ml of water remains in the faeces, the rest being absorbed by the colon. The total absorptive capacity of the gut for water is said to be of the order of 18 l/day. The absorption of water occurs down the osmotic gradients generated by the active transport of sodium, and possibly, in some areas, chloride and hydrogen carbonate ions. Sodium ions are actively absorbed throughout the intestinal tract, most importantly in the small intestine and colon. Active transport of sodium in the small intestine is linked with the active transport of glucose and aminoacids. In the colon the active transport of sodium ions provides the osmotic gradient which is essential for the absorption of water. In the small intestine and colon chloride ions are actively absorbed in exchange for hydrogen carbonate ions. Potassium ions diffuse from the extracellular fluid to the intestinal lumen down gradients of charge. Chronic diarrhoea can result in sufficient loss of potassium to cause a drop in plasma potassium concentrations – hypokalaemia. Active transport mechanisms exist for two other important ions: calcium and iron. There are carrier proteins for both these ions – calcium-binding protein for calcium and apoferritin for iron. These transport mechanisms are described further in Chap. 24 (calcium metabolism) and Chap. 4 (iron metabolism).

Vitamins are absorbed by diffusion: the water soluble ones are small molecules whilst uptake of the lipid-soluble is facilitated by their lipid carriers. Vitamin $B_{12}$, cyanocobalamine, is exceptional in that it is taken up by a process of phagocytosis in the ileum, but only if complexed with a glycoprotein secreted in the stomach.

Bile salts are absorbed in the ileum after performing their emulsifying action higher in the intestine and pass back to the liver to be recycled. Their absorption stimulates the liver to increase its rate of synthesis of more bile salts.

## Functions of saliva

### Protective functions

Saliva protects the cells of the oral mucosa. Mucosal surfaces differ from skin in being much less well protected against evaporative loss of water and so aqueous secretions are necessary to protect them from drying. Production of saliva in large quantities also helps in washing away particles or solutions remaining around the teeth or oral tissues. The protective action of the saliva is enhanced by its glycoprotein (mucoprotein) content. The potentially abrasive action of foodstuffs is moderated by covering the tissues and mixing the food with a viscous solution of glycoproteins to act as a lubricant. This lubricant fluid also assists in the formation of a bolus of food suitable for swallowing. The lubricant layer contributes to the production of speech sounds by providing a smooth surface to the resonating chambers.

Saliva contains several components which afford some protection against bacterial attack. Leucocytes in saliva and the substances they synthesise may contribute to its antibacterial activity. Even the bacteria of saliva have some protective role since 'foreign' bacteria find it difficult to establish colonies in an existing ecological unit. The enzyme lysozyme acts specifically against bacteria whose cell walls contain muramic acid. Antibodies in saliva may react with the antigens of some oral bacteria. Lactoperoxidase and thiocyanate ions constitute an antibacterial system.

The blood clotting factors in saliva may accelerate blood clotting in the mouth.

Saliva protects the teeth in a number of ways: it limits the pH drop when acid is produced in the mouth because it contains hydrogen carbonate which acts as a buffer, and it is supersaturated with respect to the apatite crystals of the enamel surface so that enamel will begin to dissolve only if the pH drops markedly. If decalcification of the tooth surface occurs, saliva is an effective re-calcifying solution. Newly-erupted teeth dissolve more readily in acid than teeth which have been exposed to saliva for some time and whose surface apatite crystals have taken up ions from saliva.

### Digestive functions

Saliva has little digestive function in the oral cavity, its role as a lubricant being much more important. Starchy foodstuffs remaining around the teeth are broken up by the action of amylase. Few other digestive enzymes are present in saliva in any quantity and the time available for digestion in the mouth is so short that they are unimportant.

### Water balance

Saliva, like the other digestive secretions, consists largely of water. This water passes into the lumen of the digestive tract, which is functionally outside the body, and will be lost unless it is reabsorbed further down. If it is lost, either by vomiting or by diarrhoea, total body water will be depleted by that amount. Some water loss occurs during panting and hyperpnoea, mainly by evaporation from the pulmonary membranes, but some is also lost from the mouth.

Reduction in extracellular fluid volume leads to reduction in saliva volume and the resulting dry mouth gives rise to a sensation of thirst. Intake of water into the mouth, even without swallowing, can abolish the thirst sensation, presumably because sensory inputs from the oral mucosa inhibit the activity of the hypothalamic osmoreceptors.

Factors controlling extracellular fluid volume can affect saliva production. Thus the secretion of antidiuretic hormone increases the permeability of the striated ducts, more water is reabsorbed, and so the final saliva is less hypotonic.

### Excretory functions

Although saliva has been described as an excretory route for several substances from plasma, it is not very efficient since saliva and its constituents are usually reabsorbed in the lower parts of the gut.

Urea, uric acid and ammonia pass into saliva and are broken down in the mouth or in the gut and lost. In uraemia the raised levels of blood urea are reflected in the saliva and give a characteristic odour to the breath. Thiocyanate is said to be excreted in saliva. The heavy metals, including lead, mercury and bismuth, appear in saliva if blood levels are raised, and may be deposited in the oral tissues.

### Solvent functions

Saliva dissolves foodstuffs and aids in their comminution and digestion. Saliva is also necessary for taste sensation: it dissolves sapid substances and spreads them over the surface of the lingual papilli to bring them into contact with the taste buds.

### The importance of saliva in relation to oral disease

The events leading to dental caries and to periodontal disease are probably initiated in dental plaque and variations in saliva are important only as they affect the fluid matrix or the bacteria of the dental

**Table 11.5 Summary of the functions of the organs of the digestive tract**

---

*Mouth*
Initial comminution of the food; initial digestion of starch by amylase activity.

*Oesophagus*
Transport.

*Stomach*
Storage of food to provide controlled flow to intestinal tract; mixing and some further comminution to form chyme; digestion of protein by pepsin in presence of secreted hydrochloric acid; production of hormone gastrin; some absorption; secretion of mucoprotein factor necessary for absorption of vitamin $B_{12}$ in ileum.

*Duodenum*
Digestion of proteins, fats, carbohydrates; production of hormones secretin and pancreozymin-cholecystokinin; activation of trypsin by secreted enterokinase.

*Pancreas*
Production of hydrogen carbonate-rich secretion containing enzymes which digest proteins and peptides (trypsin, chymotrypsin, carboxypeptidase, elastase, collagenase), RNA and DNA (ribonuclease and deoxyribonuclease), starch (amylase), and fat (lipase). Endocrine function of islets of Langerhans.

*Liver*
Production of bile containing hydrogen carbonate, bile salts, bile pigments (produced continuously but stored in gall bladder and released in response to digestive stimuli); metabolism of absorbed aminoacids, fatty acids and monosaccharides; interconversion of foodstuffs; storage of carbohydrate as glycogen; urea production; synthesis of proteins, particularly blood proteins; release of glucose into bloodstream; storage of vitamins; storage of iron; some storage of fat; production of blood cells in fetus; destruction of red cells; breakdown of haemoglobin and production of bile pigments; protective effect of large macrophages (Kupfer cells); synthesis of salts of cholic acid from cholesterol; inactivation, conjugation with glucuronide or sulphate, and excretion of steroid hormones; inactivation and excretion of other hormones; inactivation, conjugation and excretion of many drugs – detoxification.

*Jejunum and ileum*
Digestion and absorption; secretion of disaccharidases.

*Colon*
Absorption of salts and water; storage of continuously digested food to permit excretion of faeces at intervals.

*Rectum*
Storage and excretion as in colon.

---

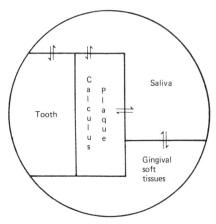

**Figure 11.4** A diagrammatic representation of the interactions between saliva, the teeth and the soft tissues of the mouth.

plaque. The dental environment may be regarded as a multi-component system as shown in Fig. 11.4. The important interactions of saliva are its antibacterial activity, its role in plaque formation, its place in the equilibrium between saliva and plaque constituents, and its possible effects on dental calculus formation.

The pellicle formed on cleaned tooth surfaces exposed to saliva is similar in composition to the glycoproteins of saliva. Formation of pellicle may be due to the presence in saliva of a specific glycoprotein which binds to apatite, or to spontaneous precipitation of salivary proteins caused by interaction between them and other surfaces, by the change in properties of glycoproteins when sialic acid is split off by neuraminidase, or by drying of saliva on the teeth.

Pellicle or plaque may be formed as a result of interactions of bacteria with saliva. Salivary bacteria produce acid by metabolising sugars: this may lower the pH locally to levels where salivary proteins precipitate at their iso-electric points. Salivary bacteria might attach to the tooth surface because they produce sticky surface carbohydrate layers from simple dietary sugars, or because they aggregate in the presence of salivary proteins or calcium ions.

The diffusion of larger molecules in the gelatinous matrix of dental plaque is probably restricted, but small molecules and ions appear to exchange readily with those of saliva. Sucrose, glucose and organic acids like lactic acid diffuse easily though plaque. Calcium and phosphate, unless bound, also diffuse easily through the matrix. Calculations based on the normal calcium and phosphate concentrations in saliva suggest that apatite should begin to dissolve when the saliva reaches pH 5.5: this has been termed the 'critical pH'. However, as plaque fluid concentrations of calcium, phosphate, and hydrogen ions may differ from those in saliva, this value of the critical pH may not be applicable to dental enamel and dental plaque. The pH at which enamel will begin to dissolve depends on concentrations of calcium and phosphate at the enamel surface.

Although dental calculus deposition depends more on conditions in dental plaque than on those in saliva, loss of carbon dioxide from saliva, or production of ammonia from urea, aminoacids and small peptides in saliva, may provide an alkaline environment in which calcium salts are less soluble. Local concentrations of calcium may be raised because of breakdown of calcium-binding proteins, or because of loss of carbon dioxide with its ability to form complexes with calcium salts. Local concentrations of inorganic phosphate may be raised by salivary phosphatases which break down salivary organic phosphates. Calcium-binding proteins from saliva may pass into the dental plaque matrix and act as seeding substances for apatite crystal growth.

## The functions of the different parts of the gut

The functions of the separate organs of the digestive tract are summarised in *Table 11.5*.

### Further reading

Davenport, H.W. *Physiology of the Digestive Tract,* 5th edn, Year Book Medical Publishers, Chicago, 1982

Sernka, T.J. and Jacobson, E.D. *Gastrointestinal Physiology – the Essentials,* Williams and Wilkins, Baltimore, 1983

Tenovuo, J. *Clinical Chemistry of Saliva,* CRC Press, Boca Raton, in press.

Chapters on saliva in textbooks of Oral Physiology or Oral Biology.

# 12

# Endocrine systems

Hormones, or endocrine secretions, have traditionally been defined as chemical substances secreted directly into the bloodstream from a specialised tissue and producing effects on distant organs. They are a pathway of communication alternative to the nervous system, differing from it in being slower, but at the same time more suitable for widespread integrated activity in a number of effector organs or cells. Endocrine systems are suitable for mediating changes in activity of many organs together, particularly when the responses are required in minutes, or hours, or days, rather than in fractions of a second. The rapid, but transient, response of the nervous pathways would be inefficient in controlling processes such as growth and reproduction, or in modifying metabolic activity, or maintaining ordered bodily rhythms.

In general, hormones increase or decrease the activity of target organs or cells; they rarely initiate activities. Although many hormones are secreted at a low basal rate, few maintain a constant level of production: their secretion is adjusted in response to nervous stimuli (usually via the hypothalamus) or to changing conditions around the secreting glands. Rhythmical variation occurs in the circulating levels of many hormones: aldosterone output varies during a 24 h cycle with levels low in the early hours of the morning and high in the late afternoon, whilst the levels of gonadotrophic hormones and oestrogens in women follow a cycle of approximately 28 days length. Other hormones are secreted in response to changes in the environment of the secreting cells: thus high concentrations of glucose in the blood cause insulin to be secreted, and decreases in blood ionised calcium levels stimulate secretion of parathormone.

The classical view of hormones and their activity has been modified over the last decade by the discovery of new chemical transmitters and a re-evaluation of the mechanisms of activity of the known hormones. A number of chemical substances have actions on cells and tissues which can be described as hormonal, although, unlike most hormones, they are not secreted by defined aggregations of cells, or they act locally rather than on distant organs. Examples include the first substance ever to be called a hormone, secretin, which, like the other hormones of the gut, is not secreted by recognisable glandular structures. The peptides released by damaged cells are not hormones in the classical sense because they act very locally. The identification of biologically active peptides in the hypothalamus, and later in other areas of the brain, has led to the realisation that some substances previously considered as hormones may act as neurotransmitters and some neurotransmitters appear to act elsewhere as hormones. This raises difficult questions about the distinctions between neurotransmitters and hormones. Many of the cells which produce biologically active peptides are derived, like nerve cells, from neuro-ectoderm, and resemble nerve cells in taking up and decarboxylating amine precursor substances. They have therefore been seen as parts of an 'amine-precursor uptake decarboxylase' system, or APUD system; the APUD cells may be descendants of a more generalised communication system which now manifests itself in the nervous system, in some parts of the classical endocrine systems and as isolated cells in other organs such as the gut.

Since the terms 'endocrine' and 'hormone' exclude many chemical substances with biological activity, other terms have also been used. The term 'paracrine' is applied to chemical substances which act locally rather than being transported in the bloodstream to sites of activity. Some of these are

**Table 12.1** Hormones derived from an aminoacid, and their sites of production.

| Hormone | Aminoacid | Site of production |
| --- | --- | --- |
| Adrenaline | Phenylalanine/tyrosine | Adrenal medulla |
| Dopamine | Phenylalanine/tyrosine | Hypothalamus |
| Histamine | Histidine | Mast cells |
| Noradrenaline | Phenylalanine/tyrosine | Adrenal medulla |
| Prolactin inhibiting hormone (PIH) | Phenylalanine/tyrosine | Hypothalamus |
| Thyroxine | Tyrosine | Thyroid gland |
| Tri-iodothyronine ($T_3$) | Tyrosine | Thyroid gland |

**Table 12.2** Polypeptide hormones and their sites of production

| Hormone | No. of aminoacids | Site of production |
| --- | --- | --- |
| Adrenocorticotrophic hormone (ACTH) | 39 | Anterior pituitary |
| Angiotensin II | 8 | Plasma |
| Antidiuretic hormone (ADH) | 9 | Hypothalamus |
| Atrial natriuretic peptide | 28 | Atria of heart |
| Calcitonin | 32 | Thyroid ultimobranchial tissue |
| Cholecystokinin-pancreozymin (CCK-Pz) | 39 or 34 | Duodenum |
| Corticotrophin-like intermediate zone peptide | ? | Pituitary pars intermedia |
| Corticotrophin releasing hormone (CRH) | 41 | Hypothalamus |
| Encephalins | 5 | Central nervous system |
| β-Endorphin | 31 | Brain and pituitary |
| Enterogastrone | ? | Small intestine |
| Gastrin | 17 or 34 | Stomach (pyloric antrum) |
| Gastric inhibitory peptide (GIP) | 43 | Small intestine |
| Glucagon | 29 | Pancreas |
| Growth hormone releasing hormone (GRH) | ? | Hypothalamus |
| Inhibin | ? | Testis (seminiferous tubules) |
| Insulin | 51 | Pancreas |
| Kinins | 8–12 | Tissues |
| β-Lipotrophin | 91 | Anterior pituitary |
| Luteinising hormone releasing hormone (LHRH) | 10 | Hypothalamus |
| Motilin | 22 | Duodenum |
| Oxytocin | 9 | Hypothalamus |
| Pancreozymin *see* Cholecystokinin-pancreozymin | | |
| Prolactin releasing hormone (PRH) | ? | Hypothalamus |
| Parathormone | 84 | Parathyroid glands |
| Relaxin | 52 | Ovary (corpus luteum?) |
| Secretin | 27 | Duodenum |
| Somatomedin C | 63 | Liver |
| Somatostatin | 14 | Hypothalamus, gut |
| Thyroid stimulating hormone releasing hormone (TRH) | 3 | Hypothalamus |
| Vasoactive intestinal peptide (VIP) | 28 | Small intestine |
| Vasopressin *see* Antidiuretic hormone | | |
| Villikinin | ? | Small intestine |

small molecules which have effects on many different cells, but particularly affect blood vessels: these are termed 'autacoids'. Another group, termed 'modulators', includes substances which influence the biological effects of other hormones or nerves but have no apparent specific direct action on cells.

## The chemical nature of hormones

Hormones are messenger substances, passing information from one cell to another, usually through the circulatory system, and often over great distances. Such substances must be readily synthesised in the body, able to act on specific receptors, and readily destroyed or inactivated when they have achieved their purpose. Most hormones are either polypeptides or cholesterol derivatives (steroids), although some are simpler still – often aminoacid derivatives – and a few are more complex – such as the glycoprotein hormones (*Tables 12.1–12.4*). Hormones which act locally can be very simple in structure, but if the target organ is distant simple molecules have the drawbacks of rapid diffusion and inactivation. Indiscriminate diffusion from the circulation of the simpler hormones is reduced by binding to larger molecules such as the plasma proteins. Cholesterol derivatives are usually poorly soluble in water: their transport in the circulation is also facilitated by binding. When hormones are carried in a bound form they are probably inactive: only the small amount of free hormone in equilibrium with the bound can attach to cell receptor sites. With the development of radioimmunoassay methods, estimation of active hormone concentrations in the blood has become possible. Because only the free hormone can pass into the interstitial fluid and be transferred into an exocrine secretion, the estimation of hormones in saliva presents a method for determining the free concentrations in blood.

The biological activity of hormones is governed not only by their blood concentrations but also by the number of receptor sites on the cells. This can vary from time to time: for example, the number of gonadotrophin receptor sites on the cells of the ovarian follicle changes during the menstrual cycle.

It is convenient to group hormonal substances as water-soluble and lipid-soluble, since there are major differences between these two groups.

### Water-soluble hormones

The water-soluble hormones include aminoacid derivatives, polypeptides, proteins and glycoproteins. They bind to cell membrane receptors and their actions are achieved by means of intracellular second messengers.

### *Aminoacid derivatives* (*Table 12.1*)

Structurally the simplest hormones, these usually act locally. A high concentration is required, therefore, only at the local site, and the dilution which results from the rapid outward diffusion of these small molecules does not prevent them from acting. However, the ready diffusibility may assist some, such as adrenaline, to produce very widespread effects.

### *Polypeptides* (*Table 12.2*)

The polypeptides range from the very small – tripeptides – up to molecules with over 80 aminoacid residues. Since peptide hormones are small, they are more readily synthesised; they are usually broken down at the receptor site, or by enzymes in the blood or in the tissues, so that their half-lives are relatively short – often minutes. Nearly all the hormones produced by nerve cells, or tissues of nervous origin, are polypeptides – with adrenaline, once again, being an exception. Many peptide hormones act very locally but some, particularly the larger ones, have target organs far from the site of synthesis. The chemical structure of most of these hormones is now known in detail and many of the smaller ones have been synthesised.

### *Proteins and glycoproteins* (*Table 12.3*)

A number of hormones are protein or glycoprotein molecules which are able to travel easily in the bloodstream without the need for carrier molecules. Although the synthesis of these molecules is obviously more complex than that of the small polypeptides, such hormones can survive for much longer periods in the circulation and can produce

**Table 12.3 Protein and glycoprotein hormones and their sites of synthesis**

| Hormone | Number of aminoacids | Sites of synthesis |
|---|---|---|
| Erythropoietin | | Plasma |
| Follicle stimulating hormone (FSH) | 236 | Anterior pituitary |
| Growth hormone | 191 | Anterior pituitary |
| Human chorionic gonadotrophin (hCG) | 231 | Placenta |
| Human chorionic somatomammo-trophin (hCS) | 191 | Placenta |
| Luteinising hormone (LH) | 217 | Anterior pituitary |
| Prolactin | 198 | Anterior pituitary |
| Thyroid stimulating hormone (TSH) | 212 | Anterior pituitary |

long-lasting effects on more distant organs. Very often they do not produce direct effects, but stimulate the secretion of other, more locally acting, hormones. The trophic hormones of the pituitary gland, thyroid stimulating hormone, follicle stimulating hormone and luteinising hormone, are good examples. These three, incidentally, are synthesised in the same cells, the basophil cells, and are made up of the same α subunit with slightly differing β subunits, providing an obvious economy in synthesis.

## Lipid-soluble hormones

Two groups of biologically active substances fit into this category: the steroids (derivatives of cholesterol) and the prostaglandins (derivatives of arachidonic acid).

### *Steroid hormones* (*Table 12.4*)

The steroids are derivatives of cholesterol, occurring either as cyclic alcohols or their fatty acid esters. The basic subunit of structure is the cyclopentano-perhydrophenanthrene ring (Fig. 12.1). This has three six-carbon rings to which are assigned the letters A, B and C, and a five-carbon D ring. The physiologically active compounds are classified into groups by their structure. The oestrogens have 18 carbon atoms – the 17 of the ring itself, plus a methyl group attached to carbon number 13. The A ring is aromatic in these compounds. In the androgenic hormones, testosterone and the 17-keto-steroids, there are 19 carbon atoms – methyl groups are

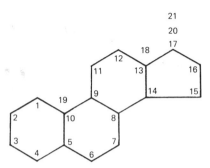

**Figure 12.1** The fundamental unit of steroid structure: the cyclopentano-perhydrophenanthrene ring, with the numbering of the carbon atoms.

attached at both the 13 and 10 positions. A side chain of two further carbon atoms attached at position 17, in addition to the two methyl groups, making a total of 21 carbon atoms, characterises the adrenocortical hormones and progesterone, whilst the bile acids have a five-carbon chain and cholesterol and vitamin D derivatives an eight-carbon chain in this position. All the steroids are synthesised along similar pathways (summarised in Fig. 12.2) and their breakdown also follows similar routes.

As lipid-soluble substances the steroids act widely in the body: they are able to penetrate cell membranes and they interact with intracellular receptor molecules. Lipid substances are relatively poorly soluble in blood plasma and so most steroids are bound to carrier proteins for transport. Breakdown of the steroids may result in the formation of more active metabolites but eventually all are broken down in the liver and/or conjugated with glucuronic acid so that they can be actively secreted into the urine by the organic acid transport system of the proximal convoluted tubule of the kidney.

### *Prostaglandins*

The prostaglandins are derivatives of arachidonic acid, one of the essential fatty acids in the diet. This is released in the cell by phospholipase $A_2$ from cell membrane phospholipids, and converted by the action of prostaglandin endoperoxide synthetases (endoperoxidases) to the endoperoxides $PGG_2$ and $PGH_2$. These two compounds are the precursors of all the prostaglandins, different cells having different synthetic pathways from the endoperoxides. The prostaglandins are considered to be modulators rather than hormones proper. They are produced in a variety of tissues – including the prostate gland which originally gave them their name. Many of the activities of the prostaglandins are related to those of the second messenger cyclic adenosine monophosphate (*see below*) with the prostaglandins either augmenting or diminishing the effect of a

**Table 12.4** Lipid-soluble hormones and their main sites of production. The position of substitutions in the steroid structure is given as C21, C19, or C17

| Hormone | Structure | Main site of production |
| --- | --- | --- |
| *Steroids* | | |
| Aldosterone | C21 | Adrenal cortex |
| Corticosterone | C21 | Adrenal cortex |
| Cortisol | C21 | Adrenal cortex |
| Cortisone | C21 | Adrenal cortex |
| Dehydroepiandro-sterone (DHEA) | C19 | Adrenal cortex |
| 1,25-Dihydroxy-cholecalciferol $(1, 25\text{-}(OH)_2D_3)$ | | Kidney |
| 17 β-Oestradiol | C17 | Ovary |
| Oestriol | C17 | Placenta, ovary |
| Oestrone | C17 | Ovary |
| Progesterone | C21 | Corpus luteum, placenta |
| Testosterone | C19 | Testis |
| *Prostaglandins* | | |
| *see* section at end of chapter. | | |

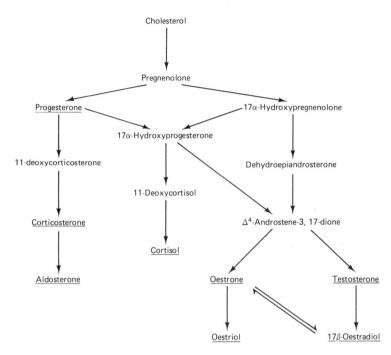

**Figure 12.2** The synthetic pathways for steroid hormone production. The principal hormones are underlined. The hormones produced in a particular organ depend upon the presence of the enzymes for the specific pathways.

hormone in stimulating cAMP release. They seem to be particularly important in their effects on the smooth muscle of various tissues, on blood platelets, and on inflammatory reactions. Salicylates, such as aspirin, inhibit prostaglandin synthesis. All the prostaglandins act locally and they have very short half-lives.

## Mechanisms of action of hormones

The mechanisms by which hormones produce their effects on cells are now becoming fairly well understood. A surprisingly small number of basic mechanisms are used to achieve diverse effects from diverse hormones. The mechanisms fall into two broad groups, those appropriate to water-soluble hormones, and those appropriate to lipid-soluble hormones. The exceptions to the general rule that only the steroids enter the cells are the thyroid hormones, which will be considered separately, and insulin, which also has some intracellular actions.

### Water-soluble hormones

The water-soluble hormones interact with receptors at the cell membrane and most of them then cause the production of a second messenger substance within the cell, which in turn produces a specific effect. In many cases the cyclic nucleotide, cyclic 3,5-adenosine monophosphate (cAMP), functions as the second messenger, but cyclic guanosine monophosphate (cGMP) may be an alternative in other instances, and the actions of many hormones involve calcium ions and the protein calmodulin.

### Mediation of hormonal activity by cAMP

Many hormones (*Table 12.5*) bind to specific receptors on the membranes of target cells causing a chain of reactions in the membrane which culminate in the activation of the enzyme adenylate cyclase on its inner surface. This enzyme converts ATP to cAMP. A range of protein kinases able to activate specific protein activities (usually enzymatic) exists in different cells. The kinases are usually in an inactive form with a so-called regulator unit. When cAMP reacts with the regulator unit, the kinase is activated and becomes able to phosphorylate the target protein so that it too becomes active (Fig. 12.3). Substrates for the protein kinase include cell membrane proteins (so that cell permeability may be altered), rate limiting enzymes, ribosomal proteins, and microtubule proteins. Calcium concentrations in the cell fluid, the cytosol, may affect the ultimate result of protein kinase activation.

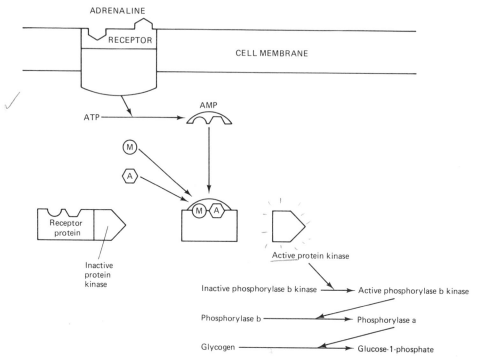

**Figure 12.3** The mechanism of action of hormones whose effects are mediated by cyclic AMP at the cell membrane as a second messenger. Adrenaline is taken as an example in this diagram: other hormones would activate different final reactions. A signifies ATP, M signifies a magnesium ion.

### Mediation of hormonal activity by cGMP

Although adenylate cyclase is located on the inner surface of the cell membrane, the comparable guanylate cyclase appears to be found only in the cytosol. It is known that cGMP concentrations are changed in some circumstances where hormonal activity is taking place, and also that cGMP can mimic the activity of some hormones. Nevertheless, there is no absolute evidence that cGMP is a second messenger in the action of any specific hormone. Since changes in cytosolic cGMP are almost always related to changes in cytosolic calcium concentrations, it may be that they interact in some mechanisms of hormone action.

### Mediation of hormonal activity by phosphoinositide derivatives and calcium ions

Many of the hormones that bind to membrane receptor sites do not cause increases in cAMP within the target cells (*Table 12.5*). Some of them may use cGMP as a direct second messenger but many of them bind to receptors which activate a phosphodiesterase in the membrane. This splits membrane phosphoinositide-4,5-phosphate to give two second messengers, inositol-1,4,5-triphosphate and diacylglycerol. Diacylglycerol then activates an enzyme known as C-kinase which in turn activates other enzymes by phosphorylating them. In addition diacylglycerol stimulates production of arachidonic acid, prostaglandin formation, guanylate cyclase activity, and sodium pump activity. Inositol-1,4,5-triphosphate causes the release of ionic calcium into the cytosol. Calcium seems to act as a third messenger. The protein calmodulin, which has been found in the cells of a very wide variety of tissues, is able to take up calcium ions and then bind to some protein kinases and activate them (Fig. 12.4).

### Mediation of the activity of thyroid hormones

Thyroid hormones are able to enter cells through the cell membranes. Tri-iodothyronine binds to receptors in mitochondria and in the nucleus (Fig. 12.5) but thyroxine itself must be converted intracellularly into tri-iodothyronine before it can bind to these receptors. In the nucleus the thyroid hormone stimulates cell protein production by affecting gene transcription. Mitochondrial activity is increased by intracellular tri-iodothyronine and so

**Table 12.5 Second messengers for hormones**

| Cyclic 3'5' adenosine monophosphate | Others such as calcium and cGMP |
| --- | --- |
| Adrenocorticotrophic hormone | Angiotensin |
| Catecholamines (at β receptors) | Catecholamines (at α receptors) |
| Chorionic gonadotrophin | Growth hormone |
| Follicle stimulating hormone | Insulin |
| Glucagon | Oxytocin |
| Gonadotrophin releasing hormones | Prolactin |
| Luteinising hormone | Prostaglandin $F_{2\alpha}$ |
| Lipotrophin | Somatomedin |
| Melanocyte stimulating hormone | Somatostatin |
| Parathyroid hormone | |
| Prostaglandin $E_1$ | |
| Thyrotrophin releasing hormone | |
| Thyroid stimulating hormone | |
| Vasopressin | |

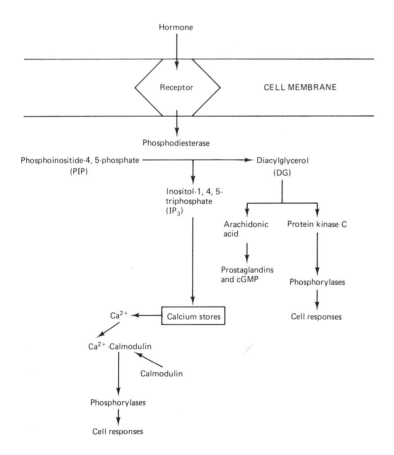

**Figure 12.4** The mechanism of action of hormones whose effects are mediated at the cell membrane through activation of phosphodiesterase. Second messengers include phosphoinositides, calcium ions, calmodulin, and cyclic GMP.

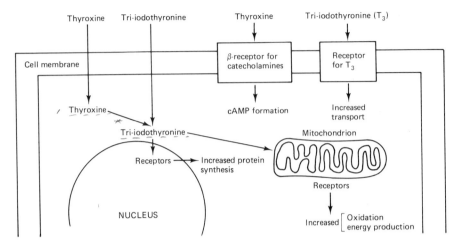

**Figure 12.5** The mechanisms of action of thyroid hormones at the cellular level. Note that thyroxine is converted to tri-iodothyronine inside the cell.

there is an increased rate of oxidative metabolism and increased energy production. In addition to these intracellular effects thyroid hormones also bind on the cell membrane to specific receptors and to catecholamine β-receptors, causing an increase in membrane permeability and membrane transport of certain substances.

## Lipid-soluble hormones

Lipid-soluble hormones, such as the steroid hormones, can enter cells readily and bind to receptors present in the cytoplasm of the cell (Fig. 12.6). When a steroid hormone combines with its receptor molecule the complex enters the nucleus where it attaches to chromatin. As a result the number of transcription sites increases and there is increased production of messenger RNA. The final result is an increase in the synthesis of particular proteins, each specific to the stimulating hormone. A good example of this is the stimulation of production of calcium-binding protein in the cells of the small intestine by an increase in circulating levels of the steroid 1,25-dihydroxycholecalciferol (a vitamin D metabolite).

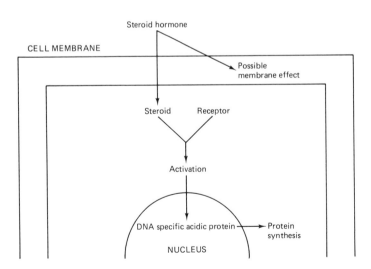

**Figure 12.6** The mechanism of action of the lipid-soluble steroid hormones within the target cells.

## Substances with hormonal activity and the sites of their secretion

The major sites of hormone secretion are shown in Fig. 12.7. This diagram, however, does not include tissues in which there are scattered hormone-producing cells. The classical endocrine glands, with recognisable aggregations of secreting cells, include the pituitary gland (with the anterior lobe or adenohypophysis, the intermediate lobe, and the posterior lobe, or neurohypophysis), the thyroid gland (together with the ultimobranchial tissue in man), the parathyroid glands, the endocrine islets of Langerhans in the pancreas, the adrenal glands (with medulla and cortex), the testes, the ovaries and the corpus luteum, and the placenta. In addition to these classical endocrine glands, hormones are produced in the hypothalamus, nerve cells, in the mucosa of the stomach and the small intestine, in the liver, and in the kidneys. Each endocrine organ will now be described, together with its hormones, classified according to their chemical nature. Alternative names for each hormone are given in brackets.

## The hypothalamus

One of the most important sites of hormone production is the hypothalamus. Its position and structure have already been discussed in Chap. 9. It consists of nervous tissue and, like most nervous tissues, produces hormones which are polypeptide in nature. Cell bodies in the paraventricular and supra-optic nuclei have axons extending down through the pituitary stalk to terminate in the posterior part of the pituitary gland. The blood supply to the hypothalamus derives from the internal carotid arteries and the arterial circle of Willis. There is an extensive capillary network in the median eminence which links up into a number of sinusoidal vessels. These pass down the pituitary stalk to break up into a second capillary plexus in the anterior part of the pituitary gland – thus constituting the hypothalamo-hypophyseal portal system. By virtue of these neural and vascular links, the hypothalamus is able to send hormonal substances to control hormone production in the anterior pituitary gland, and peptide hormones to be secreted by the posterior pituitary gland. The hypothalamus contains receptors for temperature,

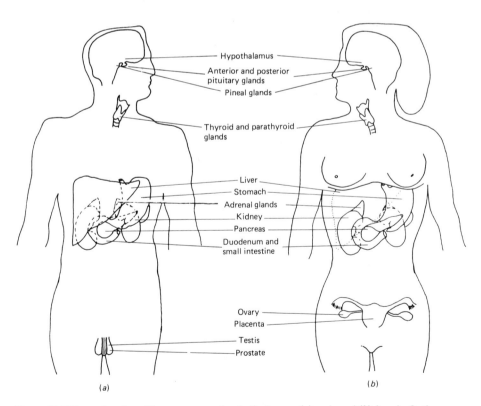

**Figure 12.7** The major sites of hormone secretion in the human (*a*) male and (*b*) female. In the female figure the liver is removed to show the right kidney and suprarenal (adrenal) gland.

osmolality and blood glucose concentrations, and receives nerve inputs from the limbic brain and from sensory pathways. It represents, therefore, the integrating centre for the activity of the nervous system and the endocrine systems, and can be seen as the controlling centre for most forms of hormonal activity (*Table 9.1*).

The hypothalamus produces at least eight polypeptide hormones. Two of these, antidiuretic hormone (ADH, or vasopressin) and oxytocin, are octapeptides synthesised in the supra-optic and paraventricular nuclei. The hormones are produced in the cell bodies, where they can be stained and seen as granules. They then combine with carrier proteins, termed neurophysins, which transport them along the axons into the posterior pituitary gland, where they are released into the bloodstream. The supra-optic nucleus was thought to produce only antidiuretic hormone and the paraventricular only oxytocin, but it is now known that both nuclei can synthesise the two hormones. The principal stimulus for the synthesis and release of antidiuretic hormone is the stimulation of osmoreceptor cells in or near the supra-optic nucleus by a rise in the osmotic activity of the plasma. Below 280 mosmol/l secretion ceases. The hormone is, however, also released during exercise and in response to stress and trauma. Angiotensin II (*see* p. 209) increases antidiuretic hormone secretion as does nicotine, but alcohol reduces secretion. It travels in the bloodstream to reach the kidneys, where it reduces the volume of urine excreted by increasing the reabsorption of water in the collecting ducts. This is achieved by increasing the permeability of the cells of the collecting ducts to water, and also by decreasing blood flow to the renal medulla so as to preserve the osmotic gradient. Both actions are mediated by cAMP. In large doses the hormone has a general vasoconstrictor activity (hence the name vasopressin), but normally this is seen only in the effect on the vasa recta in the kidney medulla. The other physiological circumstance in which the vasoconstrictive action of the hormone is important is in response to haemorrhage. In stress, antidiuretic hormone acts on the liver to raise blood glucose concentrations. It also acts in conjunction with corticotrophin releasing hormone to stimulate ACTH release. Antidiuretic hormone is rapidly inactivated in the liver and kidneys so that its half-life is only about 9 min.

The other octapeptide, oxytocin, has two established actions, both in the female: it causes contraction of the uterine smooth muscle when this becomes sensitive to the hormone towards the end of pregnancy, and it causes contraction of myoepithelial cells in the lactating mammary gland, causing the expulsion of already formed milk. In both instances the stimulus for release of the hormone appears to reach the hypothalamus by nervous pathways – stretching and contraction of the uterine muscle is one stimulus, and tactile stimulation of the nipple and areola by the suckling baby the other. Even the sound of a baby crying can trigger oxytocin release. Genital stimulation may cause release of oxytocin in the non-pregnant woman. The function of oxytocin in the male is unknown, although it may be involved in sexual arousal. The hormone is probably inactivated at the target organs, or in the liver and kidneys. Its half-life is short, like that of antidiuretic hormone.

Both the octapeptide hormones of the hypothalamus act on the smooth muscle of blood vessel walls to cause vasoconstriction: similar molecules can be synthesised with the vasoconstrictor activity greatly enhanced and the other properties reduced. One such analogue is octapressin, or felypressin, which is used as a vasoconstrictor in some local anaesthetic mixtures to prevent the rapid removal of the anaesthetic agent by the local blood flow. In most local anaesthetic mixtures adrenaline is used as the vasoconstrictor agent, but adrenaline also has actions on the heart and general circulation and so may have harmful effects in some patients.

The remaining hypothalamic polypeptide hormones are releasing or release-inhibiting hormones which pass through the portal system to stimulate or inhibit the release of hormones from the anterior pituitary gland. Four releasing and one release-inhibiting peptides have been characterised, and one other is thought to exist.

Thyrotrophin releasing hormone (thyroid stimulating hormone releasing hormone, TRH) is a tripeptide produced in the anterior hypothalamus which has more recently been found in many parts of the brain. It stimulates production of thyroid stimulating hormone from the anterior pituitary. Such a small molecule is rapidly inactivated. The stimuli for its production are a decrease in deep body temperature, or emotional or stressful stimuli. Circulating levels of thyroid hormones do not normally affect TRH production: thyroid hormone levels are controlled by a negative feedback system in the anterior pituitary by means of thyroid stimulating hormone secretion. TRH appears to adjust the sensitivity of this feedback control system rather than being directly involved in a feedback control.

Corticotrophin releasing hormone (CRH) passes in the portal circulation to stimulate the anterior pituitary gland to secrete adrenocorticotrophic hormone (ACTH). The production of β-lipotrophic hormone is also stimulated. Although the cells secreting CRH are somewhat scattered, the axons terminate in the median eminence near the blood capillaries. Production of CRH seems to depend upon nervous stimuli reaching the hypothalamus – CRH release occurs in emotion and trauma. In stress antidiuretic hormone is also necessary to enable CRH to act. The normal corticosteroid rhythm of the body is probably controlled by a diurnal rhythm of CRH release.

Gonadotrophin releasing hormone (GnRH, or luteinising hormone releasing hormone, LHRH), a decapeptide, is thought to be synthesised in cell bodies in the pre-optic and anterior areas of the hypothalamus and released from the terminals of these neurones in the posterior part of the median eminence. It stimulates the anterior pituitary gland to produce luteinising hormone and/or follicle stimulating hormone. Although a separate follicle stimulating hormone releasing hormone may exist, present evidence suggests that there is a single gonadotrophin releasing hormone. If this is so, other factors must control the response of the pituitary cells to the releasing hormone.

Somatostatin (growth hormone release inhibiting hormone, GIH) is a tetradecapeptide secreted into the hypothalamic capillary circulation. It causes inhibition of growth hormone release from the anterior pituitary gland. Feedback control of growth hormone secretion results from stimulation of somatostatin production by circulating growth factors produced as a result of growth hormone activity. Somatostatin may also be produced during sleep and in response to high circulating levels of glucose and cortisol. It is also found in nerve cells in other parts of the brain and in the dorsal horn of the spinal cord, in the mucosa of the the stomach and the small intestine, and in the pancreas. All its known actions are inhibitory: it is capable of inhibiting secretion of the intestinal hormones, gastrin, secretin, vasoactive intestinal peptide, gastric inhibitory peptide and motilin, it inhibits gastric motility and both gastric and pancreatic secretion, and it inhibits the endocrine secretion of TRH, insulin and glucagon.

Growth hormone releasing hormone (GRH) is a large peptide (44 aminoacid residues) produced by cells in the arcuate nuclei and released from their terminals in the median eminence. Lesions of the hypothalamus or of the connecting vessels to the pituitary gland can reduce growth hormone secretion because of the lack of stimulation by GRH. Increases in growth hormone secretion in exercise, in stress and during starvation may be due to stimulation of the hypothalamus and secretion of GRH.

A releasing hormone for prolactin (PRH) is produced in the hypothalamus, but has not yet been characterised.

The hormone which passes from the hypothalamus to the anterior pituitary gland, via the portal circulation, to inhibit the release of prolactin (prolactin release inhibiting hormone, PIH) is now known to be dopamine – a catecholamine which acts as a neurotransmitter within the nervous system.

## The posterior pituitary gland (the neurohypophysis)

The hormones secreted by the posterior pituitary gland – antidiuretic hormone and oxytocin – are synthesised in the hypothalamus and are described above.

## The anterior pituitary gland (the adenohypophysis)

The anterior portion of the pituitary gland contains blood sinusoids lined by cords of secretory cells classified by their staining reactions into the agranular chromophobe cells and the granular chromophile cells. Some of the chromophobe cells may be precursor cells or non-secreting cells related to the chromophil cells, but others secrete the polypeptide adrenocorticotrophic hormone (ACTH). The cells containing acidophil granules secrete the protein hormones, growth hormone and prolactin; while the cells with basophil granules secrete the glycoprotein trophic hormones for the thyroid glands and the gonads – thyrotrophic hormone, luteinising hormone and follicle stimulating hormone.

Adrenocorticotrophic hormone is a large polypeptide which has a circulating half-life of about 10 min. The sites of inactivation are unknown. During sleeping hours the level of ACTH is low, but it begins to rise at about 06.00 h and stays at high levels until about 11.00 h. The rhythm is apparently dominated by impulses from the limbic cortex; it is an important feature of the human 'biological clock'. Secretion of ACTH is stimulated by stress, emotion, and injury, as a result of the production of CRH by the hypothalamus. There is a direct feedback inhibition of secretion by circulating cortisol and, to a lesser extent, other hormones of the adrenal cortex. ACTH binds to membrane receptors on cells of the adrenal cortex. In the zona fasciculata this causes increased formation of cAMP and activates a protein kinase which phosphorylates a ribosomal protein, a phosphorylase, and a cholesterol ester esterase. This sets in motion the synthesis of pregnenolone from cholesterol, the first step in steroid synthesis. The trophic hormone also causes increased uptake of glucose and calcium ions into the cells. The steroid hormone synthesised as a result of ACTH activity is cortisol, together with lesser amounts of the other glucocorticoids. ACTH is also thought to have a permissive role in the secretion of aldosterone.

β-lipotrophin is a very large polypeptide, produced by the same cells as ACTH, which has no direct hormonal activity, but is a precursor of the peptides with morphine-like activity, the endorphins and encephalins, which are important in the central control of pain sensations.

The acidophil cells of the anterior pituitary gland produce two protein hormones, growth hormone (GH, or somatotrophic hormone, SH) and prolactin, both containing about 200 aminoacids and having molecular weights around 22 000 daltons.

The output of growth hormone is about 0.7–4.0 nmol/day in both adults and children and a plasma level of about 11 pmol/l is maintained. Concentrations rise rapidly in response to stimuli such as a reduction in blood glucose levels, an increase in circulating aminoacid levels, and in stress. Thus exercise, fasting, sleeping, and starvation, all result in increases in secretion of growth hormone. Secretion occurs in response to a releasing hormone, and is suppressed by somatostatin, from the hypothalamus. The hormone is destroyed in the liver: its half-life is about 20–30 min. Growth hormone probably does not produce direct effects on the growth of tissues, but stimulates the production of other peptide hormones in the liver and other tissues. The best known of these is somatomedin C, formed in the liver, which is responsible for the effects of growth hormone secretion on cartilage and bone growth. Growth hormone is one of the factors involved in the maintenance of blood glucose levels during fasting and, more importantly, during starvation. These metabolic effects are produced by the hormone itself, in contrast to the effects on growth.

Prolactin is closely related to growth hormone but has different antigenic properties. In conjunction with oestrogen and progesterone it causes milk secretion in the mammary glands. Its secretion is inhibited by dopamine from the hypothalamus, possibly by a negative feedback mechanism evoked by circulating prolactin levels. The hypothalamus also stimulates prolactin secretion via a prolactin releasing factor and TRH. Oestrogens stimulate prolactin production, and there seems to be a nervous pathway from touch receptors in the breast, similar to that which stimulates oxytocin production. Although blood prolactin concentrations in males are almost two thirds as high as in females, its function in males is unclear. One of its effects is to increase the number of receptors for androgens. It has also been suggested that prolactin may assist in salt and water retention. It is one of the hormones whose concentration in the blood is increased during stress: if the hormone has a role in salt and water balance this would be a factor in common with other hormones released in stress.

The basophil cells of the anterior pituitary gland produce a number of glycoprotein hormones, of which thyroid stimulating hormone (TSH) is one. They have a common α subunit and specific β subunits, with molecular weights between 25 000 and 30 000 daltons. Thyroid stimulating hormone is the main controlling factor in the secretion of thyroid hormones from the thyroid gland. Although its secretion is normally controlled by direct feedback from the concentration of circulating thyroid hormones, there is an over-riding control from the hypothalamus by TRH secretion. The hormone increases cAMP concentration in the cells of the thyroid glands, causing them to take up colloid from the follicles and transfer it to the bloodstream. At the same time there is increased glucose oxidation in the gland and an associated increase in iodine uptake and iodination of tyrosine to produce the thyroid hormones. Blood flow also increases. Thyroid stimulating hormone is broken down in the kidney and, to a lesser extent, in the liver. It has a half-life in the body of about 1 h.

Although follicle stimulating hormone (FSH) derives its name from its action in the stimulation of the ovarian follicles of the female during the menstrual cycles, it is also produced in the male and is necessary for the production of spermatozoa. The production of FSH in both sexes is controlled from the hypothalamus, probably by production of GnRH. In the female, increasing levels of oestrogens in the circulation inhibit secretion of FSH. In the male the situation is less clear: testosterone inhibits GnRH production, limiting luteinising hormone secretion, and this may also affect FSH secretion. A substance termed inhibin, thought to be a polypeptide or small protein, which inhibits FSH secretion, is secreted by the seminiferous tubules during growth and development in the male. The biological half-life of FSH is about 3 h, but the sites of destruction are not known. In the female, FSH controls the early development of the ovarian follicle and, together with luteinising hormone, the maturation of the follicles. Both FSH and luteinising hormone are necessary for oestrogen secretion to occur from the follicles. In the male FSH is necessary for the maintenance of the spermatogenic epithelium of the testes. Both FSH and testosterone are required for the production of spermatozoa.

The third of the hormones from the basophil cells of the anterior pituitary is luteinising hormone (LH, or interstitial cell stimulating hormone, ICSH). The direct stimulus to LH production is LHRH, a peptide apparently identical with GnRH, from the hypothalamus, but this secretion is in turn controlled by negative feedback from circulating levels of testosterone in the male, and oestrogen and progesterone in the female. The negative feedback link between circulating levels of oestrogen and LH secretion is only operative at moderate constant levels of oestrogen: if oestrogen levels rise markedly, as they do about 24 h before ovulation in the normal menstrual cycle (Chap. 22), a positive feedback mechanism is triggered, and LH secretion is stimulated. This results in a burst of LH secretion which is followed about 9 h later by the release of the ovum. Although, as with FSH, the site of inactivation is unknown, the half-life of the hormone is less than 1 h. Luteinising hormone in the male stimulates the interstitial cells (of Leydig) of the testes to secrete testosterone. For this reason it is sometimes called interstitial cell stimulating hormone. In the female its role is more complex. Together with FSH it controls maturation of the ovarian follicles and oestrogen production from them. A sudden increase in LH concentrations triggers ovulation, after which LH is the stimulus for oestrogen production by the corpus luteum. Although the cause of the rise in body temperature observed at or just before ovulation is uncertain, it may be due to the activity of LH. The control of the menstrual cycle by LH and FSH is discussed in detail in Chap. 22.

## The pars intermedia of the pituitary gland

The loose stroma between the anterior and posterior parts of the pituitary gland does not produce any significant hormones in the adult human but plays an important role in fetal life. The role of melanocyte stimulating hormones (MSH), secreted by the pars intermedia, is not yet established in the human. These peptides have a similar aminoacid sequence to part of the ACTH molecule and it is possible that the pigmented spots seen in the mouth in Addison's disease are due to melanocyte stimulation by the excess of circulating ACTH.

In fetal life this gland secretes corticotrophin-like intermediate lobe peptide (CLIP), which controls secretion of androgens from a special zone between the medulla and cortex of the fetal adrenal glands. These androgens are essential for the normal growth of the fetus (see p. 289).

## Thyroid gland

The thyroid gland is a bilobed gland lying on the anterior surfaces of the laryngeal, thyroid and cricoid cartilages, and extending down over the pharynx to the fourth tracheal ring. The lobes are joined by an isthmus across the second and third tracheal rings. The gland is organised in acini, or follicles, in which the colloidal exocrine secretion accumulates as an eosinophilic store. The 'inactive' gland is engaged in synthesis and storage, so that the follicles are full and their cells flattened, but the active gland uses up its store, passing the colloidal material back to the bloodstream. The gland actively accumulates some 20–40% of ingested iodine. Salivary glands share this property of active transport of iodine. The dietary requirement for iodine is between 0.8 and 1.2 μmol/day (100–150 μg), although about 4 μmol are present in the normal diet. Of this, the gland takes up about 1 μmol and converts about 0.6 μmol into thyroid hormones. Plasma iodide concentrations are around 24 nmol/l (3 μg/l). The ion is almost completely reabsorbed in the kidney tubules.

In the thyroid gland iodide is oxidised to iodine and possibly to iodate: it is then combined with tyrosine to form mono-iodotyrosine, then di-iodotyrosine, and these two molecules are condensed to form thyroxine ($T_4$) and a smaller amount of tri-iodothyronine ($T_3$). The process of secretion of the hormones was described on pp. 37–38. Small molecules such as thyroxine and tri-iodothyronine would be rapidly lost from the bloodstream were they not bound by proteins. About 70% of plasma thyroxine is bound by thyroxine binding globulin (TBG), and the remainder by a thyroxine binding pre-albumin. Should these become saturated, albumin can also bind the hormone. The free thyroxine in the blood is only some 0.02 nmol/l. Tri-iodothyronine is less well bound: although its plasma concentration is only about 2% that of thyroxine, about 0.005 nmol/l are free. Tri-iodothyronine is more active than thyroxine; indeed, thyroxine is converted into tri-iodothyronine in target cells. The two hormones act on many tissues to increase metabolic rate and oxygen usage. They are important in growth and development. The mechanism of action at a cellular level has already been discussed and their biological effects are described in Chap. 24 onwards. In the tissues the hormones are de-iodinated and de-aminated. However, the half-life of thyroxine is about 6 days. In the liver the hormones are conjugated with sulphates or glucuronides and enter the bile; as a result, about 15% of the amount of iodine taken in daily is lost through the faeces. Secretion of thyroid hormones is controlled by thyroid stimulating hormone from the anterior pituitary so that the concentration in the circulation remains fairly constant, but this control system is subject to a further control by thyrotrophin releasing hormone from the hypothalamus. Excessive production and secretion of thyroid hormones can result from the presence in the circulation of abnormal immunoglobulins which bind to the TSH receptors on thyroid cell membranes, causing cAMP production in similar fashion to TSH itself. The first of these to be discovered was given the name long-acting thyroid stimulator (LATS); this immunoglobulin is however not so clearly associated with hyperthyroidism as is the more recently described human thyroid stimulatory immunoglobulin (HTSI). Both these immunoglobulins increase thyroid hormone levels in the blood and therefore cause suppression of TSH release.

## Ultimobranchial tissue of the thyroid

The parafollicular cells, clear cells, or C cells, of the thyroid gland arise from tissue forming the ultimobranchial bodies in lower vertebrates. They secrete a polypeptide hormone, calcitonin, when blood calcium levels exceed 2.38 mmol/l. Gastrin also stimulates calcitonin secretion. Concentrations of hormone in the blood are higher in males than in females. Calcitonin acts upon bone to inhibit the transport of calcium from bone to the extracellular fluid. The duration of its effect appears to be less than 10 min. Although it was originally thought to be important in preventing blood calcium concentrations from rising too high, its role now seems to be slight except possibly immediately after food intakes.

## Parathyroid glands

The parathyroid glands are four small discs of tissue embedded in the posterior surface of the thyroid gland, usually one in each superior lobe, and one in each inferior lobe. They contain two cell types: the small chief cells which secrete parathormone, and the larger oxyphil cells whose function is unknown.

Parathormone is a large polypeptide split off a still larger precursor prohormone in the gland itself. The amount secreted is controlled by the concentration of calcium in the blood; at a plasma calcium concentration of 3 mmol/l no parathormone is secreted, but as calcium levels fall secretion of the hormone increases linearly. The half-life of parathormone is about 20 min. The gland responds directly to the blood ionic calcium concentration, and there is no known control of the gland by either the hypothalamus or the pituitary gland. Parathormone maintains the total calcium concentration around 2.5 mmol/l by increasing absorption of calcium in the small intestine and in the kidney. It can also cause mobilisation of calcium from bone and increase phosphate excretion in the kidney. In the absence of parathormone plasma calcium levels fall to 1.75 mmol/l.

## The pancreatic islet tissue (islets of Langerhans)

The pancreas is an exocrine gland with its head lying within the loop of the duodenum. It consists mainly of acini and ducts of a digestive gland, but about 1–2% of its weight is made up of small islands of endocrine tissue, 75 × 175 μm in size, and between 1 and 2 million in number. The endocrine cells are classified by their staining properties with a Mallory stain into A, B, D, and F cells. The red stained granules of the A cells contain glucagon, the blue-purple stained granules of the B cells contain pro-insulin, and the D cells contain somatostatin. The F cells produce a polypeptide whose function is as yet unknown.

Glucagon is a polypeptide structurally similar to the gastro-intestinal hormone, secretin. It is secreted by the α cells of the pancreatic islets in response to a decrease in plasma glucose concentrations, although an increase in plasma aminoacid concentrations has the same effect, as do exercise and fasting. The hormone stimulates cAMP formation in liver cells, thus causing activation of phosphorylase and the breakdown of liver glycogen to glucose. It also increases the deamination of aminoacids and the breakdown of fats. It stimulates secretion of growth hormone, insulin, and somatostatin. These latter two effects are important in producing the appropriate balance of hormones after digestion of a meal (p. 303). The hormone is inactivated in the liver by a peptidase; its half-life is short, being between 5 and 10 min.

Insulin is derived from a prohormone made up of three polypeptide chains, A, B and C. The C chain is split off as the hormone is secreted. Secretion of insulin is stimulated by increases in concentration of glucose and aminoacids in the plasma, by glucagon, by nervous impulses in the right vagus nerve, and by gastro-intestinal hormones. Adrenaline and noradrenaline inhibit insulin release. The active hormone binds to receptors in most tissues of the body except the brain, the kidneys, and the red blood cells. It increases the uptake of glucose into the insulin-sensitive cells. In the liver this is partly due to an increase in the rate of phosphorylation of glucose inside the cell which increases the concentration gradient for glucose diffusion, but the mechanism in other cells is not clear. In addition to increasing cell glucose uptake, and thus lowering blood glucose concentrations, insulin also encourages formation of glycogen, fats and proteins. It is inactivated mainly in the liver by a group of enzymes collectively termed insulinases. The half-life is about 5 min.

Somatostatin (see Hypothalamus above) is produced in the D cells. It inhibits hormone production by both A and B cells, so that production of both insulin and glucagon are decreased. Since glucagon stimulates somatostatin production, it also indirectly inhibits insulin production, and operates a negative feedback control of its own production.

The actions of insulin and other hormones affecting metabolism are described further in Chap. 24.

Although gastrin secretion may occur in the fetal pancreas in man, it does not seem to be produced in the pancreas after birth.

## The adrenal medulla

The adrenal glands are a pair of endocrine organs situated over the superior poles of the kidneys. Each gland is made up of a medulla and a cortex.

The adrenal medulla consists of interwoven cords of granule-containing cells, in close contact with preganglionic sympathetic nerve cells, and separated from each other by venous sinuses. The cells have been described as ganglion cells without axons, but containing granules of neurotransmitter substances. In man some cells contain granules of noradrenaline and others granules of adrenaline. Both hormones, or transmitters, are formed from phenylalanine and tyrosine but the final methyl group of adrenaline can only be added in the cells of the adrenal medulla. Thus sympathetic postganglionic nerves and some other nerves can synthesise noradrenaline, but not adrenaline. The gland secretes about six times as much adrenaline as noradrenaline, but the concentrations in blood are about 1.2 nmol/l adrenaline, and 0.16 nmol/l noradrenaline, the blood ratio varying between 4:1 and 8:1.

Adrenaline is transported in the bloodstream and inactivated at the target organ by the enzymes catecholamine *ortho*-methyl transferase (COMT) and monoamine oxidase (MAO). The final product of these reactions, vanillylmandelic acid (VMA) is excreted in the urine. COMT is widely distributed, particularly high concentrations being found in the liver and kidneys, but is not present in sympathetic nerve terminals, unlike MAO which is found in the mitochondria. As a result, the half-life of adrenaline is only about 2 min. Excretion of noradrenaline exceeds that of adrenaline because, although more adrenaline is secreted by the adrenal medulla, noradrenaline is a transmitter at sympathetic nerve terminals and the total daily production is therefore greater than that of adrenaline. The stimuli for adrenaline secretion are varied, but include all those which produce fright, fight, or flight reactions. Exercise and a fall in blood glucose concentrations are further stimuli. The hormone supplements the effects of the sympathetic nervous system. Adrenaline, acting at $\beta_1$ receptors, increases cardiac rate and stroke volume, thus raising blood pressure, and, acting at $\beta_2$ receptors, dilates blood vessels in skeletal muscle and cardiac muscle. It increases blood glucose concentrations by stimulation of $\alpha$ and $\beta_2$ receptors in the liver to promote glycogenolysis, and it stimulates $\beta_1$ receptors in adipose tissue to mobilise free fatty acids. It causes increased heat production. The main role of noradrenaline is as a neurotransmitter, but it is also secreted by the adrenal glands. Its actions are generally similar to those of adrenaline, except that it is usually involved in vasoconstriction rather than vasodilatation because it acts principally at $\alpha$ receptors and $\beta_1$ receptors. The excretory pathways are the same.

## The adrenal cortex

The adrenal cortex of the adult is organised in three zones. The outer zona glomerulosa is a thin layer made up of whorls of large cells which produce the steroid hormone aldosterone. Next to it is the wide zona fasciculata, made up of columns of cells separated by venous sinuses, and next to the medulla is a layer in which the columns of cells merge into an interwoven network about the venous sinuses – the zona reticulata. These inner two zones produce cortisol and androgenic, or masculinising, hormones. A small amount of corticosterone is secreted in all zones. The cells are typical steroid hormone producing cells, and secrete other hormones intermediate along the synthesis pathways, such as progesterone, oestrogens, and testosterone (Fig. 12.2). The hormones of the adrenal cortex are usually classified as glucocorticoids, mineralocorticoids, and androgens, according to their main spectrum of activities. The glucocorticoids raise blood glucose levels but have little effect on the excretion of sodium and potassium ions by the kidneys. The mineralocorticoids cause the kidneys to retain sodium ions and excrete potassium ions. This division is not absolute: the glucocorticoid cortisol, for example, although of low mineralocorticoid activity, does affect sodium retention by the kidneys because its plasma concentration is relatively high.

Cortisol (hydrocortisone) is the main glucocorticoid produced in the adrenal cortex. It is synthesised in all three zones but the production from the outer zone is relatively unimportant. The stimulus to secretion is adrenocorticotrophic hormone (ACTH) from the pituitary gland, which in turn is controlled by the hypothalamus. Stress of any kind appears to be the initiating stimulus. Plasma concentrations (*Table 12.6*) vary from a peak in the morning to low levels in the evening, following ACTH levels. Only about 5% is free, the rest being bound by an $\alpha$ globulin known as transcortin. The free fraction in plasma can be estimated by measuring the concentration in saliva. The hormone is metabolised in the liver, conjugated with glucuronide and then secreted into the kidney tubules. Some passes into the bile and is lost in the faeces. In salivary glands the hormone is converted into cortisone; whilst in plasma the principal active glucocorticoid is cortisol, in saliva cortisone is more abundant. The physiological half-life of cortisol is between 60 and 90 min.

Corticosterone is another glucocorticoid with similar physiological characteristics to cortisol. Its half-life is about 1 h.

All the glucocorticoids have similar actions: they increase the mobilisation of glucose in the liver, and they cause breakdown of fats and proteins so that gluconeogenesis can occur. Thus the plasma concentrations of glucose, aminoacids, and free fatty acids, all increase. Peripheral utilisation of glucose is decreased. These actions together help to provide and maintain a supply of glucose to tissues which cannot use other substrates at times when the supply of glucose in the blood is limited. Secretion of the glucocorticoids is increased in stress: their role in stress is to raise blood glucose levels, to increase movement of water into the plasma compartment to maintain blood volume, to maintain vascular sensi-

**Table 12.6 Hormone production in the adrenal cortex**

| Hormone | Daily production (μmol) | Plasma concentration (μmol) | Role |
| --- | --- | --- | --- |
| Dehydroepiandrosterone | 73 (male) 55 (female) | 1.56 | Androgen |
| Cortisol | 55 | 0.42 | Glucocorticoid |
| Corticosterone | 8.6 | 11.5 | Mineralocorticoid |
| Aldosterone | 0.4 | 0.16 | Mineralocorticoid |
| Progesterone | Very small | | |
| Oestrogens | Very small | | |

Note that although the secretion rates and the plasma concentrations of aldosterone are much less than those of corticosterone, it is about 200 times as potent in its mineralocorticoid effect.

tivity to the catecholamines, and to permit the catecholamines to act on fat cells to mobilise free fatty acids as an emergency energy source. The hormones inhibit or reduce inflammatory and allergic reactions. Kinin and histamine release from damaged cells is reduced, local swelling is decreased, fibroblast activity is reduced, systemic effects of bacterial toxins are suppressed, and antibody production, after an initial boost, is much reduced. In addition to these effects, the glucocorticoids alter blood cell counts: red cell counts rise slightly, lymphocyte, eosinophil and basophil counts fall, and the numbers of other white blood cells are increased.

Corticosteroids are used widely in medicine and dentistry to reduce inflammation; on the whole they alleviate the symptoms but do not treat the cause of the disease. Their use may suppress body reactions to harmful stimuli. Since they are lipid-soluble they are absorbed through the skin or through the mucous membrane of the mouth and so even surface preparations can produce systemic effects. Administration of corticosteroids suppresses ACTH secretion and over a period of time some atrophy of the adrenal cortex may occur, leaving the patient less able to react to stress. Patients on long-term steroid therapy may therefore present problems in dental practice or oral surgery.

The principal mineralocorticoid is aldosterone. The metabolic pathway for its synthesis is found only in the zona glomerulosa. Since only a small proportion in plasma is bound by plasma protein, its half-life is relatively short at about 20 min. It is inactivated and bound to glucuronides in the liver and excreted in the urine. The main stimulus for aldosterone secretion is angiotensin II, produced in the plasma from a plasma protein known as angiotensinogen as a result of the action of the enzyme renin from the juxtaglomerular bodies of the kidney tubules. A fall in blood pressure or a decrease in plasma sodium concentrations is the primary stimulus (other stimuli are discussed later in relation to renin). ACTH probably has little direct effect on aldosterone secretion, but it has a permissive effect in enhancing the response to other stimuli. The adrenal cortex is also stimulated directly by decreases in plasma potassium concentrations in excess of 20%. The hormone produces its effect of stimulating sodium reabsorption in the distal tubules of the kidneys by increasing production of proteins involved in transport of sodium ions, and increasing ATP production. As more sodium ions are reabsorbed, some potassium ions are lost in the urine in exchange. A similar effect is produced in the salivary glands, aldosterone increasing reabsorption of sodium in the striated ducts and decreasing the concentration of sodium ions in the final saliva. In the small intestine and, more particularly, the colon, aldosterone increases sodium transport promoting absorption of sodium ions and hence water absorption also. Changes in salt balance also affect water balance: increased sodium reabsorption is accompanied by an increase in extracellular fluid volume.

Although in the testis the androgenic hormone, dehydroepiandrosterone (DHEA), is a precursor of testosterone and, indeed, of oestrogens, further conversion in the adrenal gland is very limited. Control of its production in the adult is probably via ACTH. It is secreted both free and as a sulphate. In the fetus DHEA is secreted by a special intermediate zone of the adrenal gland next to the medulla and is very important in growth, particularly in the female, in whom it is the principal anabolic hormone. Its other vital role in the fetus is to provide a substrate from which the placenta can synthesis oestriol (see below and p. 282). Like the other androgenic hormone, testosterone, DHEA stimulates protein synthesis. It is inactivated in the liver and excreted as a glucuronide. Like testosterone again, the hormone can produce masculinising effects in women if secreted in excess.

## The testes

The testes are a pair of organs situated in the scrotum of the male. They are made up of a series of blind-ended tubules, the seminiferous tubules, lined with germinal epithelium which gives rise to the spermatozoa. The tubules are separated by groups of cells containing lipid granules – the interstitial cells of Leydig. These produce the characteristic androgenic steroid hormone, testosterone. The prostate gland, and probably other tissues also, convert testosterone into a more active metabolite, dihydrotestosterone. The seminiferous tubules secrete prostaglandins.

Although it is mainly produced in the testes, a small amount of testosterone is also synthesised in the adrenal glands, and some may be formed in the ovary. Production is controlled by the hormone LH from the anterior pituitary gland, itself controlled by LHRH from the hypothalamus with negative feedback control from the level of circulating testosterone. In plasma the hormone is bound by a β globulin, gonadal steroid binding globulin (GBG), and by albumin, so that only about 3% is free. The plasma concentration is 20 times as great in males as in females. Testosterone is an anabolic hormone, promoting protein synthesis. It is responsible for growth and development in the male, including the development of secondary sexual characteristics. It stimulates bone growth and red blood cell production. Psychological effects of the hormone include aggressive behaviour. In many tissues testosterone can be converted into oestrogens, but the main route for inactivation is to androsterone in the liver. It is conjugated with glucuronide and excreted in the urine. It is possible that testosterone must be converted into the dihydro- form before it can produce its effects on cells. Inhibin is a polypeptide hormone secreted in the seminiferous tubules, which inhibits FSH production in the anterior pituitary gland.

## The ovaries

The ovaries produce a number of hormones. The follicles, maturing under the influence of FSH, secrete oestrogens, and when the follicle bursts to release the ovum, the tissue remaining forms the corpus luteum which can synthesise both oestrogens and progesterone. The menstrual cycle is described in Chap. 22.

The oestrogens are the characteristic steroid hormones of the female. They include 17β-oestradiol, oestrone, and oestriol. The pathway by which they are formed is via the androgens and in some cells androgens act as substrates for oestrogen production. Progesterone is formed by a much simpler pathway, involving only two steps from cholesterol, and is an intermediate in the synthesis of all steroid hormones.

17β-Oestradiol is the main oestrogen produced in the female except during pregnancy. It is secreted by the theca interna cells of the ovarian follicles, by the corpus luteum, and, in pregnancy, by the placenta. The adrenal cortex also produces a small amount. The stimulus for oestrogen production is a combination of secretion of FSH and LH by the anterior pituitary gland – FSH by itself is not effective. There is a negative feedback control, as oestrogens in the plasma suppress LHRH secretion by the hypothalamus, and therefore suppress the secretion of pituitary gonadotrophins. The concentration of 17β-oestradiol in plasma is about 0.37 nmol/l at menstruation, rises sharply after menstruation to a peak immediately before ovulation, falls rapidly again, and then rises more slowly for some 8 days after ovulation. From this peak, concentrations decline slowly to the original level. The daily production in males is around 70% of that in females in the early follicular phase. About 70% of circulating oestrogen is bound by the plasma gonadal steroid binding globulin (GBG) which binds testosterone also. In the liver the hormones are oxidised and/or conjugated with glucuronide or sulphate and the conjugates are then secreted into the urine in both male and female.

Oestrogens from the developing follicle cause the endometrium (the mucosal lining of the uterus) to proliferate and thicken, and its glands to increase in size. They cause the glands of the uterine cervix to produce a watery alkaline secretion. Under the influence of the oestrogens the uterine muscle thickens and becomes more active and more susceptible to the stimulus of oxytocin. The ovarian follicles develop more rapidly in the presence of oestrogens and the fallopian tubes increase in activity. A number of other effects are reported, such as salt and water retention, a lowering of blood cholesterol concentrations, a slowing of the rate of red blood cell maturation, and some psychological effects. The metabolic rate in women is lower than in men but rises at menstruation and during pregnancy. It is possible that oestrogens are responsible for this. Oestrogens increase the blood flow to the oral mucosa and render the blood vessels more permeable.

Female secondary sexual characteristics result in part from an absence of testosterone, but principally from the presence of the oestrogens. The development of the breasts at puberty and the changes which occur in the breasts in preparation for, and during, lactation, all result from the actions of oestrogens (Chaps. 22 and 23).

The comments on oestrogens relate to oestrone also.

Oestriol is the third oestrogen produced in significant amounts. Its production is greatly increased during pregnancy, since it is the principal oestrogen secreted by the placenta. It is much less potent than 17β-oestradiol, having only about one tenth the activity. There are relatively large amounts of oestriol excreted in the urine by both women and men. The general comments on oestrogens apply also to oestriol.

Progesterone follows pregnenolone in the metabolic pathways of steroid hormone synthesis – so it is produced in small quantities in all steroid producing glands. Thus both the adrenal glands and the testes produce some progesterone. In the corpus luteum and in the placenta the pathways for steroid synthesis from cholesterol stop at the 17α-hydroxylase stage; the oestrogens produced in the placenta are therefore synthesised from androgens secreted by the fetal adrenal glands. In the corpus luteum secretion is stimulated by LH from the anterior pituitary. Plasma concentrations are about 10 times higher during pregnancy than in the midluteal phase of the menstrual cycle. The hormone is bound by a carrier protein in the bloodstream. Progesterone causes development of the secretory glands of the endometrium after ovulation. When the corpus luteum ceases to secrete progesterone the endometrium can no longer be maintained, and it is shed in the menstrual flow. Progesterone causes the mucous secretion of the uterine cervix to become thicker and more tenacious. In pregnancy it is important in the development of the breasts, but inhibits milk secretion. It decreases the excitability of the uterine muscle and its response to oxytocin. There is some evidence that progesterone increases the permeability of blood vessels and alters blood flow in the oral mucosa. Certainly in other tissues it causes some vasodilatation. Progesterone also has a general thermogenic effect. Progesterone is converted to pregnanediol in the liver, conjugated with glucuronide, and secreted into the urine in the kidneys.

Synthetic progesterones are used to block the burst of LH secretion in the normal menstrual cycle which causes ovulation: they are the basis of many oral contraceptive preparations.

Three very similar polypeptides termed relaxins are produced in the ovaries, possibly from the corpus luteum. They are believed to be secreted to relax the symphysis pubis and the uterus, and to soften the cervix to aid in the process of parturition (birth). Their significance in the human is uncertain.

## The placenta

The fertilised ovum becomes embedded in the wall of the uterus, and its outer layers build up to form the placenta, the nutritive organ for the growing fetus. It has a profuse blood supply, with the fetal and maternal vessels arranged in parallel. It is an active hormone-secreting organ but is best considered, not in isolation, but as forming a functional unit with the fetus itself, the feto-placental unit. The endocrine activities of the placenta and the fetus are very closely linked. The placenta produces steroid hormones, (progesterone and oestrogens), protein and glycoprotein hormones, and some prostaglandins. Placental production of its main oestrogen, oestriol, depends upon the production of dehydroepiandrosterone by the fetus, because the placenta has no alternative pathways for oestrogen synthesis. The hormones of the placenta affect both the fetus and the mother.

In addition to the steroid hormones, progesterone and the oestrogens, which have already been described, the placenta produces a protein hormone, human placental lactogen (hPL, human chorionic somatomammotrophin, hCS), which is similar to growth hormone. It stimulates development of the mammary glands and milk production, causes retention of nitrogen, calcium and phosphorus in the mother, so maintaining the maternal tissues and promoting growth of the fetal tissues, and also stimulates the laying down of fat and the switchover to a glucose-sparing type of metabolism (cf growth hormone on p. 302).

If pregnancy occurs, the control of the function of the corpus luteum in producing progesterone and oestrogens is taken over by human placental gonadotrophin (hCG), a glycoprotein hormone similar in structure to TSH, FSH and LH. The highest concentrations are observed in the blood between 5 and 10 weeks after conception; after this the placenta itself takes over the synthesis of progesterone and the corpus luteum becomes unimportant. This is the hormone assayed in human pregnancy tests.

## The mucosa of the stomach and the small intestine

A number of polypeptide hormones are produced by scattered cells in the stomach and the small intestine.

Gastrin is produced mainly by cells in the lateral walls of glands in the pyloric antrum of the stomach, although there are also some gastrin-producing cells in the duodenum. Several gastrins of different molecular size have been described. Secretion is stimulated by the vagus nerve, or by the distension of the stomach wall, or by the presence of peptides and proteins in the antrum. Alcohol stimulates gastrin secretion. In the bloodstream gastrin is sometimes found attached to a carrier protein. The main actions of gastrin are to increase the secretion of acid, and to a lesser extent, pepsin, by the glands of the stomach, to increase the motility of the stomach and to relax the gastro-oesophageal (cardiac) sphincter. As concentrations of gastric acid increase in the antrum, they inhibit further gastrin secretion. In the pancreas gastrin stimulates secretion of insulin and glucagon. It increases calcitonin secretion and it may also have some effect on the secretion of amylase in saliva.

Secretin is produced by mucosal cells in the upper part of the small intestine in response to the presence of acid chyme in the duodenum. Free fatty acids in the duodenum also stimulate its secretion. Secretin acts on the pancreas, causing it to secrete a juice rich in hydrogen carbonate and on the liver, stimulating secretion of bile. It also inhibits acid production in the stomach, and reduces gastric and intestinal motility. It may stimulate insulin secretion. The activity of secretin was demonstrated by Bayliss and Starling in 1902 – the first hormonal effect to be recognised.

In addition to secretin the cells of the mucosa of the upper part of the small intestine were originally thought to secrete a hormone stimulating pancreatic enzyme production and another causing contraction of the gall bladder. These two, pancreozymin and cholecystokinin, have now been characterised as a single polypeptide hormone. The digestion of fat and protein causes breakdown products to appear in the small intestine, and these (in particular, certain aminoacid residues and fatty acids with more than 10 carbon atoms) stimulate the hormone-producing cells to secrete CCK-Pz. The hormone stimulates enzyme secretion from the pancreas, and causes the gall bladder to contract, sending a stream of bile into the duodenum. If it were to act by itself, it would stimulate gastric acid secretion, but normally it acts as a competitive inhibitor of gastrin-stimulated acid secretion. CCK-Pz stimulates glucagon secretion and may also increase calcitonin secretion.

There are several other polypeptide hormones secreted in the upper part of the small intestine. Their role in the integrated control of the gastro-intestinal tract is still being studied. They include enterogastrone, a peptide similar to epithelial growth factor, which is able to inhibit gastrin activity in the stomach; motilin, which causes increased motility in the stomach and small intestine; gastric inhibitory peptide (GIP), which inhibits gastric secretion and gastric motility, and stimulates the release of insulin in the pancreas; vaso-active intestinal peptide (VIP), which produces a general vasodilatation in the intestinal tract and causes an increased secretion of electrolytes and water in the small intestine, but inhibits gastric acid production; and villikinin, a substance thought to cause contraction of the muscularis mucosae in the villi of the small intestine.

A substance with similar immunological reactions to glucagon, glucagon-like immunoreactive factor, GLI, has been detected in the mucosa of the stomach, duodenum and ileum. It is also known as enteroglucagon and glicentin. Like glucagon itself acting directly, enteroglucagon stimulates insulin release. It also increases gastric motility. The stimulus for enteroglucagon release is the presence of glucose or free fatty acids in the ileum.

Somatostatin is secreted in many parts of the gut and in glands such as the pancreas. Its function there is uncertain, but it is known to inhibit production of acid in the stomach, and it may be a neurotransmitter.

## The kidneys

The kidneys do not themselves produce hormones but contain or produce enzymes which activate substances with hormonal activity. The enzymes include renin, produced in the juxtaglomerular bodies in response to a fall in blood pressure, and the renal erythropoietic factor (REF), produced in response to a fall in the amount of oxygen carried in the blood. The kidneys can also convert a metabolite of vitamin D into products with greater or lesser hormonal activity.

The afferent and efferent arterioles of a renal glomerulus are close together as they enter the glomerulus and also come into contact with the beginning of the distal convoluted tubule. The tubular epithelium in contact with the afferent vessel wall has dark-staining cells which contain granular material. These cells form the macula densa. The tunica media of the wall of the afferent vessel contains epithelioid cells known as juxtaglomerular cells. Together these structures form the juxtaglomerular bodies or apparatus. A fall in blood pressure causes these organs to secrete the enzyme renin, which splits off angiotensin I from angiotensinogen. In the lungs, a non-specific converting enzyme removes a further aminoacid from angiotensin I to convert it into the active angiotensin II (Fig. 12.8a). This has a very short half-life because it is broken down by circulating peptidases. Angiotensin II causes vasoconstriction – it is the most powerful physiological vasoconstrictor known – and stimulates the adrenal cortex to secrete aldosterone. Sodium ions are therefore retained in the distal tubules of the kidneys and this results in an increase in extracellular fluid volume and the blood pressure. Angiotensin II also stimulates a sensation of thirst. Fall in blood pressure at the juxtaglomerular apparatus is sensed by stretch receptors in the juxtaglomerular cells. This is not, however, the only stimulus for renin secretion. The macula densa seems to respond to the transport of sodium and chloride ions across the wall of the distal tubule: renin secretion increases as the rate of transport of these ions falls. Decreases in potassium ion concentration also stimulate renin secretion, possibly by altering sodium and chloride movement across the macula densa. Prostaglandins, particularly prostacyclin, increase renin secretion: they may be the mediators by which the macula densa affects it. Both the sympathetic nerves to the kidney and circulating

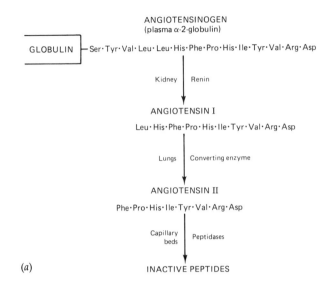

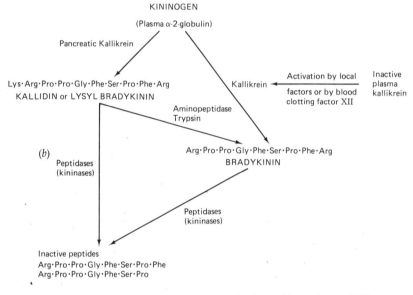

**Figure 12.8** (*a*) The mechanisms of angiotensin production and breakdown. (*b*) The mechanisms of bradykinin production and breakdown.

adrenaline increase renin secretion. Vasopressin and angiotensin II itself decrease renin release. Renin, and angiotensin production, are not limited to the kidneys: some is found in the hypothalamus, and it has been reported in many other situations.

Low concentrations of oxygen in the blood reaching the kidneys stimulate production of the renal erythropoietic factor, and this causes the formation of the glycoprotein, erythropoietin, in the blood. Erythropoietin production is favoured by alkalosis. Androgenic hormones stimulate production and oestrogens inhibit it. The hormone is a glycoprotein with a half-life of about 5 h. It stimulates stem cells in the bone marrow to form pro-erythroblasts and therefore increases red blood cell production.

Although vitamin D has traditionally been considered as a dietary factor, it has always been recognised that the active vitamin is also synthesised in the skin from a provitamin, using the energy derived from ultraviolet light. The active vitamin is cholecalciferol, vitamin $D_3$. This is hydroxylated at the 25- position in the liver and then passes to the kidneys in the circulation. In the kidneys the 25-hydroxycholecalciferol is further hydroxylated, either to 1,25-dihydroxycholecalciferol, $1,25(OH)_2D_3$, or to 24,25-dihydroxycholecalciferol, $24,25(OH)_2D_3$, or to the inactive 1,24,25-trihydroxycholecalciferol. The amounts of the different metabolites produced is governed by the blood calcium concentration. Low calcium levels stimulate and high calcium levels depress the formation of $1,25(OH)_2D_3$. This effect is probably mediated by parathormone. $1,25(OH)_2D_3$ stimulates the production of calcium-binding protein in the cells of the small intestine and probably also stimulates calcium transport. By increasing absorption of calcium, the hormone raises the concentration of calcium in the blood. A similar effect may be produced on bone cells by $24,25(OH)_2D_3$, which also inhibits parathormone secretion.

## The liver

The liver is involved in many metabolic transformations, of which the hydroxylation of cholecalciferol mentioned above is but one. Somatomedin C is one of the few characterised substances with hormonal activity produced in the liver.

Somatomedin C is a polypeptide produced in the liver in response to circulating growth hormone in children. Growth hormone does not act directly on tissues to cause growth, but stimulates the formation of polypeptides which act as growth factors. The somatomedins are examples of these. Somatomedin C promotes the uptake of sulphate ions into proteoglycans, and of thymidine into DNA. It stimulates the synthesis of RNA and of protein molecules.

## Autacoids

The autacoids are locally-acting substances, producing effects mainly upon blood vessels. They are important in inflammation and in the response to injury.

Histamine is produced by decarboxylation of the aminoacid histidine. It is usually found in mast cells, and is released when the tissue is damaged, or in the course of antigen-antibody responses. It causes an increase in the permeability of blood capillaries, and therefore an outflow of plasma proteins and fluid, and also produces vasodilatation of the arterioles, but constriction of the venules. It is taken up and inactivated by eosinophils. In the stomach histamine stimulates acid secretion, acting on the $H_2$ receptors of the parietal cells. The effect is mediated by cyclic AMP.

A number of small polypeptides, termed kinins, are cleaved from plasma proteins by enzymes termed kallikreins, during tissue activity or tissue damage. The best known of these is bradykinin (Fig. 12.8b). They increase blood flow by causing vasodilatation, and also increase capillary permeability to some extent.

## Prostaglandins

These derivatives of arachidonic acid (Fig. 12.9) produce a number of local effects. Their actions are often linked with that of cAMP but no clear pattern of activity emerges. Thus they mimic the action of TSH, ACTH, LH, and PTH on target tissues – all activities involving cAMP – but they antagonise the lipolytic effects of hormones, and the actions of ADH, which also involve cAMP. They increase intestinal motility. Ovulation is associated with an increase in prostaglandin activity.

Both the prostaglandins E decrease acid production in the stomach. They induce and increase inflammatory responses, and they act as pyrogens – they raise the hypothalamic set point for temperature regulation. They lower blood pressure. Prostaglandin $E_1$ prevents blood platelet aggregation and degranulation, while prostaglandin $E_2$ stimulates these processes.

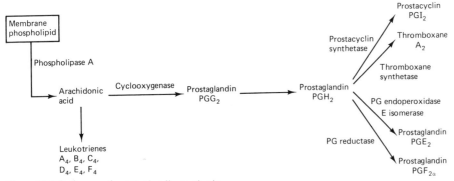

**Figure 12.9** Pathways of prostaglandin synthesis.

In hypertension prostaglandin A lowers blood pressure and has diuretic and natriuretic effects.

Although prostaglandin $F_{2\alpha}$ increases intestinal motility, its main effects seem to be on the female genital tract, where it can cause ovulation. Its effect on contraction of uterine smooth muscle has led to suggestions that it may be the main factor causing parturition to begin.

Prostacyclin is found in the intima of the blood vessels. It relaxes vascular smooth muscle and inhibits the aggregation of blood platelets. Its presence in the maternal aorta and pulmonary vessels and in the placenta is thought to be essential for the proper development of the fetal and neonatal circulation. It causes the release of renin in the juxtaglomerular bodies of the kidneys.

Thromboxane $A_2$, produced by the blood platelets, is important in both normal and abnormal blood clotting. It is a vasoconstrictor and it promotes aggregation of blood platelets.

## Further reading

Hardy, R.N. *Endocrine Physiology*, Arnold, London, 1981

Tepperman, J. *Metabolic and Endocrine Physiology*, 5th edn, Year Book Medical Publishers, Chicago, 1987

# Section III
## Physiological control

# 13

# Mechanisms of control

## Feedback and control systems

The fundamental concept in physiology is that of control. This includes control of the internal environment so as to permit cell and organ function to continue at an optimal level, and control in the sense of the direction of both conscious and unconscious functions on which, however indirectly, the survival of the individual and the species depends. Control of the internal environment means the maintenance of an environment for the cells of the body, stable in temperature, in osmotic activity, in pH and in the concentrations of ions, nutriments and metabolites. The term homeostasis is used to describe this maintenance of a stable environment. In this chapter the essential elements of a control system will be considered; this outline will form a framework for the consideration of homeostatic mechanisms and other control systems in subsequent chapters.

Most control systems are termed feedback systems, since information about any changes which are occurring must be fed back to some controller which can cause an appropriate action, usually one to restore the original condition. Systems of this kind are negative feedback systems – the active change reversing the initial change. Negative feedback systems are inherently stable, always resisting change. They are typical of biological systems.

The second type of control referred to above is that necessary for action, and is therefore more characteristic of conscious activity. This type of control includes positive feedback mechanisms, where a detected change induces a response designed to increase the magnitude of the change itself. It also includes the type of control known to engineers as servo control, in which a small induced change is amplified by various mechanisms to provide a final much larger change. Positive feedback systems are inherently unstable, accentuating changes imposed upon the organism.

The control of a central heating system provides an everyday example of a negative feedback system. Heaters are set to maintain a given temperature in a room. Any departure from the set temperature causes the heaters to operate more actively if temperature falls, or less actively if temperature rises. Positive feedback can be illustrated in a supermarket price war. One firm cuts prices, others to maintain sales follow suit. The first firm is then compelled to cut prices further if it does not wish to lose customers: a cycle of price cutting is set up. The cycle ends, however, when firms can no longer cut their prices further and go out of business. Servo control is exerted in steering a ship: a small, effortless movement of the wheel produces a much larger, more difficult movement of the rudder and this in turn causes the whole ship to change direction.

## A basic control system

The basic elements of a control system are summarised in a block diagram (Fig. 13.1). The variable factor to be controlled must be measured by some sensor – in biological terms, a receptor. Some receptors monitor the variable directly, others indirectly. Thus the carotid body chemoreceptors assess directly the partial pressure of oxygen dissolved in the blood. The chemoreceptors of the medulla oblongata measure blood carbon dioxide, but they do this by measuring the hydrogen ion concentration in the cerebrospinal fluid, a related variable.

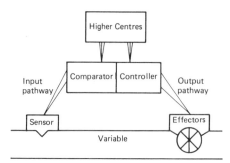

**Figure 13.1** Outline of a model control system.

The information must be fed back to some controlling centre along an input pathway, usually a nerve, although in some instances this is not necessary because the sensor and the controlling centre are the same cell.

The controlling centre, sometimes called the comparator, is usually in the brain or spinal cord. Here the information from the sensor or sensors is compared with some internally set level and any departure from the set level causes appropriate instructions to be sent out. Sometimes the comparison of observed values with the set values may occur at the sensors themselves. A room thermostat provides one example, since the required temperature is set in a device which is the actual temperature sensing device and is also the control switch for the heating system. The hypothalamus acts both as a sensor and a controller for the osmotic activity of the extracellular fluid. The term controlling centre is therefore preferable for the area in which incoming information is translated into outgoing instructions and the term comparator reserved, if necessary, for the particular structures where the comparison takes place, whether in the control centre or elsewhere.

The output pathway may be via nerves or via hormonal messengers in the bloodstream. It may be so short as to have no separate identity, or it may be a clearly defined pathway of considerable length.

The effector organ is usually muscle or a gland, capable of changing the controlled variable.

These components make up the simplest possible feedback circuit. The circuit may be complicated by the duplication or multiplication of its components – several sensors, several input pathways, hierarchies of control centres, multiple output pathways, several effector organs. In the control of blood pressure, for example, the effector may be the heart or the blood vessels. In the control of rising body temperature, blood vessels may dilate and sweat glands may secrete.

Further complications arise in the possibility of control of the controlling centre itself, either by alteration of its comparator function or by directly altering its activity.

## Direct and indirect sensors

Two kinds of sensor are distinguished – misalignment detectors and disturbance detectors. Misalignment detectors monitor the controlled variable either directly or indirectly, disturbance detectors respond to changes likely to affect the controlled variable. A room thermostat is a misalignment detector, an outdoor thermometer a disturbance detector. Disturbance detectors give early warning of an impending change in the controlled variable. Physiologically, skin temperature receptors are disturbance detectors, whilst hypothalamic deep temperature receptors are misalignment detectors. An example of the interaction of the two types of receptor can be seen in the control of blood osmolality. Osmoreceptors in the hypothalamus act as misalignment detectors to provoke a sensation of thirst which can be abolished by drinking. Intake of water will not immediately affect blood osmolality; the oral and pharyngeal receptors can function as disturbance detectors to cause suppression of the osmoreceptor signals so that the body is not overhydrated by continuing to drink beyond the necessary amount.

## Variation of the set level

The set level of a feedback system in the body may be altered in a number of ways. For example, in exercise the temperature setting of the hypothalamic control centre is set at a much higher level.

Activity of the limbic brain may alter the setting of the blood pressure controlling system in the medulla oblongata by means of impulses passing through the hypothalamus. Stiffening of the arterial walls with increasing age reduces the sensitivity of the carotid sinus stretch receptors: as a result blood pressure is maintained at a higher level in the old than in the young.

The concentration of cortisol in the blood is monitored by the hypothalamus and, as the concentration falls, the hypothalamus secretes corticotrophin releasing hormone (CRH) into the portal bloodstream: when this reaches the anterior pituitary gland the secretion of adrenocorticotrophic hormone is stimulated. This in turn stimulates the adrenal cortex to secrete cortisol. The sensitivity of the hypothalamus to cortisol is reduced by stress: and so higher concentrations of cortisol are maintained in the bloodstream.

Normally the regular alternation of breathing, inspiration and expiration is controlled by the respiratory centres in the medulla oblongata but this control can be voluntarily over-ridden during breath-holding, or during speech.

## Mathematical analysis of control systems

The study of feedback systems in engineering and electronic applications has led to the development of mathematical expressions for their analysis. They deal mainly with the manner in which the mechanisms act, rather than the mechanisms themselves. Thus a system may be proportional in that the response is directly proportional to the magnitude of the departure from the set level. In its simplest form such a response will begin rapidly, when the discrepancy is greatest, and slow down as the discrepancy diminishes, finally reaching the set level fairly slowly. This kind of response is seen in the control of the concentration of carbon dioxide in the blood: since the body can tolerate small excesses of carbon dioxide relatively easily, the slow but close adjustment to the set level is permissible. However, if the concentration of blood oxygen falls rapidly, a much more dangerous situation arises, and a rapid adjustment must be made. Rapid adjustments cannot allow for the diminishing error in the system and so the correction overshoots, corrects the other way, overshoots again, and ends in a series of rapid alternate corrections seen as an oscillation about the set level which gets smaller and smaller until it dies away. In more complex responses the rate of correction of the variable is speeded up by having a 'derivative' control: the response then depends not on the magnitude of the departure from the set level but on the rate of change of the controlled variable from the set level. Again, a potentially dangerous fall in blood oxygen concentration stimulates a response whose speed depends not only on the absolute change in oxygen concentration but also on the rate at which it is changing.

Characterisation of a system as being 'proportional' or 'derivative' is merely a way of describing how it responds: it does not reveal the mechanism of the response although it may indicate something of the way in which the components, such as the sensors and the controlling centre, must operate.

## Positive feedback systems

The majority of biological systems, and the majority of examples given, are negative feedback systems. Positive feedback systems, with their natural instability, are much less common. In fact, if positive feedback systems continued indefinitely, the results would be incompatible with biological survival: the organism can exist only by maintaining some degree of internal homeostasis. All the examples of positive feedback systems in physiological situations, therefore, have a cut-off point, beyond which the positive feedback ceases to operate, and other controls take over. A good example of positive feedback, though not strictly of control, occurs in the initiation of the action potential. A change in membrane permeability permits sodium ions to enter a nerve cell; this results in depolarisation; depolarisation increases permeability; more sodium ions enter; depolarisation increases; and so the process goes on until suddenly the permeability to sodium ions decreases again. A positive feedback control system is exerted in the initiation of childbirth. Movement of the fetus causes stimulation of receptors in the cervical segment of the uterus; nervous impulses reaching the hypothalamus cause oxytocin to be released from the posterior pituitary gland; oxytocin causes uterine contraction; uterine contraction stimulates receptors in the uterine wall; more oxytocin is released; and the process continues until birth actually takes place.

## Servo control systems

The distinction between a positive feedback system and a servo system is summarised in Fig. 13.2. In the servo system a small induced change is amplified by various mechanisms to produce a final much larger change. Biologically the initial small change is often the imposition by a higher centre of a new set level on the feedback system in order to achieve a change in a controlled variable; normally in a feedback system the response is to the change in a controlled variable. In muscle contraction the muscle spindles may be used as a servo system. In order to close the jaws upon a tough bolus of food the system must be reset to a new value appropriate to the closed position. The amount of force exerted by the masseter and medial pterygoid muscles is relatively small until the teeth meet the resistance of the bolus. If the muscles contract with the same force throughout the stroke, the bolus is scarcely penetrated by the teeth. The stretch receptors in the muscle signal that shortening of the muscle is occurring as programmed; muscle activity continues as before. However, when impulses pass to A$\alpha$ motor neurones innervating the extrafusal muscle fibres a simultaneous output to the A$\gamma$ motor

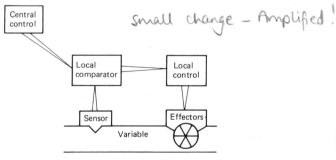

**Figure 13.2** Outline of a servo control system. The central control sets the level at the local comparator and/or control centre: the local circuit then adjusts the variable.

neurones reaches the contractile fibres within the muscle spindle itself. As their contractile poles shorten the stretch receptors are elongated; this in turn results in increased drive to the A$\alpha$ motor neurones to reinforce the contraction of the extrafusal fibres. Only when the bolus is penetrated do the extrafusal fibres shorten sufficiently to reduce the discharge of the spindle receptors, and hence reduce further drive to the A$\alpha$ motor neurones. The activity of a negative feedback regulator has been adjusted by a servo control which has altered the set level.

# Reflexes

A reflex is defined as an involuntary response to a peripheral stimulus. This definition is not sufficiently exclusive and a number of supplementary conditions have to be laid down. Some activities are difficult to classify as involuntary. In order to aid in definition one can add that a response must be inborn, and that it must be present in all members of the species. There are problems over complex activities like the maintenance of an erect posture, or movements such as walking where conscious control is exerted over a series of automatic actions. The proviso that the response must be inborn eliminates learned responses such as are seen in conditioned reflexes. General usage applies the term much more generally than does scientific usage, referring to many learned responses as reflexes. Drivers of motor cars are said to have good, or rapid, reflexes: such responses are obviously not manifest in all members of the species. In learning responses, the individual repeats particular thought pathways sufficiently often to omit, apparently, intervening steps: the co-ordination of two feet, two arms and two eyes becomes 'automatic' and the individual sequence of activities is not broken down into its component parts. Such learned responses are termed conditioned reflexes: they are discussed later. Another proviso, which is perhaps understood in the term involuntary, is that the nervous pathway should not involve the cortex of the brain; however, an input signal to the cortex giving perception of the stimulus simultaneously with and later than the reflex response does not invalidate description of a response as reflex. Withdrawal of the hand from a very hot object is achieved without the need for information to reach the cortex even though a painful sensation is almost simultaneously evoked. Not all reflexes are limited to the nervous system; there are also hormonal reflex responses. Some hormonal reflexes have already been mentioned above; others will occur in later chapters. For the moment, however, only the properties of entirely neuronal reflexes will be described.

A number of responses involve enteroreceptors rather than exteroreceptors: the use of the term peripheral stimulus in the definition of a reflex includes all stimuli outside the central nervous system. Reflexes controlling the process of breathing, or the beating of the heart, are triggered by internal stimuli. Many of these reflexes occur within the autonomic nervous system: indeed, the autonomic nervous system is so named because it is involved almost entirely in reflex activity (*see* Chap. 9).

## The reflex arc

The reflex arc is made up of an afferent pathway from a receptor and an efferent pathway to an effector: it is similar in outline to a negative feedback control system (Fig. 13.3). Indeed the whole range of regulatory reflexes are themselves negative feedback control systems. However, the pathway outlined in the diagram is appropriate not only for regulatory reflexes but also for two other types of reflex: the nociceptive reflex in which a painful or harmful stimulus causes the organism to withdraw itself or attempt to remove the stimulus, and the other type of reflex in which the organism produces a change of position or state in order to ensure its survival. The simplest entirely neuronal reflex arc consists of two neurones only. Although a single neurone reflex may appear to be impossible, the branching of a sensory nerve fibre can permit a sensory impulse travelling towards the spinal cord to pass antidromically down another branch and produce a response at a normally sensory termination. Such an axon reflex is observed in the control of the skin circulation (*see* p. 235). Leaving this

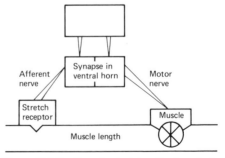

**Figure 13.3** Outline of a stretch reflex in a similar format to *Fig. 13.2*.

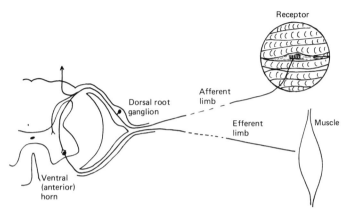

**Figure 13.4** Outline of the components of a stretch reflex. The components are obviously not to scale but are represented at sizes to render them identifiable – the receptor is placed in a circle to indicate that it is a low-power magnification.

anomalous example aside, however, it is even debatable whether a true two-neurone reflex can exist. The stretch reflex – contraction of a muscle in response to being stretched – is described as a two-neurone reflex, but if all the concurrent actions which permit that muscle to contract are included, many more neurones are involved.

However, what is normally described as the simplest reflex takes the following form (Fig. 13.4). The reflex pathway will be described in terms of a spinal sensory nerve, the spinal pathways and spinal motor nerves; reflexes involving cranial sensory and motor nerves follow analogous routes. In many homeostatic reflexes the effector pathway is hormonal. A sensory nerve enters the spinal cord by a dorsal root or its cranial equivalent. The cell body of the spinal sensory afferent is in the dorsal root ganglion (the cranial equivalent is the nucleus of the cranial nerve). This is the afferent limb of the reflex arc. The neurone then synapses with an anterior horn cell – the cell body of a motor nerve – and the motor axon then leaves the cord in the ventral root of the spinal nerve, as the efferent limb of the reflex arc. Such a simple reflex – and probably only the stretch reflex is as simple as this – is usually accompanied by synergistic reflexes or other additional reactions. A muscle cannot easily shorten unless its antagonist muscle relaxes; contraction of an extensor muscle usually requires relaxation of a flexor muscle, or vice versa. This is achieved by stimulation of other interneurones (internuncial or messenger neurones) which inhibit the motor nerve to the opposing muscle. The axons of such internuncial neurones may stay in one segment of the spinal cord or project up or down to other segments to activate additional pathways.

Because most reflexes occur at spinal level and are modified or inhibited by impulses from higher in the spinal cord or from the brain itself, damage to the spinal cord above the level of the reflex pathway or damage to the brain may leave a spinal reflex intact or even uncover a fundamental response normally masked by control from the higher levels. Normal reflex activity may persist during general anaesthesia but be abolished by local anaesthetics which block the sensory pathways. The complex responses which prevent the biting of the lips or cheeks during mastication may fail to operate when the mandibular nerve is blocked after carrying out treatment in the lower dental arch.

## Properties of reflexes

The general properties of reflexes are those of nerve-synapse chains. Thus reflexes show a latency, a finite time from stimulus to response. The rate of conduction of an impulse along a nerve is rapid in comparison with that of synaptic transmission. Reflex latency is therefore proportional to the number of synapses along the reflex pathway although the length and diameter of the axons involved is also important. From the latency it is possible to calculate the number of neurones and synapses involved. Very recent work suggests that the jaw jerk reflex – closing together of the jaws by the action of the masseter muscle after it has been stretched by a sharp downward tap on the chin – can be achieved by a two-neurone, one-synapse pathway (Fig. 13.5).

Like synapses again, reflexes show fatigue. Although nerves eventually fatigue, the usual cause of a pathway ceasing to conduct impulses is the depletion of neurotransmitter granules at the presynaptic terminals of the nerves. The rate at which a reflex begins to show fatigue is of the same order as that at which synapses cease to conduct.

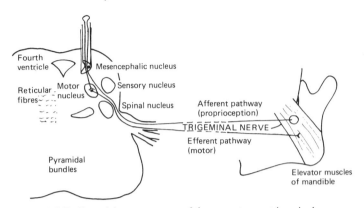

**Figure 13.5** Outline of the components of the masseter, or trigeminal, stretch reflex usually known as the jaw jerk reflex. The mandible on the right is not on the same scale as the section through the pons. The four nuclei of the trigeminal nerve (the mesencephalic, motor, sensory and spinal nuclei) are labelled. Note that the sensory fibres do not have cell bodies in a separate ganglion as in the spinal reflex pathway.

Synapses which repeatedly conduct impulses adapt to permit transmission to occur more easily than it does in synapses used less frequently. This adaptation is called facilitation: repeatedly evoked reflexes show a similar adaptive property.

A number of other properties of synaptic pathways are also exemplified in reflexes although not specific to them. Reflexes may show summation or occlusion; recruitment and after-discharge may be observed. Summation and occlusion have already been discussed on pp. 76 and 81. Recruitment is the term used to describe the increasing intensity and scope of a reaction resulting from the addition to the original response of responses mediated through longer neural pathways. Contraction in a muscle may be amplified and extended by the involvement of other motor units. The contraction of a motor unit after a monosynaptic reflex may be maintained by impulses from the stimulated sensory neurone activating other neuronal pathways with several synapses which eventually reach the same effector motor neurone. This is known as after-discharge. The term is also used to describe delayed activity in other motor neurones outside the original monosynaptic reflex.

### Levels of reflex activity

The simple reflexes occur at spinal level or, when cranial nerves are involved, at the levels analogous with the spinal segments. The more complex a reflex, the more levels are involved: many of the reflexes vital to life which involve several body systems are integrated within the medulla oblongata of the brain, or higher still, in the pons. Thus integration occurs in the medulla for the control of the heart, the control of the circulation, the control of blood pressure, and the control of breathing. Here, too, integration of these vital activities with each other is achieved. Reflexes like coughing, sneezing, vomiting and swallowing have centres in the medulla. Reflexes involving visual stimuli pass through the midbrain. Most complex of all, the reflexes which maintain the internal environment constant in osmotic activity, in chemical composition, and in temperature, have their centres within the hypothalamus. Whether the fundamental biological drives of hunger, thirst and sexual activity, and the manifestations of emotion may be considered as reflexes is debatable: if they are, these are examples of reflexes triggered within the limbic cortex.

Studies on men with spinal or brain injuries have demonstrated these different levels of reflex control. Transection of the spinal cord results first in a loss of tone in the muscles innervated below the level of the transection. There is then a return of reflex activity in these muscles although voluntary activity will not be possible in the absence of intact descending motor nerves. The withdrawal, or flexor, response returns first. Stimulation of a limb causes it to be withdrawn by the activity of the flexor muscles. This is demonstrated in the Babinski reflex, a curling of the foot in response to touching the sole. The withdrawal reflex is particularly noticeable after nociceptor (harmful) stimuli. Increase in force of the stimulus may result in an extensor response – stretching of the limb to push away the offending stimulus source. The push obtained by the leg during walking may be a response of this kind to the pressure on the sole of the foot. Greater stimulus strength may also lead to a crossed extensor reflex – withdrawal of the stimulated limb but a push from the opposite limb to support the body as its weight is thrown over to that side.

Transection of the cord will leave intact the reflexes which cause emptying of the bladder as it is stretched by urine production and emptying of the rectum as faecal material accumulates within it. Sweating may occur locally. Normally these responses are controlled by impulses from higher centres or are modulated by other reflexes.

## Conditioned reflexes

The human brain is extremely powerful, but achieves this power, in part, by limiting its immediate window of awareness to a relatively narrow field of attention. This is made possible by relegating as many control functions as possible to levels of mental activity below those of consciousness. It is these control functions which were referred to at the beginning of the chapter as learned or conditioned reflexes. This type of reflex is not unique to man: most animals can learn patterns of activity of varying complexity. The classical example is that in the dogs studied by Pavlov. He observed that dogs fed regularly at a time signalled by the ringing of the bell of a nearby church associated the sound with feeding and began to salivate at the sound of the bell. The presence of food in the mouth is an unconditioned stimulus to salivation: learning is not apparently required to establish the pattern of reflex activity. The bell associated with food intake became a conditioned stimulus. Even if the dogs were not actually fed, the sound resulted in increased salivation. Conditioned reflexes, once established, are strengthened by frequent repetition – a process termed reinforcement. However, if the bell stimulus was used sufficiently often without being accompanied by feeding, the reflex weakened and eventually disappeared. Dissociation of the conditioned and unconditioned stimuli leads to extinction of the reflex.

At their simplest, conditioned reflexes involve a conditioned stimulus linked to an unconditioned stimulus producing a physiological reflex; at their most complex, they include patterns of learned behaviour such as those involved in driving a car, when the final activity is far removed from a physiological reflex.

The information gained from experiments with conditioning has led to the application of these principles to education and learning. The stimulus-response theories of teaching are based on such principles. Aversion therapy is another application. Unfortunately, the same principles have also been used to develop methods of political or social indoctrination.

## Further reading

Horrobin, D.F. *Principles of Biological Control,* Medical and Technical, Publishing Co., Aylesbury, 1970

Burton, A.C. *Physiology and Biophysics of the Circulation,* Year Book Medical Publishers, Chicago, 1972

Coxon, R.V. and Kay, R.H. *A Primer of General Physiology,* Butterworths, London, 1967

Selkurt, E.E. *Physiology,* 5th edn, Little Brown, Boston, 1984

# 14

# Posture, locomotion and voluntary movement

A number of terms should be explained before discussing these concepts. Muscles of the skeletal system are usually described according to their function as flexor or extensor. A flexor muscle contracts to produce bending at a joint – the biceps muscle, for example, bending the arm at the elbow joint – whilst an extensor muscle has the opposite action of straightening the line of bones at a joint – as the triceps muscle straightens the arm. Slightly different terms are used in the case of the muscles of mastication which may be elevator or depressor to the mandible, or may move it forward (protrusor) or back (retrusor). All these movements involve changes in the plane of one muscle in relation to another; rotational or twisting movements also occur, as in the cervical spine when the head is turned. Indeed, because of the position of insertion of muscles into the bones, rotational movements may be included in the flexor or extensor movement. The principal muscle involved in a movement is the agonist (or prime mover, or sometimes, protagonist); muscles assisting its action are synergist, and those giving the opposite movement are antagonist.

The motor neurone passing from the central nervous system (spinal cord or the nuclei of the cranial nerves) to the muscle is often termed the lower motor neurone, a convenient clinical term to distinguish these neurones from those whose cell bodies and processes are entirely in the spinal cord or brain, damage to which causes different clinical effects. The lower motor neurone, as the nerve supply to the motor unit, is the only pathway by which impulses from any source in the central nervous system can reach the motor unit. It is therefore sometimes called the final common pathway for motor activity.

Several reflexes which operate at a spinal or brainstem level have been thought to be implicated in posture and locomotion. These are:

(i) *The stretch reflex.* Stretching of a muscle stimulates the spindle stretch receptors and causes contraction of the muscle by a simple reflex usually involving only two neurones directly.

(ii) *The tendon jerk.* This is the same as the stretch reflex, but is initiated by tapping the muscle tendon to produce the stretching of the muscle.

(iii) *The flexor response.* In response to a noxious stimulus, the limb is flexed to remove it from the point of stimulation. Although the reflex is seen with noxious stimuli, it is argued that, in locomotion, the touch and pressure stimuli to a supporting limb may initiate flexion.

(iv) *The crossed extensor reflex.* This response may be associated with the flexor response, particularly if the stimuli are intense. Flexion of the stimulated limb is accompanied by extension of the limb on the other side, the contralateral limb.

(v) *The scratch reflex.* Repetitive stimulation of touch receptors causes a pattern of limb movement involving alternate extension and flexion. Since locomotion is a pattern of alternate extension and flexion of the limbs, analogies have been sought between the scratch reflex and locomotion.

These simple reflexes may be involved either by themselves or as part of a larger pattern. The more complex and more specific reflexes involved in posture and locomotion will be dealt with in the context of the whole process.

# Posture

Most of our understanding of neurophysiology comes from animal experiments, of which the vast majority have been carried out on quadrupeds. Many of the ideas about human posture and locomotion, therefore, are based upon observations on quadrupeds, although some information comes from experiments on other primates and more has been gained from a few experiments on human subjects. The human, except as an infant, is a biped and differs from the quadrupeds in posture and in the manner of locomotion.

The most important single difference is that the quadruped has four supporting limbs, and the vertebral column with its muscles and ligaments is unimportant, whilst man has only two supporting limbs and depends upon the rigidity of the vertebral column to maintain the upright posture. Reflexes relating activity of front and hind limbs have relevance to adult man only when equilibrium has been lost completely.

Posture may be considered to involve three processes. The first may be termed tonic, or static, posture; the second is the correction of small changes in body position, and is given the name of 'righting', or righting reflexes; and the third is the maintenance of a general postural position during movement – statokinetic reflexes. Another term sometimes used is 'attitude': the spatial relationship of the body structures to each other (Fig. 14.1). This is a part of all three aspects of posture but is probably most important in the statokinetic responses.

## Static posture (stance)

Tonic or static posture in man is the maintenance of an upright vertical posture above the two supporting limbs. It may be achieved by the fixation of joints by simultaneous contraction of the extensor and flexor muscles. This is true in the lower limbs, as it is in all four limbs of a quadruped: but in addition the vertebral column must be fixed both intrinsically and extrinsically so that pelvis, abdomen and thorax are positioned above the pelvic basin. This combination of activities is referred to as the positive supporting action, and it seems to be initiated by proprioceptive impulses from the small interosseous muscles of the feet. It is replaced by the negative supporting action when flexion of a limb causes excitation of proprioceptive nerves from the plantar or volar muscles (on the sole or upper side of the foot). Although the positive supporting action seems necessary for the adoption of stance, it is probable that stance is maintained with very little effort in the muscles except when an unusual stance is adopted. Even then the body adopts a position in which the centre of gravity is above the centre of ground contact. Thus wearing high-heeled shoes which tilt the plane of the feet upwards from toe to heel causes the shoulders to be held back to counteract the forward thrust of the pelvis induced by the leg position (Fig. 14.2). Animal experiments demonstrate that a truly normal stance cannot be established or maintained unless the cortex and basal ganglia are intact. Spinal pathways are insufficient of themselves for standing. However, transection in the upper part of the pons results in decerebrate rigidity which can permit standing. Stance with normal muscle tone requires at least the forepart of the midbrain, and the positive and negative supporting reactions need the basal ganglia and the cortex to be intact.

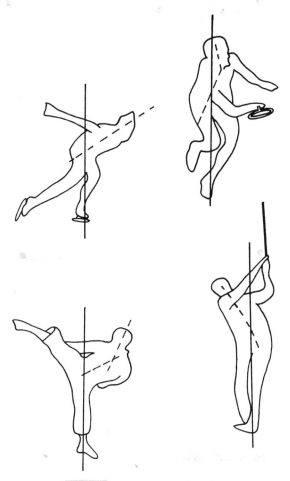

**Figure 14.1** Attitude – the balancing of body structures during movement. The vertical line through each figure passes through the centre of gravity. The dashed line gives the adjusted position of the line originally vertical in the standing upright position. The centre of gravity is at the intersection of the solid and the dashed line. Note how the body weight is distributed in different positions.

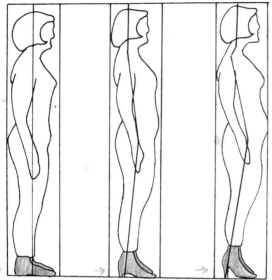

**Figure 14.2** Posture – the balancing of body structures in the upright standing position at rest. A vertical line through the centre of gravity is drawn on the first figure. As the line through the legs moves away from the vertical because of the increased tilt as the heels become higher, the body position is adjusted to bring the centre of gravity back over the feet. The sections of the original line through legs, trunk, neck and head, are shown in their original positions as the posture is adjusted. The effect is exaggerated in the horizontal direction to show it more clearly.

## Righting reflexes

Righting reflexes in man involve correction of sway from the vertical and vary in the degree of the response according to the amount of displacement. The idea that normal posture is maintained by an infinite series of small righting reflexes producing an oscillation about a mean posture vertical is no longer accepted and it is thought that a static posture requires little contraction of either extensor or flexor muscles.

The first consideration with righting reflexes is the nature of the sensory mechanisms which initiate them. Four groups of receptors are important in man and it is usually said that at least one of the first two groups is essential for the maintenance of posture.

(i) *Visual receptors.* One of the most important sensations in the maintenance of posture in man is vision. If the position of a fixated image of a static object on the retina changes, first the eye muscles contract to maintain the original position, then the neck muscles adjust the head position, and finally the trunk is moved to support the head. Impulses suggesting a change in body position from the vertical will pass to the visual cortex but also to the vestibular nucleus.

(ii) *Vestibular apparatus.* In man this is another very important sensory organ in the maintenance of posture. The otolith organ acts as an absolute indicator of head position in respect to gravitational pull. The semicircular canals sense rate of change of the position of the head and therefore provide early warning of the extent of head movement in relation to the starting position. The organs provide accessory control of the direction of gaze of the eyes. Thus the vestibular nuclei connect with the oculomotor nuclei, with the cerebellum and with spinal motor neurones.

(iii) *Proprioceptive sensors in muscles, tendons and joints of the neck.* Any local reflex stabilizing the line of sight will move the head in relation to the body and hence produces a change in attitude. In order to correct this, the position of the neck structures is monitored. The body can then follow the head position. These reflexes seem to be important in animals but their relevance to human posture is unclear. One suggestion is that the possible sequence of events begins with the perceived movement of a retinal image, the eye level is corrected by moving the head, and this movement stimulates labyrinthine and neck receptors to align the body to support the head. In man these reflexes may be simply a fraction of the larger input of proprioceptive impulses from trunk and legs – although in the infant they probably play a role similar to that in other quadrupeds.

(iv) *Proprioceptive sensors in muscles, tendons, ligaments and joints throughout the body, but particularly around the vertebral column and in the legs; touch and pressure receptors in the feet.* These sensors are responsible for the recognition of the spatial position of the body parts. The slight corrections of posture needed to counter sway originate in signals from the muscle stretch receptors, from ligament and joint receptors, and are manifested in changes in length and tension principally in the extensor muscles on the side away from which sway has occurred together with relaxation of those on the other side.

## Statokinetic reflexes

The term attitude has already been defined briefly, but since it refers more to the situations involving the statokinetic reflexes, it is given a fuller definition here. The term is used to describe the relationship of the body parts to each other in such an arrangement that the postural load is evenly distributed, and the different muscles of the body are in an equilibrium appropriate to the activity being carried out. The statokinetic reflexes maintain attitude during movement.

The postural reactions during movement are based upon the same general reflexes as are involved in the static posture and righting reactions. Whereas in postural control the gravitational receptors of the utricle and saccule are important in maintaining the upright stance, in movement the rotational receptors of the semicircular canals become the more important. Similarly, the eye reflexes, which are supplementary in standing posture, become much more important in maintaining attitude during movement. Finally, the proprioceptive sensors of muscles and joints, particularly those of the head and neck, which initiate the classical righting responses, are equally important in the control of attitude during movement.

## Locomotion

Locomotion is the movement of the body, by a co-ordinated series of muscle activities, in a particular direction whilst the relation of the body as a whole to the ground is maintained.

There are several reactions which have been investigated in animals which come midway between righting reflexes and locomotion proper. These are seen in relation to the application of an overbalancing force. A slight push against one side or one limb of an animal usually induces sway, or the transfer of weight to the contralateral limb. An increase in the force of the push leads to a stepping reaction in which, after weight has been shifted from the nearside limb, it is moved sideways to be placed closer to the contralateral limb, which, in turn, is moved further away. In effect, the animal has removed itself from the push by a sideways step. If the movement of the leg on the pushed side is prevented, the contralateral leg produces a hop. These are reactions to a pushing force. Another reaction possibly related to locomotion is the magnet reaction. This is a positive supporting reaction observed when a light touch stimulus is given to the paw, and the limb is extended to increase the force against the touching object. Such a reflex could be useful in regulating the strength of the positive supporting reaction when the animal is moving across rough terrain. In its extreme form the magnet reaction may induce an animal to follow a light touch stimulus moving away from it.

The process of locomotion in man, as exemplified in walking, consists in raising one leg by flexion and producing extensor thrust in the other so that the body falls forward. The flexed leg is then extended to correct the fall and the weight is transferred to it. The forward component of the gravitational pull on the tilted body is used to provide energy for forward movement. Movements at the three major joints – ankle, knee and hip – are involved. Sensory information from the muscles, ligaments and joints is necessary for control of the movement. Although posture in animals cannot be maintained without a cerebral cortex, locomotion is possible without it. A rhythmic pattern of alternating extension and flexion can be produced at spinal cord level, and is analogous in many ways to the scratch reflex. Such a pattern, however, cannot maintain locomotion proper because it does not include any means of regulating body posture or of co-ordinating activity of muscles other than those of the limbs directly concerned. There appears to be a neural pattern organised centrally, but subject to modification by afferent impulses. Further, the initiation of locomotion is usually a voluntary act and so cortical impulses are necessary for this. It is clear that the process is subject to central control, probably at the level of the cerebellum and vestibular nuclei, and is not simply a rhythmic alternation of afferent impulses and local reflexes which maintain the muscle contraction at a set level.

## Muscle movements

Control of muscle movements can be exerted at all levels from the cortex to the spinal cord at the segment of origin of the muscle (*see* Fig. 14.3).

### Cortical control

The cortex is responsible for the initiation of voluntary movement. The conscious initiation of movement is a process originating in the formulation of the will to move in the frontal cortex. Except in the case of relatively simple movements such as rapid movements involving very few muscles, impulses pass from the frontal cortex to the basal ganglia. There may be a two-way traffic of impulses between these ganglia and the frontal cortex as the pattern of movement is determined, but the next part of the process consists of signals back to the cortex in the motor and premotor areas. Stimulation of points on the motor cortex in the precentral gyrus causes localised muscle responses on the contralateral side. Stimulation of the supplementary motor area next to the motor cortex produces complex diffuse responses. Ablation of the motor cortex causes an immediate loss of voluntary control of the appropriate muscles, affecting particularly the distal musculature and the muscles involved in speech. After a recovery period, which may extend for some months, control of gross movement returns but control of delicate skilled movement of the digits is lost permanently. Although control of accurate finger movements is lost, the more general arm movements are still controllable. Equally, although

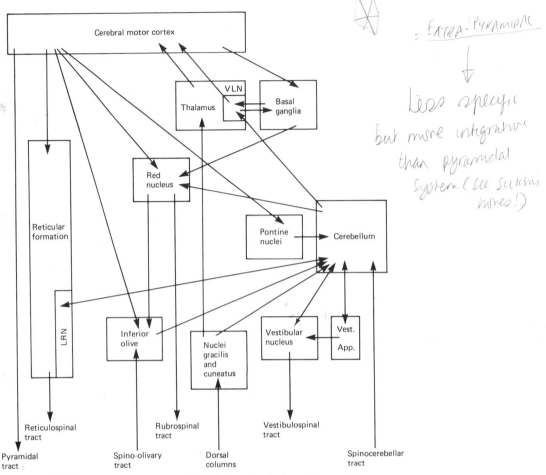

**Figure 14.3** The pathways of motor control. LRN, lateral reticular nucleus; VLN, ventrolateral nucleus of the thalamus. Vest App, the vestibular apparatus.

toe movements could not be controlled, walking would still be possible. There is no obvious loss of motor control if the supplementary area only is excised. In monkeys, damage to the motor cortex abolishes the control of fine skilled movements, but it appears to be the execution of the movement that is affected, rather than the actual learned skill, since the ability to carry out the movement may return if the damage to the cortex is limited and the adjacent cortex is able to take over control. In general, the increased complexity of the human cortex seems to render such recovery unlikely.

From the cortex the principal motor fibre pathway is in the pyramidal tracts. Fibres pass from the cortex to the internal capsule, then the middle cerebral peduncle, and then decussate in the pyramids at the pontine level to pass to the corticospinal tract. Lesions of this tract have the same effects as lesions of the motor cortex.

Other fibres pass to what are called the **extrapyramidal tracts** – that is, any tracts other than the pyramids. From the cortex fibres travel to the reticular formation and thence in the reticulospinal tract (Fig. 14.3). This system seems to be concerned with synergistic effects. It is less specific but more integrative than the pyramidal tract system. A bulbar **inhibitory centre** has been described in the lower pons and medulla which causes inhibition of all reflex movements whether flexor or extensor. Above this, in the midbrain and extending down to the bulbar area, is a **flexor reflex facilitatory centre**.

### Basal ganglia

Yet another group of cortical fibres travel to the basal ganglia – the caudate nucleus, the putamen, and the globus pallidus (Fig. 14.3). The role of the basal ganglia is to store and then provide a template for a particular co-ordinated activity. This storage process may be shared with frontal cortex: fibres pass in both directions between the frontal cortex and the basal ganglia, and when a muscle action is desired the plan of execution is devised between the frontal cortex and the basal ganglia. The evidence suggests that their role is in the co-ordination of activity and of different muscle groups to achieve a desired movement. On the whole they are not involved in the integration of that movement with the postural reflexes. This integration is achieved lower down in the system in relation to the local reflexes by the reticular centres referred to earlier and, more importantly, in relation to posture as a whole, by the cerebellum. The basal ganglia are responsible for the integration of the particular response into the other concomitant motor activities.

## Cerebellar control

The cerebellum is the largest mass of nervous tissue concerned solely with muscular control. It has been described as exerting 'provisional control' over reflex and voluntary motor activity. Its role is to provide a means of allowing voluntary and reflex activity to go on simultaneously, and to ensure that the intended result of voluntary activity is achieved as far as possible. In other words, the cerebellum provides or controls the platform on which voluntary activity is superimposed. Thus it is possible to kick a football when running without falling over. The parts of the cerebellum have been identified as having particular functions (Figs 9.11 and 9.12). The cerebellum can be regarded as divided into five functional areas: a medial area which includes the vermis, two lateral areas, and two intermediate areas between them and the medial. The intermediate lobe of the cerebellum is concerned with the integration of impulses from the cerebral cortex and those from muscles themselves. The lateral lobes and the hemisphere control highly skilled learned activities. Once a particular pattern of activity is learned it persists as a kind of template in this part of the cerebellum, and the pattern can then be repeated in an unmodified form without intervention from the cortex after its initiation. Thus although the cortex must initiate the act of striking a nail with a hammer, the actual process is achieved without conscious control and cannot easily be stopped in mid-stroke. This is a learned motor activity – a conditioned reflex. It is only when the pattern of activity fails to meet external conditions, as when one walks a step too many down a staircase, that consciousness may intervene and modify the pattern or switch in a new pattern. The integrative role of the cerebellum is achieved by virtue of the many afferent and efferent fibres reaching and leaving it. The initiation of a simple learned pattern of movement after the pattern has been invoked by the basal ganglia, takes place in the intermediate lobe of the cerebellum, where the pattern has been integrated with any minor necessary postural adjustments and learned. The continuation of the movement is directed by an on-going control in which the desired and actually achieved movements are compared and aligned. This control activity takes place in the intermediate cerebellum. Finally, the movement is terminated by activity in the lateral and intermediate zones. More complex movements, where postural adjustment is an important part of the movement, are controlled by the anterior lobe and the vermis – the medial zone of the cerebellum.

There are many afferent pathways from exteroreceptors and proprioceptors and from the cerebral cortex. Thus the inferior peduncle carries inputs from the dorsal horn of the spinal cord, the nuclei gracilis and cuneatus, the inferior olive, and from the vestibular nucleus and the medullary reticular nuclei; the middle peduncle carries inputs from the cerebral cortex via the pontine nuclei; and the superior peduncle carries some spinal, tectal and tegmental inputs. Efferent fibres pass via the thalamus to the cortex, to the vestibular nuclei, to the reticular formation, and to the motor nuclei for the head and neck. The fibres to the cerebral cortex help to integrate the motor activity with the sensory information. The projections to the spinal cord are generally ipsilateral whilst those to the cortex are contralateral.

## Vestibular and reticular control

The remaining higher control areas are the vestibular nuclei and the motor areas of the reticular formation extending from the bulbo-pontine nuclei to the midbrain. The reticular formation receives inputs from visual, somatosensory and labyrinthine receptors. It sends signals to motor neurones whose axons pass in the reticulospinal tract by the ventral funiculus and the ventral part of the lateral funiculus to the median longitudinal fasciculus. The role of the reticular formation is in part integrative. Although it can excite flexor responses, many of the fibres of the reticulospinal tract neurones are in fact inhibitory to the lower motor neurones. The vestibular nuclei are also integrative in function, particularly in the integration of impulses arising from the spinal cord and cerebellum. The vestibulospinal pathways carry motor signals to both $\alpha$ and $\gamma$ motor neurones. Experimental stimulation of lateral vestibular nucleus shows that this part of the vestibular area will cause excitation of extensor muscles.

## Local control

Finally, control is also exerted at a local level. Three types of receptor appear to be necessary: some kind of fast-reacting receptor responding to rate of change of length of a stimulated muscle, capable of providing advance information on the likely result of the degree of stimulation; some receptor of length in the antagonist muscle to prevent overstretching; and some receptor of tendon or ligament tension to prevent damage to these structures. The first two functions are performed by the muscle spindles, the last possibly by the Golgi tendon organs and ligament and joint receptors. The reflexes providing a local control are the stretch receptor reflex, the crossed extensor reflexes and the inhibitory Golgi tendon organ responses, and the effects of flexor response on extensors or vice versa. Local responses of this kind may be relatively simple with few synapses and interneurones, staying at one segmental level in the spinal cord, or may be more complex, with interneurones passing up or down to other levels. An interneurone whose functional connections have been relatively well explored is the Renshaw cell. This is stimulated by an axon of the α motor neurone. At the synapse, acetylcholine is the transmitter, as at the muscle terminal of the α motor neurone. The Renshaw cell receives other spinal and supraspinal inputs. One of its axons passes to the α motor neurone again and inhibits it. This provides a negative feedback to the muscle and may be important in limiting the duration of muscle activity resulting from α motor neurone excitation. Obviously the inhibitory signal will be at least one synaptic delay interval after the stimulation of the motor nerve. Other axons of the Renshaw cell may terminate on interneurones, and it is thought to cause presynaptic inhibition of inhibitory interneurones to the motor nerves to antagonist muscles. The Renshaw cell by virtue of its connections may act as a local integrator of motor impulses from nerves at its own level of the spinal cord, from other segmental levels, and from the brainstem and midbrain areas.

It was thought for some time that the final direct control of muscle activity occurred either by simultaneous activation of α and γ motor neurones, or by sequential activity with the γ motor neurone preceding the α motor neurone activity by a very short interval. This would be a servo control mechanism, or a follow-up servo control. The γ motor neurone was thought to cause the intrafusal fibres to adopt a length appropriate to render the spindle in a stretch equilibrium at the intended final length of the whole muscle. The α motor neurone then stimulated contraction of the extrafusal fibres. If this contraction were appropriate to the intended length, the contracted muscle spindle would not be excited by stretch. If the contraction of the whole muscle were insufficient, the contracted muscle spindle would be stretched, and the sensory fibres would carry impulses back to the spinal cord to produce further reflex activation of the α motor neurone. This circuit is described as a γ loop. In recent years the importance of this reflex pathway has been questioned.

## Further reading

Davson, H. and Segal, M.B. *Introduction to Physiology, Volume 4,* Academic Press, London, 1978

Robertson, T.D.M. *Neurophysiology of Postural Mechanisms,* Butterworths, London, 1965

There are a number of books dealing with movement including the term 'kinaesthetics' in their titles – interesting but rather specialised!

# 15
# Control of pressure and flow in the cardiovascular system

The control of the circulation may be viewed in two different ways: control to ensure an adequate distribution of blood to the body as a whole, and control to adjust that distribution to provide for the differing demands of the different parts of the body. At the same time blood flow to the priority organs – the brain and the heart – must be maintained. A peripheral adjustment of pressure may well result in a reduction in flow through the vessels concerned whilst vasodilatation and an increased volume flow of blood may be at the expense of pressure and velocity of flow either in the local circulation or by differential effects elsewhere.

## The control of arterial blood pressure

Pressure depends upon two factors: the flow produced by the pump, and the resistance of the flow pathway. The blood pressure, therefore, is the product of the cardiac output and the resistance of the circulation to flow. The latter is termed the total peripheral resistance (TPR). Blood pressure and cardiac output are both directly measurable; the peripheral resistance is not. It is the ratio of pressure to the cardiac output and is often given in arbitrary peripheral resistance units, or PRU, obtained by dividing pressure in mmHg of mercury by flow in ml/s. Thus the peripheral resistance in the systemic circulation is normally around 1.2 PRU (a mean pressure drop of 13.3 kPa, or 100 mmHg of mercury, divided by 83 ml/min, or 5 l/min) and that in the pulmonary circuit, 0.2 PRU (assuming a mean pressure drop of 2.2 kPa, or 17 mm). In more conventional units one PRU is 133 kPa/s/m$^{-3}$.

If blood pressure increases, the transmural pressure in the major arteries is increased and the vessel walls stretch. Similarly, a decrease in pressure permits the vessel walls to recoil. Blood pressure can, therefore, be monitored by stretch receptors in the walls of the blood vessels. It is probable that such stretch receptors are found in many parts of the circulation including the walls of the cardiac chambers themselves. However, the best known and most important are found in the walls of the arch of the aorta and in the walls of the internal carotid arteries just after the bifurcation of the common carotid vessels (Fig. 15.1). These stretch receptors

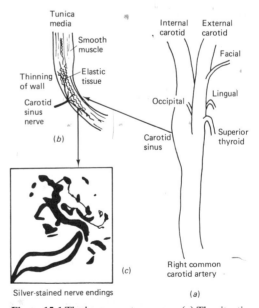

**Figure 15.1** The baroreceptor organs. (a) The situation of the carotid sinus at the bifurcation of the common carotid artery; (b) a low power section through the wall of the carotid sinus to show the thinning of the vessel wall and the position of the nerve terminals; (c) a higher power section through a nerve terminal to show the actual endings.

are branched nerve terminals with bulbous swellings – almost like the flower-spray type of nerve endings. The internal carotid artery is dilated to form a swelling termed the carotid sinus and the walls of the sinus are more distensible than those of the vessel before and after this point. The carotid sinus and the arch of the aorta, then, can be regarded as sensor areas for blood pressure measurement, baroreceptor areas. Pressures below 9 kPa produce no response from these sensors but from that pressure up to about 20 kPa their response is almost linear. Any recording of their normal activity reflects the regular cycle of systolic and diastolic pressures with a burst of action potentials at each systole superimposed upon the regular frequency of action potentials proportional to the diastolic pressure.

The input pathways of the blood pressure control system (Fig. 15.2) are a branch of the glossopharyngeal nerve from each carotid sinus and a branch of the vagus nerve from the aortic arch. These nerves are called depressor nerves since an increased input of action potentials along them to the medulla reduces the rate and strength of the heartbeat.

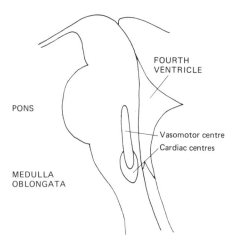

**Figure 15.3** The situation of the blood pressure control centre in the medulla oblongata. Midline sagittal section.

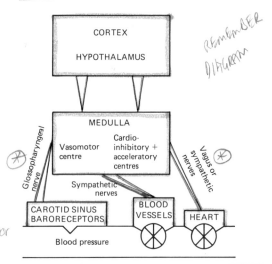

**Figure 15.2** Outline of the blood pressure control system. The carotid sinus system is shown: the aortic arch receptors send their input through the vagus nerve to the same centres.

The impulses pass to a controlling centre within the medulla oblongata (Fig. 15.3). Cells here can be grouped into areas termed the vasomotor centre, which controls dilatation or constriction of the arterioles, and the cardiac centres, which control the activity of the heart. The 'set level' for blood pressure probably depends as much on the sensitivity of the receptors as on the controlling centre, and there is evidence that in hypertension – a state in which the blood pressure is continuously above normal – the sensitivity of the receptors is reduced.

Adjustment of set levels in physiological systems is often achieved by altering the sensitivity of the receptors.

The output pathways of the system are the sympathetic nerves to the smooth muscle of the arteriolar walls, and the cardiac branches of the vagus and sympathetic nervous systems.

The effector systems are two-fold. Since the blood pressure is proportional to the cardiac output and the peripheral resistance, variations in one or both of these may be used to maintain blood pressure. Cardiac output may be increased by a diminution in the vagal restraint of the sinoatrial node or by an increase in the sympathetic acceleration of the rate initiated by the sinoatrial node. Sympathetic stimulation will produce positive inotropism (an increase in the force of contraction of the heart) as well as positive chronotropism (an increase in the rate at which it beats). The ventricles both fill and empty more completely so that their end diastolic volume is greater and the end systolic volume less. The stroke volume may also be increased by an increase in venous return resulting from sympathetic venoconstriction. Peripheral resistance may be increased by sympathetic stimulation of arteriolar constriction. It may be decreased by a decrease in sympathetic stimulation and hence a relaxation of the arterioles – vasodilatation (the passive word dilatation, rather than the active word dilation, is used to describe this change in calibre, because active contraction of the muscle walls can only reduce the diameter of the vessel). Thus sympathetic stimulation of the blood vessels has two effects: constriction of the arterioles to raise peripheral resistance and constriction of vessels on the venous side to increase return of blood to the heart and raise cardiac output.

The feedback control system is in fact much more complex than this (Fig. 15.4). There is evidence of

*The control of arterial blood pressure* 231

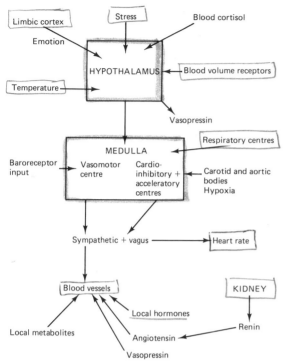

**Figure 15.4** Factors other than the baroreceptor reflex which influence the control of blood pressure and the state of dilatation of the blood vessels.

other pressure receptors in both the atria and the ventricles as well as in other blood vessels. Other factors can modify the response to changes in blood pressure. The cardiovascular control areas receive inputs from the receptors in the carotid and aortic bodies which monitor the partial pressure of oxygen in the blood. These peripheral chemoreceptors will be dealt with more fully in Chap. 17 in relation to respiration. They respond to decreases in partial pressure of oxygen or to decreased blood flow. In general impulses from them cause an increase in vasomotor activity. Impulses from the aortic bodies usually cause tachycardia, but those from the carotid bodies may cause bradycardia. In the medulla itself increased partial pressures of carbon dioxide and decreased partial pressures of oxygen at first stimulate the cardiovascular control areas, but then depress nervous activity.

The feedback control system is able to maintain a fairly constant arterial blood pressure by these apparently simple components of a feedback loop. This system by itself is able to adjust the blood pressure during changes in posture, or after changes in the calibre of blood vessels, or when the venous return to the heart is varied. It does not, however, adjust the differential flow to different parts of the body, except inasmuch as the effects of sympathetic stimulation are selective in different vascular beds,

and it does not adjust flow in response to external stimuli or changes in external environment. Differential flow can be controlled entirely locally, or alternatively, may be directed by some higher centre. The activity of the medullary centres may be adjusted by interaction with other medullary or pontine centres involved in controlling other activities, such as respiration. Some integration of body systems occurs, therefore, even at medullary level. There are, in addition, controls exerted by higher centres. Biofeedback treatments make use of voluntary control of blood pressure by conscious cortical direction. Normally, however, there is no direct cortical control but the hypothalamus, which is the main area of the brain concerned in the integration of hormonal and nervous responses, and the area through which manifestations of limbic brain activity are channelled, exerts a general control over the medullary cardiovascular centres. The responses of anger and fear, and the responses to exercise, involve complex adjustment of the functioning of the cardiovascular system. Part of the response to loss of blood or a decrease in body fluids is cardiovascular (Chap. 16), and the control of body temperature, centred in the hypothalamus, demands the balancing of peripheral vasoconstriction and vasodilatation with the need to maintain the blood pressure.

The sympathetic system with its combination of activity through nervous pathways and the release of adrenaline from the adrenal medulla can control both the overall blood pressure and also flow to different organs. Both adrenaline and noradrenaline act differentially via α- and β-receptors (p. 142). Their separate effects on the cardiovascular system, however, are different. Noradrenaline is almost always vasoconstrictor. It therefore causes an increase in both systolic and diastolic blood pressures, and the resultant stimulation of the baroreceptors slows the heart to a greater extent than the noradrenaline can accelerate it, and bradycardia results. Adrenaline has a major vasodilator effect via β-receptors in skeletal muscle and liver which reduces the effect of vasoconstriction elsewhere, and peripheral resistance falls. Diastolic pressure therefore falls but the increase in cardiac rate and output causes the systolic pressure to rise. Activity of the sympathetic nervous and hormonal outputs can be controlled through the hypothalamus, although even in the absence of hypothalamic integration, the baroreceptor control circuit automatically adjusts the blood pressure back to normal values. Thus, when adrenaline is injected and a more rapid heart rate temporarily established, any rise in blood pressure is usually transient because compensation occurs via the baroreceptor reflex. Hypothalamic control may be seen as over-riding but exerted only when circumstances outside the cardiovascular system demand it.

*[Handwritten at top: INTRINSIC CONTROL — ① STARLINGS ② LAW OF LAPLACE : larger spherical body will ③ HEART CONTRACTS MORE c̄ GREATER RESISTANCE need more force to reduce circumference.]*

In addition to the general mechanisms for controlling blood pressure, there is also an intrinsic control on the activity of the pump itself. This is seen, for example, in the Frank-Starling principle. Under experimental conditions the force of contraction of the cardiac pump is directly related to the degree of stretch of the cardiac muscle fibres. This observation parallels that embodied in the length-tension curve of skeletal muscle. Translating the observation into terms relating to cardiac function, the force of expulsion of blood from the heart is proportional to the end diastolic volume or, less directly, to the magnitude of the venous return. Thus an increase in venous return, or an increase in central venous pressure, results in an increase in cardiac output. The relationship between force of expulsion from the heart and the end diastolic volume is not however constant: it depends upon the degree of sympathetic stimulation, since sympathetic activity increases the contractility of the heart – the positive inotropic effect. It is not possible therefore to relate cardiac output directly to venous return over long periods of time: but the Starling effect allows the rapid adjustments necessary as venous return changes from beat to beat and min to min. This is important in balancing the outputs of the two sides of the heart. Since the heart consists of two pumps operating in series in the one continuous circuit of the systemic and pulmonary circulations, each side of the heart must pump the same volume of blood to prevent the blood accumulating in one part of the circuit. Small discrepancies from beat to beat can be accommodated by the elasticity of the pulmonary vessels but the discrepancies are smoothed out over several beats by the operation of the Starling mechanism. There are, however, situations which appear to invalidate the Frank-Starling principle. Thus the heart is actually at its largest in a resting recumbent subject, because of the balance of hydrostatic pressures, and the stretch of the cardiac muscle fibres at end diastolic volume is then such as would be expected to give a maximal force of contraction. In intense exercise, with a much greater venous return, the heart is smaller – but the level of sympathetic activity, and hence the heart rate and the contractility of the cardiac muscle, is greater, so that cardiac output is greatly increased. Another anomaly in the Starling relationship (see p. 121), arises because contractility increases only up to a certain diastolic ventricular volume, and then decreases again.

Another intrinsic control is provided by the geometry of the heart chambers. As the cardiac volume increases, relatively less contraction of the fibres is required to expel the same volume of blood. This benefit is limited by the Law of Laplace which predicts that in a larger spherical body, relatively more force will be needed to reduce the circumference of the body.

A third factor, not easily explained, is the apparent ability of the heart to contract more forcibly against a greater resistance. This too may be linked to the stretching of the muscle fibres.

*[Handwritten: ✱ EXAM Qu]*

### The control of cardiac output
*[Handwritten: EXERCISE / EMERGENCY SITUATIONS]*

The control of cardiac output as such has not yet been mentioned. All the elements of the control of cardiac output have however been outlined. Cardiac output is the product of heart rate and the stroke volume. Both of these are controlled by the autonomic nervous system and the hormone adrenaline. Further, the stroke volume depends upon the venous return. Control of cardiac output is important when extra demands are made upon the cardiovascular system: the more important of these are described in the chapters on exercise and emergency situations. In general, cardiac output responds to the demands of the body so that during exercise it increases linearly with the workload from a resting level of 6 l/min to between 25 and 35 l/min.

### The control of blood flow and of the distribution of blood throughout the body

Variations in flow to different organs are achieved by nervous stimuli controlling the constriction or dilatation of arterioles, arteriovenous shunts and precapillary sphincters. In general these are central effects mediated through the sympathetic nervous system: differences in activity represent the differential effects of stimulating α- or β-receptors. The sympathetic nerves are usually vasoconstrictor and are probably tonically active in most blood vessels. Most blood vessels show rhythmical contraction and relaxation but how far this is intrinsic and how far it is due to a neural drive is unclear. Variation in the calibre of the vessels is normally produced by altering the rate of sympathetic discharge. Contraction is active, dilatation passive. Dilatation also occurs as a result of stimulation of sympathetic β-receptors and in a few isolated examples parasympathetic nerves may be responsible for vasodilatation. In the salivary glands the increased blood flow necessary to generate a secretion is achieved by a parasympathetic vasodilatation synchronous with the stimulus to the secretory cells. The vasodilatation which fills the cavernous sinuses and hence erects the penis or the clitoris is due to parasympathetic impulses. Vasodilatation in the gut during digestion and absorption may be due to the presence of local metabolites or it may be that vasoactive intestinal peptide (VIP), which is released during the digestive process, causes vasodilatation in the gut as it can in other tissues.

There is some degree of intrinsic stabilisation of blood flow to particular regions or organs despite changes in the perfusing pressure. This is termed autoregulation: it is due mainly to the ability of the walls of the blood vessels, like smooth muscle fibres elsewhere, to respond to stretching by contraction. The vasodilatory effect of local metabolic products may also be a factor in autoregulation.

The major control of local blood flow in most organs, however, is chemical: the concentrations of oxygen, carbon dioxide, hydrogen ion, and other metabolites in the blood control blood flow in individual vascular beds. As a general rule, a decrease in oxygen concentration or an increase in the other factors causes vasodilatation. Thus, the blood supply to an organ is usually governed by its metabolic need. The importance of the different factors varies in different organs and so they are now considered separately.

## Brain

The brain is extremely resistant to factors which may reduce or even increase its blood supply. The important need is for constancy. This huge concentration of nerve cells is always operating at near a maximum level and its demand for oxygen, and therefore blood, is near maximal all the time. Brain blood flow remains fairly constant at around 54 ml/100 g of tissue/min, about 750 ml/min. There are sympathetic nerves to the cerebral blood vessels but their influence on blood flow is negligible. Oxygen tensions are important but relatively little increase in brain blood flow is produced by a fall in oxygen tension and the fall has to be to about 6.3 kPa before any change is observed. The most dramatic influence is an increase in tension of carbon dioxide – but even that rarely manages to double brain blood flow. The effect probably occurs via an increase in hydrogen ion concentration since a fall in pH produces similar results (see also pp. 253–254). When carbon dioxide is blown off from the lungs by hyperventilation, the low partial pressures of carbon dioxide in the blood cause constriction of the cerebral vessels and dizziness and loss of sensation may result. Sensitive new methods have shown that blood flow to the brain is not uniform, and changes in blood flow related to the activity of different parts of the brain have been observed. A special factor which influences brain blood flow is intracranial pressure. Rises in intracranial pressure cause a fall in brain blood flow, and vice versa. Since the brain is enclosed in a rigid bony box any increase in intracranial pressure might cause compression of brain cells. Although the effect of a rise in pressure is cushioned by the cerebrospinal fluid, the production of cerebrospinal fluid itself is increased by the increased blood pressure.

## Lungs

The arterioles of the lungs are insensitive to the normal vasomotor stimuli. In fact, the principal vasomotor stimulus to the pulmonary vessels is hypoxia, which here causes vasoconstriction. In every other local circulation hypoxia, if it has any effect at all, causes vasodilatation. If the blood in the lungs is hypoxic it implies that gas exchange is not occurring in that part of the lung: this untypical response allows blood to be shunted to better oxygenated areas.

Sympathetic stimulation of the pulmonary veins causes constriction. This has little effect upon the pulmonary circulation as such, but the reduction in capacitance forces an extra volume of blood into the left side of the heart and allows the cardiac output to be rapidly increased.

The volume of blood in the lungs may also be affected by the intrathoracic pressure; during inspiration the reduced intrathoracic pressure draws blood into the pulmonary vessels whilst forcible expiration may reduce the blood volume from around 1 litre to some 200 ml.

## Liver

The liver has two separate supplies of blood, one from the portal vein, carrying nutrients absorbed in the small intestine but having had some of its oxygen removed in its passage through the vessels of the gut, and the other from the hepatic artery, which is the main oxygen supply for the liver. Only the branches of this latter vessel are subject to control in the liver, and one of the factors controlling them is the flow of blood from the portal circulation.

The liver is principally affected by sympathetic vasoconstrictor nerves although their effect is not of great magnitude. Normal blood flow to the liver is high – about 1500 ml/min – reflecting its intense metabolic activity, although only about a quarter of this blood supply comes from the hepatic artery. The liver accounts for about one fifth of the body's oxygen usage at rest. When an organ operates at this level of metabolic activity there is virtually no room for an increase in activity: as a result, mechanisms for enhancing the blood supply as a whole are lacking; however, a reduction in delivery of blood by the portal vein is counterbalanced by dilation in the hepatic circuit, probably because metabolites are not being removed so efficiently. On the other hand, if the need for blood and oxygen elsewhere is increased, as in exercise, or in emergency situations, the maintenance of blood flow to this metabolically, but not vitally, important organ may be reduced. This reduction is achieved by sympathetic vasoconstriction.

## Spleen

The effect of sympathetic nerves on the spleen in many carnivores is to produce a marked vasoconstriction. In these animals the sinuses of the spleen concentrate blood cells and it is an important reservoir from which cells can be mobilised in emergency situations. This is not the case in the human. The human spleen is important in early life as a haemopoietic organ (p. 61) and throughout life as a source of lymphocytes and a site of destruction for red blood cells. Some vasoconstriction is seen, therefore, in the human spleen when sympathetic activity is high – as in exercise, or in emergencies – but this is simply due to the general redistribution of blood away from less essential organs to provide a greater blood flow to the heart and the muscles. Although cholinergic drugs produce a vasodilatation in the spleen it has no parasympathetic nerve supply.

## Kidney

The most characteristic feature of blood flow to the kidney under normal circumstances is its autoregulation: above a critical pressure of about 9 kPa blood flow remains constant. As pressure rises, the resistance to flow rises in parallel, probably because the stretching of the vessel walls causes a myogenic contractile response. This is also seen in other tissues, but to a much lesser degree. It is likely that in the kidney, with its closely-fitting capsule, the effect is enhanced by pressure on the vessels resulting from an increase in extravascular fluid. When vasoconstriction does occur, as a result of enhanced sympathetic activity either through sympathetic nerves or circulating adrenaline, it is normally relatively short-lived: the explanation for this is not known. When blood pressure falls drastically in the sequence of events known as shock (p. 324ff) circulating adrenaline causes massive vasoconstriction in the kidney, and, if this persists, damage to the cells and necrosis may follow.

## Gastro-intestinal tract

The principal control of gastro-intestinal blood vessels is via sympathetic α-receptors and hence sympathetic vasoconstriction. In emergency situations, or when the demand for blood and oxygen for vitally important tissues is increased, blood is shunted away from the gastro-intestinal tract by the selective action of the sympathetic nervous system or circulating adrenaline. Both β-receptors and acetylcholine receptors, which might conceivably cause vasodilatation, have been identified in the gastro-intestinal tract but their physiological roles are debatable.

## Heart

One of the most interesting features of the blood flow to the heart is its anomalous relation to systole and diastole. Since systole involves cardiac muscle in contraction, it results in compression of the coronary vessels causing increased venous outflow into the chambers of the heart while flow in the arteries ceases. During diastole arteriolar flow is at its greatest.

When the heart is subjected to an increased load in exercise, or in emergencies such as those involving blood loss, the blood flow to the heart muscle is increased. The exact mechanism by which this occurs is uncertain. Coronary blood vessels respond to hypoxia by vasodilatation but it is not clear whether this is a direct response to hypoxia or whether it is simply a response to accumulation of metabolites. Since cardiac muscle already, in 'resting' conditions, removes most of the oxygen from the blood, hypoxia is a relative term in this context. Carbon dioxide and hydrogen ion accumulation do not produce noticeable effects on coronary blood vessels but other metabolites produce very marked effects. The major control of the calibre of coronary blood vessels is chemical; nervous control is relatively unimportant. Adrenaline, however, causes a vasodilatation by a direct effect on the vessels as well as by an indirect effect due to the increase in heart activity.

## Skeletal muscle

Blood flow increases in actively exercising muscle. Four different controlling systems for skeletal muscle blood vessels have been described. There is normally sympathetic vasoconstrictor tone via sympathetic nerves and α-receptors. Inhibition of this tone can increase blood flow two to three times. Sympathetic nerves supplying cholinergic receptors cause vasodilatation and can increase blood flow four to five times. Signals to these nerves arise from the hypothalamus and seem to bypass the medullary centres. Sympathetic cholinergic vasodilatation is thought to be the mechanism by which blood flow to muscle is increased just before exercise and also the cause of the fall in peripheral resistance which precedes fainting. β-receptors supplied by some sympathetic fibres are relatively unimportant, although they are also stimulated by adrenaline. Finally, the vasculature of skeletal muscle is very sensitive to metabolic products which can cause vasodilatation, increasing blood flow six- to tenfold. Hypoxia, however, is not very effective as a stimulus. Exercise and its effects on all the body systems is discussed further in Chap. 26.

As in cardiac muscle, arterial flow is reduced and venous flow increased during contraction of skeletal muscle fibres. The compression of the veins by contraction of skeletal muscle is an important factor in promoting the return of blood to the heart.

## Skin

The control of the skin blood vessels is almost entirely nervous, mediated through the sympathetic nervous system operating via α-receptors. Vasodilatation seems to be entirely passive – a release of vasoconstrictor tone. Although the pathway appears simple, the control is exceedingly complex. The blood vessels of the skin participate in the general control of blood pressure through the peripheral resistance, a control exerted through the medullary centres; in the control of body temperature and in the response to local skin temperature as a result of the hypothalamic temperature controlling centres; and in emotional responses such as blushing originating in the limbic brain and modulated by the hypothalamus. Local chemical stimuli such as hypoxia and the accumulation of metabolites cause vasodilatation as in reactive hyperaemia, the flushing of the skin after a period of occlusion of the blood supply to an area.

In the skin of the face, hands, and feet there are many arteriovenous anastomoses situated close to the surface. These provide a shunt pathway by which blood flow to the skin can be greatly increased without the blood necessarily having to traverse the capillary networks. In cold conditions both these anastomoses and the capillary beds close down under sympathetic stimulation although if the skin of the face, hands or feet are suddenly exposed to intense cold the anastomoses dilate, apparently by an axon reflex of the type described below. As a result of this the skin appears red rather than white. In some skin areas the arteries and veins are arranged in a countercurrent system which helps to maintain the gradient of temperature in the tissues and in the blood (p. 105).

In addition to these skin responses which may be general or local, there are some specific responses to potentially harmful stimuli. The white line reaction and the triple response are examples. If a line is drawn on the skin with a blunt instrument relatively light pressure shows first a blanching as the blood is pressed from the local capillaries, and then after about 15 s the white line, or tache, results from the contraction of precapillary sphincters along the line and a draining of the blood from the capillaries and venules. A firmer pressure gives in about 10 s a red line and – later – perhaps min later – a local swelling surrounded by a diffuse red mottling. This threefold reaction is termed the triple response. The red tache is due to dilation of capillaries. The swelling results from minor damage to tissue causing the release of histamine and, therefore, increased vascular permeability and an outflow of plasma from the vessels to cause local oedema. The red flare is thought to be due to an axon reflex. This term describes the stimulation of a sensory nerve and antidromic conduction of the impulse down another nearby branch of the same nerve. The antidromic impulse is thought to cause the release of a peptide from the sensory nerve terminal. The peptide, called substance P, causes arteriolar dilatation.

## Gingivae and oral mucosa

The oral mucosa behaves very similarly to the skin. The colour of the mucosa, like that of skin, is a sensitive indicator of the blood flow through the area and the state of oxygenation of the blood. The colour changes in gingivitis are due to the vascular changes of an inflammatory response.

## Tooth

The blood supply to the tooth pulp is subject to constraints rather like those governing the blood supply to the brain. The tooth itself forms a rigid container so that any expansion of the pulp results in pressure and possibly cell damage. Further, both arteriole and venule reach the pulp through the very small orifice of the root apex in the fully-formed tooth. Hence, dilatation of the arteriole will compress the venule and impede venous drainage. Measurements of blood pressure in human teeth give values of 4.3 kPa systolic and 3.5 kPa diastolic. Noradrenaline and adrenaline raise the pressure from a mean of 3.7 kPa to around 4.0 kPa, whilst acetylcholine causes it to fall to 3.2 kPa. Application of heat to the tooth causes vasodilatation and a rise in pressure. Cold causes a reduction in pressure. Any operative procedures such as conservation reduce the intradental blood pressure. Increases in blood flow to the tooth almost always cause pain because of the inability of the pulp to expand in its cavity.

# The effects of systemic hormones and locally produced vasoactive substances

Blood pressure and blood flow may also be modified by hormonal influences acting both in the body as a whole and more locally. Adrenaline has already been discussed. At least two hormones have been found to affect blood pressure. Vasopressin, or antidiuretic hormone, in pharmacological rather than physiological doses, causes vasoconstriction and raises blood pressure. Physiologically, it also causes vasoconstriction in the renal medulla, reducing blood flow and helping to maintain the gradient of osmotic activity which increases towards the pyramids – a factor of importance when maximum osmotic activity is needed in order to excrete a concentrated urine and retain as much water as possible (Chap. 16). Vasoconstriction more generally, with a consequent rise in blood pressure, could obviously be of importance if blood volume were greatly decreased and it is interesting that in surgical shock and after blood loss, vasopressin is released from the hypothalamic neurones into the posterior pituitary gland. Another polypeptide hormone, angiotensin II, is the most potent physiological vasoconstrictor known. It is produced by the enzyme renin secreted into the bloodstream by the juxtaglomerular organ of the kidney in response to a fall in blood pressure (Fig. 12.8). In addition to stimulation of baroreceptors, the juxtaglomerular body is also stimulated by low sodium levels, prostaglandins and sympathetic activity.

Locally produced chemical substances also cause changes in the local blood supply. Some of these are physiological transmitter substances but others are substances produced when cells are damaged. The vasodilator effect of substance P has already been mentioned as mediating part of the triple response. The kinins are a group of small polypeptides split off from larger plasma proteins, usually by locally produced enzymes. The best known member of the group is bradykinin, a peptide produced from a plasma protein by the action of the enzyme kallikrein which is present in many secreting glands (see Fig. 12.8(b)). Bradykinin and the related kallidin cause vasodilatation and hence increase the supply of blood to an organ. Prostaglandins of the PGA and PGE groups have local vasodilator effects. Prostacyclin, the prostaglandin produced by the endothelial walls of blood vessels, is vasodilator, but thromboxane $A_2$, the prostaglandin released from platelets, is vasoconstrictor (see Chap. 27).

Damage to cells releases cell contents and, of these, adenosine diphosphate in particular is capable of producing local vasodilatation. Histamine has vasodilator activity and also increases the permeability of the capillaries. A number of other small polypeptides released by tissue damage or in inflammation can cause vasodilatation. By increasing local blood flow all these substances would contribute to three of the traditional cardinal signs of inflammation – redness (rubor), warmth (calor) and swelling (turgor). The fourth sign, pain (dolor), also arises from substances released from damaged cells. All the vasoactive substances mentioned here can also cause a sensation of pain or itching.

When a blood vessel is damaged a number of mechanisms co-operate to cause constriction of the vessel wall, reducing blood flow and bringing the edges of the cut together. These are dealt with more fully in Chap. 27, but they include a local response of the smooth muscle of the vessel wall, which constricts in response to the physical stimulus of injury, and the vasoconstrictive action of locally released substances such as thromboxane $A_2$ from damaged cells and serotonin from the platelets. Vasoconstriction is also enhanced by the damage to the endothelium which interrupts prostacyclin production. During haemostasis ADP is said to be vasoconstrictor – a role at variance with its reported activity in inflammation.

## Further reading

Burton, A.C. *The Physiology and Biophysics of the Circulation,* 2nd edn, Year Book Medical Publishers, Chicago, 1972

Folkow, B. and Neil, E. *Circulation,* Oxford University Press, Oxford, 1975

Rushmer, R.F. *Structure and Function in the Cardiovascular System,* Saunders, Philadelphia, 1972

# 16

# Control of the osmotic activity of the blood and of the blood volume

## Control of the osmotic activity of the blood

There is a close inter-relation between the salt content, the osmotic activity, and the volume of the blood. Indeed, because all the body fluid compartments are in equilibrium, changes in any of these blood variables will also affect other body fluids. Thus, although this chapter considers control of blood osmotic activity separately, the real-life situation is much more complex than that presented here.

## Control of water loss

The body controls osmotic activity in the blood by adjusting water balance (Fig. 16.1). The sensors are the osmoreceptors located in the hypothalamus, principally in the supra-optic nucleus. The osmoreceptors are cells particularly sensitive to osmotic changes, shrinking as the extracellular fluid becomes hypertonic, and swelling as it becomes hypotonic. The normal osmolality of the blood plasma is around 283 mosmol/kg, with a range of some 20 mosmol/kg on either side of this. This is equivalent to an osmolarity of about 300 mosmol/l (osmolality refers to the number of osmotically active particles in a kilogram of solute, osmolarity to the number in a litre of solute).

The cells which synthesise antidiuretic hormone are nerve cells with axons extending into the posterior pituitary gland. They are probably not the actual osmoreceptor cells but cells in communication with them. They produce and store the antidiuretic hormone in granules in the cell body, and then transport it in combination with a protein

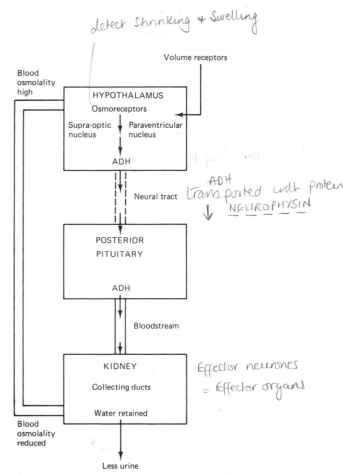

Figure 16.1 The mechanisms of control of blood osmolality.

carrier, termed a neurophysin, along their axons, down through the pituitary stalk to the posterior pituitary gland. Here it is released into the bloodstream in response to an increase in blood osmolality. The hormone passes via the bloodstream to the effector organ, the distal nephrons of the kidneys.

The hormone binds to receptor sites on the serosal walls of the cells of the distal nephrons and cyclic 3,5-AMP is released to diffuse across the cells and increase the permeability of the mucosal walls of the cells. The calculated effective pore size of the cell walls increases from around 0.8 nm to 4.0 nm. Although the effective change is one in diffusion of water, other small molecules such as urea will also diffuse more readily. Since the collecting duct is passing down through regions of increasing osmotic activity, water is reabsorbed and a more concentrated urine results. Although antidiuretic hormone decreases the rate of blood flow through the vasa recta and thus helps to maintain the osmotic gradient in the medulla, there is a natural limit on the amount of concentration which can be achieved. In man the maximum level is 1400 mosmol/l. About one third of this osmotic activity is due to the urea content and the remainder to the salt content. One consequence of this limit of concentration is that sea water cannot be used as a water source for man because its salt content is too high.

By these means the osmotic activity of plasma is raised or lowered by the excretion of more or less water without changing the total salt content. This leads to concomitant changes in blood volume which must also be compensated. Where a conflict arises between the need to maintain blood volume and the need to adjust osmolality, the blood volume is protected at the expense of osmolality. There is a minimum acceptable blood volume below which cardiac operation is inefficient.

Activity of the hypothalamic osmoreceptors, or the secretion of antidiuretic hormone, is modified by many other influences. Sleep or wakefulness, stress, emotional factors, external temperature, levels of circulating corticosteroids and adrenaline, all affect either the level of osmoreceptor activity or the amount of ADH secreted.

### Control of water intake

A second control system operates rather differently. The water content of the body may be adjusted not only by varying the loss of water but also by varying the intake. Water intake, however, is a conscious act, not a reflex activity. Nonetheless it may be stimulated by the sensation of thirst; and one effect of increased plasma osmolality is to stimulate a sensation of thirst via the hypothalamus. Removal of water from the tissues by the high plasma osmotic activity may also result in dryness of mouth, which causes a sensation of thirst. The thirst areas in the hypothalamus are then responsible for the conscious initiation of drinking, or fluid seeking. Drinking will not immediately reduce plasma osmolality, but the stimulus of fluid in the mouth inhibits thirst centre activity so that the drinking does not continue until the eventual change in plasma osmolality occurs. Experimental evidence suggests that there are also receptors in the oesophagus, stomach, and possibly the small intestine, which can contribute to the control of water intake before the actual re-establishment of the normal level of plasma osmolarity. These sensors function as disturbance detectors (p. 216).

The sensation of thirst is also stimulated by renin, by angiotensin and by hyperthermia. In hyperthermia thirst and hunger show contrary patterns: thirst is stimulated while hunger sensations are inhibited. More usually, the intake of food is associated with drinking; this may be due to habit or some primitive association of the appetites.

### Control of blood volume

It is essential if blood pressure, and therefore the circulation of blood, is to be maintained, that a sufficient volume of blood be present in the circulatory system. The control of blood volume occurs principally in the kidneys although other mechanisms exist and will be discussed further in Chap. 27.

Blood volume may be sensed by stretch receptors or by pressure receptors – themselves stretch receptors serving a different function. Such receptors may be situated in blood vessels or may be positioned to monitor extracellular fluid volume – since plasma and total extracellular fluid volume are closely related via the exchange across capillary walls. The low pressure side of the circulation, the capacitance vessels, normally contains the greater proportion of the blood volume, and so stretch receptors on this side of the circulation are well placed to monitor blood volume. Such sensors are located in the walls of the great veins and in the walls of the atria. They can respond to both increases and decreases in blood volume. In addition there may be arterial volume receptors; certainly some receptors which respond mainly to blood pressure also respond to increases in blood volume (although not to decreases). The juxtaglomerular bodies in the renal arterioles respond to both increases and decreases in blood volume.

Inputs pass by autonomic afferents to the brain and reach the hypothalamus (Fig. 16.2). Here an increase in blood volume results in an inhibition of antidiuretic hormone release whilst a decrease in blood volume stimulates release of the hormone.

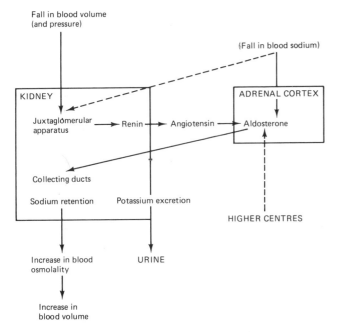

**Figure 16.2** The mechanisms of control of blood volume.

Antidiuretic hormone then acts as described earlier. Its action leads to a decrease in plasma osmolality unless salt is also retained. However, a second mechanism exists to control salt output and retention in parallel with the control of water volume.

A decrease in blood volume causes a decrease in blood pressure. This has little direct effect on filtration in the kidney, even though a fall in filtration pressure might be expected to reduce the filtration fraction, because the pressure in the glomerular capillaries is regulated by the intrinsic balancing of afferent and efferent arteriolar diameter to maintain a constant pressure difference. However, cells in the walls of the afferent arterioles which form part of the juxtaglomerular apparatus, respond to a fall in the pressure of the blood flowing through the arteriole by causing the secretion of the enzyme renin. This splits off a peptide, angiotensin I, from a precursor in blood plasma, angiotensinogen, one of the β globulin plasma proteins. The peptide is converted into an active form, angiotensin II, in the lungs by a rather non-specific enzyme known as converting enzyme. Angiotensin II causes vasoconstriction (see p. 236) but also causes the adrenal cortex to secrete the hormone aldosterone. This steroid hormone reaches the cells of the distal nephrons of the kidneys, enters them, and stimulates the active pumping of sodium ions from the tubular fluid back into the interstitium and thence into the bloodstream. This sodium retention will usually be accompanied by water retention or will by itself increase plasma osmolality and cause water to be redistributed in the body compartments so that the plasma volume, and indirectly the blood pressure, are restored. Increase in plasma osmolality results in the release of antidiuretic hormone, as described in the previous section, and so more water is retained by the kidney.

Blood volume is always restored by increasing fluid intake as well as fluid retention: the indirect method of increasing fluid retention by causing retention of salt is not by itself sufficient to restore the original volume. Stimulation of receptors by a decrease in blood volume is another stimulus for the hypothalamic thirst centre; a similar correction exists for loss of blood volume as does for increase in plasma osmolality. Further, angiotensin II is a potent stimulator of the sensation of thirst. The hypothalamus contains not only the thirst centres, the osmoreceptors, and the antidiuretic hormone synthesising neurones, but also produces corticotrophin stimulating hormone. This passes to the anterior pituitary to stimulate the secretion of adrenocorticotrophic hormone – which has a short-lived effect on the secretion of aldosterone.

Recently another mechanism thought to control blood volume has been studied. Some cells in the right atrium and, indeed, in the left atrium as well, contain granules of material which is released when the volume of the atrium is increased. Since this is on the low pressure, capacity, side of the circulation, it is particularly affected by blood volume. An increase in blood volume, then, is now thought to cause release of a diuretic and natriuretic hormone which increases the blood flow in the vasa recta, diminishes the osmotic gradient through the kidney medulla, and causes the excretion of a hypotonic urine. This action is aided by an inhibition of renin release, leading to a reduction in sodium reabsorption. The net effect is a reduction in the blood volume. This mechanism is likely to be supplementary to the main controls of blood volume, rather than a major influence.

## Salt appetite

Animals whose sodium chloride intake is restricted actively seek out salt-containing foods and their oral taste receptors may possibly become more sensitive to salt. This implies that there is some means whereby a particular food drive can be initiated: presumably via the limbic brain or the hypothalamus.

In man, salt balance is achieved with an intake of between 2 and 10 g/day in the diet. Most diets are well within this range, and salt intake depends more upon individual taste preferences than on any salt appetite. When salt intake falls below 1 g/day a temporary excess of excretion in relation to intake occurs and weight loss is observed temporarily. Over a few days a new equilibrium is established with increased aldosterone secretion and sodium retention. An increased salt intake (30–40 g/day) is corrected by the opposite adjustments to establish an equilibrium of intake and excretion of sodium; but the increased salt intake results in a weight gain which may be as much as several kilograms, and this weight gain persists even after equilibrium is achieved.

## Other systems interacting with the control of blood volume

The control system described for blood volume is made more complex in its responses by the interactions with other body control systems which are affected by blood volume. Clearly there are interactions with the control system for blood pressure and there must be neural links for this purpose between the hypothalamus and the medulla oblongata. Changes in blood volume can affect blood pressure which, in turn, affects blood flow and the tissue gas equilibria. There may, therefore, be routes by which impulses from the blood gas chemoreceptors may influence control of blood volume. Hormones such as adrenaline may affect the system since adrenaline causes changes in blood vessel diameter and cardiac activity. Temperature may have varying effects. Even mental and emotional stimuli, which affect the sensitivity of the hypothalamic receptors, may be important.

## Control of blood potassium concentrations

Plasma potassium concentrations are normally stabilised between 3.5 and 5 mmol/l. Even on a potassium-free diet hypokalaemia (low blood potassium levels) develops slowly because of the enormous reservoir of potassium in the cells. The normal diet contains an appreciable amount of potassium: this must be excreted if it is not to accumulate in the body. An increase in plasma potassium concentrations stimulates the secretion of aldosterone by the adrenal cortex. This increases the secretion of potassium ions into the tubular fluid in the distal nephron. Potassium ions are reabsorbed in the proximal convoluted tubule, secreted in the distal nephron. Since this potassium is exchanged for sodium, taking the place of hydrogen ions, an acidosis may develop – a hyperkalaemic acidosis. The increased retention of sodium ions may raise blood sodium concentrations and cause oedema. With a low sodium diet, the reverse applies: potassium ions are lost from the body, together with hydrogen ions. Similarly, in hypokalaemia the relative lack of potassium ions means that more hydrogen ions can exchange with sodium ions in the distal nephron and an alkalosis may develop.

## Other mechanisms of the control of kidney function

In addition to the mechanisms already discussed, there are direct nervous effects on the kidneys even though the kidneys can still function adequately when denervated. Sympathetic nerves are presumably involved in the autoregulation of blood pressure in the glomeruli, and in the distribution of blood flow between the cortex and medulla. Sympathetic stimulation causes the release of renin from the juxtaglomerular apparatus. This may play a role in the correction of reductions in blood pressure since sympathetic activity is invoked by the vasomotor centre in the medulla oblongata when the input from the baroreceptors of the carotid sinus and aortic arch is reduced because of low blood pressure.

Kidney function is also a vital part of the control of the pH of body fluids: this is described in Chap. 18.

## Control of the urinary bladder

### The control of urination

Urination, like defaecation (described on p. 280), is a process involving not only the organ containing the material to be voided, but also the muscles of the pelvis, the abdomen and even the thorax.

The primary stimulus for urination is distension of the bladder. Although the smooth muscle of the bladder wall elongates plastically in response to continued stretch, receptors in the trigone do initiate impulses up to a centre for micturition, and indeed, to consciousness. The reflex occurs at spinal level but is normally inhibited by supraspinal control, of which at least part is conscious in man. At about 150 ml distension of the bladder, impulses are generated in the pelvic nerves to the first, second, and third sacral segments of the spinal cord. Here the nerves synapse with the pelvic and pudendal efferent nerves, which cause relaxation of the internal and external sphincters respectively. The pelvic nerves stimulate contraction of the detrusor muscle of the bladder wall and fluid is forced into the urethra. Contraction of the detrusor muscle is increased by further impulses in the pelvic nerves resulting from stimulation of the pudendal nerve afferents from the urethra and impulses in the hypogastric nerve efferents triggered by hypogastric nerve afferents signalling distension of the urethra. The presence of fluid in the urethra causes relaxation of the external sphincter by a pudendal nerve reflex. This reflex is enhanced or inhibited by higher centres and its activity is extended in time by the length of the various reflex pathways involved. Thus stimulation of the anterior pons, or parts of the hypothalamus and limbic brain, will enhance the reflex and cause activation at lower levels of bladder tension, whilst stimulation of part of the midbrain, other parts of the hypothalamus and limbic brain, and of the cortex itself, can inhibit micturition. Spinal section above the sacral centre renders micturition an involuntary act.

Micturition involves contraction of pelvic and abdominal muscles to increase bladder pressure and enhance the effect of detrusor contraction, and also fixation of the diaphragm and thoracic muscles to increase abdominal and/or pelvic pressure. Thus micturition may produce similar although smaller thoracic effects to those seen in defaecation (p. 280).

## Further reading

Pitts, R.F., *Physiology of the Kidney and Body Fluids,* 3rd edn, Year Book Medical Publishers, Chicago, 1974

Valtin, H. *Renal Function: Mechanisms Preserving Fluid and Salt Balance in Health,* 2nd edn, Little, Brown & Co, Boston, 1983

Vander, A.J. *Renal Physiology,* 3rd edn, McGraw Hill, New York, 1985

# 17

# Normal control of respiration and the control of blood oxygen and carbon dioxide concentrations

## Normal control of respiration

### Control of the normal rhythm of inspiration and expiration

The normal rhythm of respiration is generated in the medulla oblongata as a result of the alternating activity of neurones driving the inspiratory muscles – the diaphragm and the external intercostal muscles – and neurones which, in the resting or relaxed subject, inhibit their activity. As the depth of respiration is increased, these latter neurones, or others in the same group, initiate activity in the internal intercostal muscles and then in the abdominal muscles to assist in expiration. Finally, other accessory muscles may come into play (p. 128). The rhythmic activity of breathing, then, results from the activity of a neuronal cycle-generator in the brainstem, and not, as in the circulatory rhythm, from a cycle generated by the muscle pump itself.

The nerve cells directly involved in controlling respiration are situated in two groups, one primarily controlling inspiration and the other primarily controlling expiration but with some neurones also serving inspiratory muscles (Fig. 17.1). The first group is the dorsal group, lying in the tractus solitarius. These neurones show rhythmic activity and they have axons, or are linked to axons, which travel in the phrenic nerve to the diaphragm, whose contraction results in inspiration. Branches of these neurones also pass to stimulate neurones in the other (ventral) group.

The second group is found in two adjoining nuclei – the nucleus ambiguus more cranially and the nucleus retroambigualis more caudally. Neurones of the nucleus ambiguus send branches to travel in the vagus nerve to reach the accessory muscles of respiration on the same side of the body. Neurones of the nucleus retroambigualis serve the intercostal muscles and assist in both inspiration and expiration. The nerves to the expiratory internal intercostal muscles pass only to the contralateral muscles but those to the inspiratory external intercostals pass to both contralateral and ipsilateral muscles.

Impulses from the dorsal 'inspiratory centre' neurones stimulate the ventral 'expiratory centre' neurones, and these in turn then inhibit those of the dorsal group. When the activity of the dorsal group increases, more of the ventral group are stimulated, and active expiration is added to the passive response produced by inhibition of the inspiratory neurones. In either case a cycle of activity is set up. When damage to the medulla involves the respiratory neurones respiration can still continue by voluntary activity, but only in the conscious subject. The few unfortunate individuals with such lesions (a condition known as Ondine's syndrome) must live in an artificial respirator so that breathing can continue while they sleep.

The activity of the respiratory centres is adjusted by a higher centre, situated in the pons, formerly referred to as the pneumotaxic centre but now known to be a part of the nucleus parabrachialis medialis (Fig. 17.1). This nucleus receives inputs from the vagus nerve, the glossopharyngeal nerve, and from higher centres including the cerebral cortex. Many of the control mechanisms which adjust respiratory rate and depth operate through this nucleus. The respiratory centres are also affected by activity in nearby chemoreceptors which are sensitive to the pH of cerebrospinal fluid and hence indirectly to blood partial pressures of carbon dioxide.

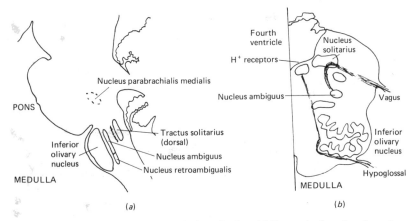

**Figure 17.1** The centres for the control of respiration. (a) Parasagittal section through the brainstem to show the nuclei involved in respiration, (b) section through the medulla oblongata of the level of cranial nuclei X and XII. Only one side of the medulla is shown. The positions of the respiratory nuclei and the receptors for the pH of cerebrospinal fluid are shown.

## Control of the depth of respiration and the adjustment of lung ventilation

Although control of lung ventilation is primarily concerned in maintaining blood oxygen and carbon dioxide concentrations at favourable levels, there are in addition some within-system reflexes to protect the pulmonary apparatus and to balance rate and depth of ventilation to achieve the desired results in the most economical manner. Since not all the inspired gas undergoes exchange with the blood (p. 129) deeper respiration is more efficient than shallower. Two further constraints appear at high rates of respiration: as the tidal volume is increased the work necessary to overcome the elastic recoil of the lungs and thorax increases, and at high rates of respiration the increase in volume flow increases the work necessary to overcome airway resistance. At very high rates of respiration the time available for each breath may be insufficient to allow enough air to be drawn into the lungs with each breath.

A number of sensory endings in or around the lungs transmit information to the medulla via the vagus nerves. The best known of these are stretch receptors in the thorax which are stimulated by inspiration; as stimulation increases, impulses pass up the vagus to inhibit the inspiratory neurones. This is the inflation, or Hering Breuer, reflex. The rhythm of respiration is initiated by the inspiratory and expiratory neurones themselves, but this reflex modifies the pattern of inspiratory depth and frequency to maintain a balance between them. This sensory input is only one of the factors influencing the activity of the respiratory neurones. A deflation reflex, in which afferent fibres of the vagus nerve show increased activity towards the end of expiration and then stimulate the inspiratory neurones, has also been described – but this potential reflex appears unimportant in normal breathing. From time to time, the rhythm of breathing is interrupted by deeper inspirations. These seem to be intermittent corrections to adjust discrepancies arising from the manner in which the body endeavours to maintain optimum efficiency of breathing in terms of rate and depth. Their occurrence is an example of a paradoxical reflex: a positive feedback reaction to inspiration. Sometimes they are apparent as sighs.

Higher centres may also control inspiration and expiration, as in speech (Chap. 20) and in sudden shock – when prolonged inspiratory spasm may occur.

## Control of breathing to protect the respiratory apparatus

Other stimuli may modify the breathing pattern. The reflex reaction to irritation of the nasal mucosa is a sneeze. It consists of a deep inspiration followed by a rapid uncontrollable expiration. Stimulation of chemoreceptors in the trachea or in the bronchi may cause coughing or increased respiration in an endeavour to blow away the irritant. This is seen with some gaseous anaesthetics which initially cause rapid shallow breathing. Mechanical irritation in the trachea or bronchi usually causes coughing. In coughing the glottis is closed and the pressure in the thorax raised by the action of the expiratory muscles. Sudden release of the alveolar air through constricted bronchi gives a high velocity airflow which may reach as high as 500 mph. The reflex is sensitised by excessive inflation and deflation so that it may become paroxysmal. Irritants such as dust and cigarette smoke increase mucus production from the protective glands of the airways, damage the cilia on the cells lining the airways, and may cause coughing.

The airways may be protected by the inhibition of inspiration. Breath-holding may be voluntary or may occur as part of another reflex. Thus inspiration is inhibited during swallowing and vomiting. The breath is held when diving into water and during weeping. Voluntary breath-holding is terminated by the stimulus of increased carbon dioxide and decreased oxygen concentrations in the blood so that preliminary hyperventilation increases the time during which the breath can be held. The period may also be extended by making use of the swallowing reflex. Another means of protecting airflow is by sniffing. This consists of one or more rapid inspirations which are limited to the nasal cavity. Such inspirations help to sweep back nasal mucus which may then be swallowed or expectorated.

### Other involuntary modifications of the breathing pattern

Some changes in respiration are involuntary but their significance is difficult to identify. These include the sigh, the yawn and the hiccough. Sighing is a slow deep inspiration followed by a slow expiration which results in a decrease in blood carbon dioxide concentration and an increase in blood oxygen concentration. The stimulus is unknown. The yawn and the hiccough are even more curious. Yawning is a long deep inspiration through the open mouth followed by a slow expiration, often accompanied by a stiffening or stretching of the trunk. Although it is associated with sleep, weariness or boredom the actual stimulus is unknown. The hiccough is an intermittent spasmodic constriction of the diaphragm. Although it may be caused by many conditions affecting thoracic or abdominal organs there is usually no identifiable stimulus. Treatments are completely arbitrary – swallowing, rebreathing from a paper bag, a sudden fright or a sharp stimulus like drinking vinegar and sugar.

Although snoring might be considered as another respiratory modification, it is in fact only a noisy version of normal breathing due to relaxation of the soft palate and the posterior pillars of the fauces which are then able to vibrate in the passing airflow.

### Respiratory manoeuvres which affect cardiovascular regulation

Reference has already been made (p. 124) to the interaction of pressures in the thorax and abdomen and the effect of thoracic pressure on the cardiac output. Two deliberate manoeuvres demonstrate the relationship between pressure in the thoracic cavity and the cardiovascular system. Valsalva's manoeuvre is the name given to expiration against a closed glottis. It occurs normally during defaecation when it is used to increase thoracic pressure to assist in raising abdominal pressure. This action decreases or prevents venous return from the abdomen and consequently cardiac output drops. The brief increase in systemic blood pressure due to the increase in thoracic pressure is therefore followed by a decrease in cardiac output and a fall in systolic pressure. This results in reduced baroreceptor activity and a reflex increase in heart rate together with vasoconstriction. Diastolic pressure therefore rises. When the pressure in the thorax is released the blood pressure rises, overshoots the normal level and finally stabilises back to it. Muller's manoeuvre is an inspiration against a closed glottis. This lowers thoracic pressure so that blood accumulates in the thoracic vessels. It is sometimes useful in preventing fainting.

## Control of blood oxygen and carbon dioxide concentrations

The controls of the respiratory system so far discussed are those which operate to maintain a regular rhythm of respiration and those which act to protect the respiratory apparatus. The main controlling factors in respiration, however, are not directly related to the respiratory apparatus, but rather to the attainment of the objectives of respiration – the supply of oxygen to the tissues of the body as a whole, and the removal and disposal of the carbon dioxide produced by the tissues.

The form of the association curve for oxygen and haemoglobin indicates that the alveolar oxygen pressure has to fall to 10 kPa (75 mmHg) before the amount of oxygen carried in the blood falls noticeably. This represents a fall from 15% oxygen in the alveoli to around 11%. Again, mixed venous blood normally contains around 75% of its haemoglobin saturated with oxygen, so that an effective oxygen reserve exists. Small changes in the partial pressure of oxygen in the blood do not normally, therefore, provide an index of the respiratory need. The amount of carbon dioxide in the blood is much more directly related to its partial pressure. The main stimulus modifying the basic respiratory rhythm, therefore, is that due to carbon dioxide concentrations either directly or via the pH change associated with carbon dioxide carriage. Respiration may indeed cease temporarily (a condition known as apnoea) when carbon dioxide has been blown off from the lungs, as after hyperventilation. Conversely, a mixture of gases containing 6% carbon dioxide, or even the alveolar gases forcibly expired by another individual, can initiate respiration when this has ceased for some reason. The basis of the 'mouth-to-mouth' method of resuscitation is the mechanical effect of blowing air in and out of the lungs, but its effect is assisted by the carbon dioxide in the gas. The oxygen content of expired air is almost sufficient to ensure adequate oxygenation of the blood. In a similar fashion, respiration can sometimes be increased by rebreathing air in a paper bag.

## Controlled variables

The variables controlled and monitored in the control of respiration are the partial pressures of oxygen, carbon dioxide, and the concentration of hydrogen ion in the blood. The controls for these three variables are closely interlinked with each other and each has influences on the others. However, it is conceptually easier to consider them separately, looking at the individual components of each control system (Figs 17.2 and 17.3). Carbon dioxide and oxygen concentrations in the blood will be dealt with here, and the control of blood pH in the next chapter.

Several terms are used to describe conditions in which the supply of oxygen is reduced or the amount of carbon dioxide inhaled or present in the blood is

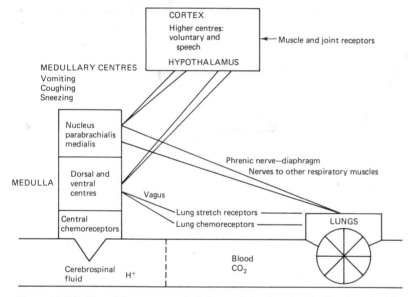

**Figure 17.2** Outline of the components of the control system for blood carbon dioxide concentrations.

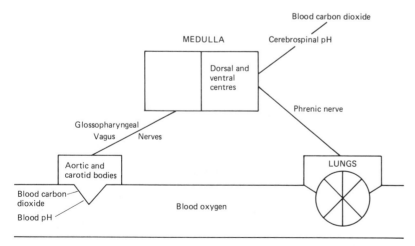

**Figure 17.3** Outline of the components of the control system for partial pressure of blood oxygen. This figure should be taken in conjunction with *Figure 17.2*, since the modifying influences of the higher centres have not been included in this second diagram.

increased. The term hypoxia indicates that insufficient oxygen is reaching the tissues. It was defined more loosely on p. 129, as a low oxygen tension in the inspired gas, but this usage is actually a contraction of the term hypoxic hypoxia. Four types of hypoxia are distinguished: anoxic, anaemic, stagnant, and histotoxic. Anoxic hypoxia (perhaps better, hypoxic hypoxia) occurs when the inhaled gas mixture is low in oxygen concentration or is at low pressure, so that the partial pressure of oxygen is low. Anaemic hypoxia results from the haemoglobin content of blood being low, so that less oxygen is carried in a bound form. If the blood supply to a tissue is reduced, or stops completely, as a result of disturbance to the arterial inflow or the venous outflow, stagnant hypoxia ensues. In histotoxic hypoxia the oxygen is available in the blood, but cannot be utilised by the tissue because of some poisoning of the metabolic pathways, as in cyanide poisoning. Strictly, anoxia indicates a complete lack of oxygen, but the term is sometimes used instead of hypoxia. The term hypoxia is often used without qualification when hypoxic hypoxia is meant. Another term can be used to refer to low concentrations of oxygen in the blood: hypoxaemia (or anoxaemia). This may be associated with a change in skin colouration towards the blue of de-oxygenated haemoglobin – cyanosis. Cyanosis is seen when the concentration of de-oxygenated haemoglobin in the blood exceeds 0.78 mmol/l (5 g/dl). Thus cyanosis is not seen in severe anaemia, when insufficient haemoglobin is present, or in toxic hypoxia, such as that due to cyanide or carbon monoxide poisoning. When there is marked cutaneous vasoconstriction as in very cold conditions, or in shock, cyanosis may be masked by the skin pallor.

An increase in carbon dioxide concentration is termed hypercarbia or hypercapnia – the two terms are synonymous.

When the supply of oxygen is much reduced and at the same time carbon dioxide concentrations are high, a state of asphyxia is said to occur. This may result either from rebreathing of air in a confined space or from obstruction of the airways.

### Sensors of blood carbon dioxide concentrations

Two major groups of sensory cells are responsible, directly or indirectly, for monitoring blood carbon dioxide concentrations. One of these is the central chemoreceptor area, consisting of cells within or close to the respiratory neurones (or centres) on the ventrolateral surface of the medulla oblongata near the roots of the vagus and glossopharyngeal nerves (Fig. 17.1). These cells are in contact either with cerebrospinal fluid or tissue fluid in equilibrium with it. Carbon dioxide in the blood forms carbonic acid and the hydrogen ions resulting from dissociation of this acid are largely buffered by the haemoglobin and plasma proteins of the blood. Large changes in carbon dioxide concentration are therefore necessary to produce significant changes in the pH of blood. Carbon dioxide, however, diffuses readily from the blood into the cerebrospinal fluid. Carbonic acid formation changes the pH of cerebrospinal fluid much more readily because it has a low protein content and therefore relatively little buffering capacity. It is this pH change in the cerebrospinal fluid which stimulates the central chemoreceptors and activates the nearby respiratory controlling neurones. These chemoreceptors are responsive to small changes in carbon dioxide levels but their response to changes in blood carbon dioxide concentrations is relatively slow because the changes are sensed not in blood directly but only after it has equilibrated with cerebrospinal fluid. Hydrogen ions themselves do not readily cross the blood-brain barrier, so that changes in blood pH affect the cerebrospinal fluid only after some delay. The central chemoreceptors are not therefore sensitive to changes in blood pH.

The pH of cerebrospinal fluid is very closely controlled by two mechanisms: respiration, and the adjustment of the blood flow to the brain. A downward shift of 0.05 pH units from the normal 7.32 of cerebrospinal fluid results in a ten-fold increase in ventilation and almost a doubling of cerebral blood flow. However, it is with the respiratory changes that this section is concerned.

The second group of cells are those which are situated in the two small accumulations of sensory cells lying alongside the carotid arteries and supplied directly with blood from small branch arteries. These are the carotid bodies, about 1–2 mm in diameter and weighing about 2 mg (Fig. 17.4). They are made up of sensory cells and nerve terminals. They contain high concentrations of transmitter substances such as dopamine and adrenaline. Similar small accumulations of cells are found near the aorta and are termed aortic bodies. These organs respond directly or indirectly to changes in blood carbon dioxide concentrations. Although their response is more rapid, in that they are monitoring arterial blood, they are probably less sensitive in that the changes they observe for a given increase in carbon dioxide partial pressure are less than those perceived by the central receptors. As a result, they account for only some 30% of the total response to an increase in blood carbon dioxide concentrations. The cells appear to be more sensitive to hydrogen ion concentrations than to actual partial pressures of carbon dioxide: so the response to carbon dioxide may be indirect. The real importance of these cells, however, is in the sensing of partial pressures of oxygen in the blood, as described below.

# Control of blood oxygen and carbon dioxide concentrations

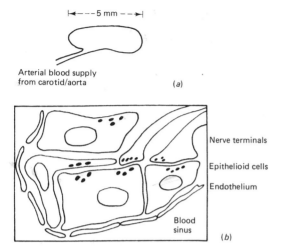

**Figure 17.4** The peripheral chemoreceptors. (*a*) Gross appearance of a carotid body; (*b*) section through part of a carotid body.

## Input pathways

The input pathway from the central chemoreceptors is to the cells of the medullary respiratory centres.

Inputs to these cells also arrive from the carotid bodies via the glossopharyngeal nerves and from the aortic chemoreceptors via the vagus nerves.

## Sensors for blood oxygen concentrations

The sensors for blood oxygen concentration lie in the peripheral chemoreceptor organs, the carotid and aortic bodies (Fig. 17.4). These structures have an unusually high blood flow, equivalent to 2 l/min/100 g weight. Because of their small size the blood leaving these organs is almost as oxygen-saturated as the blood entering them. The cells take up oxygen directly from the dissolved oxygen fraction – their rate of metabolism and the flow of the blood is such that the bound oxygen is almost untouched. No activity is observed in the chemoreceptor nerves if the partial pressure of oxygen is above 15 kPa and the impulse traffic in the nerves increases markedly when the partial pressure falls to around 9 kPa. Thus the response to a decrease in oxygen tension in the blood is delayed in that a large fall is necessary to stimulate a response, but it is rapid in that, once stimulated, the response is immediate. The stimulus is directly from arterial blood.

The response of the peripheral chemoreceptors is to dissolved oxygen in the blood. This means that changes in the total oxygen content or in the amount of haemoglobin-bound oxygen do not influence the chemoreceptors unless the dissolved fraction also changes. Thus the chemoreceptors are not stimulated by the hypoxia of anaemia, where less haemoglobin carries less oxygen, or the hypoxia of carbon monoxide poisoning, where the haemoglobin carries carbon monoxide in preference to oxygen. However, inadequate perfusion of the aortic bodies, such as occurs with a fall in cardiac output, reduces their oxygen supply and will produce both increased respiration and also increased heart rate and vasoconstriction. The carotid bodies are less affected by a reduction in cardiac output and do not give the same circulatory response.

## Controlling centre

In common with many other reflex systems involving a considerable degree of integration, the control system for respiration is situated in the medulla oblongata of the brainstem (*see* above). The functioning of the central control is modified by changes in blood oxygen concentrations even when the concentrations are above those necessary to stimulate the peripheral chemoreceptors. The sensitivity of the medullary centres to the indirect stimulus of carbon dioxide at the central receptors is related to the blood oxygen concentrations. Thus the response to increases in blood carbon dioxide depends upon the concentration of oxygen in the blood. Asphyxia, therefore, in which oxygen concentrations begin to fall as carbon dioxide levels rise, is a more potent stimulus to respiration than is hypercarbia alone.

In addition to the controls from higher centres already mentioned, there are complex interconnections with centres controlling blood pressure and flow, since the respiratory and cardiovascular systems are both parts of the nutrient-supplying and metabolite-removing mechanisms of the body. There are also links with a number of other centres, controlling reflexes which may involve parts of the respiratory system. These include swallowing and vomiting. Reflexes which may involve pressure changes in the thorax – such as defaecation – are also linked to the respiratory centres.

In many animals panting is a response to an increase in deep body temperature. This response is initiated from the hypothalamus. In man, temperature control does not involve the panting reflex, but the hypothalamus can change the pattern of activity of the respiratory centres as a result of limbic brain activity. Thus the increased rate of breathing in anticipation, in excitement or in fear, originates in the hypothalamus. The rapid inspiration in response to a sudden painful stimulus – sometimes indicating the answer to the question 'Does it hurt?' - may well involve hypothalamic pathways.

The most obvious involvement of higher centres, however, comes in the conscious control of breathing. Voluntary breath-holding and voluntary deep breathing are easily achieved. Control of breathing also occurs at a less conscious level in the production of speech, in singing, and in the playing of wind instruments. Breathing has to be regulated in exercise, more particularly in swimming. Control of respiration has even been utilised to help otherwise paralysed subjects to carry out a number of tasks such as typing. In speech, singing, and the playing of wind instruments expiration is controlled and inspiration takes place only between phrases or sentences. Speech may leave the speaker breathless or may have the same effects as overbreathing (hyperpnoea).

### Output pathways

The principal output pathway is via the phrenic nerves to the diaphragm, but nerves to the other muscles of respiration will be involved to a greater or lesser extent.

### Effector organs

The effector organ is best described as the respiratory unit because the lungs, thorax, diaphragm and abdominal muscles, even the neck and nasal muscles, act as a unit to modify ventilation and hence the controlled variables of blood carbon dioxide and blood oxygen concentrations. The effector can produce two responses: an increase in the rate of breathing and an increase in the depth of breathing. Either will increase ventilation although the dead space volume means that an increase in depth may be more efficient than a similar increase in rate. In response to increased levels of carbon dioxide in the blood, a parallel increase in rate and depth of respiration is usually seen.

### Response to the composite stimulus of carbon dioxide excess and oxygen lack

In the laboratory it is possible to isolate the responses to carbon dioxide excess and those to oxygen lack by causing the subject to rebreathe various gas mixtures with and without carbon dioxide absorbers in the circuit. The effects of asphyxia can also be studied. The differences seen in the laboratory in the responses to hypercarbia, asphyxia and hypoxia are shown in Fig. 17.5. These responses are also seen to some extent in anaesthetised patients and provide useful guides to their state. In normal life, however, these responses will not be seen as separate entities and the control systems interact almost completely.

In hypercarbia the subject shows a rapidly increasing rate and depth of breathing and feels uncomfortable. This is even more marked in asphyxia where the subject feels as if choking for breath. In both these conditions the discomfort is obvious. In hypoxia the onset of the condition is insidious and cyanosis may be apparent before any increase in ventilation is observed. The subject may drift into unconsciousness without any sensation of discomfort. As the amount of oxygen available to the brain cells slowly decreases, their activity is reduced. After an initial period of loss of sensation, actual brain damage will occur. In hypoxia the ventilatory response sometimes shows first as an increase in depth of breathing, without the concomitant increase in rate observed in hypercapnia or asphyxia. This is a more efficient way of increasing oxygen supply because increased alveolar ventilation is achieved more readily by increasing tidal volume than by increasing respiratory rate.

It is rare for a purely hypoxic or purely hypercapnic response to be evoked in real life. There are exceptional circumstances, such as the fall in oxygen partial pressures experienced when an individual is suddenly removed from a low to a high altitude (see p. 253), but normally hypoxia is associated with hypercarbia. The systems of control are therefore modified.

Although the respiratory rhythm can apparently be maintained by the rhythm generator, in practice respiration may cease in the absence, or in the presence of lower than normal, levels of carbon dioxide. This necessary stimulus for the maintenance of respiration is sensed at the central receptors in the medulla. Similarly, increases in the partial pressure of carbon dioxide stimulate respiration largely by their effect at the central receptors The peripheral receptors respond mainly to hypoxaemia but the fall in blood oxygen levels, as an isolated stimulus, must be large if it is to produce an effect. However, decreases in the partial pressure of oxygen in the blood sensitise the cells of the central control system so that they respond more readily to signals from the central receptors enabling even small increases in the partial pressure of carbon dioxide in the blood to stimulate respiration. Thus asphyxia produces an almost maximal response from the controlling centres. Decreases in pH of the blood may also affect the peripheral chemoreceptor response. Major decreases in the blood oxygen concentration, however, may reduce the response of the central chemoreceptors, since hypoxia depresses the activity of nerve cells.

Although central and peripheral chemoreceptors act together to increase respiration when the partial pressure of carbon dioxide in the blood rises above 5.3 kPa, a contrary, or paradoxical, effect may be seen when the blood is acidified in other ways. Stimulation of the peripheral receptors increases respiration, driving off carbon dioxide through the lungs and hence reducing the carbon dioxide concentration in the cerebrospinal fluid. This in turn reduces its hydrogen ion content. The central receptors are less stimulated, and they oppose the response to the peripheral receptor stimulation.

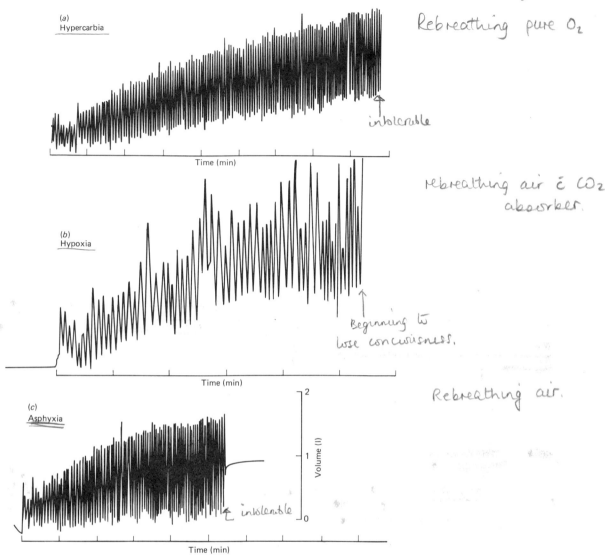

**Figure 17.5** The effects of raised partial pressures of carbon dioxide and lowered partial pressures of oxygen on respiration. (a) Rebreathing pure oxygen; (b) rebreathing air from a spirometer incorporating a carbon dioxide absorber; (c) rebreathing air – asphyxia. The experiments in (a) and (c) were terminated by the subjects when breathing became intolerable; (b) was terminated as the subject began to lose consciousness. Note that although the vertical volume scales are the same in each recording, the time trace in (b) is different.

Quite apart from interactions at the receptor level and at the medullary level, there are interactions between the state of the blood in regard to carbon dioxide and oxygen content simply because the changes in ventilation evoked by one stimulus will affect the concentration of the other. Thus hypoxia may stimulate ventilation, drive off carbon dioxide, and remove the carbon dioxide stimulus to maintain ventilation. This is seen in the response to hypoxia at high altitudes described in the next chapter.

## Further reading

Comroe, J.H. *The Physiology of Respiration,* 3rd edn, Year Book Medical Publishers, Chicago, 1987

Dejours, P. *Respiration,* Oxford University Press, Oxford, 1966

Robinson, J.R. *Fundamentals of Acid Base Regulation,* 5th edn, Blackwell, Oxford, 1975

Widdicombe, J. and Davies, A. *Respiratory Physiology,* Arnold, London, 1983

# 18
# Control of blood pH

## The normal blood pH and the variables which set it

The pH of blood, and hence extracellular fluid generally, must be maintained at a reasonably constant level. In general mammals maintain the pH of the internal environment between 6.9 and 7.8. Although mammalian enzymes will function outside these limits *in vitro*, in the body corrective mechanisms to restore the pH would produce conditions too extreme for the animal's survival. In man the normal range of pH of body fluids is between 7.35 and 7.45. Arterial blood is normally around 7.40, mixed venous blood around 7.37.

Since the principal blood buffer is the carbonic acid-hydrogen carbonate system, the pH may be calculated by the Henderson-Hasselbalch equation for that system: that is, the pH is equal to:

$$\text{dissociation constant of the weak acid} + \log \frac{\text{concentration of salt}}{\text{conc. of undissociated acid}}$$

that is,

$$pH = 6.1 + \log \frac{[\text{hydrogen carbonate}]}{[\text{carbonic acid}]}.$$

$$= 6.1 + \log \frac{[HCO_3^-]}{[ApCO_2 + H_2CO_3]}$$

where $ApCO_2$ is the solubility of carbon dioxide in water times the partial pressure of carbon dioxide.

The ratio $\dfrac{[HCO_3^-]}{[ApCO_2 + H_2CO_3]}$ is usually 20:1 giving a pH of 7.4.

The changes in blood pH consequent upon altering the two variables in this equation if no other buffering mechanisms are present can be calculated readily. A change in the partial pressure of carbon dioxide from 5.3 kPa to 10.6 kPa would cause a fall in pH to 7.1, whilst halving the partial pressure to 2.65 kPa would raise the pH to 7.7. In reality, other blood buffers cushion these pH changes, giving final values around 7.2 and 7.6. The buffering of hydrogen ions by these other buffers increases relatively the amount of hydrogen carbonate, since

$$CO_2 + H_2O \rightarrow HCO_3^- + (\text{buffered } H^+)$$

In a similar fashion, changing the ratio of variables by doubling the hydrogen carbonate concentration from 24 mmol/l to 48 mmol/l moves the pH by 0.3 units to pH 7.7. Changes in blood hydrogen carbonate concentrations usually arise from the addition of acid, which effectively removes hydrogen carbonate, or alkali, which increases its concentration. Both these effects are lessened by the presence in the blood of other buffering mechanisms.

If, when hydrogen carbonate concentrations are changed, the partial pressure of carbon dioxide can be adjusted appropriately, then no change in pH should occur. Thus, addition of acid to the blood causes the loss of hydrogen carbonate

$$H^+ + HCO_3^- \rightarrow H_2CO_3 \rightarrow H_2O + CO_2$$

but this can be compensated by increasing respiration and lowering the partial pressure of carbon dioxide. The pH can therefore remain constant. However, there is a penalty in that the buffer capacity of the system has been proportionally reduced: production of more hydrogen carbonate will be necessary to replenish the buffer capacity.

The two parts of the ratio may be adjusted by the body. Different organs are involved: the partial pressure of carbon dioxide depends upon the balance of oxidative production in the tissues and the loss from the body by expiration, whilst the concentration of hydrogen carbonate depends upon the extent of its effective reabsorption in the kidney after the plasma has been filtered in the glomerulus. The respiratory system operates under a control which is set to maintain a partial pressure of around 5.3 kPa in arterial blood. Increases above this value result in increased ventilation, driving off carbon dioxide. If ventilation is reduced, carbon dioxide accumulates in the blood. Hydrogen carbonate concentrations are controlled by the kidneys, which normally reabsorb some 24 – 28 mmol/l of cations in excess of anions, a deficit made good by hydrogen carbonate. An effective increase in reabsorption increases blood hydrogen carbonate concentrations, an effective decrease causes hydrogen carbonate loss in the urine. The Henderson-Hasselbalch equation governing blood pH is sometimes given as a functional description rather than a numerical equation, by rewriting the final ratio as one of renal activity to pulmonary expiration – kidney over lungs.

## Causes of changes in blood pH

*metabolic & respiratory*

Changes in blood pH occur as a result of the ingestion of acid or alkali, as a result of metabolic disturbance, or as a result of a failure to match respiratory excretion of carbon dioxide to its metabolic production. The first two disturbances are referred to as metabolic, the third as respiratory. Changes in blood pH are strictly referred to as 'acidaemia' and 'alkalaemia'; the words 'acidosis' and 'alkalosis' refer to the amount of buffer base (i.e. hydrogen carbonate) present in relation to the normal concentration in the blood. The two sets of terms are often used loosely as being interchangeable although they describe different variables.

Ingestion of acids or alkalis may occur directly, or substances ingested may give rise to acids or alkalis in the course of digestion and metabolism. Thus, ammonium chloride is converted to urea and hydrochloric acid. Cations which readily form insoluble salts, such as calcium, also may cause an acidaemia because the cations remain in the undigested faeces in combination with phosphate and hydrogen carbonate (derived, via the gut secretions, from the plasma), and so soluble chlorides replace the blood buffering anions. The reverse occurs with the salts of weak organic acids, such as citrate and lactate: the acid anions are metabolised and then hydrogen carbonate has to be retained to balance the excess cations, resulting in alkalaemia. The metabolism of fruits which contain the salts of weak acids, or the ingestion of large amounts of sodium hydrogen carbonate may therefore result in a metabolic alkalaemia: the loss of gastric acid during vomiting is, however, a more usual cause of this condition.

Metabolism by itself usually results in the production of acid. Complete oxidation of fats and carbohydrates yields water and carbon dioxide only. Anaerobic metabolism results in production of lactic acid. Oxidation of proteins also yields water and carbon dioxide, but any sulphur- or phosphorus-containing proteins will in addition yield sulphuric or phosphoric acid. The total daily production of carbon dioxide is around 13 000 mmol/day. In the blood this will generate an equivalent amount of hydrogen ions which will then be buffered mainly by haemoglobin. Excretion of the carbon dioxide through the lungs will result in the elimination of these hydrogen ions. The main acid produced by metabolism, therefore, is exploited to provide the main buffering system against pH change in the body. Very little phosphoric acid is produced, since in practice most of the phosphorus ingested is already present as phosphate. Further, phosphate itself can act as a buffer by virtue of the possible transition from monohydrogen phosphate to dihydrogen phosphate. Some 50 – 100 mmol of sulphuric acid are formed each day, the amount relating closely to the amount of methionine and cysteine in the diet. This acid must be excreted through the kidneys, presenting some problems which will be discussed later.

Respiratory acidaemia or alkalaemia results when respiration for some reason does not maintain the blood carbon dioxide levels in proportion to the hydrogen carbonate concentrations. Hypoventilation is defined as ventilation inadequate to remove carbon dioxide from the body as fast as it is produced. It can be voluntary, as in breath-holding, or it may be due to obstruction of the airways, paralysis of the respiratory muscles, or the actions of drugs or toxins on the respiratory centres. When the inspired air contains so much carbon dioxide that the metabolic carbon dioxide cannot be eliminated through the lungs a technical state of hypoventilation exists as defined above. In hypoventilation carbon dioxide is retained, blood carbon dioxide levels increase, and the plasma pH falls. If carbon dioxide concentrations increase markedly it acts as a narcotic and depresses the activity of the respiratory centres – a positive feedback situation which leads to respiratory failure. Hyperventilation is defined as ventilation in excess of that required to remove carbon dioxide as fast as it is produced. Such a definition excludes increased ventilation when this is a response to an increase in carbon dioxide production – as in exercise. Hyperventilation may occur voluntarily as in physiological experiments, or before swimming underwater, or involuntarily as a result of emotional stress or hysteria, or in response to low partial pressures of oxygen at high altitudes. The loss of carbon dioxide from the blood results in alkalaemia, or a rise in pH. Overanxious or hysterical patients may hyperventilate, induce an alkalaemia, and then show symptoms of tetany. This hyperexcitability of motor nerves is due to a reduction in ionic calcium concentrations resulting from the increase in calcium binding capacity of plasma proteins in alkaline conditions. Another adverse effect of hyperventilation may be potentially dangerous to the underwater swimmer: removal of carbon dioxide from the blood can lead to cerebral vasoconstriction, and this together with hypoxia due to apnoea can cause loss of consciousness underwater. The hypoxia during underwater swimming also causes bradycardia.

## Control of blood pH

The pH of the blood is controlled by adjusting the ratio of hydrogen carbonate concentration to that of carbon dioxide and carbonic acid. Changes in pH are apparent as changes in this ratio: control of pH consists of the restoration of the ratio by respiratory and renal adjustments. However, the first stage of control is by the purely physicochemical process of buffering, a process which permits the presence and transport of acid or alkali with relatively little pH change.

### Buffering in the blood and body fluids

The carbon dioxide generated by metabolism is hydrated in the red blood cells with the assistance of carbonic anhydrase. This reaction generates hydrogen ions and hydrogen carbonate ions. Some carbon dioxide combines with the terminal groups on the haemoglobin molecules to form carbamino-compounds and more hydrogen ions. When the respiratory quotient is around 0.7, that is, when 7 molecules of carbon dioxide are generated for every 10 molecules of oxygen consumed (as when fat is the sole metabolic substrate) all the hydrogen ions generated can be buffered by the histidine groups of the haemoglobin exposed as it gives up oxygen. In other words, haemoglobin is a stronger base than is oxyhaemoglobin and its additional buffering power is equivalent to 7 moles of hydrogen ion for every 10 moles of oxyhaemoglobin giving up oxygen. The hydrogen carbonate ions are exchanged across the red cell membranes for chloride ions (p. 111). With greater values for the respiratory quotient, as would be observed with any mixture of substrates, or with protein or carbohydrate as a sole substrate, hydrogen ions in excess of the buffering capacity of the red cells are produced and are buffered elsewhere or cause a change in pH.

Other acids in the plasma are first diluted, a process taking place automatically as diffusion occurs rapidly in the plasma, more slowly throughout the extracellular fluid compartment, and finally throughout the intracellular fluids also – a final dilution into some 45 litres of fluid.

Dilution by itself can only lessen the ionic challenge slightly; buffering is needed to prevent large shifts in pH. This is achieved first in the blood, by the buffering power of hydrogen carbonate, haemoglobin, plasma proteins, and phosphate, in that order. Hydrogen carbonate (about 20 mmol/kg of cells and 28 mmol/kg in the plasma) contributes some 65% of the blood buffering capacity. It is an effective buffer because it combines with hydrogen ion to form carbonic acid which can be dehydrated and blown off as carbon dioxide. Such buffering assumes that more hydrogen carbonate can be manufactured rapidly to restore the buffer base content of the blood. Haemoglobin constitutes a further 28% (although the main role of haemoglobin is in buffering the hydrogen ions generated during carriage of carbon dioxide), plasma proteins about 6%, and phosphate about 1% of blood buffering. Non-hydrogen carbonate buffering accounts for some 42 mmol/kg in the blood cells, but only some 16 mmol/kg in the plasma. The blood is capable of buffering around 140 mmol of strong acid without a lethal drop in pH. This is greatly in excess of the 50–100 mmol of strong acid formed in total in 24 h.

Within the extravascular fluid, buffering is due almost entirely to hydrogen carbonate.

In the intracellular fluid compartment, hydrogen carbonate buffering is relatively unimportant (10 mmol/kg). Cell proteins play a larger part (30 mmol/kg), but the major cell buffer is phosphate (140 mmol/kg). The buffering of hydrogen ions by the intracellular buffers results in a release from the cells of other cations in exchange. Although potassium is the cation in greatest concentration in cells, the process of exchange is controlled in some way to ensure that sodium ions are preferentially lost.

The relative importance of the body fluids in buffering can be illustrated by experiments in which acids and alkalis (10–20 mmol/kg body weight) were infused into animals. With acids about 20% of buffering occurred in the blood, about 25% in the extravascular fluid and about 50% outside these compartments. With added alkali only about 32% of the buffering was outside the vascular and the extravascular fluid compartments, the remainder being shared between these two compartments roughly in proportion to their volumes – about 1:3.

Although intracellular buffering has already been mentioned, there is in fact a further extracellular component in that part of the buffer reserve. This is bone, a tissue made up of hydroxyapatite crystals in a collagenous matrix. The surface crystals partake in ion exchange with the fluids around them. About one third of the sodium ions in the body are located in bone, and about one third of these exchange with extracellular sodium ions in 24 h. Some of them exchange readily with hydrogen ions. Bone can, therefore, buffer acids added to the extracellular fluids. As acidaemia increases, however, calcium ions also are lost, and excreted in the urine. Effectively, decalcification of bone occurs in prolonged acidaemia. Alkali can also be taken up by bone: hydrogen carbonate and carbonate both exist in apatite crystals, and the amounts of these ions in bone increases in alkalaemia and decreases in acidaemia. Buffering, or hydrogen ion exchange, in bone is probably the major component of the so-called intracellular buffering.

## Respiratory control of blood pH

Buffering counters the effects of acid and alkali on the pH of the blood: lungs and kidneys actually remove the disturbing factors. The lungs can excrete acid in the form of carbon dioxide; they have no direct effect upon concentrations of hydrogen carbonate or alkali (Fig. 18.1).

If the plasma hydrogen carbonate remains at normal levels the respiratory system sets the levels of partial pressure of carbon dioxide in the blood and hence its pH. Carbon dioxide partial pressures in excess of 5.3 kPa stimulate respiration, removing the main acid produced in metabolism. Other acids eliminate some hydrogen carbonate by reacting with it to produce carbon dioxide and water, but also themselves stimulate respiration via their effect on hydrogen ion receptors in the carotid and aortic chemoreceptors. Alkalis increase the hydrogen carbonate concentration but the associated rise in blood pH reduces the stimulus to ventilation so that carbonic acid concentrations rise, and the ratio of base to acid remains constant. A rise of 0.1 units in blood pH doubles ventilation, whilst a fall of 0.1 units halves it. The effect of acidaemia is illustrated by the 'air hunger' seen in uncontrolled diabetic patients with acid blood pH due to ketoacid production.

As the partial pressure of carbon dioxide increases in the blood, a parallel increase occurs in cerebrospinal fluid, and the cerebrospinal fluid ratio of hydrogen carbonate to carbonic acid falls, pH falls, and the intracranial receptors stimulate ventilation. In these circumstances the central chemoreceptors act in concert with the peripheral chemoreceptors stimulated by a lower blood pH and a raised carbon dioxide concentration. If the raised partial pressures of carbon dioxide are maintained, however, the cells of the choroid plexuses (which are rich in carbonic anhydrase) hydrate carbon dioxide, and actively transfer hydrogen carbonate to the cerebrospinal fluid while the hydrogen ions are transferred back into the blood. This raises the cerebrospinal fluid hydrogen carbonate concentration, restoring its pH by a non-respiratory mechanism. The preservation of the pH of cerebrospinal fluid is more important than the maintenance of low partial pressures of carbon dioxide in the blood. Thus respiration is stimulated by the peripheral receptors but not now by the central receptors. This paradoxical response is also seen when the pH of the blood is disturbed by acids other than carbon dioxide, or by alkalis.

The interaction between the prime purpose of the respiratory system in its maintenance of oxygen supplies to the tissues and the removal of carbon dioxide, and its secondary role in adjusting blood pH, is seen in the response of the body to a sudden translocation to high altitude. At high altitude atmospheric pressure is less, and so the 21% of oxygen in the air exerts a correspondingly smaller partial pressure. If this is sufficiently low the arterial chemoreceptors respond to the lower partial pressure of oxygen in the blood by stimulating respiration. This results in the blowing off of carbon dioxide and a removal of the normal stimulus to respiration. Loss of carbon dioxide causes alkalaemia and so the smaller amount of carbon dioxide diffusing into cerebrospinal fluid causes this to become more alkaline also. There is therefore a conflict between the peripheral chemoreceptors responding to oxygen lack and the central ones responding to carbon dioxide lack. In the short term this may produce periodic respiration, with periods of apnoea when no respiratory stimulus exists alternating with respiration driven by oxygen lack. The inputs from the two groups of receptors remain opposed for several days until the cerebrospinal fluid hydrogen carbonate concentrations fall sufficiently to restore its pH at least partially. The blood remains alkaline and hypoxic drive continues to be

Excess bicarbonate — Exceeds renal threshold (actually exceeds amount of hydrogen ion secreted to tubular fluid)

→ Urine

Excess hydrogen ion — Forms carbonic acid with bicarbonate and is excreted as carbon dioxide

→ Lungs

- Secreted by kidney tubules
- Causes bicarbonate reabsorption
- Renders urine acid
  - Buffered by phosphate
  - Combines with ammonia to form ammonium ions
- Causes potassium retention

→ Urine

**Figure 18.1** The elimination of acid and base from the body.

reduced by this factor. Whilst the cerebrospinal fluid is alkaline there is cerebral vasoconstriction; if this is severe the subject may experience light-headedness, headache, lassitude, nausea and lack of appetite, and possibly vomiting. The kidneys excrete an alkaline urine containing hydrogen carbonate in order to restore the blood pH by adjusting the ratio of hydrogen carbonate to carbon dioxide. Within a week or so the plasma pH is normal, with a lower than normal partial pressure of carbon dioxide and a lower than normal hydrogen carbonate content. The hypoxic drive to respiration is then unopposed. The most probable explanation of this adaptation is a re-setting of the sensitivity of the central chemoreceptors.

## Renal control of blood pH

After the rapid preliminary adjustment by buffering and the slower adjustment by respiratory elimination of carbon dioxide, the final stage of adjustment of blood pH occurs in the kidneys, which are able to excrete acid as hydrogen ions and alkali as hydrogen carbonate (Fig. 18.1). This process is relatively slow: a hydrogen carbonate load of some 10 mmol/kg body weight takes some 24 h to excrete, and the final redistribution of ions involved in buffering may take between 3 and 7 days, whilst with a similar acid load the kidney may excrete only about a quarter of the acid within 24 h and the whole process of restoration to normal takes about a week. Long term correction involves not only hydrogen carbonate and hydrogen ions, but also those ions which have been exchanged in the process of buffering – chloride, lactate, sodium, potassium, and even, perhaps, calcium.

The excretion of acids other than carbonic acid can only be achieved through the kidneys. There is no problem over excretion of anions, which are simply not reabsorbed from the glomerular filtrate. The hydrogen ions of the acids do not in fact reach the kidney, because they are buffered or have been exchanged with ions from bone crystals. The pH of the glomerular filtrate is about 7.4. The hydrogen ions have been replaced mainly by sodium ions and, to a lesser extent, by potassium ions. The body cannot afford to lose these ions: this difficulty is overcome by the generation of new hydrogen ions which can be actively secreted into the tubular fluid in exchange for sodium ions. The source of these hydrogen ions is carbon dioxide produced by the metabolism of the cells of the kidney tubules. This is readily hydrated with the assistance of the enzyme carbonic anhydrase, present in high concentration in these cells (Fig. 18.2). The secretion of hydrogen ions into the kidney tubule is controlled by the partial pressure of carbon dioxide in the plasma, since this limits the exchange of carbon dioxide between the cells and the surrounding fluids, and thus drives the reaction in the direction of hydration of the carbon dioxide. The hydrogen carbonate produced passes back into the plasma, restoring that broken down by the acids originally. In the plasma the alkalinity of this hydrogen carbonate causes the buffered hydrogen ions to be restored to the blood. These hydrogen ions react with the hydrogen carbonate, generating carbon dioxide which is then blown off in the lungs. In similar fashion, the alkalinity of the blood causes bone to give up hydrogen ions.

In the tubular fluid the secreted hydrogen ions react with hydrogen carbonate to form carbonic acid and then carbon dioxide. The cells of the kidney tubules are impermeable to hydrogen carbonate but freely permeable to carbon dioxide, which therefore diffuses in. Once inside it is re-hydrated and provides more hydrogen ions for secretion and hydrogen carbonate ions to be transferred to the blood. Effectively, then, hydrogen carbonate has been reabsorbed. Every hydrogen ion secreted is associated with a sodium ion reabsorbed and a hydrogen carbonate ion transferred to the plasma. This process can continue until all the hydrogen carbonate in the tubular fluid has been removed. After that, the hydrogen ions remain in the tubular fluid, making it more acid. The hydrogen ions are buffered by monohydrogen phosphate, but this provides only limited buffering because the amount is small and limited to that derived from the diet. The secretion of hydrogen ions occurs in both proximal and distal convoluted tubules, although the pH falls to a greater extent in the distal convoluted tubule.

Within the tubular fluid there is a limit to the acidity which can be achieved because there is a maximum gradient against which hydrogen ions can be secreted. There may also be a limit on the ability of the cells to tolerate an acid tubular fluid. The lowest pH observed in urine is around pH 4.6. The tubular fluid does, of course, contain buffers. The principal buffer is phosphate, but creatinine also provides some buffering. At pH 4.6 all the phosphate is present as dihydrogen phosphate and two thirds of the creatinine is in its acid form. On a normal diet some 20-30 mmol of acid are excreted as free or buffered hydrogen ions, but this increases ten-fold in diabetes mellitus. Excretion of acid in this way restores the blood base content because every hydrogen ion secreted is generated from carbonic acid which simultaneously generates a hydrogen carbonate ion, and it is this hydrogen carbonate ion which is transferred to the plasma. Hydrogen carbonate is added to the blood, not simply transferred from the tubular filtrate. Thus blood pH is restored both by secreting acid and by restoring base.

Hydrogen ions can still be excreted even when the gradient against this process seems to be too great because the pH of the urine may be prevented from

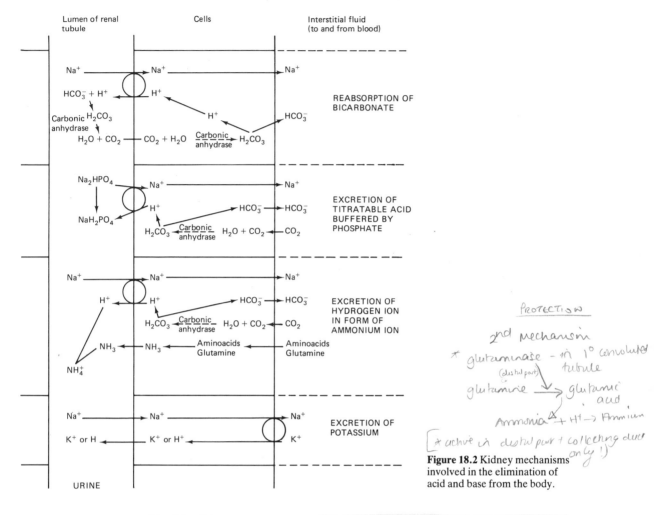

Figure 18.2 Kidney mechanisms involved in the elimination of acid and base from the body.

falling further by a second mechanism. The cells of the proximal convoluted tubules contain a glutaminase which converts the aminoacid glutamine into the deaminated acid glutamic acid, and generates ammonia as a result. Although this enzyme is produced in the proximal convoluted tubules, it is inactive in this part of the nephron because the cell pH is too high. The enzyme diffuses to the distal parts of the nephron where the intracellular pH is more appropriate and so it is actually in the distal convoluted tubule and the collecting ducts that the ammonia is produced. This ammonia combines with hydrogen ions (Fig. 18.2) to form ammonium ions. More ammonia can be produced by oxidative deamination of aminoacids. Ammonia production therefore allows the exchange of hydrogen ions for sodium ions to continue without the pH of the urine being lowered still further. Normally some 30–50 mmol of acid are secreted as ammonium ion each day and again, in severe acidosis, this amount can rise ten-fold.

The excretion of alkali in the urine has already been dealt with in passing. Hydrogen carbonate is normally reabsorbed as a result of hydrogen ion secretion: only when the amount in the tubular fluid exceeds the hydrogen ion production in the tubular cells will hydrogen carbonate be excreted in the urine. Thus the kidneys preserve the body store of hydrogen carbonate. The rate of reabsorption depends upon the partial pressure of carbon dioxide in the plasma. Normally the kidneys operate as if there is an effective threshold of 28 mmol/L in the plasma beyond which concentration hydrogen carbonate will be excreted in the urine.

So far the adjustments to blood pH carried out by the kidneys have been described as involving the ions of hydrogen, sodium, ammonium and hydrogen carbonate only. The situation is more complex in practice because exchange of sodium ions can take place with potassium as well as hydrogen ions. This is particularly evident in the distal parts of the tubules. There exists, therefore, a competition between hydrogen and potassium ions – as hydrogen ions are lost into the tubular fluid so potassium ions will be retained in the cells, and vice versa. Production of an alkaline urine is associated with loss of potassium ions from the body. Lack of potassium in the renal cells will result in acidification of the urine and an over-addition of hydrogen carbonate to the plasma. Diets high in potassium content are associated with the production of an alkaline urine. The critical factor in potassium secretion is probably the intracellular pH unless the balance of exchange is disturbed by the necessity to control sodium excretion (Chap. 16).

## Further reading

Davenport, H.W. *The ABC of Acid-Base Chemistry,* 6th edn, University of Chicago Press, Chicago, 1974

Pitts, R.F. *Physiology of the Kidney and Body Fluids,* Year Book Medical Publishers, Chicago, 1974

Robinson, J.R. *Fundamentals of Acid Base Regulation,* 5th edn, Blackwell, Oxford, 1975

# 19

# Mastication and deglutition

## Mastication

The term mastication describes the process of chewing. It does not include incision (the biting off of suitably sized portions of a larger piece of food), the ancillary movements of tongue and cheeks which transfer and sort the food, and the crushing of food against the hard palate by the tongue, although all these processes are stages in the intake and processing of food in the mouth. As food is taken into the mouth and processed to prepare it for swallowing there is involvement not only of the muscles of mastication, but also the muscles of the cheeks, the tongue and the lips together with the teeth, the hard palate, and the temporomandibular joints. The rare occurrence of tongue, cheek or lip biting demonstrates the efficiency with which the procedure is performed, and the sensitivity of the reflexes which govern the whole process.

## The masticatory apparatus

### The muscles of mastication and the movements of the mandible

The basic process is the movement of the mandible and its associated structures in relation to the maxillae. The mandible bears against, and is attached to, the glenoid fossa and the articular eminence of the temporomandibular joint, both parts of the temporal bone. At rest the articulating portion of the mandible, the condyle head, rests in the articular fossa, separated from it by the synovial disc and the upper and lower synovial cavities (Fig. 19.1). Although, in theory, a purely hinge movement is possible at the joint, the act of closing the teeth of the two jaws together involves a small backward movement of the condyle, and only half the maximum opening of the mouth is possible using a hinge movement without forward movement of the condyle. In normal mouth opening, there is a composite of a hinge movement and a sliding forward of the condyle head onto the articular eminence.

In the upright subject with the jaw musculature at rest, the mandible is normally maintained in such a position that the teeth of the two jaws are separated by a space of about 2 mm. Although this rest position may be by no means as constant as was once thought, there is certainly a freeway space, as it is

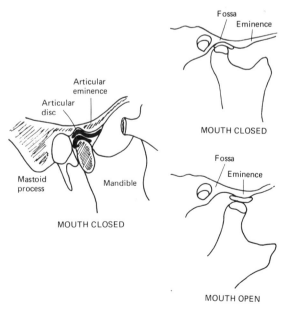

**Figure 19.1** The temporomandibular joint, and its forward translation as the jaws open.

called, and any gross change in the resting distance between the basal bone of the mandible and maxilla caused by dentures or dental work will either be visually obvious or result in muscular discomfort. Measurement of the preferred vertical dimension obtained by separating the jaws by known amounts suggests that there exists only a narrow range of tolerated inter-jaw distances. It was thought that the mandible was held in a sling of muscles with a constant degree of tone in the masseter and medial pterygoid muscles supporting the jaw against gravitational pull (Fig. 19.2). Evidence for tone in these muscles, or in the temporalis, lateral pterygoid, or suprahyoid muscles has not been obtained.

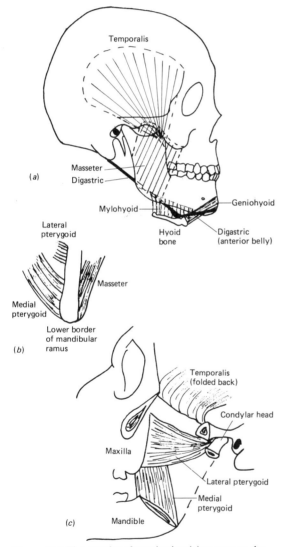

**Figure 19.2** The muscles of mastication (*a*) as seen on the skull; (*b*) left, the lower border of the mandible in vertical section to show the sling of muscles; (*c*) the pterygoids seen after cutting away the ramus of the mandible.

The alternative view is that the mandible sits indeed in a muscle sling, but that the physical properties – length, elasticity, etc. – of the muscles, tendons, ligaments, joint capsule, and surrounding structures, are such that the rest position can be maintained without the need for muscular activity. Only when the mandible moves from this neutral position do muscle and proprioceptive receptors come into play, signalling the discrepancy and activating muscular correction. This would be a situation analogous to that in the control of posture where muscle contraction is not necessary to maintain an upright stance but only to correct sway from the vertical. The correction in jaw position is made by adjusting tension in the elevator muscles only.

The muscles of mastication are the elevator muscles, masseter, medial pterygoid, anterior temporalis, and the depressor muscles, the suprahyoid group and the anterior belly of digastric (Fig. 19.2). The lateral pterygoid acts to pull the mandible forward, and the posterior temporalis together with the suprahyoid muscles retrudes it. Muscle spindles have been found in all these muscles but the number in the elevator muscles is relatively large (approaching the number in muscles which move the fingers) whilst there are very few in the depressor muscles. The afferent fibres from these spindles enter the pons in the motor root of the trigeminal nerve and their cell bodies are in the trigeminal mesencephalic nucleus. There do not seem to be any tendon receptors associated with the masticatory muscles.

The human temporomandibular joint capsule contains many receptors, usually free nerve endings, but with some more complex endings resembling those of Ruffini or Golgi. Some of the receptors are pain receptors but many are proprioceptive. Other sensors associated with the process of mastication are the mechanoreceptors of the periodontal ligaments of the teeth, touch and pressure receptors on the palate, the gingivae, lips, cheeks and tongue, and pain receptors in the periodontal ligaments or in the other structures just mentioned.

## The occlusion of the teeth

The teeth of the upper and lower jaws probably come into contact during most chewing cycles. In the perfect dentition the cusps of the teeth in one jaw fit into the fossae or interdental spaces of the teeth in the other jaw, and the planes of contact match exactly in 'normal occlusion' with the mandible approximating to the maxilla by movement in a sagittal plane only (Fig. 19.3). The buccal cusps of the mandibular teeth bite into the fossae of the maxillary teeth. Rotation of the mandible with slight retrusion of one condyle to a fixed position and the other condyle moving forward should result

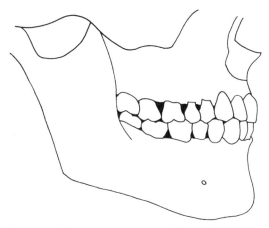

**Figure 19.3** The normal occlusion of the teeth.

in a slight separation of the jaws as the teeth slide evenly on the inclined planes of the cusps. In practice the occlusion is rarely perfect and cuspal guidance as the jaws approximate leads to deviation in the path of mouth closure. Facets are ground upon the teeth by chewing movements and at worst such cuspal disharmony leads to pain in muscles or joints. There is some elasticity in the system: the teeth move in the periodontal ligament, intruding into the sockets with forces below 1 N, but beyond that flexing the bone itself and finally, atforces around 15 N causing pain. The movements of chewing produce a net force on the teeth which causes them to tilt and move towards the mesial: this is the cause of 'physiological mesial drift'. The surfaces of the teeth in a perfect dentition should lie on the surface of a solid described by the movement of the mandible. Two curves have been described on plane sections of this solid: the curve of Monson and the curve of Spee (Fig. 19.4); although the curves actually specified by Monson, Spee, or Villain do not in fact match particularly well those seen in the living occlusion.

Forces exerted upon the teeth are sensed by mechanoreceptors in the periodontal ligament.

## The tongue

The tongue is a muscular structure intimately involved in mastication. Although in the open mouth it appears to lie in the floor and at the rear of the oral cavity, when the mouth is closed it fills the volume delineated by the floor of the mouth, the gingiva-covered alveolar processes of the mandible and maxillae, the teeth and the palate. Only a thin film of saliva separates it from these tissues. It may press slightly between the teeth and the impressions of the tooth crowns can sometimes be seen on its sides. The tongue is covered by a mucosa which contains receptors for touch, pressure and tempera-

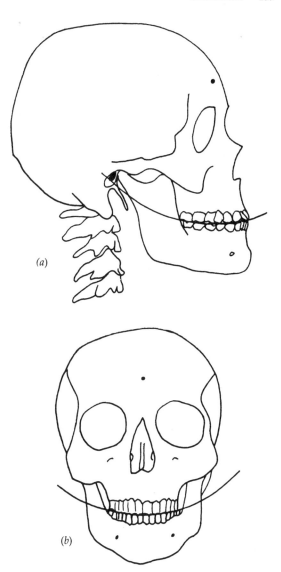

**Figure 19.4** The occlusal curves. (*a*) The curve of occlusion as seen in lateral view; (*b*) the curve of occlusion as seen anteriorly. Various names have been given to these curves; but the descriptions given by Monson, Spee and Villain do not in fact correspond closely with the curves observed in living subjects.

ture, and for the special sense of taste. It is essential for stereognosis, the sensing of the size and shape of objects, in the mouth. In eating it is involved in the manipulation of the food portion incised, and it is the prime mover in sucking and swallowing. As food is masticated the tongue keeps it passing outwards past the masticatory surfaces into the cheek pouches, and receives it passed back from the cheek pouches. The degree of comminution of the food is monitored by the tongue, and it selects the food to be swallowed. The tongue is made up of the intrinsic muscles – the horizontal and vertical muscles, with their origins and insertions into connective tissue – and the extrinsic muscles attaching it to other structures – genioglossus, hyoglossus, and styloglossus (Fig. 19.5). The intrinsic muscles change the shape of the tongue whilst the extrinsic muscles change its position. The muscle is skeletal muscle and in theory under voluntary control – although in practice its actions are largely reflex. When patients are asked to move the tongue away from a particular part of the mouth, they respond by moving the tip only and the rest often rolls frustratingly into the area from which the dentist was trying to remove it. Accidental damage to the tongue during dental work is almost always on the lateral aspects. The muscles of the tongue contain many muscle spindles.

## Other oral structures

The hard palate is sensitive to touch, pressure, pain and temperature. Its rigid mucosa-covered surface is very important in assessing the texture of food. Patients wearing upper dentures which cover the palate often complain of loss of taste sensitivity and at least part of this is a loss of appreciation of texture rather than actual taste. In conjunction with the tongue, the hard palate is important in stereognosis. It is against the palate that the tongue generates suction in drawing liquids or air into the mouth. The ridges, or rugae, of the anterior palate enable the tongue to crush and break up softer foods. This is less important in man than in some other animals. There are mucous glands and taste buds in the palatal mucosa.

The lips assist in the first taking up of food. In sucking up liquids they may help to generate suction along with the tongue, although they are also drawn back by the suction the tongue has produced. In sucking and in suckling they provide a seal around the bottle or straw or teat. In normal mastication they provide an anterior seal to the oral cavity, although the movement of the chewed food usually occurs to the sides of the mouth and little reaches the labial sulci (or the vestibule) unless the mouth is overfilled. Then the orbicularis oris muscle contracts in response to stretch to force the food back between the anterior teeth into the oral cavity proper.

The cheeks are made up of the buccinator muscles and provide the lateral walls of the oral cavity. In the resting mouth they lie closely approximated to the alveolar processes and the posterior teeth of maxilla and mandible. as a result the cheeks apparently 'fall in' when the posterior teeth are extracted. The muscular wall is relatively thin and the potential space between it and the dento-alveolar complex can be expanded under pressure to form the cheek pouch. Food forced between the teeth by the tongue passes into the pouch, stretching it to a greater or lesser degree. The stretched buccinator then contracts, squeezing the food back into the oral cavity proper.

## The process of food intake and comminution
### Incision

Food is taken into the mouth by the conscious act of incision. The protrusor muscles guide the mandible forwards onto the articular eminence as the depressors pull the anterior part downwards in the hinge-like part of the movement. Stretch receptors and joint receptors signal that the opening is sufficient to admit the object. The size of the mouthful is fairly constant in the individual, but may range from 10 ml to 50 ml. A normal mouthful of fluid is almost 20 ml. The elevators then move the

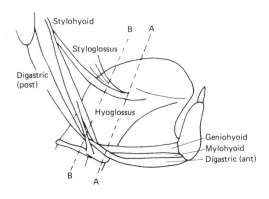
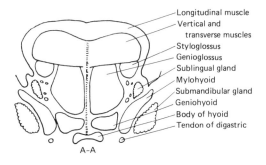
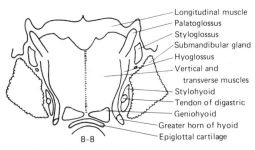

**Figure 19.5** The muscles of the tongue.

anterior part of the mandible and therefore, the incisor teeth, upwards with a hinge movement of the condyle on the eminence. When the incisors of upper and lower jaws meet in protrusion the remainder of the closing movement is achieved by the action of the retrusor posterior fibres of the temporalis. If the food cannot be cut by the incisors, the whole head may be twisted to assist in breaking or cutting. The head can be used with the upper incisors acting as gouges, as when a bone and attached meat is gnawed. Sometimes the food is stabilised against the upper teeth and the lower incisors repeatedly pulled alternately upwards and backwards by the retrusors and forward and downward by the protrusors. In eating food from a spoon the lips and teeth are used and the lip muscles brought into play. In all these activities of food intake the initiation and end of the movements is consciously willed but the intervening muscle movements are a complex interplay of excitation and inhibition of motor neurones in the mesencephalic nucleus. Antagonist inhibition and the balancing of flexor and extensor responses seen in limb and trunk movements are replaced in masticatory movements by the protrusor/retrusor and elevator/depressor interactions.

## Mastication

When the food is in the mouth the process of mastication begins; although, if stereognostic impulses identify the particles as being too large, there is further intra-oral incision, or even crushing between the canine teeth. Mastication consists of a series of rhythmical chewing cycles. Each of these is made up of a rapid upward (closing) stroke bringing the teeth into contact with the food, followed by a slower power stroke causing the teeth to crush through the food; the teeth are then separated again by an opening stroke which has a slow initial phase and a final rapid phase. The masticatory rhythm appears to be controlled by a rhythm generator situated in the brainstem either in or close to the motor nucleus of the trigeminal nerve. Stimulation by impulses from higher centres or from peripheral receptors initiates activity in the rhythm generator. This activity can then be modified by further sensory impulses. In animals electrical stimulation of the cortex or corticobulbar pathways produces rhythmic chewing movements but little is known in man of possible higher controls of mastication – except that the process can be initiated voluntarily. Reflexes which modify the basic rhythm include a servo-mechanism for the control of elevator muscles, a positive feedback mechanism to increase the force exerted when resistance to jaw approximation is met, and a protective mechanism to prevent damage to teeth, periodontal ligaments and oral mucosa. The last is demonstrated by the abrupt cessation of the closing movement when a hard particle is sensed between the teeth.

A number of reflexes which may play a part in the modification of masticatory activity have been studied in detail. These are:

(i) *The jaw jerk.* If a subject sits or stands upright with the jaw muscles at rest, and a sharp downward tap is administered on the chin, the masseter muscles contract and the body of the mandible moves upwards so that the teeth of the two jaws close together (*see* Fig. 13.5). The force of contraction can be increased by biting on an object, or decreased by the subject contracting the jaw depressor muscles against a restraining force under the chin. The jaw jerk is a mono-synaptic reflex resulting from the stretching of the muscle spindles in the masseter muscles. It has a latency of 7–12 ms. After the reflex, there is a short period during which activity in the masseter muscle ceases to be measurable – the masseteric silent period. This seems to be a manifestation of spindle unloading. The activity of the depressor muscles, such as the digastric, is reduced during the jaw jerk, but there is no depressor equivalent to the jaw jerk – an upward tap on the chin has no reflex effect – possibly because of the depressor muscles have very few muscle spindles.

(ii) *The unloading reflex.* If, as the teeth are brought together through a resistant object, there is a sudden decrease in resistance, the closing movement is halted by decreasing masseteric and increasing digastric activity. Such a halt is probably due to the sudden inactivation of the masseter muscle spindles.

(iii) *The jaw opening reflex.* Mechanical or electrical stimulation of the lips, the oral mucosa, and the teeth, causes inhibition of elevator muscle activity. In man it has not been possible to demonstrate any increase in depressor activity although in animals a depressor response leading to jaw separation has been shown. Actual tooth contact also causes a silent period in masseteric activity, but, again, in man no activation of the depressor muscles has been found.

The process of mastication has so far been discussed as if it were due simply to bilateral contraction and relaxation of the muscles of mastication. It is, in fact, much more complex. In addition to the vertical movement there is also a grinding movement, in which the mandible is moved from side to side mainly by the action of the lateral and medial pterygoid muscles but with some activity also in the depressor and elevator muscles. In man, the grinding is usually a protrusion-retrusion movement with the condylar heads remaining in the glenoid fossae. As mastication proceeds the food is moved into the cheek pouch on one side by a lateral movement of the tongue tip and side. This simultaneously creates a space beside the tongue into which food can be squeezed from the other cheek pouch by the buccinator muscle. The process is controlled by pressure receptors on tongue and cheeks, and joint receptors around the temporomandibular joints. When the tongue sensors give appropriate input information, the process of mastication ceases, and that of swallowing begins. However, although there is some adjustment of the number of chewing cycles to the consistency of the food, in any one individual the number of chewing strokes between intake of food into the mouth and the act of swallowing is fairly constant. The number of strokes was found in one study to be greater in men than in women, and greater in adults than in children.

### The importance of mastication

In a modern Western diet, there is little need for mastication of most foodstuffs. Swallowing fish, eggs, rice, bread or cheese without chewing does not seem to inhibit their digestion. Experimental evidence shows no residue from these foods. Chicken and stewed meats leave little residue if not chewed but meat generally, potatoes, peas and carrots were not fully digested unless chewed. Mastication does, however, provide a control of the size of food particles to be swallowed. Large particles may damage the walls of the oesophagus or even of the stomach while food which is too finely ground may pass rapidly through the gut and remain undigested. In terms of oral health, the stimulation of the gingival tissues resulting from mastication, particularly of coarse or fibrous foods, improves both keratinisation and local circulation. There is also, apparently, some emotional satisfaction from the rhythmic movement of the jaws in chewing. This may be due to its rhythmical nature, or it may be a specific example of a primitive emotional response to oral sensations.

### The forces of mastication

The jaw elevator muscles are extremely powerful; acrobats and trapeze artists use them to support the whole body. Measurements made by biting as hard as possible on some electrical or mechanical transducer (a gnathodynamometer) show static forces of 440 N between posterior teeth, falling to around 150 N between premolars or canines. Males can exert some 520 N, females some 340 N. The muscles respond to training and by chewing on resistant objects these values can be increased by about 30%. Measurements on Eskimo women, who chewed sealskins to soften them for use as clothing, yielded values from 1450 to 1700 N. The upper value for biting force is limited not only by muscular strength but also by the pain threshold in the periodontal ligaments. More realistic values recorded during actual eating are around 80 N with a normal Western diet. Edentulous subjects wearing dentures were able to produce forces around 64 N.

It is uncertain to what extent these forces measured between the teeth are experienced at the temporomandibular joint. In man the bony structure of the joint and articular surfaces appears relatively slight, certainly in comparison with that of the carnivore. Assumptions that the joint was load bearing have been questioned. The late disappearance of the condylar cartilage appears inappropriate if the joint is load-bearing, and various vector analyses of the forces of mastication show that it would be mechanically possible for the joint to be no more than an attachment of the mandible to the skull. The most recent work, however, supports the idea of a load-bearing joint.

## Swallowing or deglutition

The mouth is the entrance of the digestive tract. When food has been sufficiently masticated, it is swallowed by a process of voluntary and involuntary activity of muscles. The process is also known as deglutition. Liquids and semi-liquids do not need comminution in the oral cavity and are therefore passed on immediately by a slightly different mechanism, called variously, drinking (from a vessel), suckling (from teat or nipple), or sucking (from a tube or from a large volume of liquid).

### Sucking and drinking

Fluid may be taken through the oral cavity into the oesophagus down a gradient of pressure. The tongue is pressed against the palate and the tip moved to create a negative pressure. In sucking directly from a fluid the pressure is assisted by lip movements, or may be exerted through a tube with the lips forming a seal around the tube. Suction continues until only the posterior part of the tongue remains in contact with the palate, and the anterior part of the mouth is filled with fluid. The tip of the tongue is then brought up to touch the anterior palate immediately behind the incisors and a wave of contraction passes backwards as the posterior part of the tongue drops down, forcing the fluid back. The soft palate is raised to seal off the nasopharynx by meeting the pharyngeal wall and the adenoidal pad. The epiglottis deflects the fluid stream to either side as the larynx is raised below it. Fluid may pass down the oesophagus under the influence of gravity, but there is some peristaltic motion engendered, beginning with constriction of the superior constrictor of the pharynx. This is seen more obviously when gravity cannot assist: it is possible to drink in the upside down postion when the transport of fluid must be due solely to peristaltic waves. The sequence of events begins with the voluntary inward suction but when the volume of fluid in the anterior part of the oral cavity seems appropriate, the swallow proper is initiated. This initial action may be voluntary or involuntary, and it can be voluntarily inhibited. The remainder of the process is involuntary; it is controlled by a centre in the medulla which integrates muscle movements and also the breathing pattern. In a long continuous fluid intake such as the 'yard of ale', the movements of the tongue are largely inhibited, the head is tipped back, and the fluid is literally poured down the oesophagus. The tongue makes reflex wave-like movements which may assist in the process. The position is similar to that of the sword swallower, who aims to present as straight a line as possible through the oral cavity down into the oesophagus. Drinking from a vessel is similar to sucking except that less suction pressure is required, and the lips are used to form a tissue-liquid or tissue-vessel seal.

## Suckling

In breast-feeding the mother's nipple and part of the areola are pulled into the mouth by sucking. The whole of the lower jaw with the tongue attached is raised and lowered alternately by the masticatory muscles. In X-ray films this has been described as looking like a boat rocking upon waves. The tooth pads are not approximated. The nipple reaches as far back as the junction of hard and soft palates. Although suction is created as the mandible is lowered and pressure is exerted as it rises, the stimulus of movement is sufficient to cause the milk 'let down' reflex whereby oxytocin causes contraction of myoepithelial cells and milk is expressed from the nipple.

With feeding bottles the jaw is again raised and lowered, but the teat is held between the tongue and the tooth pad of the upper jaw. The tongue then compresses the teat with a pressure wave so that the contents are forced into the mouth.

In both cases a swallowing action follows.

## Swallowing of solid food

When the food is sufficiently comminuted – the degree of comminution varies in different individuals – the swallowing process begins. Three stages are described:

(i) *Bolus formation.* The tongue forms the pasty mass of saliva and food into a ball or bolus so that it lies in a hollow on the dorsal surface and touches the hard palate in the anterior part of the oral cavity. The soft palate is raised by the actions of levator and tensor palati to seal the nasal cavity. The teeth are brought together to provide a seal around the mouth (pressures of 4–7 kPa have been measured between the teeth during swallowing). The tongue cups itself to produce pressures of 10 kPa in the midline as the mylohyoid muscle undergoes a wave of contraction carrying the bolus backwards. This movement may be assisted by a negative pressure produced in the back of the mouth as the posterior part of the tongue which has been heaped up to protect the pharynx now falls back to the floor of the mouth. All stages of this process so far can be inhibited voluntarily, and the process can be begun voluntarily.

(iia) *The oral phase of swallowing.* When the bolus touches the back of the mouth and the pharyngeal wall, a new stage of the swallow begins. This is an entirely reflex mechanism, one not subject to voluntary control. Although there is variation in the sensitivity of different areas in the back of the oral cavity, swallowing is probably initiated by general stimulation of these areas rather than stimulation of any one specific area. The respiratory tract is protected in several ways. The nasopharynx is closed off by the soft palate. The larynx and oesophagus are raised and the oesophageal opening dilated by relaxation of the cricopharyngeal sphincter. This brings the larynx up behind the backward-sloping epiglottis. Contraction of mylohyoid and any suction produced by tongue movement carry the bolus back to the oesophageal opening. The bolus is split by the projection of the epiglottis and flows round it on either side. The epiglottis begins to move back but does not in fact cover the larynx in a cap-like fashion until the swallow is almost complete. During a rapid sequence of swallows it may in fact stay in this position. The vocal cords are approximated and inspiration inhibited. Swallowing can therefore be used as a device to increase a period of voluntary breath-holding.

The duration of a swallow from the beginning of bolus formation to the end of the oral phase is about 1 s.

(iib) *The pharyngeal phase of swallowing.* As the bolus enters the pharynx, the superior constrictors begin a peristaltic wave which passes down the oesophagus. Breathing is re-established with an inspiration. The cricopharyngeus muscle, which forms a sphincter like ring, relaxes to allow the bolus to pass into the oesophagus and then contracts.

(iii) *The oesophageal phase.* The peristaltic wave begun in the pharynx continues down the oesophagus, generating a gradient of pressure which carries the bolus onwards. Further contractions of the striated muscle of the upper oesophagus drive on any food left behind by the initial peristalsis. These are reflexly controlled by the vagus nerve, via motor endplates, in response to the stimulation of receptors by the presence of food in the oesophagus. The waves of contraction continue through the lower third of the oesophagus, which is made up of smooth muscle innervated by a myenteric plexus supplied by the vagus nerve. At rest the oesophagus is normally relaxed except for a sphincter like lower zone which prevents reflux from the stomach. This relaxes to allow the bolus to pass and then contracts again. Simultaneously the stomach relaxes – a state termed receptive relaxation. The oesophageal phase of swallowing lasts for several seconds.

Integration of the whole swallowing pattern takes place in a medullary centre in the reticular formation.

### Variations in the swallowing pattern

The main variant on the usual swallowing pattern is the 'tooth apart' swallow. Instead of bringing the teeth of the two jaws together to provide a firm outer wall against which the tongue can act, the subject keeps them apart in the rest position and uses cheeks and lips as the outer wall. This means that the tongue thrusts forward between the anterior teeth and the lips are often compressed by action of the circumoral muscles to assist in providing the rigid wall. This swallowing pattern is often seen in subjects who have or have had a habit of sucking thumbs or fingers.

A different problem is seen in the cleft palate patient, who may simply force food or drink out through the nasal cavity when attempting to swallow. This can be prevented by a prosthesis replacing the palatal deficiency.

### Frequency and volume of swallowing

Swallowing is not confined to meals since there is a slow continuous secretion of saliva into the mouth, and this must be swallowed as it accumulates. Studies of swallowing frequency suggest that a subject swallows about 600 times in 24 h – only a quarter of these being during food intakes. The sleeping subject produces little saliva and swallows only some 50 times during the 8 h of sleep.

Estimates of the volume of a single swallow vary from just over 20 ml in male subjects to about 14 ml in females and as little as 5 ml in small children.

The frequency of swallowing means that it is a muscular activity of some importance to the oral structures. The teeth adopt a position in the horizontal plane which lies at the null point of the various forces acting upon them. These are the tongue, cheek and lip forces. Subjects with a tooth apart swallow are much more likely to show protrusion of the upper teeth because of the powerful tongue thrust outwards towards the weaker lips. The effect on the lower teeth is more variable.

### Gagging or retching

If the swallowing reflex is inhibited when the posterior part of the oral cavity is stimulated, another reflex is initiated. The mouth is held open and the bolus or the material present at the back of the oral cavity is expelled by movements originating in the posterior part of the tongue. This reaction is known as gagging or retching. It may occur during the taking of impressions for dentures, particularly in the upper jaw, and in very sensitive patients impression taking may prove impossible unless the back of the mouth is anaesthetised. If retching fails to remove the stimulus from the oropharyngeal area, continued stimulation can invoke vomiting.

### Further reading

Jenkins, G.N. *Physiology of the Mouth,* 4th edn, Blackwell, Oxford, 1978

Lavelle, C.L.B. *Applied Physiology of the Mouth,* Wrights, Bristol, 1975

Osborn, J.W., Armstrong, W.G. and Speirs, R.L. *Companion to Dental Studies, Volume 1, Anatomy, Biochemistry and Physiology,* Chapter 12, Blackwell, Oxford, 1982

# 20

# Speech or vocalisation

## Speech sounds

Speech is the production of intelligible sounds by modification of the airflow in the respiratory and oral passages. Language is the organisation of speech sounds to convey information to others of related culture: some primitive roots of language, such as the infant's response to its mother, seem to be common between races. In man the sounds of speech are usually produced in the first instance by expiration through a slit whose length, width, and ability to vibrate, can all be controlled. This slit is the glottis: it includes both the true and false vocal cords. The sounds can subsequently be modified by resonance and interposition of barriers into the airflow. Some languages also include clicks, in which sudden local air movement is produced by muscle activity in the oral cavity.

Sound is the sensation arising from longitudinal vibrations of molecules in a rhythmical manner in a direction radial to the vibrating source. In a sound transmitted through air in one direction, the air molecules vibrate back and forth in the direction of the sound, and therefore produce alternations of pressure at the receiver. The frequency with which they vibrate gives the quality known as pitch to the sound; and the extent of forward and backward movement about the mean position is the amplitude, perceived as pressure change or loudness. Media other than air can also transmit sound but the speed of transmission and the loss of sound energy in transit will vary with the medium. The sequence of vibrations spreading from the source is called a sound wave; it may be represented graphically as pressure changes (or longitudinal molecular displacements) against time.

Sounds are characterised by their main frequency (pitch) and, since few sound sources generate only the main frequency, by the combinations of multiples of the main frequency, or harmonics, which accompany it and provide tone, or timbre. Different singers singing the same note are recognisable by the different combinations of the harmonics produced by their vocal apparatus. The loudness of a speech sound is governed by the force of expiration from the lungs, i.e. the velocity with which the airflow passes the vocal cords. With clicked sounds the amount and speed of the muscle movement provides variation in loudness. In intelligible speech the duration of a note and the manner of initiating and stopping the sound wave will also be important in characterising the sound.

In normal speech the adult male voice uses frequencies between 100 and 150 Hz, whilst the female voice uses frequencies between 200 and 300 Hz. The upper harmonics of an 's' sound may reach 9000 Hz. Until this century the limits of frequency recorded for singers were 55 and 2048 Hz, but by special training some singers have now reached 27 Hz and 4096 Hz – although the highest and lowest notes called for by composers are well within this range. The loudest recorded sounds reached by shouting are around 110 db, and although the limit for normal intelligible speech is about 180 m, sounds produced by the human voice have been heard at distances up to 8–9 km.

## The control of speech

The control of speech is a complex voluntary activity. Actual control of the movements of vocalisation is initiated in the motor cortex. However, the area anterior to the motor cortex just above the Sylvian fissure (Fig. 9.2), known as Broca's area, controls word formation, and integrates the activity of the muscles of speech and respiration through a nearby cortical area. Broca's area is connected by the arcuate fasciculus to Wernicke's area in the superior temporal gyrus. Lesions of the fasciculus and of Wernicke's area produce a condition known as fluent aphasia, in which the patient can produce normal intelligible speech sounds made up of a mixture of jargon and nonsense. Wernicke's area is essential for the understanding of both written and spoken language, and for the production of intelligent speech. When speech is abnormal, writing is also usually affected. Control of respiration by Broca's area and the neighbouring cortical areas is an over-riding voluntary control of the respiratory rhythm generated in the pons and medulla oblongata. In respiratory terms, speech is a controlled expiration, each syllable representing an outflow of air, and each punctuation mark or pause a halt to expiration and a time for inspiration (Fig. 20.1). A subject speaking for a long period may show signs of hyperventilation. The need to take a breath limits the length of phrase that can be sustained, although this can be increased by training. The enhanced expiration during speech causes increased loss of water vapour from the lungs and the oral mucosa.

Two processes are normally distinguished in the production of speech: phonation, which is the production of an airflow and the establishment of a fundamental frequency, and articulation, which is the modification of the airflow by resonance or by various degrees of stoppage to produce vowels and consonants.

## Phonation

The apparatus of phonation includes the lungs and the respiratory muscles, the trachea and the larynx. The larynx consists of cartilage and fibrous tissue, with a number of muscles which can adjust its position or its dimensions. The cartilages are the cricoid cartilage, forming a ring at the base with its widest portion posteriorly, the thyroid cartilage, forming the anterior wall with its lower horns attached to the cricoid, and the two arytenoid cartilages attached to the posterior part of the cricoid, but able to pivot and glide upon it.

The vocal cords, or ligaments, are the thickened edges of the vocal folds which project into the larynx and are attached anteriorly in the midline to the thyroid cartilage, and posteriorly to the inner projections of the arytenoid cartilages (the vocal apophyses). The vocal cords contain elastic tissue; the vocal folds are made up of fibro-elastic tissue and the lateral crico-arytenoid and thyro-arytenoid muscles. A functionally separate strand of the thyro-arytenoid muscle lateral to the vocal cord is sometimes called the vocalis muscle. Above the vocal cords are a further pair of folds, the false vocal cords, whose function, it is thought, is to provide a mucous secretion to lubricate the true cords. They may also touch the true cords and shorten the vibrating length to increase the maximum frequency of sound obtainable. In normal respiration the vocal cords are relaxed, and airflow occurs freely through the triangular space between them. The position and tension of the cords can be changed externally by movements of the arytenoid cartilages produced by the lateral and posterior crico-arytenoid muscles, the tranverse arytenoid muscles and the lateral parts of the thyro-arytenoid muscles (Fig. 20.2). Muscles which alter the position of the anterior and posterior insertions will also affect the cords: the infrahyoid group of muscles stabilise the thyroid cartilage, and the cricothyroid and the posterior crico-arytenoid muscles stabilise the arytenoid cartilages. The recurrent laryngeal nerve provides a motor supply to

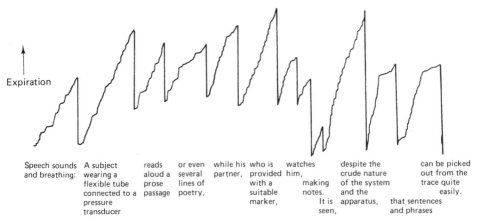

**Figure 20.1** Stethograph recording of the chest movements during speech. The arrow indicates the direction of expiration. Note how breathing is controlled as the subject reads the sentence printed below the trace – each phrase is a series of expirations, and inspiration occurs only at the pauses or the punctuation marks.

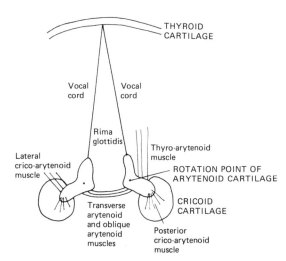

**Figure 20.2** The vocal cords, the cartilages of the larynx, and the muscles which control the length and tension of the cords. Rotation of the arytenoid cartilages about their pivot points changes the width of the gap between the cords – the rima glottidis – and alters the tension of the cords.

all the muscles except the cricothyroid, which is supplied by the external laryngeal nerve. Both nerves are branches of the vagus.

The cricothyroid muscle pulls the upper border of the cricoid upwards, and its posterior fibres draw the cartilage backwards. This increases the distance between the thyroid and arytenoid insertions of the vocal cords and therefore tenses them. Although relaxation may be achieved without antagonist muscle action because of the elasticity of the cords, the thyro-arytenoid muscles can relax the cords still further by their contraction. The width of the space between the vocal cords – the rima glottidis – is adjusted by the posterior crico-arytenoid muscles, which rotate the arytenoid cartilages to open the slit, and the lateral crico-arytenoid muscles which are their antagonists. The posterior part of the slit is closed by the action of the transverse arytenoid muscles which bring the cartilages together.

The vocal cords produce sounds by alternately stopping and releasing the airflow. During sound production, therefore, they must be approximated along their lengths and the slit between them opened and closed at such a speed as to provide puffs of air at the appropriate mixture of frequencies. The dominant frequency depends upon the length, the tension, and the thickness of the cords. Before puberty the vocal cords of the male are only slightly longer than those of the female, but at puberty the increase in size of the thyroid cartilage and the vocal cords is sufficient to cause the male cords to double in length, dropping the voice by about an octave, whereas the female vocal cords

increase in length by only a third, to produce a 2–3 tone difference. In the adult male the cords are about 15 mm long, in comparison with about 11 mm in the adult female. The frequency of the sound produced by the cords increases as length and thickness decrease. Thus the voice is not a natural string resonance, but a driven frequency, and depends on muscle activity as well as airflow. Indeed, the frequency of impulses in the fibres of the recurrent laryngeal nerve is sometimes proportional to the sound frequency produced. However, this clonic theory of sound production does not explain the production of high frequency notes – since at high frequencies of stimulation, tetanus of the muscles occurs. Two other explanations have been put forward, both of which envisage a steady state of contraction of the vocalis and other muscles. In the myo-elastic theory the air pressure is thought to force open the slit between the cords against the closing force of the muscle contraction: the tension in the muscle and the elastic tissue close the slit as the pressure is dissipated. This theory sees the muscle tension as determining the sound frequency. A third theory, the aerodynamic theory, is similar, but envisages the closing force coming not from the tissue but from the suction generated at the edges of the jet of air. The individual quality of the human voice may be related to the other turbulence patterns created as air passes at high velocity through a narrow slit. At lower frequencies the frequency of vibration is controlled by tension in the cords and by altering the thickness or shape of the cords; at high frequencies the cords are stretched to increase the tension by tilting the back of the cricoid cartilage downwards or by a relative tilting of the thyroid cartilage. At still higher frequencies, tension is increased further by increased muscle contraction without a further increase in length. Increases in loudness, or pressure, tend to increase the frequency. At the upper end of the soprano range the vocal cords remain in contact over about a third of their length and only the remaining portion vibrates. This may be assisted by the the contact of the false vocal cords which restricts vibration.

The loudness of the sound produced depends upon two factors – the air pressure produced by expiration, and the degree of closure of the rima glottidis. A firmly closed rima will produce loud sounds, a rima which allows air to escape without vibrating will produce proportionately less sound. The two mechanisms are used together in normal speech. Whispering is achieved by leaving part of the rima glottidis open throughout the production of the sound. Consequently, the intelligibility of whispering depends mainly on the subsequent articulation of the sound and very little on the process of phonation.

Control of the muscles of vocalisation is very precise; to produce the range of different notes possible, the cords must be varied in length by about 1 μm for each step in change of frequency. The receptors involved in this control are muscle spindles and spiral nerve endings in the laryngeal muscles and the joint receptors in the joints between the cartilages. There is also the control from higher centres which sets the frequency, and the feedback from the ears which identifies and corrects the frequency to some internal standard – although 'perfect pitch' is presumably a learned standard. Alternatively the sound produced may be matched to some external sound. The fact that the voice is never heard only through air, as others hear it, means that speakers can rarely recognise their own recorded voices. A good singer produces sounds to match those heard; but the untuneful are not able to match the airborne sound with that transmitted through bone and tissue. This seems true even of accent and language intonations: regional or foreign speakers are often unaware of their own accent.

In addition to the modification of airflow past the vocal cords, phonation also includes the selective amplification or suppression of particular sound frequencies by the resonators of the vocal tract. These are of two types: the pipe resonators represented by the tube from the glottis to the mouth, or the glottis to the nose, and the Helmholtz resonators, which are resonating chambers with narrow outlets. The changes due to the pipe resonators are demonstrated by the difference in the sound of the voice produced during an upper respiratory tract infection. The curious quality of speech from a cleft palate patient is another example. The Helmholtz resonators of the vocal tract are seven in number: the vestibule between the true and false cords, the larynx up to the root of the tongue, the pharynx, the oral cavity between the tongue and the hard palate, the labial cavity or vestibule between the lips and teeth, the nasal cavity, and the paranasal sinuses. The oral cavity can be divided into a posterior and an anterior cavity on a basis of the tongue position. The effect of each potential resonator in modifying speech depends upon the adjustment of the airflow from one to another and its limitation by movable structures. Since the limitation of airflow in various ways is considered as articulation, sound production by varying the size and the shape of the resonating chambers is intermediate between phonation and articulation.

## Articulation

### Vowels

Vowels differ from consonants in that they are produced by unstopped airflow. They are formed by shaping resonating chambers to select particular frequencies of voice. This involves constriction

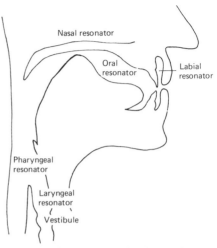

**Figure 20.3** The resonating chambers which amplify certain frequencies at the expense of others. The paranasal sinuses are not shown. The tongue is shown humped up at the rear of the mouth, separating the oral chamber from the pharyngeal. Variations in the tongue position will vary the relative sizes of these two chambers, and give each vowel sound its characteristic pair of formant frequencies.

along the airway, but no complete blocking of airflow. The structure producing the constriction is termed an articulator. Of the seven potential resonating chambers (Fig. 20.3), only three need be considered in detail – the nasal cavity separated from the others by the soft palate (or velum) and the oral and pharyngeal chambers separated from each other by the tongue. In English and in most European languages the entry to the nasal chamber is closed for all vowel sounds, whereas in French a number of nasal vowels are used. The pharyngeal chamber amplifies the lower frequencies, and the oral chamber the upper frequencies, the frequencies selected being dependent upon the size of each chamber. Thus for a short 'a' the main oral frequency is about 1300 Hz and the main pharyngeal frequency is about 720 Hz. These two frequencies are termed the formants of the vowel. As the tongue is rolled forward to the hard palate, the oral chamber becomes smaller and the oral formant rises still higher, whilst the pharyngeal chamber becomes larger and the low formant drops further. Thus, for a short 'i', the oral formant is around 2400 Hz and the pharyngeal formant is around 280 Hz. If the tongue rises towards the back of the mouth and the soft palate a different combination of frequencies will be obtained. Vowels may be open or closed, depending upon the degree of narrowing of the connection between the chambers. The volume of the vestibule between the lips and the teeth may be added to that of the oral resonator by rounding and slightly projecting the lips: this will lower the frequency of the oral formant.

## Consonants

Consonants are produced by briefly stopping the airflow. They may be characterised by two features – the manner of the stopping, and the position of the barrier (*Table 20.1*). A structure forming a barrier is termed an articulator.

The airflow may be stopped momentarily and then released, as in plosive sounds. In the plosives the soft palate closes the nasal cavity, and airflow is blocked completely by one of several possible stops in the oral cavity. Nasal sounds are similar except that the nasal passage is left open and air leaks through it. The sounds 'b' and 'm' illustrate the difference between a plosive and nasal sound with the oral stop in the same position. Lateral sounds are those in which air is allowed to escape on either side of the tongue which is acting as the barrier to airflow. When the articulating structure vibrates against the other wall of the barrier a rolled, or vibrant, sound is produced. Sometimes the barrier to airflow is left very slightly open, either as a flat slit ('f'), or as a more rounded opening ('w'). This type of consonant is a fricative. A combination of a fricative and a plosive gives an affricative such as 'ch' or 'j'.

The position of the barrier to airflow gives another classification – bilabial, labiodental, interdental, linguodental or alveolar (apico-alveolar), alveolopalatal or prepalatal, palatal or mediopalatal, velar, and glottal. The last of these includes not only the aspirate 'h', but also the break between vowels which the Londoner and New Yorker may substitute for 't' - as in be'er for better.

In addition to these classifications it is often possible to distinguish between strong and weak forms of consonant, depending upon the air pressure necessary to produce them. Thus 't' and 'd', 'p' and 'b', are sounded by the same combinations of plosives and positions but with different air pressures. Air pressure on the tongue, one of the main articulators, in fact varies between 3.4 and 16 kPa during speech. In English a weak form is usually associated with vibrations of the vocal cords and is therefore termed voiced, whereas the stronger consonants include no tone from the vocal cords and so are unvoiced. The difference between 'w' and 'wh' shown in *Table 20.1* has largely disappeared from English: few people now distinguish between 'witch' and 'which', although the latter is correctly pronounced 'hwitch'.

## Clicks

The click is used in English speech solely as an expression of disapproval – although a double click is sometimes used in non-verbal communication – in contrast with the wide variety and usage of clicks in African languages. The sound is produced by closing the oral passage at two points simultaneously, reducing the pressure between the two points, and then allowing air to re-enter rapidly to give the click. No phonatory airflow is necessary, and respiration may be unaltered.

## The effects of dental variations and dental procedures on speech

The quality of speech is affected by changes in the volume of the resonating chambers, and in the position and size of the articulators. Thus speech differs in sound if the palatal arch is high or if it is flat; if the tongue is large or if it is small; if there is obstructive growth of lymphoid tissue in the oro-nasopharynx; if there is a palatal cleft; or if the musculature of the soft palate is deficient or functionally abnormal. The consonants will be affected if the articulators cannot produce the normal barriers to airflow: if the lips are short, if the jaw relationship is abnormal, if teeth are missing or misplaced, or if tongue to palate contact is difficult.

Table 20.1 The form and position of the stops used in producing the sounds of the consonants. V indicates voiced, S unvoiced

|  | Plosive | | Nasal | Lateral | Rolled | Fricative | | Affricative | |
|---|---|---|---|---|---|---|---|---|---|
|  | V | S | V | V | V | V | S | V | S |
| Bilabial | b | p | m |  |  | w | wh |  |  |
| Labiodental |  |  |  |  |  | v | f |  |  |
| Interdental |  |  |  |  |  | that | thin |  |  |
| Alveolar | d | t | n |  |  | z | s |  |  |
| Prepalatal |  |  |  |  | r | v | sh | j | ch |
| Palatal |  |  |  | l |  | y |  |  |  |
| Velar | g | k | ng |  |  |  |  |  |  |
| Glottal |  |  |  |  |  |  | h |  |  |

Racial or familial peculiarities may affect the nature of speech sounds. Thus the mandibular retrognathism fairly common among the English aristocracy may be instrumental in producing some aspects of their speech considered as affectation – including the lisp. The change in effective palatal position, and even tooth position, in the upper denture wearer may change the sounds whose production involves these structures. Edentulous subjects cannot pronounce the dental consonants properly. The sounds most commonly affected by dental abnormalities and by dental treatment are the dental, alveolar and prepalatal consonants with 's' and 'th' being particularly susceptible to change.

## Further reading

Malmberg, B. *Phonetics*, Dover, New York, 1963
Appropriate chapters in textbooks of oral physiology/oral biology.

# 21

# Control of digestion

Three types of control of the individual units of the digestive tract can be distinguished. Chronologically first is the cephalic phase of control, in which such stimuli as sight or sound or smell or the thought of food cause increased movement or secretion in the gastro-intestinal tract (Fig. 21.1). The oral sensations of taste and touch are sometimes included as psychic or cephalic stimuli. The input pathways for these reflexes are sensory nerves and the controlling centre is in the cortex. Outputs may be through the medulla oblongata via centres controlling specific digestive organs, or directly through the vagus nucleus. Examples of such reflexes include the secretion of saliva, the increase in gastric motility and secretion in response to psychic stimuli, and the contraction of the gall bladder.

The second type of control is intra-organ or within-organ control: the adjustment of movement and/or secretion in response to the nature of the food material in the organ itself. Examples are the control of salivary secretion by the sensation of taste, the adjustment of rate of emptying of the stomach in response to the texture or osmotic activity of the gastric contents, and the stimulus to defaecate as the rectum is distended. These reflexes may be nervous or hormonal; they may be truly

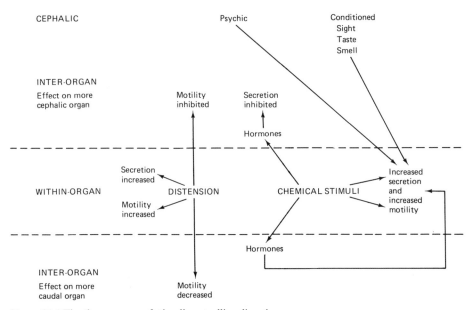

**Figure 21.1** The three groups of stimuli controlling digestive organs.

272  Control of digestion

local with only local nerve pathways involved, or they may involve central nuclei (as in salivary secretion), or they may involve local secretion of hormones which then circulate eventually to reach the target organ (as in the gastric secretion of acid).

The third group of controls are inter-organ controls. These are responsible for the co-ordination of activity throughout the system. Thus distension of an organ may cause sphincters ahead to open, or it may inhibit movement and close sphincters behind – thus maintaining movement through the gut. Presence of food material in one organ may stimulate secretion of mucus in the next part of the digestive tract.

The digestive reflexes will be described in order from the intake of food down to final defaecation. Mastication and swallowing are dealt with in Chap. 19. At the upper end of the digestive tract conscious control and reflexes involving the extrinsic innervation dominate gut activity; lower down mechanical factors become important in stimulating both extrinsic and intrinsic reflexes; lower still hormonal control becomes the important feature; and progressively the reflexes involving the intrinsic innervation take over control. Finally, at the lower end as at the upper, conscious control is established. The gastrointestinal tract is not isolated from the rest of the body: it receives motor and secretomotor and inhibitory fibres from the parasympathetic and sympathetic nervous systems, and its blood supply is a part of the cardiovascular system as a whole. Circulating hormones affect the activity of the gut. The blood supply, and hence activity, of the gut may be influenced by emotion, or exercise, or blood loss. Equally, the motor and secretory activity may be modified by changes in general sympathetic activity. Stomachs exposed surgically or by trauma, can blush with embarrassment or blanch with fear. Constipation may be a symptom of emotional doggedness, or diarrhoea a sign of nervous haste.

## The control of salivary secretion

### The external stimuli and the afferent pathway (Fig. 21.2)

#### Psychic stimuli

Stimuli to the sense organs of the head, and, specifically, the mouth, are often associated with food intake. The stimulation of sight or hearing causes salivation only by a conditioned reflex. The sight of food, or the sound of cooking, or listening to a description of food, is a psychic stimulus rather than a local or direct stimulus.

A small but real increase in salivary flow occurs in response to thoughts about, or suggestions of, food. The sensation of mouthwatering, however, is probably due to increased awareness in the oral

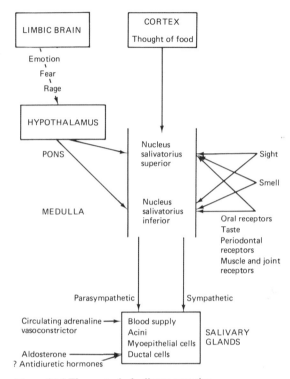

**Figure 21.2** The control of salivary secretion.

area, rather than to increased salivary flow. Foods which are disliked by a subject may inhibit salivation.

#### Local stimuli (cephalic stimuli, excluding psychic stimuli)

Local stimuli for the secretion of saliva may be taken as all the cephalic stimuli, with the exclusion of the psychic, or conditioned, stimuli. They include smell and taste, touch and irritation of the oral mucosa, proprioceptive impulses from the masticatory muscles and the temporomandibular joint, and impulses from the pressure sensors of the periodontal membranes of the teeth.

Olfactory irritants, in contrast to non-irritating olfactory stimuli, appear to cause a direct rather than a conditioned salivary reflex.

An increase in salivary flow is observed whenever the sense of taste is stimulated. Acid stimuli are the most effective, sweet and salt less so, and bitter least of all. Many experiments suggest that the stimulation of taste sensors in the posterior part of the tongue, served by the glossopharyngeal nerve, results mainly in a flow of saliva from the parotid gland, whose secretomotor fibres travel in the same nerve. Chorda tympani inputs have a greater effect on the glands innervated by the chorda tympani secretomotor fibres.

Unilateral stimulation of smell, taste, touch, or proprioceptive sensors causes a predominantly unilateral secretion of saliva.

## Interorgan stimuli

Interorgan stimuli have little influence on salivary secretion. Irritation of the oesophagus causes a reflex salivation, but irritation of the stomach results in salivation only as a part of the nausea and vomiting reflex. Salivary secretion is not inhibited by digestive activity elsewhere in the gut.

## Central control

Impulses pass along the axons from the various sensors to reach the brain and spinal cord, and after one or more synaptic links, efferent secretomotor impulses are generated.

The cell bodies of the preganglionic secretomotor neurones are the sites at which incoming impulses are finally integrated to generate suitable secretory responses. In the sympathetic system these cell bodies lie mainly in the region of the second thoracic nerve. The spinal reflexes for salivation are under the control of fibres passing downward from the medulla oblongata and higher areas such as the hypothalamus.

The cell bodies of the preganglionic parasympathetic fibres are in the nuclei of the facial and glossopharyngeal nerves. A group of neurones in the reticular formation, extending from the facial nucleus to the nucleus ambiguus, gives salivary responses on stimulation and has been named the nucleus salivatorius. It is made up of the nucleus salivatorius superior, where stimulation causes secretion mainly from the ipsilateral submandibular gland, and the nucleus salivatorius inferior, where stimulation causes secretion mainly from the ipsilateral parotid gland.

The secretomotor cell bodies, in addition to receiving inputs from the efferent fibres of the reflex loops, also receive inputs, both excitatory and inhibitory, from other parts of the brain. Some examples of this may be seen in the conditioned reflexes mentioned previously. Other examples are the stimulation and inhibition of secretion by hypnosis, the inhibition of secretion by sleep, and the inhibition of secretion which causes a dry mouth as a result of fear or excitement. The flow of unstimulated saliva is high in infancy and decreases up to the age of five years – the period during which the brain is maturing and increasing evidence of inhibition by higher centres becomes apparent. Hypothalamic activity can also affect salivation: secretion is increased when the posterior hypothalamus is stimulated, and overt salivation occurs in conjunction with the sham rage, eating and vomiting responses which can be produced by hypothalamic stimulation.

A feeding reaction which includes a salivatory response is produced on stimulation of the cortex at the lower end of the fissure of Rolando and the junction of the frontal and anterior sigmoidal gyri. Stimulation of the rhinencephalon, including the amygdaloid nucleus, has similar effects.

## The efferent pathway

The flow of saliva is controlled entirely by nervous activity. Hormones may modify the secretion but do not themselves initiate a flow of saliva. Pharmacological agents such as pilocarpine can cause salivation by mimicking the action of transmitter substances such as acetylcholine and noradrenaline.

Control is exerted mainly by the parasympathetic nerve supply but also by the sympathetic. The parasympathetic fibres originate in the glossopharyngeal and facial nerves. Sympathetic nerve fibres synapse in the superior cervical ganglion and the fibres pass from there to run with the blood vessels to the salivary glands.

It is now usually accepted that stimulation of parasympathetic nerves in man causes production of a copious flow of saliva whilst sympathetic stimulation selectively causes secretion of protein and glycoprotein. However, such observations depend mainly on experiments with drugs which can selectively block different receptors; the effects of the two types of stimulation are probably never separated in real life.

Five possible effects in the glands can occur as a result of nervous stimulation – initiation of secretory activity, increase in blood flow to maintain secretion, synthesis of new secretory products, changes in activity of the duct cells, and contraction of the myoepithelial cells. Secretory activity may be stimulated by either parasympathetic or sympathetic impulses or by the two systems acting in concert. The innervation of the mucous and serous cells appears to be similar; but variations in secretion may occur as a result of the stimulation of different receptor sites on the effector cell and hence activation of different secretory pathways. When both sympathetic and parasympathetic stimulation are employed their effects are synergistic. Parasympathetic nerve fibres are vasodilator and the sympathetic fibres vasoconstrictor. Once secretion has begun the glands produce bradykinin, a possible vasodilator, which might over-ride the nervous control of the blood vessels. Recently VIP (vasoactive intestinal peptide) has been suggested as an additional neurotransmitter which could cause vasodilatation during secretion. It is not known whether the secretory products of the cells are continually being synthesised or whether they are synthesised in response to nervous stimuli. Sympathetic stimuli, which cause the cells to discharge their secretory products, do not appear to initiate their synthesis at the same time. The presence of cholinesterase in or near ductal cells suggests that they have a parasympathetic innervation, but the effect of autonomic activity on duct cell function has not been explored. The myoepithelial cells of human salivary glands probably respond to both parasympathetic and sympathetic stimulation, and possibly also to bradykinin. However, the role of the myoepithelial cells in the expulsion of saliva from the gland is still debatable.

## The transduction of the nerve stimuli
(Fig. 21.3)

Stimulation of the secretomotor nerves results in the release of the appropriate neurotransmitter substances: acetylcholine from the parasympathetic nerves, noradrenaline from the sympathetic. Both substance P and vasoactive intestinal peptide (VIP) can stimulate secretion from salivary cells and there may be other possible neurotransmitters from secretomotor nerves. Circulating adrenaline can also act on the glands. It has been shown that even a single nerve impulse in the chorda tympani nerve can release more than enough acetylcholine to initiate secretion. The neurotransmitters bind to membrane receptor sites on the acinar cells, possibly on some ductal cells, and certainly on the muscle cells of the blood vessels. They are inactivated, respectively, by acetylcholinesterase or by monoamine oxidases. Monoamine oxidases are present in high concentrations in salivary glands and can probably inactivate all the circulating catecholamines which normally reach them.

Acetylcholine binds to muscarinic receptors (Fig. 21.3) and causes an increase in turnover of membrane phospholipids, generation of inositol tri-phospate and the release of calcium from the mitochondria and the microsomes where it is stored into the cytosol. There is also an influx of calcium from outside the cell through calcium 'channels'. It is the increased concentration of calcium ions in the cytosol that causes the opening of calcium-dependent potassium channels in the basal and basolateral portions of the cell membranes and begins the process of secretion described on p. 36. Stimulation of acinar cells by acetylcholine results in an increase in intracellular cyclic GMP but the significance of this is unclear. The increase in calcium concentrations inside the cells may also contribute to the process whereby protein is exported from the cells by exocytosis. Fluid-containing vacuoles have been observed near the Golgi apparatus of acinar cells after acetylcholine stimulation: their contents and function are unknown.

Catecholamines may bind either to α receptors or to $\beta_2$ receptors (Fig. 21.3). Their effect at α receptors is similar to that of acetylcholine at the muscarinic receptors. At $\beta_2$ receptors, however, they produce a different pattern of stimulation. They stimulate production of cyclic AMP causing

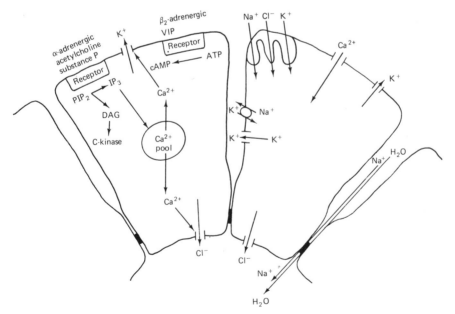

**Figure 21.3** The control of salivary secretion at the cellular level. This diagram shows in an abbreviated form the same second messenger mechanisms as *Figs. 12.3* and *12.4*. VIP vasoactive intestinal peptide, cAMP cyclic adenosine monophosphate, $PIP_2$ phosphoinositide 4,5-phosphate, $IP_3$ inositol-1,4,5,- triphosphate, DAG diacyl glycerol.

calcium release but also activate a protein kinase, which itself causes a release of intracellular bound calcium. The free calcium ions bind to calmodulin and the complex then initiates the movement of secretory vesicles and granules along pathways delineated by microtubules and microfilaments towards the luminal cell membrane. Here the increased concentrations of ionic calcium immediately beneath the cell membrane are able to alter the relative charges on the plasma membrane and the vesicle membrane to permit their fusion and the loss of the vesicle contents by exocytosis. Cyclic AMP has been demonstrated to play some part in initiating synthesis of new protein for secretion but this does not seem to occur at the same time as secretion. Although cyclic AMP is associated with secretion of protein by the cells, calcium is also necessary, and secretion can occur when calcium concentrations are raised even in the absence of cyclic AMP.

## Factors affecting the concentration of salivary constituents

### Flow rate

The main factor affecting the composition of saliva from individual glands is the flow rate (Fig. 21.4). The effects on salivary constituents have been mentioned in Chap. 11 but can now be summarised. As flow rate increases the concentrations of total protein, amylase, sodium and hydrogen carbonate increase. There is no change with flow rate for potassium above 0.3 ml/min. Fluoride also shows very little change with changes in flow rate. As flow rate increases the concentrations of magnesium, phosphate, albumin, aminoacids, urea, uric acid, and ammonia decrease. However, at fast flow rates the concentrations of urea and other small non-polar molecules increase again. The concentrations of chloride, calcium and protein-bound carbohydrates at first fall with increasing flow rate but above about 0.3 ml/min they all increase.

If a constituent is in markedly different concentrations in parotid and submandibular saliva, the concentration in mixed saliva will be influenced by the relative contribution of each gland to the flow. Thus at slow flow rates submandibular saliva constituents will predominate whilst at fast flow rates the parotid saliva constituents will become progressively more important. In the unstimulated secretion the composition of the saliva from the accessory glands may be important. One example of these effects is the fall of calcium concentrations in mixed saliva as flow rate increases. This occurs despite the increasing concentrations of calcium in the individual secretions and is due to the lower concentrations of calcium in parotid saliva which forms a greater proportion of the mixed secretion at faster flow rates. Another example is the change in balance between glycoproteins and amylase in mixed saliva as the flow rate increases.

### Duration of stimulation

Maintenance of a constant stimulus to the salivary glands for periods greater than 3 min whilst keeping the flow rate constant results in a reduction of the concentrations of many components (Fig. 21.5). However, after an initial fall the concentrations of calcium, hydrogen carbonate and protein begin to

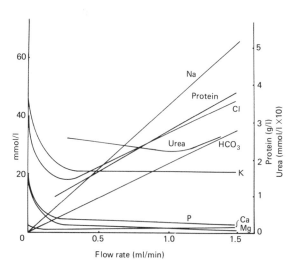

**Figure 21.4** Variations in the concentrations of salivary constituents at different rates of flow of saliva – in this case parotid saliva.

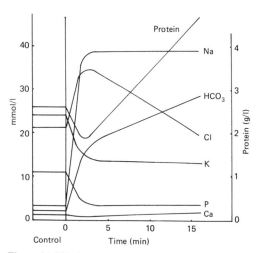

**Figure 21.5** Variations in the concentrations of salivary constituents with varying periods of stimulation of the glands while flow rate was maintained constant – parotid saliva.

rise again. Sodium and iodide concentrations are not affected by duration of stimulation after the first few min. Phosphate, magnesium and potassium concentrations fall at first but then remain steady. Chloride concentrations, after an initial increase, fall throughout the period of stimulation. Protein concentrations, after falling briefly, normally increase over the first few min of stimulation, but on prolonged stimulation (an hour or more) the concentrations fall. Saliva collected during the first 3 min of stimulation may be discarded to eliminate this source of variation.

### Nature of the stimulus

Protein concentrations in parotid and submandibular saliva vary with the stimulus, salt tastes provoking particularly high concentrations. Variations in composition of mixed saliva may arise from possible differences in stimulus sensitivity between the glands.

### Time of day

There are circadian rhythms in the concentrations of protein, amylase, sodium, potassium, calcium, chloride, phosphate, organic phosphate, thiocyanate and the steroid hormones in saliva (Fig. 21.6). The flow rate itself varies at different times of day but the changes in concentrations of the main ions do not seem to be related to the changes in flow rate. In general, protein concentrations are high during the daytime and seem to be related to mealtimes. Sodium and chloride concentrations, and to a lesser extent, calcium concentrations are high in the early hours of the morning whilst potassium and phosphate concentrations are highest

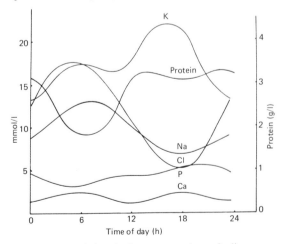

**Figure 21.6** Variations in the concentrations of salivary constituents at different times of day – submandibular saliva. The data have been analysed and a mixture of sine curves fitted to give a 'best fitting' smooth curve.

in the afternoon. The circadian variations in steroid hormone concentration are directly related to the plasma circadian variations.

### Plasma concentrations

The concentrations in saliva of urea, uric acid, aminoacids, nitrate, glucose and, when plasma levels are abnormally high or low, potassium and calcium ions, are related to plasma levels. Phosphate concentration is not. There is some relationship between the sodium and chloride concentrations in saliva and plasma but this is more likely to be due to hormonal activity than to any simple secretory mechanism. The concentrations of steroid hormones in saliva are directly related to the concentrations of free steroids in plasma.

### Hormonal effects

There is no actual evidence that the water content of saliva is affected by the circulating levels of antidiuretic hormone. Increased permeability of the striated duct cells would therefore permit more reabsorption of water, and might explain the reduced salivary secretion observed.

Transport of sodium and potassium ions in the striated ducts is affected by aldosterone, which increases the reabsorption of sodium. Chloride reabsorption will be concomitantly increased.

The effects of other systemic hormones on the composition of saliva in man have not been explored. There are minor differences in ion and protein concentrations in saliva from males and females, and at different times in the menstrual cycle, but these are probably not significant and cannot be specifically ascribed to the events of the cycle. Both testosterone and thyroxine increase salivation. Salivation is often increased during pregnancy and at the menopause there is often xerostomia (dry mouth) resulting from a decrease in secretion.

A number of locally-acting substances can be sialogogues (substances which increase salivation). Bradykinin and its precursor, kallidin, stimulate secretion by increasing the blood flow. Substance P and two active alkaloids, physalaenin and eledoisin, have sialogogue activity.

### Effect of age

The rate of secretion of saliva is higher in infants than in adults, and there are minor differences in concentrations of components, possibly due to a delay in the maturation of the striated ducts. Saliva flow rate in older subjects – over age 60 years – is often reduced but this is almost certainly due to the medication received by most subjects in this age group. Sodium concentrations are lower than usual in older subjects.

## Influence of diet

The size and activity of the salivary glands are influenced by the degree of stimulation, either mechanical or gustatory, to which they are normally subjected.

Increases in the amount of calcium or phosphate in the diet do not appear to affect salivary concentrations of these ions, probably because salivary phosphate levels are not related to those in plasma and, although salivary calcium concentrations are related, the plasma concentration is normally maintained within very narrow limits. There is a transient increase in fluoride concentrations in both plasma and saliva after a dose of fluoride.

The buffering power of saliva from subjects on predominantly protein diets is reported to be higher than that of saliva from subjects on predominantly carbohydrate diets. High protein intakes raise plasma urea concentrations and may therefore raise salivary urea concentrations. People on high protein diets, with low carbohydrate intakes, have less amylase in their saliva.

# Control of the oesophagus and gastro-intestinal tract

## Control of the oesophagus

Swallowing begins as a voluntary act but ends as an involuntary one. The upper third of the oesophagus is composed of striated muscle but even this is subject to involuntary control via the vagus nerve. Lower down, the striated muscle gives way to smooth muscle and the arrangement of muscle fibres and the nerve plexuses becomes more typical of the alimentary tract. The sole control seems to be the intra-organ control exerted through the plexuses. The cardiac (gastro-oesophageal) sphincter responds to the movement of materials into the digestive tract, relaxing as the food bolus is swallowed. Sensory impulses in the fifth, ninth and tenth cranial nerves pass to the medulla oblongata and stimulate the vagal efferents to the oesophagus.

## Control of the stomach

The stomach is stimulated both by psychic influences and by the presence of food in the mouth (Fig. 21.7). Thus, an appetite secretion is described, consisting of acid and pepsin. Patients with gastric or duodenal ulcers used to be recommended to keep the stomach partially filled and to avoid any particularly appetising foods to try to diminish the acid secretion and its effect. Once food is smelled and tasted the gastric secretion and motility increase. Contact of the partially masticated food with the stomach wall causes a number of intra-organ reflexes. In man histamine stimulates a flow of acid secretion with a smaller amount of enzyme secretion; it is possible therefore that the food swallowed may damage the surface layer of the gastric epithelium, causing histamine release and stimulating secretion. The food certainly stimulates endings of parasympathetic nerves causing impulses to pass to the medulla and cause secretomotor activity in the vagus efferent fibres. Hypoglycaemia (low blood sugar levels) stimulates secretomotor activity in the vagus nerve. Stimulation of the stomach by the vagus causes secretion of pepsinogen, acid and mucus. The vagus also sensitises receptors in the gastric mucosa within the body of the stomach and, more particularly, in the antrum. These receptors respond to the presence of peptides and proteins in the chyme causing the secretion into the bloodstream of the polypeptide hormone, gastrin. This stimulates the gastric glands to secrete an acid-rich digestive juice. Normally, when the antrum becomes acid, the secretion of gastrin is inhibited. Thus acid production by the gastric glands is adjusted to match the food intake and composition.

The movements of the stomach are controlled by local stimuli, operating either directly by distension

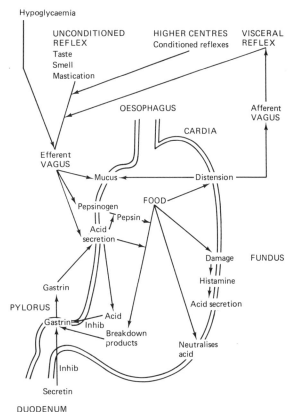

**Figure 21.7** The control of the secretions of the stomach.

of the smooth muscle, or by physical or chemical stimulation of the mucosa. Mucosal receptors respond to decreasing particle size, to increasing osmotic activity (due to breakdown of larger molecules into smaller), decreasing viscosity (again due to the reduction in molecular size during digestion) and increasing volume. The results of these stimuli are an increase in gastric motility and an increase in the rate of gastric emptying. Distension stimulates secretion as well as motility; the other stimuli mentioned may also affect secretion since solutions of high osmotic activity inhibit it. Below 200 mosmol secretion is stimulated. The activity of the stomach is reduced when it contains aminoacids or fats. The practice of taking an alcoholic drink before eating may be rationalised as alcohol is a stimulant of gastric secretion; so also is caffeine.

The stomach is also subject to inter-organ reflexes. The influence of food in the mouth has already been mentioned. Filling and distension of the next part of the gut (the small intestine) inhibits gastric activity. Thus, as the duodenum fills, it inhibits its own further filling. Irritation of the duodenum may have the same effect as distension. The duodenum also has chemical sensors responding to fat, to carbohydrate, to acid, and possibly to osmotic activity. All these reduce gastric activity. Fats are particularly potent in this respect and may act by causing secretion of enterogastrone, a hormone related structurally to the epithelial growth factor found in saliva. This hormone inhibits gastric motility and also inhibits the action of gastrin. The presence of hydrochloric acid, and also proteins and aminoacids, in the duodenum causes the production of the hormone secretin, which, in addition to its main effect on the pancreas, causes inhibition of gastric activity. Some secretion of gastrin takes place from duodenal cells. Finally, cholecystokinin-pancreozymin is secreted in response to the presence of fats and soaps, and this hormone has a slight inhibitory effect on gastrin. Enteroglucagon is secreted in the ileum in response to the presence of glucose and fatty acids: it inhibits gastric motility and may be part of an ileogastric reflex. It also stimulates insulin secretion.

There are other external factors which affect the stomach. These include circulating adrenal hormones, such as cortisol, which reduce gastric secretion generally. The effect of hypoglycaemia in increasing secretion of acid and pepsin is probably mediated centrally and may involve hypothalamic monitoring of blood glucose levels.

## The act of vomiting

Food taken into the stomach, or even passed to the duodenum, is usually digested or eliminated via the small and large intestines. However, irritation of the stomach or duodenum by toxins, or by abrasive or obstructive material, or irritation of the back of the pharynx or some parts of the oesophagus, can initiate the reflex series of events termed vomiting. Impulses pass – possibly by sympathetic afferents – to the medulla, where a series of responses are integrated. Impulses reaching the cortex are interpreted as nausea. The medullary neurones can themselves initiate vomiting, or may be stimulated to do so by hypothalamic areas as a result of emotional responses in the limbic brain. The process begins with an increase in salivation and then a deep inspiration. After inspiration the glottis is closed and the soft palate seals off the nasal cavity. The pressure in the abdominal cavity is raised by contraction of the external muscles (which would produce an abdominal expiration were the glottis not closed) and the muscle walls of the oesophagus and the stomach relax, together with the gastro-oesophageal and pyloric sphincters. Powerful contractions of the small intestine and the pylorus in the high pressure of the abdominal cavity help to raise gastric pressure above oesophageal pressure, and the gastric contents are forced upwards. These contractions are not of themselves propulsive but act to prevent the intestinal contents also being forced upwards by the increase in intra-abdominal pressure. There are usually a number of contractions of the abdominal muscles, raising the semi-liquid column up to the cricopharyngeal sphincter, until sufficient pressure is generated to allow it to spill over through the mouth. Closure of the glottis and the raising of the soft palate prevent the vomitus reaching the larynx or the nose unless the reflex is blocked by drugs. Inspiration is not resumed until the nasopharynx is clear.

## Control of the small intestine and associated structures

The chyme reaching the small intestine is there exposed to the action of the pancreatic secretion and of the bile, the secretion produced in the liver and stored in the gall bladder. Control of the digestive process in this part of the gut, then, involves control of the liver, gall bladder and pancreas, as well as that of the duodenum, jejunum and ileum (Fig. 21.8).

The small intestine, like the earlier parts of the digestive tract, is subject to control by cephalic influences. The cephalic stimuli already mentioned act via vagal pathways to cause an increase in the movements of the intestine and in the secretion of mucus by its glands. There is also some stimulation of pancreatic enzyme secretion and contraction of the gall bladder sends a flow of bile into the duodenum. There is no evidence that the composition of gastric or pancreatic secretions is modified to match the food to be digested.

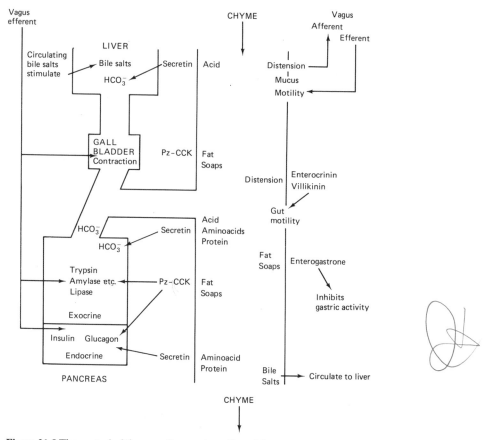

**Figure 21.8** The control of the secretions and motility of the small intestine and its associated glands, the liver (and gall bladder) and the pancreas.

The presence of chyme in the duodenum has many effects. Distension of the duodenum, sensed by vagal afferents, increases movement and secretion of mucus. It may also cause rhythmic contraction and shortening of the villi; possibly because of the production of a hormonal substance termed villikinin. There seems to be some doubt as to whether these movements of the villi are produced by the contractions of the muscularis mucosae. Chemical stimuli on the whole cause secretion of hormones which then affect glands producing digestive secretions. A neutral chyme in the early part of the duodenum may stimulate acid secretion in the stomach; but more usually the chyme is acid and stimulates the production of secretin in the duodenum. This has its main effect upon the pancreas, causing the production of an alkaline secretion composed mainly of sodium hydrogen carbonate, but also stimulates the liver to increase its secretion of bile. Elsewhere secretin inhibits gastrin-stimulated acid secretion in the stomach although, in contrast, it may stimulate the secretion of pepsinogen. It decreases duodenal motility.

Lipids and soaps in the duodenum cause the secretion of cholecystokinin-pancreozymin and enterogastrone. The name of the hormone cholecystokinin-pancreozymin indicates the two aspects of its activity, 'pancreozymin' referring to the stimulation of the pancreas to produce a secretion rich in enzymes, and 'cholecystokinin' indicating that it causes contraction of the gall bladder. It also stimulates intestinal motility but inhibits gastric motility. It stimulates the production of bile salts in the liver – as secretin probably increases liver hydrogen carbonate secretion. The other chemical stimulants in the small intestine are the bile salts themselves. They stimulate bile salt production in the liver after their reabsorption in the ileum.

This account is simplified in that it describes the effects of each hormone separately. A single type of foodstuff is rarely present in the gut by itself and so the effects of the hormones and the nervous reflexes are usually mixed. Thus gastrin helps stimulate the production of pancreatic enzymes whilst secretin and cholecystokinin-pancreozymin enhance each other's activity on the pancreas. Gastrin and the intestinal hormones are co-operative in their effects on the pancreas but antagonist in their actions on the gastric glands.

The liver produces its secretion continuously but may be additionally stimulated by secretin and by the reabsorbed bile salts – a kind of positive feedback loop resulting from the enterohepatic circulation.

## Control of the large intestine

The presence of chyme in the colon causes enhanced motility and secretion.

Inter-organ controls include the gastro-colic reflex: entry of food into the stomach after a delay of 1–10 min causes increased motility, hyperaemia, and secretion of mucus.

## Control of the anus and rectum

### Defaecation

The progress of food and waste materials through the digestive tract is largely automatic. Only the intake and exit of these materials to and from the tract are voluntarily controlled – at least, in the normal adult. As described earlier, there are some three or four mass peristaltic movements through the colon daily. These fill the rectum or increase the tension within it. When internal rectal pressure is sufficiently high relaxation of the internal anal sphincter occurs by a local reflex. The external anal sphincter, however, is under voluntary control via the splanchnic nerves. Damage to the spinal cord may interrupt this voluntary control so that in a 'spinal man' defaecation is an involuntary reflex act.

If stimuli from the rectum are permitted to act by conscious volition, there takes place the defaecation reflex, controlled by a network of nerves in the medulla oblongata. A powerful peristaltic movement sweeps down the colon and rectum and may empty the colon as far back as the splenic flexure. The diaphragm descends, the glottis is closed, and the chest muscles contract. Abdominal contraction raises intra-abdominal pressure to 12–25 kPa. The mass of the faeces is important – a large volume is more easily expelled. Greater pressures can be achieved in a squatting position. The reflexes controlling defaecation and those controlling urination interact with each other – the two activities cannot be performed simultaneously. The rise in abdominal pressure has other consequences. A rise in arterial pressure occurs as pressure is exerted in the abdomen but this is followed by a fall in arterial pressure as the abdominal pressure prevents the return of venous blood to the heart. The sudden changes in intracranial pressure consequent upon these arterial pressure changes may even be fatal in ill subjects.

As a consciously controlled activity, defaecation may be subject to emotional variation. Constipation and diarrhoea may have emotional origins.

## Further reading

Davenport, H. W. *Physiology of the Digestive Tract,* 5th edn, Year Book Medical Publishers, Chicago, 1982

Sernka, T. and Jacobson, E. *Gastrointestinal Physiology,* Williams and Wilkins, Baltimore, 1983

Sanford, P.A. *Digestive System Physiology,* Arnold, London, 1982

Chapters on saliva in oral physiology/oral biology textbooks.

# 22

# The menstrual cycle, pregnancy and lactation

## The menstrual cycle

The human female experiences from menarche to menopause a cyclical variation in hormone levels, in the state of the generative organs, and in the release of ova (Fig. 22.1). The cycle has a rhythm of approximately 28 days, a lunar month, and is therefore termed 'menstrual'. It varies from 21 to 40 days, but is usually fairly constant for any one individual. The cycle is normally only interrupted by pregnancy and lactation – although it may be replaced by a pharmacological cycle in which ovulation is inhibited. The normal sequence of events occurring as a result of the hormonal changes begins with a proliferation of the epithelial lining of the wall of the uterus and the underlying endometrium. Associated with these changes is a thickening of the epithelial lining of the vagina and a cornification of its surface layers. The uterine muscle also increases in thickness. This pre-ovulatory part of the cycle is therefore called the proliferative phase. The uterine changes represent a preparation for the implantation of the fertilised ovum. The ovum is released at approximately mid-cycle (12–21 days), and travels down the fallopian tube to the uterus. After ovulation the endometrium enters a secretory phase, during which the glands become coiled and secrete more actively, while the endometrium as a whole becomes oedematous. Should fertilisation not occur, the endometrium degenerates and sloughs off, and the cells, secretions and blood from the raw surface are lost in the discharge of menstruation.

The hypothalamus controls the cycle by stimulating a cyclical release of the gonadotrophin hormones, follicle stimulating hormone (FSH) and luteinising hormone (LH).

The first part of the cycle, from the beginning of the menstrual flow until the release of the ovum, is related to events taking place in the ovarian follicle and is therefore termed follicular; the second part is termed the luteal phase, because it is related to events occurring after ovulation, when the follicle becomes the corpus luteum.

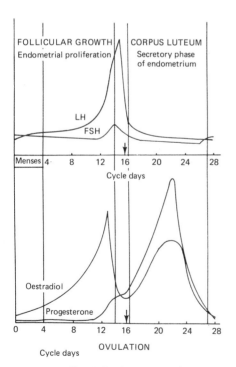

**Figure 22.1** Circulating hormone levels during the menstrual cycle.

## The follicular phase = First Part of Cycle

The rise in concentration of FSH in the blood at the beginning of the menses (the period of 3–7 days during which menstrual flow takes place) stimulates several follicles to start developing but as the concentration decreases again, all but one of these follicles atrophy. Early in the follicular phase the granulosa cells (G cells) of the follicle have many receptors for FSH but few for LH. Although FSH concentrations in the blood are decreasing, it is able to stimulate the G cells to proliferate and to form oestrogens from androgens. The thecal cells of the follicle are able to synthesise androgens from cholesterol and so provide substrate for the G cells. The oestrogen promotes the development of more FSH receptors and so increases the sensitivity of the G cells to the hormone. The base of the follicle becomes progressively more vascular, and so the oestrogens readily gain access to the bloodstream. The oestrogen produces a number of effects: locally it stimulates the development of the uterine and vaginal epithelium, and the development of the endometrial cells and the glands, and it causes increased vascularisation of the tissue. As the secretion of oestrogen increases, the hypothalamic receptors reverse their activity and initiate a positive feedback: LHRH is secreted, stimulates LH secretion causing the G cells with a now increasing number of LH receptors to secrete more oestrogen, and this now stimulates more LHRH production. The end results of this process are high levels of oestrogen and of LH. At the mid-follicular stage the G cells produce folliculostatin, an inhibitor of FSH secretion. Just before ovulation the LH causes the follicular cells to produce progesterone. As progesterone secretion increases there is an apparently unrelated rise in local concentrations of prostaglandin $F_{2\alpha}$ and an increase in proteolytic activity in the follicular fluid. These together cause the follicle to rupture and the ovum to be released into the fallopian tube. The production of progesterone probably causes the 0.3–0.5°C increase in body temperature which persists until the next menses. Progesterone causes the endometrial glands to become coiled and to commence secretion.

## The luteal phase

The follicle now fills with luteal cells, yellowish in colour because of their high lipid content. These are, in fact, granulosa and thecal cells, now transformed into typical steroid-producing cells, resembling those of the adrenal cortex. They synthesise progesterone and, to a lesser extent, oestrogen. The rising progesterone and oestrogen levels inhibit gonadotrophin secretion from the pituitary gland by their action on the hypothalamus. The mechanism is a negative feedback – LH stimulating progesterone secretion, and progesterone inhibiting LH secretion. If fertilisation does not occur, the corpus luteum degenerates, possibly under the influence of prostaglandin $F_{2\alpha}$. The blood concentrations of progesterone and oestrogen fall as the corpus luteum ceases to produce them. This causes the production of prostaglandin $F_{2\alpha}$ which results in an intense vasoconstriction in the coiled spiral arteries. Deprived of blood flow, the endometrium begins to degenerate, its surface sloughs off, and the damaged raw capillaries bleed. As this begins, the secretion of FSH from the pituitary gland rises again and more follicles begin to mature.

## Sperm production in the male

The human female is born with some 800 000 oocytes already developed and most of these will be lost by degeneration over many years. Only some 400 are actually ovulated in the course of the menstrual cycles. The human male produces millions of millions of spermatozoa during his lifetime, each taking about 74 days to mature. There is no obvious cyclical variation in hormones or tissues with time.

The Leydig cells of the testis have receptors for LH: their response to circulating LH is to produce testosterone from cholesterol. The testosterone circulates and reaches the hypothalamus where it inhibits LHRH secretion. It is testosterone that maintains the secondary sexual characteristics of the male. The Sertoli cells, which line the seminiferous tubules, have receptors for testosterone and for FSH. In response to FSH the cells begin to produce an androgen-binding protein, whose further synthesis is then promoted by testosterone itself. The developing germ cells are passed between Sertoli cells into the seminiferous tubules, where they mature within the micro-environment provided by the processes of the Sertoli cells.

## Pregnancy

If fertilisation occurs and the fertilised ovum is implanted into the uterine wall, the corpus luteum, instead of involuting, continues to produce both oestrogen and progesterone. By the end of the first week after fertilisation, the tissue supporting the embryo, the placenta, is producing chorionic (placental) gonadotrophin – hCG (Fig. 22.2). The stimulating effect of this on the cells of the corpus luteum replaces the diminishing effect of the pituitary gonadotrophins. As the placenta becomes larger, and its cells more developed, it produces both oestrogen and progesterone itself, and so the need for the hCG and luteal hormones decreases. Fetal DHEA is, however, necessary for oestrogen synthesis in the placenta (p. 208). By the end of the

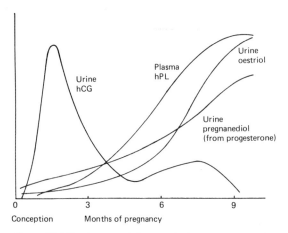

**Figure 22.2** The hormonal changes during pregnancy. hCG human chorionic gonadotrophin, hPL human placental lactogen.

first six weeks the corpus luteum is no longer necessary. The chorionic oestrogen and progesterone stimulate uterine growth and acting in concert they stimulate growth but inhibit secretion of the mammmary glands. The hormone placental lactogen (hPL), also produced by the placenta, aids in the development of the breasts. Oestrogen stimulates growth of the uterine muscle and would cause it to contract were it not for the inhibitory effect of progesterone. Other placental hormones have already been described (p. 208).

The changes in hormonal activity during pregnancy, like those of puberty, may cause an increased susceptibility to gingivitis (inflammation of the gums) and possibly also the hyperplastic enlargements of the interdental papillae termed epulides. Although there seems to be an increased susceptibility to dental caries during pregnancy, it is unlikely to be due to the hormonal changes; it is more probable that the many small meals, eaten in response to the increased appetite and lessened capacity of pregnancy, are more cariogenic. The increased susceptibility to caries used to be thought to be due to calcium being lost to the fetus: but since dental enamel is not exposed to the bloodstream calcium cannot be transferred from enamel to blood. In any case, the observation is of an increase in a bacterial disease, not a simple removal of calcium.

The hormonal changes so far described are due to changes in the gonadotrophins, the gonadal hormones, and the hormones of the placenta. The production of other hormones is also affected by pregnancy, probably in order to maintain the homeostasis of the mother as the fetus makes increasing demands upon her body constituents. Pregnancy is associated with increased appetite and food intake, possibly due to the action of prolactin on the satiety and feeding centres in the hypothalamus. Thyroxine and cortisol are produced in greater quantities during pregnancy, and in late pregnancy, aldosterone secretion increases. All these hormonal changes could assist in the control of salt and water balance.

## Events leading up to the beginning of labour

Birth is thought to be initiated by a local increase in prostaglandin $F_{2\alpha}$ triggered by the increasing concentrations of oestrogen. Two other mechanisms have, however, been postulated and may play some role in the process. Some stimulus may block progesterone secretion while maintaining oestrogen secretion. This theory envisages a positive feedback response, with some stimulus increasing oestrogen secretion and this oestrogen then stimulating the hypothalamus to increased LHRH production, increasing the output of pituitary gonadotrophins and causing more oestrogen secretion – and so on. Although the amount of oestrogen present seems to be important in labour, that by itself is insufficient support for the hypothesis. The state of fetal development is now thought to be the important factor. As the fetal production of adrenocorticotrophic hormone increases, the fetal adrenal glands are progressively stimulated to produce more dehydroepiandrosterone. This is used as a substrate for oestriol synthesis by the placenta and the increase in synthesis of oestriol blocks the synthetic pathways for progesterone. Such a switchover to oestrogen production could cause the release of prostaglandin $F_{2\alpha}$. Prostaglandin $F_{2\alpha}$ may sensitise the uterus to oxytocin, stimulating contraction. An older theory envisaged movements of the fetus in the uterus stimulating local contraction and this being sensed by nerve fibres which communicate with the hypothalamus, the hypothalamus then producing oxytocin which caused more contraction and a further stimulus to the hypothalamus. The final result was expulsion of the fetus in the process of birth. This mechanism could well be important in the maintenance of labour but is probably not the initiating process.

A peptide hormone, relaxin, is produced in the pregnant uterus up to the time of birth: it permits widening of the pubic symphysis and therefore the pelvic opening, facilitating the actual process of birth.

## Lactation

Even when birth has occurred the mother still shows hormonal changes. The baby continues to need nutriment, and this is supplied by the mother until the infant is weaned.

The mammmary glands develop under the influence of chorionic oestrogen and progesterone, but together these hormones inhibit milk secretion. After birth the placental hormones are no longer produced and the glands can therefore secrete. During pregnancy the number of prolactin-producing cells in the anterior pituitary gland increases, possibly as a result of stimulation by oestrogens. When oestrogen inhibition ceases, prolactin can stimulate the mammary glands to produce milk proteins and lactose. Other hormones are necessary for lactation: thyroid hormones, hormones of the adrenal cortex, and growth hormone. Oxytocin is the final hormone in the milk delivery chain: it causes contraction of myoepithelial cells and the expulsion of the milk from the ducts into the suckling baby's mouth. The process of suckling itself provides the stimulus for prolactin and oxytocin secretion. Touch receptors around the nipple send a neural signal to the hypothalamus to initiate secretion of the hormones.

Prolactin may inhibit the releasing factors for luteinising hormone; the continuation of lactation seems to inhibit the re-establishment of the menstrual cycle.

## Further reading

Austin, C.R. and Short, R.V. (eds) *Hormonal Control of Reproduction,* Vol. 3 of *Reproduction in Mammals,* Cambridge University Press, Cambridge, 1984

Johnson, M. and Everitt, B. *Essential Reproduction,* 2nd edn, Blackwell, Oxford, 1984

Rhodes, P. *Reproductive Physiology,* Churchill, London, 1969

Tepperman, J. *Metabolic and Endocrine Physiology,* 5th edn, Year Book Medical Publishers, Chicago, 1987

# 23

# Growth and development

The processes by which a fertilized ovum develops finally into an adult human being are still not completely understood. Although measurements of hormone concentrations in blood and tissues are yielding more information about the stimuli for various developmental changes, there is virtually no information on how the time scale and sequence of development and growth are programmed and realised. The factors which govern organ size and shape are still poorly identified. Molecules known as organisers probably diffuse between cells to control differential growth and cell membranes probably interact with each other in delineating tissue boundaries – but no organiser molecules have yet been isolated and the mechanisms of membrane interaction are obscure.

Although the mechanisms are not known, it is possible to say that genetic factors play a major role in control. Homozygous twins reared under different conditions have similar body shapes even if actual size is different. The rate of physical development of a child, in the absence of major environmental differences, lies between those of its parents. Environment influences the extent to which the inherited potential is displayed. Some environmental factors may be particularly important at critical times in development and less so at other times. Thus malnutrition is most harmful in delaying growth and development during the first five years of life. However, the effect of malnutrition is to delay rather than to prevent, and if full nutrition can be restored the individuals may finally achieve normal development. On the other hand, tissues whose cells do not continue to divide after birth, such as nervous tissue, may not recover from malnutrition occurring during fetal life. In general, malnutrition affects size rather than shape.

Since the human race shows sexual dimorphism, development is affected by the presence or absence of the Y chromosome, females having two X chromosomes and males an X and a Y. The sex of the child is, therefore, governed by the chromosome carried by the spermatozoon. Even in abnormal karyotypes such as XXY, XXX, XXXY and XO, the presence of the Y chromosome is the deciding factor in the sex of the individual.

## Measures of growth and development

The most obvious way to assess or express development is by age. However, chronological age is not necessarily related to either physical or mental development. The age of the fetus may be expressed either as post-menstrual, that is, dated from the last menstrual flow, or as post-conception. The latter is approximately two weeks later than the former. The term intra-uterine (i.u.) should mean post-conception and will be used in that sense here. Birth occurs at approximately 40 weeks post-menstrual or 36–38 weeks post-conception. Fetal age is often expressed chronologically when the assessment has been made dimensionally, since tables are available relating crown-rump lengths to the ages of average embryos. Birth is a clearly defined event in time and hence age can easily be expressed in relation to it. The next defined point in life is puberty, or the beginning of adolescence. In Europe it occurs in girls between the chronological ages of 8 and 13 years and in boys between 9.5 and 13.5 years. These are the ages at which enlargement of the sexual organs is first observed, an event which in girls precedes the first menstrual period, or menarche.

Size, or height, is not a particularly reliable indicator of development but is used. As mentioned above, it is one of the principal ways of determining the age, or developmental age, of the fetus. The crown-rump length or the crown-sole length can be measured on X-ray pictures, and ultrasonic scanning can be used to measure head size, crown-rump length and abdominal circumference.

Other measures of development are available. The growth of the legs and feet is used as an indicator between birth and 18 months; after that the hand and wrist may be X-rayed to reveal the development of the individual bones which calcify in a specific sequence (Fig. 23.1). The mean ages at which the different stages are observed are termed bone ages. Growth in the female is always some 25% ahead of that in the male during the first ten years of life, and so female bone age is ahead of male bone age at the same chronological age.

Developmental age may also be assessed by the appearance of the teeth or, preferably, by X-rays of the teeth and jaws. The calcification and eruption of the deciduous teeth occur at similar rates in males and females, but calcification and eruption of the permanent teeth are earlier in females. The correlation between the age assessed from dental development and that assessed from the hand and wrist bones is not very strong, suggesting that the growth of these tissues is separately programmed and that there is not a close relationship between the development even of similar tissues.

The best correlation between the various measures of development is that between bone age and the age of the menarche in girls – but bone age does not correlate well with the age at which puberty begins in either sex.

## Tissue growth

Tissues may grow by an increase in the number of cells, or an increase in the size of cells, or both. In normal development some tissues reach their definitive complement of cells early, and thereafter the number stays constant; other tissues are regenerative and have a continuing turnover of cells; whilst still others can show the property of regeneration only if the tissue is damaged or has to perform a much greater workload. Non-regenerative tissues include nerve, muscle and adipose tissue. Epithelia of the skin and the digestive tract are characterised by continuing death and replacement of cells; bone is constantly being laid down and resorbed. The liver and the kidneys are not normally subject to high rates of cell turnover but can show cell proliferation in response to increased metabolic load when large amounts of tissue are lost due to injury. An increase in organ size due to cell division is termed hyperplasia; an increase in organ size due only to increases in the size of cells is termed hypertrophy.

In the non-regenerative tissues growth occurs early in life. There are three phases: one of cell division, one of cell growth associated with a slower rate of cell division, and a final stage of cell growth only. Nerve cells stop dividing at about 18 weeks i.u. life and no new cell nuclei appear thereafter. The axons and dendrites can grow, and damage to one of

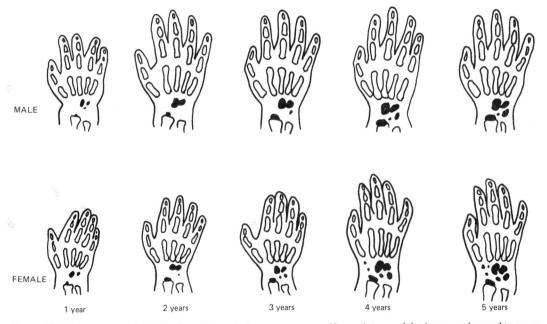

**Figure 23.1** The sequence of calcification of the wrist bones up to five years of age. X-ray pictures of the bones of the hands and feet are used to assess a bone age in the same way as X-ray pictures of the jaws may be used to assess dental age.

these cell extensions is repaired by outward growth from the cell body. The supporting cells of the nervous system, the glial cells, appear to be capable of division throughout life. The growth and maintenance of sympathetic nerves is dependent upon nerve growth factor, a peptide produced in many tissues including the salivary glands, which is taken up by the axons and transported back to the soma, or body of the cell. Muscle cell numbers are also established at an early age, probably soon after 30 weeks i.u. life. Developing skeletal muscle is a tissue in which cells fuse together during embryonic development so that the final cells have many nuclei. Some cells remain as reserve, or satellite, cells: these are mononuclear with very little cytoplasm around the nucleus. Growth of muscle cells occurs by the synthesis of new protein to increase the contractile protein content of each cell. Muscle growth in response to the stimulus of increased usage involves incorporation of some satellite cell nuclei into the existing fibres in addition to the increase in protein content. Since growth occurs by increased protein synthesis, it is dependent upon the anabolic hormones, testosterone, adrenal androgens, and growth hormone. Insulin is also anabolic in its effects. The third tissue in which the number of cells is determined before birth is adipose tissue. Like the cells of skeletal muscle, fat cells increase the size of the tissue by increasing the volume of their cell contents: in this case by storing fat. Loss of stored fat from within the cells decreases the tissue size, just as decrease in the actin and myosin content of disused skeletal muscle leads to wasting of the muscle. It is probable that in adipose tissue, again as in muscle, there are less well-differentiated cells which will not develop into mature adipocytes unless excessive fat deposition occurs. Some writers dispute whether fat cells are non-regenerative and believe that new cells can be produced in these conditions. The adipose tissue of the fetus includes a relatively large proportion of brown adipose tissue, whose forming cells resemble those of the more usual white adipose tissue but whose mature cells differ from them both in appearance and function. All the non-regenerative tissues increase in size by hypertrophy.

Growth of the regenerative tissues may be continuous and responsive to tissue changes due to cell death or damage: there are, in addition, a number of polypeptide hormones which influence the growth of different tissues. These include epidermal growth factor, fibroblast growth factor and the somatomedins produced in response to circulating growth hormone.

Growth of bone and other hard tissues must be considered separately, since in these tissues the main bulk of the tissue is extracellular, and consists of inorganic salts which may be deposited and, except for dental enamel, removed by normal cell activity. Bone is formed in two ways: by deposition into a cartilage precursor, or by development in specialised fibrous tissue sites. After the initial formation, growth may occur either by cartilaginous growth followed by replacement with bone, or by remodelling as a result of activity of the cells of the periosteum in removing or building up layers of bone on the surface. The periosteum is a double layer of cells surrounding the bone; the outer layer is mainly fibroblasts, but the inner, or cambial, layer consists of cells which can lay down or resorb bone. In many bones, such as those of the limbs, cartilage cells form the shape of the fetal bones. The primary ossification centre appears in the centre of the shafts of these bones, and the cartilage is resorbed and replaced by bone. Shortly before birth, secondary centres of ossification form in the ends of the cartilaginous structure, and then bone is laid down in the articulating caps – the epiphyses. This leaves a plate of cartilage at each end of the bone; and it is in these plates that growth in length of the bone occurs. Control of such growth is largely exerted by somatomedin C produced in the liver in response to growth hormone. In addition there is control by testosterone, possibly the adrenal androgens, oestrogen and, in a permissive sense, by thyroid hormones. As growth progresses the epiphyseal plate of cartilage becomes thinner and finally the epiphyses fuse with the shaft, or diaphysis. This normally takes place towards the end of puberty, under the control of the gonadal hormones, the thyroid hormones, and growth hormone, acting in concert. Subsequent development depends upon the muscular forces acting on the bone which cause growth in response to muscle pull. Gravitational forces may supplement this. This remodelling includes bone deposition where there is mechanical stress, and resorption in its absence. The means by which stress is translated into growth is not yet known: it is possible that the piezo-electric forces generated between bone crystals may stimulate bone deposition. The final form of most bones, therefore, is one adapted to their function. The jaws differ from the limb bones in a number of ways. The mandible is not formed in a cartilaginous precursor, but develops from primary centres in fibrous tissue. The secondary centres are cartilaginous but only one of these, the articular cartilage, is near the end of the bone, and it is covered, not by a bony cap, but by a layer of fibrocartilage. This cartilage is not fully converted into bone until early adulthood – around 20 years. The jaws have a layer of bone in which the teeth are embedded, the alveolar process; this is appropriately placed to support the teeth, and its thickness and survival depend on the functioning of the teeth themselves. The whole of the alveolar process may disappear after the loss of the permanent teeth.

The hard tissues of the teeth are laid down in a

complex of mesoderm and ectoderm. The first sign of the deciduous teeth appears at some six weeks after conception. Dental enamel reaches its final form before the tooth erupts, and once formed cannot be altered morphologically by the body: it is subject only to wear, or attrition, in the processes of cutting and grinding the food. Acids may dissolve it, and dental caries may attack it. The cells which produce the enamel die or are lost on the completion of their task. Dentine is more adaptable. Whether new odontoblasts develop in response to stimulation is doubtful, but existing odontoblasts can certainly lay down more dentine. The dentine of deciduous teeth is resorbed by osteoclast cells like those of bone. The third dental hard tissue, cementum, lies on the root surface and it behaves more like bone in being resorbed or added to in response to the stresses placed upon it.

## Fetal growth

At implantation the fertilised ovum has reached by cell division a size of about 150 cells. The blastocyst, as this is now called, develops into two structures – the outer layer forming the placenta, and the core of cells the embryo. The placenta is the organ in which exchange of nutrients from mother to fetus occurs. A parallel system of blood flow, with the vessels forming loops in the villi which extend into the maternal blood of the placenta, provides for efficient exchange between the two circulations. Exchange is selective: the placenta appears to block partially the transfer of fluoride ions to the fetal circulation, so that administration of fluoride during pregnancy is not very effective as a caries-protective measure for children. During the next eight weeks the embryo differentiates to form a recognisable human fetus, with head, limbs, nervous system, and a beating heart. Cell specialisation occurs; and organs and limbs develop by differential growth and the diffusion of 'organiser' substances. The rate of growth in length or weight is at first slow (Fig. 23.2). From now until about 18 weeks i.u. the rate of growth in length increases, whilst the rate of growth in weight reaches its peak at about 32 weeks i.u. Most of the increase in weight up to 24 weeks is due to the build up of cell cytoplasm, particularly in muscle, but from 28 weeks onwards there is accumulation of fat – from about 30 g up to about 430 g at birth. At least 10% of this fat is in the brown adipose tissue which is found throughout the body but particularly around the heart and kidneys and between the scapulae. The brown fat is sometimes called multilocular fat because the fat cells have many small fat droplets around a central nucleus, in contrast to the white fat cells which have a large single droplet pushing the nucleus to one side of the cell. The presence of many mitochondria indicates that the cells are capable of producing large amounts of energy. They do this by direct oxidation of the triglycerides present in the fat droplets. They are innervated by sympathetic nerves through β receptors. Adrenaline is much less effective than noradrenaline in stimulating the metabolic breakdown of brown fat. Up to about 28 weeks the placenta is growing more rapidly than the fetus and provides the principal direct source of its energy; after this the placental growth rate declines and the fetus begins to use some of its stored energy sources. The total fat store provides a compact source of energy for the early weeks after birth. In addition, the brown fat enables the infant to generate metabolic heat relatively easily: this is important in temperature control for the infant, because the shivering reflex does not develop until later. This oxidation of brown fat is therefore termed 'non-shivering thermogenesis'.

How far the maternal and placental hormones influence the growth of the fetus is unclear. Growth hormone is not necessary for fetal growth. On the other hand, the fetal thyroid gland begins to secrete between the 13th and the 18th week i.u.: thyroid hormone secretion is one of the principal factors controlling fetal growth. The thyroid hormones are essential for the synthesis of protein, particularly in nervous tissue. However, hypothyroid babies can be of normal length at birth. Another hormone with anabolic activity is insulin: this may also contribute to growth in the fetus.

The gonads at first are similar in appearance in the two sexes, but under the influence of the Y chromosome the testes become identifiable in the male after 7–8 weeks i.u.; in the absence of a Y chromosome the female ovary can be recognised at 9–10 weeks i.u. By 9 weeks the Leydig cells of the testes are differentiated and they begin to secrete testosterone under the stimulation of the placental

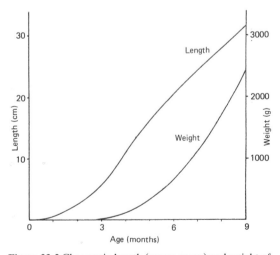

**Figure 23.2** Changes in length (crown-rump) and weight of the fetus.

gonadotrophin hormones. The testosterone acts on the cells of the external genitalia to cause the development of the penis and scrotum; in the absence of this hormone the genitalia develop as those of the female by about 12 weeks i.u. The growth of the penis and scrotum after 13 weeks depends mainly at first upon the stimulus from placental gonadotrophin; but the hypothalamus becomes 'male' at 12–14 weeks i.u. under the influence of the testosterone, and then begins to control the production of pituitary gonadotrophins. This occurs because the hypothalamic cells are able to take up testosterone and convert it to oestrogen. This causes the cells to become 'male'. Oestrogens themselves are protein-bound to such an extent in the bloodstream that they cannot be taken up by the hypothalamic cells. Testosterone secretion continues to increase up to birth. When the Sertoli cells of the testes develop they secrete Mullerian duct inhibiting factor, which causes the precursors of the fallopian tubes in the male to degenerate. At about 10 weeks i.u. the fetal adrenal cortex develops a special zone near the medulla and this continues to grow until birth, when it disappears. Its growth and secretion are controlled by a fetal pituitary trophic hormone, corticotrophin-like intermediate-lobe peptide (CLIP). This zone secretes androgens only – dehydroepiandrosterone (DHEA) and its sulphate. These androgens control the growth of the fetus in the female; they act with testosterone in the male. In addition DHEA is an essential substrate for oestriol production in the placenta which lacks the enzymes of the precursor pathways in oestriol synthesis. A spurt of follicle stimulating hormone production at 18–20 weeks i.u. initiates development of the ova. The fetal adrenal glands also secrete cortisol in an amount proportional to that secreted by the adult and the rate of secretion remains fairly constant throughout childhood.

Once the difference between male and female development is established the growth rate of the female fetus begins to exceed that of the male: thus by 20 weeks i.u. the development of the female fetus is approaching being 3 weeks ahead of that in the male. Although the deciduous teeth develop at the same rate, the permanent teeth are more advanced in the female. The greater fore-arm length to body height ratio of the male is evident *in utero*, as is the relatively longer index finger of the female. The pelvic opening is larger in the female at birth. Despite the greater rate of development of the female fetus, the newborn male is usually slightly heavier than the female. After 32–34 weeks i.u. the growth of the fetus slows down as it becomes progressively more confined. This reduction in rate of growth is directly related to the size of the placenta: it enables a child inheriting genes from a father of large body size to be borne by a mother of small body size.

# From birth to puberty

At birth there is a complete change in metabolism since the fetus no longer depends upon the maternal blood supply for nutrients, but now as a neonate has to absorb water, salts and nutrients from first the colostrum and then the milk. A change is detectable in the dental tissues – the layer of hard tissue formed at birth is differently calcified and can be seen in the enamel of the deciduous teeth and sometimes in the cusps of the first permanent molars as an accentuated incremental line known as the neonatal line. Otherwise, however, most of the processes of growth and development continue as if to some pre-set programme independent of the actual time of birth.

The rate of growth increases now that the restraining influence of the placenta is absent. Girls grow faster than boys up to puberty, reaching half their adult height by 1.75 years, as opposed to 2.0 years for boys (Fig. 23.3). The eruption of the deciduous teeth takes place at much the same age in the two sexes, but girls are ahead of boys in the eruption of the permanent dentition, usually gaining their first permanent molars some two months ahead (just before the age of 6 years) but

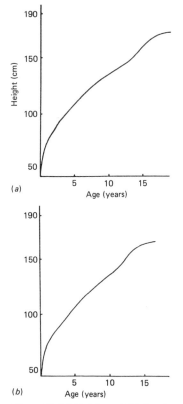

**Figure 23.3** (*a*) Male and (*b*) female growth curves up to early adulthood.

progressively increasing the difference until at about 10 years there may be as much as a year's difference in the dates of eruption of the canine teeth.

Thyroid hormones, important in fetal growth, still play a part in growth during childhood but their role may be seen as one of permitting the action of other hormones, particularly growth hormone. Growth hormone, through the action of somatomedin, is the most important factor governing bone growth, and therefore height, during this stage of development. Growth of the jaws is also under the control of this hormone. Growth hormone cannot produce its full effect in the absence of thyroid hormone or of an adequate supply of nutrients. Thus without thyroid hormones growth of the jaws is inhibited and the eruption of the teeth delayed. Differences in the final height of adults cannot be related to the amount of secretion of growth hormone in childhood in normal subjects: the final height appears to be genetically controlled. A further anabolic factor during childhood is the maintenance of insulin secretion since growth hormone acts on the cells of the endocrine pancreas, making them more sensitive to metabolic stimuli.

The secretion of testosterone falls to low levels after birth, except for a brief spurt of secretion during the first six months of life, the significance of which is unknown. Oestrogen secretion, and that of the gonadotrophins, likewise remains low until puberty as does the secretion of adrenal androgens. Prolactin and cortisol are secreted at levels proportional to those in the adult male.

Different tissues grow at different rates, depending, apparently, upon their individual programming. Overall differences appear in different parts of the body. Thus, leg length increases more rapidly than that of the trunk up to puberty: this produces a later sexual dimorphism because the later timing of puberty in the male allows more time for leg growth. Changes in the numbers of blood cells occur (Fig. 23.4) as a result of an increase in blood volume soon after birth, although the rate of production of different blood cells also changes; despite the increase in blood volume the blood proteins increase in concentration and relatively more globulins are produced. Red blood cells are produced more slowly immediately after birth as the changeover from fetal haemoglobin to adult haemoglobin production occurs: since the rate of destruction is unchanged, the red blood cell count falls until the new pattern of cell production is established. The antibodies of the ABO blood groups appear in the blood during the first year of life.

The development of reflex and control activities in the brain follows the pattern of general morphological development; the development of intelligence is more difficult to relate to physical development and estimates of mental age may differ from those of physical age. However, the more rapidly growing and developing child is also more intelligent than its peers and the slow developer may also be intellectually slower. These differences appear to persist into adult life regardless of the final body size or shape.

## Puberty and afterwards
### Changes associated with puberty

Puberty is defined as the time at which there is a sudden physiological increase in the size of the testes, prostate gland, and seminal vesicles, in the growing male, and of the uterus and vagina in the growing female. The appearance of pubic hair, which used to be regarded as the first external sign, is usually preceded, often substantially, by the increase in size of the testes in the male, and the

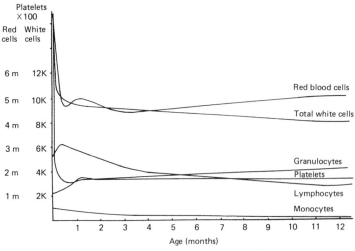

**Figure 23.4** Changes in blood cell concentrations in early life.

pubertal growth spurt in the female. The female growth spurt may or may not be coincident with the first changes in appearance of the breasts. In both sexes the changes occur as a result of changes in the sensitivity of hypothalamic cells to the circulating gonadal hormones. The timing of these changes does not seem to be related to the state of bone development or to other chronological changes in growth such as the development and eruption of the teeth. In addition to the changes in the reproductive organs themselves, morphological changes at puberty produce the secondary sexual characteristics – differences in the growth of bone, muscle, adipose tissue and hair. Although most of these can be explained teleologically in terms of traditional sex roles, the differences in hair distribution are difficult to explain. Adult males have more facial and body hair but they lose hair in the temporal region and are more likely to become bald. Pubic hair in the female has a flat upper border whilst it forms an upward-pointing triangle in the male.

## Hormonal changes at puberty

During childhood the feedback control systems of the pituitary gonadotrophins operate at the relatively low blood concentrations of the gonadal hormones. Thus a decrease in blood concentrations of oestrogens, for example, causes the hypothalamus to produce luteinising hormone releasing hormone (LHRH), which stimulates production of follicle stimulating hormone (FSH), and this in turn stimulates production of more oestrogen (Fig. 23.5). Rising concentrations of oestrogen in the blood inhibit the hypothalamic production of LHRH. A constant low concentration of oestrogen is maintained. At puberty some unknown process of maturation causes the hypothalamic cells to become much less sensitive to oestrogens, so that more oestrogen is needed to inhibit LHRH production: blood concentrations therefore rise. In the male a similar system operates in relation to the pituitary production of luteinising hormone (LH) which controls testosterone production from the testes. Pituitary hormones are usually released in spurts of secretion, and they show a circadian rhythm with more secretion occurring during the night. The release of LH in the male at puberty occurs at first in short bursts during the night only but later takes place during the day also. In the female in late puberty the hypothalamic receptors suddenly reverse their response to circulating oestrogens to initiate the sequence of events leading up to menstruation and the establishment of a regular pattern of ovulation and menstruation.

Puberty may be affected by the changes in secretion of a number of hormones involved in growth and metabolism although these are not primary factors. Thus relatively high levels of

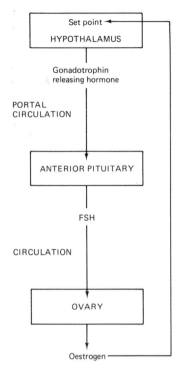

**Figure 23.5** The hormonal feedback loop at puberty – the set point shifts to a higher setting at puberty.

growth hormone during childhood continue on through puberty, with only slight peaks at birth and when puberty commences; cortisol secretion increases, but only to remain consistent with body size; insulin secretion is unchanged; and prolactin secretion remains constant in the male. Pathological alterations in secretion rates of these hormones may affect pubertal development. The pubertal growth spurt results from the increased secretion of testosterone in the male and of adrenal androgens in the female.

## Pubertal changes in girls

The occurrence of puberty in girls is chronologically some two years ahead of that in boys. There are two major changes in hormonal activity which appear to be independent: the change in hypothalamic sensitivity to oestrogens which causes the increase in secretion of FSH and oestrogens, and a change in the output of androgens from the adrenal cortex. No trigger for the rise in androgen secretion has been found and adrenocorticotrophic hormone secretion in general is unchanged. The secretion of androgens reaches a peak during puberty and then slowly declines to pre-pubertal levels by about age 60 in both males and females. Oestrogens cause the breasts to begin to develop, one of the earlier external signs of puberty, and stimulate development of the uterus and the vagina with its associated glands. Breast development occurs between 10.5 and 15 years in the average West European girl. Oestrogens also stimulate selectively the cartilage cells of the hips causing widening of the pelvis; there is also an increase in size of the pelvic outlet. The first menstrual period, or the menarche, occurs in W. European girls between 12.8 and 13.2 years. It results from a further change in the sensitivity of hypothalamic cells to oestrogens and the timing is not related to the development of the breasts or of the pubic hair. It always follows the peak in the increase in growth rate observed at puberty. The adrenal androgens (principally dehydroepiandrosterone, DHEA) are responsible in the female for the growth spurt of puberty, the development of body hair, and the final maturation of the skeleton, including the development of the skull and jaw bones. The growth spurt in girls begins at an age of about 10.5 years, peaks at about 12 years, and is complete at about 14.5 years. This earlier growth spurt means that orthodontic treatment can be initiated earlier in girls. At the peak of the growth spurt the annual increase in height may be as much as 6–11 cm/year, in comparison with the average of 5 cm/year before puberty. At this stage girls may be taller and stronger than their male contemporaries and their athletic performance often reaches a peak. Androgens are also responsible for the growth of the pubic hair, which appears relatively late in puberty (between 11 and 14 years) and only some 9 months before its appearance in the male. The axillary hair and the axillary apocrine sweat glands also develop late in puberty. The apocrine sweat glands are analogous to glands producing secretions acting as sexual signals in other animals: the function of the axillary hair may be to retain and concentrate the odours of these secretions. The final filling out of the breasts may be due to a small rise in prolactin secretion late in puberty. The hormonal changes of puberty in girls sometimes produce effects on the soft tissues of the mouth, particularly if the establishment of regular menstrual periods does not proceed smoothly. In these circumstances, the gingivae (or gums) are much more susceptible to inflammation and there may be excessive growth of interdental tissue to form an epulis.

## Pubertal changes in boys

In boys the first external sign of puberty is the enlargement of the testes at between 11.5 and 15 years. This is accompanied by changes in the external appearance of the scrotum and growth of the prostate gland and seminal vesicles. Then the penis enlarges (12–14 years) and the pubic hair appears (12–15 years). All these changes occur as a result of the change in the sensitivity of hypothalamic cells to testosterone, increased secretion of LH and FSH by the anterior pituitary gland, and a large increase in circulating testosterone levels. The nerve pathways involved in ejaculation are present from an early age, but until prostatic secretion is sufficient to provide a volume for ejaculation and sperm production has been initiated, ejaculation cannot actually take place. In fact, ejaculation usually occurs for the first time spontaneously, during sleep, about a year after the beginning of the enlargement of the penis, perhaps in response to the bursts of LH secretion occurring during sleep. The ejaculate contains few sperms at first but the numbers increase later.

The growth spurt of puberty in the male begins about 12.5 years of age, reaches a peak at about 14 years, and is complete by about 16 years. The spurt is greater than in the female, the annual increase in height being between 7 and 12 cm/year, in comparison with a prepubertal growth rate of around 5 cm/year. It usually begins about one year after the enlargement of the penis. The increased rate of growth is produced by testosterone and, to a much lesser extent, by the adrenal androgens, whose secretion shows a similar pattern of changes to those in the female. As a result of the pubertal growth spurt the average W. European male is some 13 cm taller than the female, although before puberty the sexes are very similar in height. The very large increase in testosterone secretion causes increased growth and development of some tissues preferentially. Thus the cartilage cells of the shoulders respond to testosterone, increasing the width of the shoulders and the size of the chest. The cells of the pelvic cartilages do not increase their rate of growth in response to testosterone, and so the main skeletal difference between males and females is established. The number calculated by multiplying the distance between the acromial processes by three, and subtracting the distance between the iliac crests, provides a good index of the male or female nature of the skeleton, male skeletons usually giving values above 75 cm. Growth of bone and muscle will not be complete unless growth hormone is also present.

The influence of testosterone on the growth of hair is interesting inasmuch as different sites seem to have differing sensitivity, either in terms of chronology of receptor development, or in terms of the priming exposure to testosterone necessary to establish a sufficient number of receptors. Pubic hair appears between 12 and 15 years, whilst axillary hair and the related axillary glands do not appear until about 14 years. Facial hair appears next, spreading in a fixed sequence from the corners of the upper lip across that lip, and then beginning on the cheeks and the midline of the lower lip, before eventually covering the sides and lower third of the face. After this the body hair develops to a much greater extent than that of the female. The amount of facial and body hair seems to be genetically controlled rather than directly dependent upon the concentration of testosterone in the blood. There is a change in the appearance of the breast in males – the areola around the nipple doubles in size (in the female it triples) – although it is unusual for breast enlargement (termed gynaecomastia in the male) to occur. The change in the voice of the male at puberty is a result of the growth spurt in the larynx; the increase in the size of the female larynx at puberty is less dramatic.

## Male-female differences after puberty

The changes which take place as a result of the growth spurt are in some instances similar in the two sexes, in others different. The critical factor is usually the presence or absence of testosterone. It seems appropriate at this point to make some comparisons of the less obvious secondary sexual characteristics. Adult height is partially determined by the pubertal growth spurt: some 30% of the variation between individuals results from the growth of the pubertal years. Height is governed largely by the length of the long bones: the epiphyseal closure which terminates their growth in length is finally controlled by the gonadal hormones, in addition to the on-going control by growth hormone and the thyroid hormones. The difference in height between males and females therefore results from two factors, the differing age at puberty, and the differing effect of their hormones on the time of closure of epiphyses. The growth rate of different parts of the body varies: the legs reach their peak rate before the trunk, and the trunk before the shoulders and the widening of the chest. Since before puberty the legs are elongating more rapidly than the trunk, and during puberty this relationship is reversed, the later puberty of the male results in a greater ratio of leg length to trunk length. Although the skull reaches its maximum size early in life (by 8 years it is 90% of its final size), there is a small increase in size at puberty, mainly by apposition, which increases the bone thickness by about one sixth, but also by development of the air sinuses. It is at this time that the frontal air sinuses and the supra-orbital ridges develop and the development is greater in males. The facial appearance also changes as a result of the elongation of the maxilla and, to a greater extent, the mandible. Facial muscles become more prominent. All these changes are more marked in the male – an important consideration in orthodontic assessment.

Muscle growth elsewhere is stimulated more by testosterone than by the adrenal androgens, one effect of the testosterone being to increase the amount of contractile protein incorporated by each cell, and thus to give a greater strength per gram of muscle. As a result of this the male can lift heavier weights and run faster. A similar effect is seen in cardiac muscle: the male heart is relatively larger and so systolic blood pressure is higher and the resting heart rate is lower. Because of this development of the heart, and the greater volume of the lungs in the larger thoracic cavity, the male can not only run faster, but can maintain muscular activity for longer periods. The increased blood levels of testosterone at puberty also cause the production of more red blood cells giving the blood a greater oxygen-carrying capacity in the male. The only tissue whose rate of growth slows at puberty is adipose tissue. In the male the decrease is so marked that fat is actually lost and the ratio of fat in the body decreases markedly but in the female the decrease is less and no fat is actually lost. The decrease in growth of adipose tissue varies in different parts of the body, being greatest in the limbs. The developmental pattern of the female confers a number of advantages over the male, chiefly in relation to a greater ability to survive in adverse conditions. Smaller body size is often associated with a greater ability to resist cold environments, and the extra insulation of a fat layer assists in this. The increased ratio of fat to carbohydrate gives the female a better long-term energy store and allows her to survive starvation more effectively. Again, the smaller heart size and lesser amount of muscle reduce the cost of survival. The male appears to be designed for short-term muscular activity suitable for fighting or hunting, and the female for long-term survival. The female adaptations also render her more efficient in the protection of developing embryos and able to survive the cost of childbearing. Male adaptations are for personal or family survival, female for survival of the species.

Much study is now being devoted to the process of aging and discussion continues as to whether body tissues have a programme of aging as well as one of growing. The main change which can definitively be described is an increase in the cross-linking and rigidity of collagen fibres. There is an increase in calcium content of some calcified tissues like bone, cementum and dentine, which can lead to brittleness. Bone may also, particularly in post-menopausal women, show porosity as a result of resorption: this process is said to be reversed by high doses of fluoride.

## Further reading

Austin, C.R. and Short, R.V. (eds) *Embryonic and Fetal Development,* Vol. 2 of *Reproduction in Mammals,* 2nd edn, Cambridge University Press, Cambridge, 1982

Sinclair, D. *Human Growth after Birth,* 4th edn, Oxford University Press, Oxford, 1985

Tanner, J.M. *Foetus into Man,* Open Books Publishing Ltd, London, 1978

Tepperman, J. *Metabolic and Endocrine Physiology,* 5th edn, Year Book Medical Publishers, Chicago, 1987

# 24

# The control of metabolic processes

## The control of general metabolism

The control of metabolism can be seen either in relation to the body economy as a whole, considering how the adult human maintains a moderately steady metabolic balance over relatively long periods of time, or at a more detailed level, considering how the hour-by-hour variations in metabolite availability are dealt with, and how the metabolic pattern changes in response to decreased supplies of substrates to the tissues and increased demands from them. Growth and development have been dealt with earlier. The three main classes of nutrient, carbohydrate, fats, and proteins, must be considered. Of the inorganic ions, sodium and potassium were discussed on p. 240, iron on p. 62, and calcium is given a separate treatment at the end of this chapter.

Most adult human subjects maintain, or strive to maintain, a constant weight. This implies that there is a rough balance between the intake of food, on the one hand, and the output of energy as work or heat, and the elimination of waste products, on the other (Fig. 24.1). In Western societies, with no real lack of food and yet little usage of energy in physical labour, changes in weight often represent increase or decrease in the amount of stored fat. More rarely, changes in muscle mass may affect overall weight gain or loss. Even though with increasing wealth and cultural sophistication dietary patterns shift from bulk carbohydrate diets to the more carnivorous diets, high in protein and fat, changes in body weight usually reflect carbohydrate intake and metabolism. Although the control of metabolic processes is not limited to the control of carbohydrate metabolism, or, even more narrowly, to the control of blood glucose concentrations, these subjects are of major importance.

## Food intake

It is not known what physical or mental signals cause people to eat. The sensation of hunger may be related to blood glucose concentrations sensed in the hypothalamus (and possibly in the liver also), to the degree of stretch of the walls of the stomach or other parts of the gastro-intestinal tract, or simply to the normal pattern of eating.

As one of the primitive drives in all animals, hunger is a sensation associated with limbic brain

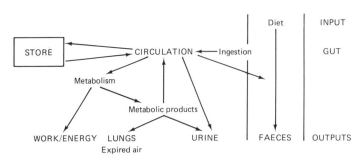

**Figure 24.1** The balance of intake and output as related to food and energy.

activity. Most manifestations of limbic brain activity are channelled through the hypothalamus and so this part of the brain is also associated with hunger drives. The hypothalamus has two centres related to feeding: a satiety centre in the ventromedial nucleus and a hunger centre laterally. The hunger centre is thought to be active except when its activity is inhibited by the satiety centre. This contains 'glucostat' cells whose activity is controlled by the amount of glucose available in the blood. When the glucose level in the blood is high, the cells show increased activity, the individual feels sated, or 'full', and the activity of the hunger centre is reduced. The reverse occurs when blood glucose levels are low. The glucostat cells are responsive to insulin, so the hormone increases their uptake of glucose (see p. 301).

The normal rhythm of food intake may be a genuine expression of human body rhythms, or it may be a manifestation of habit.

The hypothalamic receptors are well situated to monitor the glucose supply to the brain, which in normal circumstances relies on glucose as its energy source. However, it would be advantageous to stimulate eating if the supply of other nutriments to the rest of the body is deficient. The level of fat stores in the body may be monitored in some way, and this may control the desire to eat. The hypothalamus may monitor the levels of free fatty acids or certain prostaglandins in the blood as a way of determining the adequacy of the fat stores. Experimental investigations suggest that any control exerted as a result of changes in these blood constituents is very slow in operation.

The desire to eat is affected by growth, exercise, external temperature, and illness. Appetite is an emotional response to food. Lack of appetite (anorexia) or even distaste for food, may effectively suppress the need to eat. Equally, excessive appetite may lead to unnecessary eating in the absence of hunger, a condition known as hyperphagia. Appetite may be hypothalamic or limbic in origin, and may be pharmacologically stimulated or depressed.

Actual patterns of eating vary, although there is a tendency for many smaller meals to replace the three main meals of earlier custom. Thus, if every food intake is counted as a meal, many individuals today eat six meals, and some as many as twelve, in the course of a day. Small meals tend to be meals of appetite rather than of hunger: and hence are often composed of the sweeter carbohydrates – the kind of frequent sugar intake linked with a high incidence of dental caries.

## Food utilisation

All the foodstuffs may be oxidised to yield energy. This will appear as work, or heat, or stored energy. The maintenance of body temperature depends upon the heat produced in metabolism. If man were a totally efficient machine, converting all the energy produced into work instead of only some 25% of it, he would then die of hypothermia.

Complete oxidation of carbohydrate yields 17.22 kJ/g (4.1 kcal/g) and of fat 39.06 kJ/g (9.3 kcal/g). Since protein is not completely oxidised in the body, it does not yield as much energy as fat, being roughly the same as carbohydrate at 17.22 kJ/g (4.1 kcal/g). The energy available from diets may be calculated from these figures, which are the basis of 'calorie charts'. There is a small expenditure of energy involved in the gastro-intestinal and metabolic processing of these foodstuffs: some 4% of the energy from fat, some 6% of the energy from carbohydrate, but the large proportion of 30% of the energy from protein.

The energy produced in the body may be measured directly in a calorimeter which for this purpose consists of a sealed, insulated room through which pipes carry water whose inlet and outlet temperatures can be measured. More usually an indirect method is used. Since the energy is derived from oxidative processes, the consumption of oxygen is proportional to the energy produced. Each mole of oxygen produces an average of 453.5 kJ, the exact value depending upon the nature of the food catabolised. An estimate of this may be gained from the respiratory quotient – the ratio of carbon dioxide produced to oxygen used, being 1 for carbohydrate and 0.7 for fat.

The basal metabolic rate is the amount of energy produced by a subject in complete mental and physical rest in a room at a comfortable temperature, and some 12–14 h after the last meal. For an average individual this is around 7.0 MJ/day, but it is directly related to body size expressed in terms of surface area. Hence it is usual to quote the basal metabolic rate (BMR) as $J/m^2$ of body surface/h. An accepted normal value would be 168 $kJ/m^2/h$. Tables are available of appropriate values for male and female subjects of different ages and sizes. Several hormones influence the basal metabolic rate. The thyroid hormones, thyroxine and tri-iodothyronine, increase the oxygen consumption of most tissues, with the exception of adult brain tissue. Soon after birth the blood-brain barrier is sufficiently developed to exclude the thyroid hormones. Despite this, the thyroid hormones have been detected in the adult brain and they do increase overt brain activity. They also shorten reflex times peripherally. The increase in oxygen usage and energy production due to thyroid hormones results in many instances from increased activity of the $Na^+$-$K^+$-ATPase membrane pumps: increased transport of sodium ions results in greater energy usage and calorigenesis. This mechanism has been demonstrated in the salivary gland response to thyroxine. In children deficient in thyroid hormones, the incidence of dental caries is sometimes increased: this may be related to the decrease in saliva production resulting from decreased tissue metabolism. The extent of enhancement of the basal metabolic rate by thyroid hormones in tissues generally depends upon the initial rate: if the initial rate is low, it is considerably increased, but if the rate is high, very little increase is observed. The catecholamines, adrenaline and noradrenaline, also increase the metabolic rate, but only in conjunction with thyroid hormones.

The extra energy production associated with thyroid hormones means that more substrate is used. In the absence of increased carbohydrate intake, thyroid hormones stimulate the breakdown of protein and lipid stores. Urinary nitrogen excretion is increased. Liver glycogen is broken down to provide glucose. The increased metabolism results in sensations of hunger. Body temperature rises slightly, initiating heat-losing activities like cutaneous vasodilatation and sweating. Heart rate and pulse pressure increase as a result of the combined activities of thyroid hormones and catecholamines.

## Metabolic substances and storage

Now that overall metabolism has been considered, the individual nutrients and their place in metabolic balance may be described. Proteins are ingested and converted in the gut into aminoacids which are absorbed and then, together with the aminoacids from the 70–80 g of body protein which are broken down each day, constitute the aminoacid pool from which new proteins are synthesised. The aminoacid pool is maintained from these exogenous and endogenous sources, but it is also in exchange with the common metabolic pool. Aminoacids may undergo transamination, forming a ketoacid and giving up an amino group to another ketoacid, such as α-ketoglutarate, oxaloacetate or pyruvate (forming glutamate, aspartate and alanine respectively). In the liver, oxidative deamination occurs by a process of iminoacid formation and hydrolysis to ketoacids with a release of ammonia. This may be taken up by other aminoacids – such as glutamic acid, which then becomes glutamine – or converted to urea. Ammonia itself is a toxic substance and must be converted to urea for excretion. This process takes place only in the liver. In the normal adult the loss of nitrogen as urea in the urine should balance the intake of nitrogen as protein. On the recommended dietary intake of around 56 g/day, the urine should contain some 0.32 moles of urea per day. Young children and athletes in training are in positive nitrogen balance, excreting less nitrogen than they take in the diet. This is due to the activity of the hormones concerned with growth, the anabolic hormones. If the catabolic hormones of the adrenal cortex are present in excess, or if there is a deficiency of insulin, a negative nitrogen balance will be established. During starvation, when protein is catabolised to provide energy, or during immobilisation for long periods causing muscle wasting, more urea nitrogen is excreted than protein nitrogen is ingested. However, protein is used as an energy source only when carbohydrate and fat stores are depleted. Protein starvation may result in a lack of essential aminoacids. There is evidence that protein-starved children are more susceptible to dental caries, although the variations in protein intake in Western diets are unlikely to affect children's caries experience.

Lipids provide the most energy in relation to weight. Since fat is used primarily as an energy source, the intake of about 100 g/day (in Western diets) must be balanced against the breakdown for energy production. Some fat is stored but most of the fat stores of the body actually derive from glucose rather than fatty acids. Lipid is an exceedingly compact form of energy storage, not only because it provides more k/g, but also because, being hydrophobic, it can be stored without water. Storage of energy as fat is nine times as economical as storage in the form of glycogen – a 70 kg man would have to be virtually double this weight if all his energy substrate were stored as glycogen. The total fat content of the body, including all lipid material such as cell membranes, amounts to between 12 and 18% in men and between 18 and 24% in women; stored fat amounts to about 10% of body weight. Fats are stored in special cells known as fat cells or adipocytes. These form layers subcutaneously or around vital organs such as the kidneys and the heart. The cells characteristically contain a single large drop of fat; the nucleus and cytoplasm are squeezed into a thin envelope around the fat globule. Fat is mobilised as fatty acids and glycerol by a lipase which is activated by the catecholamines via a pathway involving cAMP. Glucagon and vasopressin can activate the same system. Growth hormone, glucocorticoids, and thyroid hormones also activate lipolysis, but by different routes. The metabolism of fats in many cells is disturbed if insufficient glucose is present, so that the fats are converted into ketone bodies and an acidosis develops. This is seen in starvation, in insulin deficiency, and in subjects ingesting a high fat, low carbohydrate diet. Ketone production can be recognised by the presence of acetone in the expired air.

Carbohydrate is quantitatively the greatest energy source in most diets, about half the energy requirement from the diet coming in the 300 g or so of carbohydrate eaten each day. Under normal conditions a balance will be maintained between carbohydrate intake and energy output, although over short periods of observation the picture is confused by the ready storage of excess carbohydrate and the mobilisation of stores. Since food intake occurs at intervals, and energy expenditure is continuous, a storage system is important. The form of carbohydrate absorbed and present in the bloodstream is mainly glucose. This may be stored as glycogen, principally in the liver, but also in muscle and, to a lesser extent, in other tissues. It may also be converted to fatty acids and stored as fat in the fat depots. If carbohydrate intake over longer periods of time is not balanced by energy production, changes in body weight will take place.

## Obesity

Subjects whose weight is substantially above that considered normal for their age and height are termed obese. The causes of obesity are debatable but certain general statements can be made. Body build is partly determined genetically: the general classification of individuals into ectomorphs, who are thin and bony, mesomorphs with a relative predominance of muscle tissue, and endomorphs, who are relatively large and rounded, usually with a higher proportion of adipose tissue, describes characteristics which tend to be inherited. Obesity, however, is outside this classification and could affect individuals in any of the categories, though with a preference for endomorphs. The amount of adipose tissue in the body, and therefore the potential for obesity, is probably determined early in life; feeding in infancy may influence the development of the fat cells and no new cells will be produced after some fairly early age. Beyond this point the factors influencing long-term weight increase in the human are three in number. First, and the major factor in initiating obese changes, is hyperphagia, over-eating, or more strictly, the intake and absorption of more energy substrate than the individual is using, resulting in excessive storage as fat. Second is the possibility of a metabolic factor. Obese subjects have higher levels of blood glucose and free fatty acids, and appear to be less responsive to insulin. These may be metabolic differences from the normal, or may be adaptations resulting from obesity. Thirdly, obese subjects may be eating in response to psychological or social pressures rather than hunger drives – obesity may be primarily a psychological problem.

Weight changes can result from causes other than an increase in fat stores. Most short-term changes in weight are due to loss or gain of water – many so-called slimming aids do not affect the problem of fat accumulation but simply cause a loss of fluid. This is usually rapidly replenished. Secondly, it is possible for the athlete or sportsperson in training to build up muscle mass, with a consequent weight gain (*see* p. 319).

## Control of metabolism at a tissue level

Each of the three of the major groups of biological molecules can be oxidised to provide energy: carbohydrate most conveniently, fat most compactly, and protein most reluctantly, since protein provides the structural units of the body.

All tissues need energy in order to live: the energy must be provided by materials within the cell, or materials transported to the cell in the blood. In the bloodstream materials travel in their simplest form – monosaccharides (mainly glucose), aminoacids, fatty acids, ketone bodies, and glycerol. These are the forms most readily used to provide energy, and so the blood serves as the main source of energy-producing compounds. Because its capacity is limited it cannot function for more than a few minutes without its content being replenished from digested or stored material. The human being is omnivorous, but behaves more like the carnivore than the herbivore in eating at intervals rather than continuously. The problem of providing a continuous flow of energy substrates is solved in part, as with herbivores, by storing undigested food in the stomach and then digesting and absorbing the meal over several hours. The carnivore's practice of eating large meals and diverting the substrates to stores is the other part of the solution. Stored materials are used more particularly during the sleeping period – up to 75% of liver glycogen may be used in this way.

Control of metabolism can be exerted at a molecular level, by the presence of particular enzyme pathways and by the availability of substrates, or at a physiological level, by the action of nerves or hormones on cells and on particular metabolic pathways.

Within particular tissues the combinations of enzymes result in preferred pathways of metabolism. Muscle, for example, metabolises glucose, or glycogen, or fatty acids, or ketone bodies. The choice is dictated by the speed of the muscle response and by the availability of the substrate. Brain, nervous tissue, red blood cells, the kidney medulla, and the testes normally use only glucose as a substrate. The last three, which have small energy requirements and poor complements of oxidative enzymes, use only anaerobic oxidation of glucose and produce lactate as an end product. The brain can utilise ketone bodies – after some 4 h fasting some 25% of its energy needs are met from this source, and in prolonged starvation that proportion rises to around 75%. Nonetheless, a sudden interruption of the glucose supply to the brain causes deterioration in function within a minute, and irreversible damage in a few minutes.

The unit of energy within the body is the high energy phosphate bond of ATP. Complete oxidation of glucose (aerobic oxidation) yields 38 mol of

ATP for every mol of glucose used. Anaerobic oxidation is much less efficient, yielding only 2 mol of ATP/mol of glucose, or, if stored carbohydrate is used, 3 mol/mol from glycogen. A six-carbon fatty acid completely oxidised yields 44 mol ATP/mol.

## Carbohydrate metabolism

In the resting state after a meal most tissues will use glucose from the blood as their main energy substrate. Glucose enters cells by diffusion from the blood. In muscle (of all three types) and adipose tissue this diffusion is promoted by a membrane carrier and the uptake of glucose is increased in the presence of insulin. Once inside the cells, glucose enters the glycolytic pathway and is converted to glucose-6-phosphate by hexokinase (Fig. 24.2). Glucose-6-phosphate is converted eventually to pyruvate, the key enzyme in the sequence being phosphofructokinase. This enzyme is inhibited by ATP, an end product of the reaction, and so the rate of the reactions depends upon the energy used. The rate is also limited by the substrate availability, since hexose monophosphates stimulate phosphofructokinase activity. If inhibition of phosphofructokinase slows down the sequence of reactions, glucose-6-phosphate begins to accumulate and inhibits hexokinase activity. However, accumulation of the glucose-6-phosphate also allows it to be converted to the storage carbohydrate glycogen through a series of reactions with a rate-limiting step catalysed by glucosyltransferase. This enzyme exists in two forms, the D form which is inactive *in vivo* unless very high concentrations of glucose-6-phosphate are present, and the I form, which is independent of glucose-6-phosphate stimulation. The inactive D form may be converted to the active I form by a phosphatase whose activity is inhibited by glycogen – thus providing a negative feedback control for glycogen synthesis.

The two metabolic pathways for glucose in skeletal muscle, then, allow the cells to use it for energy by glycolysis and to store it as glycogen for future use.

In liver cells the glycolytic pathway for glucose is relatively unimportant as an energy source. An enzyme found only in the liver, glucokinase, has a very high affinity for glucose, converting it to glucose-6-phosphate independently of any hexokinase activity. The glucose-6-phosphate is converted to glycogen in a manner similar to that seen in skeletal muscle. Glucose-6-phosphate also serves as a step in the pathway to pyruvate, which is converted into acetyl CoA, and does not pass into the citric acid cycle to be metabolised to carbon dioxide and water (Fig. 24.2). Acetyl CoA can then be converted into fatty acid, esterified to triglyceride, and secreted to the blood as VLDL (very low

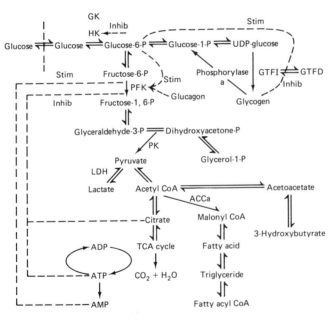

**Figure 24.2** The metabolic pathways for glucose, the points of control, and the crossovers to fat synthesis and ketone production. Enzymes at key points are abbreviated as follows: HK hexokinase, GK glucokinase, G6P glucose-6-phosphatase, GTFI glucosyltransferase I, GTFD glucosyltransferase D, PFK phosphofructokinase, LDH lactic dehydrogenase, PK pyruvate kinase, ACCa acetyl CoA carboxylase a.

density lipoproteins). In the blood the triglyceride is carried to adipose tissue where it is stored as fat. The liver, then, uses blood glucose to produce stores of carbohydrate within itself and stores of fat in the adipose tissue.

In the cells of adipose tissue glucose is taken up from the blood and passes into the glycolytic cycle in the same way as in the liver, forming acetyl CoA, then fatty acyl CoA, and, with the glycerol also formed from glucose, triglycerides. Since in these cells ATP is formed to a much lesser extent than in muscle, phosphofructokinase is not inhibited and the reactions are driven by the concentration of glucose in the cells.

## Lipid metabolism

As fats are not water-soluble they must travel in the blood either bound to protein as VLDL or as chylomicra. The normal 0–4.5 g/l of triglycerides in plasma can rise as high as 20 g/l after a fatty meal. The non-esterified fatty acids amount to about 2.0-4.5 g/l normally and the ketone bodies (acetoacetic acid and β-hydroxybutyric acid) amount to less than 10 mg/l of plasma. The free fatty acids are usually bound to albumin. Lipoprotein lipase is essential to permit entry of triglycerides to cells: it is found in the endothelial lining of blood vessels in liver, adipose tissue, and skeletal muscle. It splits triglycerides into fatty acids and glycerol which can be taken up by the cells. In adipose tissue these are then synthesised into new storage triglyceride.

In a similar fashion in the liver the triglycerides of the chylomicra arriving from the digestive tract via the lymphatic duct and the bloodstream, are converted into free fatty acids and glycerol. The liver itself does not store triglyceride to any extent in the human, but resynthesises it and exports it as VLDL back to the adipose tissue. Free fatty acids themselves are taken up by the liver and pass into two possible pathways – the esterification pathway to triglyceride, or the oxidative pathway to ketone bodies. The esterification pathway is used preferentially, but at high concentrations of fatty acid in the blood, its capacity is exceeded and the oxidative pathway takes up the excess. The liver is thus a key site for lipid metabolism: it takes up some 10–15% of triglyceride and up to 50% of the free fatty acid from the digestive tract, it forms triglyceride from glucose and from free fatty acids, and it exports triglyceride, free fatty acid, glycerol, and ketone bodies in the blood.

Muscle uses fatty acids for sustained energy production. In moderately intense work a mixture of glucose and fatty acids is used for energy production, but as the work is prolonged the respiratory quotient falls, showing that more fat is being used. The more intense the activity, the more glucose is used and the respiratory quotient approaches unity.

The availability of free fatty acids or ketone bodies increases production of acetyl CoA which both inhibits the oxidation of pyruvate, and forms citrate which inhibits phosphofructokinase – thus stopping glycolysis and glucose usage.

## Metabolism of protein

Aminoacids are taken up by the liver and by skeletal muscle cells: they are used principally to make new proteins or to replace proteins which have been broken down. In the liver there is some conversion of aminoacids into ketoacids (leucine, isoleucine, phenylalanine and tyrosine are ketogenic) which the liver uses as its own source of energy. The liver is also capable of converting aminoacids into glucose. Both these processes are dependent upon the amount of substrate in the blood.

The cells of adipose tissue take up aminoacids and form fatty acids from them by deamination and then transference into the acetyl CoA pathway. This provides the adipocytes with another source of triglyceride.

## Metabolism of stored substrate: glycogen metabolism

The total body store of glycogen amounts to about 70 g in the liver, and about 120 g in the whole skeletal muscle mass. Other cells contain small amounts, but this is not significant in the whole body total. Glycogen is purely a storage medium for carbohydrate. It may be broken down to glucose to maintain circulating glucose levels, or to glucose-6-phosphate and oxidised. Circulating glucose by itself could maintain activity for only a short period of time, but the liver by providing 180–300 g from its reserve can provide a supply of glucose for 12–24 h. The daily intake of glucose is more than sufficient to satisfy the glucose-dependent tissues – some 120 g/day for the brain and about 40 g/day for the anaerobic tissues. Overnight, liver glycogen is almost exhausted in maintaining the blood glucose levels. Skeletal muscle uses its own glycogen as a fuel reserve, but this reserve is not available to other tissues because muscle cells have no glucose-6-phosphatase to allow the conversion back to glucose and the possibility of export from the cells.

White muscle (fast muscle) has a poor blood supply, few mitochondria, low activity in its tricarboxylic acid cycle, and a high capacity for glycogen degradation. The cells store relatively large amounts of glycogen. White muscle is used for violent, short-term effort: glycogen utilisation can occur very rapidly, but the poor blood supply and the low activity of its hexokinase mean that anaerobic glycolysis is inadequate as an energy supply. In most muscles the size of the glycogen store is sufficient for only about 10 s worth of activity.

Red muscle, or slow muscle, has a good blood supply, many mitochondria, and a high capacity for the aerobic oxidation of carbohydrate or fat. Sufficient glucose or fat is provided from the blood for moderate mechanical activity; but if the need for energy exceeds that capacity, glycogen can be oxidised over a relatively long period. If the energy requirement exceeds the oxygen supply, glycogen is broken down anaerobically to lactate and ATP. The rate of glycogenolysis then becomes very high, since ATP is produced anaerobically at only one tenth of the aerobic rate. Glycogen reserves are therefore depleted rapidly and the extra energy production cannot be maintained.

Breakdown of glycogen is controlled by the enzyme phosphorylase. This is present in two forms, a and b. Phosphorylase b is normally of very low activity, but can be activated by cAMP and by phosphate. On the other hand, its activity is inhibited by ATP and by glucose-6-phosphate. None of these compounds affect phosphorylase a to any extent. Phosphorylase b is converted to phosphorylase a by phosphorylase b kinase. This in its turn is an active form of an inactive molecule, the conversion being due to a cAMP-dependent protein kinase. Phosphorylase b can also be activated by ATP and calcium ions (Fig. 24.3).

for sustained muscle activity – indeed it is the main energy substrate for cardiac muscle.

Mobilisation of fat can raise the concentration of free fatty acids in the blood to some five times the resting level, providing a wide variation of substrate availability. In contrast, blood glucose levels have a relatively restricted range of about 30% on either side of the resting level. Normally a fall of 30% in blood glucose levels is the maximum that can be tolerated, because of the dependence of the brain on this particular energy source. A third circulating energy source, however, is also derived from fat. The conversion of the mobilised fatty acids to ketone bodies in the liver can raise their concentration more than 20 times: this then provides an important energy source for the brain and for skeletal muscle. Ketone bodies inhibit glucose usage. They may also inhibit lipolysis in a feedback response.

### The generation of glucose from other substrates-gluconeogenesis

The relatively small size of the glycogen store, and the dependence of the brain and a few other tissues on glucose for energy production, means that further sources of glucose may be necessary, particularly if the liver glycogen store is depleted. Glucose is not available from skeletal muscle glycogen for other tissues. The liver, however, has the capacity to produce glucose from a number of

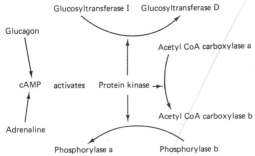

**Figure 24.3** The activation of protein kinase by cyclic 3,5-adenosine monophosphate affects the activation of glycogen-forming and glycogen-splitting enzymes in opposite directions, and also causes inactivation of acetyl CoA carboxylase.

### Metabolism of stored substrate: fat metabolism

The triglycerides in the adipocytes of adipose tissue provide an alternative energy store. These cells contain a hormone-sensitive lipase which converts triglyceride into glycerol and free fatty acids which can then be secreted into the bloodstream. The total store of about 17 kg in a 70 kg man is sufficient to support metabolism for about 40 days in the absence of any food intake. Not only does fat provide a compact energy store, but it is a suitable substrate

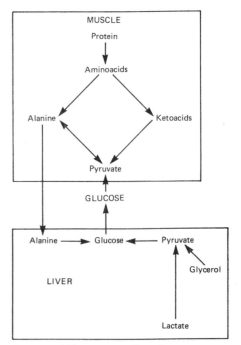

**Figure 24.4** Gluconeogenetic pathways.

other substrates – lactate from anaerobic glycolysis in the tissues, glycerol from the breakdown of triglycerides, and aminoacids, particularly alanine and glutamine, from the breakdown of skeletal muscle proteins (Fig. 24.4). Fatty acids cannot be converted to glucose because the conversion of pyruvate to acetyl CoA can proceed in only the one direction.

The process of gluconeogenesis is controlled at the molecular level by substrate availability – the presence of lactate, glycerol and aminoacids in the liver cells. The key points in the metabolic pathways are at the oxidation of pyruvate and at the phosphofructokinase step. Gluconeogenesis is stimulated if oxidation of pyruvate is inhibited and if a fall in fructose-2,6-biphosphate concentrations drives the conversion of fructose-1,6-biphosphate to fructose-6-phosphate.

## Hormonal control of metabolic pathways

There are six hormones which directly affect general metabolism in the adult: thyroid hormones, insulin, glucagon, adrenaline, glucocorticoids and growth hormone. The effects of thyroid hormones have already been described. Of the remainder only insulin is normally anabolic in the adult, promoting synthesis of molecules. Growth hormone in the absence of sufficient insulin is catabolic, particularly when levels of glucocorticoids are raised. The other four hormones are catabolic (*Table 24.1*). Other hormones do affect metabolism but their effects are more specialised and are described elsewhere.

### Insulin

The effect of insulin in promoting the uptake of glucose into the cells of insulin-sensitive tissues has already been described. Tissues which are not insulin-sensitive include intestine and kidney, the red blood cells and the brain. The blood-brain barrier is not permeable to insulin: however, the hypothalamus and nearby areas, outside the barrier, are insulin-sensitive. In muscle cells insulin increases glucose uptake: the increased availability of glucose results in enhanced glucose usage and also synthesis of glycogen, because of the concomitant stimulation of glucosyltransferase activity, due to the conversion of the inactive D form of glucosyltransferase to the active I form. This ability of insulin to activate glucosyltransferase can permit a rapid re-starting of glycogen synthesis when the stimuli to glycogenolysis in muscle are removed. In liver cells insulin does not actually increase the permeability to glucose, but it does increase the gradient for glucose diffusion by stimulating glucokinase activity. The parallel decrease in glucose-6-phosphatase activity in these cells causes glycogen synthesis to be the preferential pathway for glucose metabolism. This switch is aided by the conversion of glucosyltransferase to the I form and its activation.

The uptake of glucose into fat cells is stimulated, increasing glycolysis, but also increasing the rate of triglyceride synthesis by stimulating the conversion of acetyl CoA into long chain acyl CoA. Similarly, fatty acid synthesis and esterification to triglyceride is stimulated in the liver cells. Triglycerides are more readily taken up from the plasma into the fat cells because insulin stimulates lipoprotein lipase activity. Simultaneously insulin lowers cAMP in the adipocytes and therefore decreases the activity of the intracellular triglyceride lipase.

In muscle cells the enhanced uptake of aminoacids and the stimulation of the conversion of tRNA-bound aminoacids into proteins results in protein synthesis.

Table 24.1 The metabolic effects of hormones. Some of these effects may be indirect. A gap indicates no effect or no consistent direct effect

| Effect | Insulin | Adrenaline | Glucagon | Glucocorticoids | Growth hormone |
|---|---|---|---|---|---|
| General | Anabolic | Catabolic | Catabolic | Catabolic | Catabolic in absence of insulin, otherwise anabolic antagonised |
| Entry of glucose into cells | Increased | | | | |
| Glycogen synthesis | Increased | Dec but inc from lactate | Dec (liver) | Inc (liver) | |
| Glycogenolysis | Decreased | Inc | Inc (liver) | Inc (liver) | Inc (liver) |
| Gluconeogenesis | Decreased | Inc | Inc | Inc | Inc |
| Aminoacid uptake | Increased (muscle) | | Inc (liver) | Inc (liver) | |
| Protein synthesis | Increased (muscle) | | Dec | Inc (liver) Dec (elsewhere) | Inc (with insulin) Dec (low insulin, high cortisol) |
| Lipogenesis | Increased | Dec | Dec | Dec | |
| Lipolysis | Decreased | Inc | Inc | ?Inc | Inc |
| Ketogenesis | Decreased | Inc | Inc | ?Inc | Inc |

Gluconeogenesis is depressed because insulin lowers the availability of aminoacids and glycerol from the plasma.

### Growth hormone

When insulin levels are low and cortisol is circulating, growth hormone stimulates the breakdown of stored triglycerides by triglyceride lipase in adipose tissue and increases the amount of cAMP in these cells. The resultant increase in plasma free fatty acids stimulates liver metabolism, producing triglycerides and ketone bodies. Muscle cells respond to the increase in plasma free fatty acids and ketone bodies by switching to these as fuel and decreasing glucose usage. The brain can use ketone bodies in place of glucose and their presence inhibits brain glucose usage. Ketone bodies also stimulate gluconeogenesis in the kidney cortex. Growth hormone has a direct anabolic effect on muscle cells, stimulating RNA and protein synthesis even in the absence of insulin.

### Adrenaline and glucagon

The main actions of both adrenaline and glucagon on the target cells are to activate adenyl cyclase and increase cAMP concentrations. In normal metabolism glucagon maintains the fuel supply when no new fuel is reaching the blood from the digestive tract. Adrenaline is the hormone of exercise and stress, which increases the fuel supply to match the extra metabolic need. Thus glucagon mobilises the depot storage fuels – liver glycogen and fat in adipose tissue – whilst adrenaline also enhances the use of the intracellular glycogen stores in muscle. The increase in cAMP activates phosphorylase, and hence glycogenolysis, in both liver and muscle. In adipose tissue it activates triglyceride lipase to increase the export of free fatty acids to the plasma. This also occurs in slow red muscle where acyl CoA can be used as fuel. Gluconeogenesis is stimulated in the liver by the raised plasma concentrations of free fatty acids and by the increased cAMP. Entry of aminoacids into the liver cells is stimulated by glucagon.

### Glucocorticoids

The glucocorticoids promote protein synthesis in liver cells, affecting particularly the enzymes involved in gluconeogenesis. In muscle, on the other hand, protein synthesis is inhibited, so that aminoacids are lost to the plasma. Lipolysis is stimulated and glucose utilisation inhibited by cortisol. Glucocorticoids stimulate growth hormone production. This may be important after trauma in preparing the body to begin synthesis of repair proteins when the stress situation has been resolved.

## Changes in hormone secretion and general metabolism in normal and abnormal situations

Metabolic activity changes according to the supply of metabolic substrates from foodstuffs and according to the demands of exercise, stress or trauma. These last are dealt with in Chap. 26 and 27, but the situation of starvation, which is associated with similar hormonal and metabolic changes, is included here. The metabolic modifications resulting from the normal cycle of feeding and fasting will now be examined.

### Food intake and the control of metabolism when the food is in the gut

The intake of food is followed by a phase in which the availability of metabolic substrates in the blood exceeds the demands of the tissues. The hormonal responses immediately before and during food intake should therefore prepare the body for a phase in which excess metabolites can be converted into storage forms.

The psychic stimuli to the digestive tract are mediated through the vagus nerve: this nerve also innervates the β cells of the pancreatic islets and initiates insulin secretion – a first step in the transition to the metabolite storage phase. The presence of food in the stomach and intestine is associated with the secretion of gastro-intestinal hormones (p. 278, 279). The intake of glucose stimulates secretion of gastric inhibitory peptide (GIP), which passes to the pancreas causing the B cells to secrete insulin (Fig. 24.5). The presence of glucose (and fatty acids also) in the ileum stimulates the release of enteroglucagon: this too increases insulin release. Aminoacids stimulate the release of cholecystokinin-pancreozymin (CCK-Pz) which stimulates both insulin and glucagon secretion in the pancreas. Both gastrin and secretin may potentiate the response of the insulin- and glucagon-secreting cells. Most stimuli for insulin release also stimulate somatostatin production.

Absorption itself does not seem to be under hormonal control although somatostatin is said to slow or block absorption of glucose. This could be a negative feedback control to limit the rate of glucose absorption in relation to the increase in blood-concentrations. Thyroid hormones cause more effective absorption of glucose and other nutrients, probably because of a general stimulation of metabolic processes. This incidentally helps to provide more substrate for those metabolic processes. Thyroid hormones are sometimes described therefore as raising blood glucose concentrations; but they also increase the motility of the gut, reducing the transit time and hence the time available for absorption, and they increase usage of

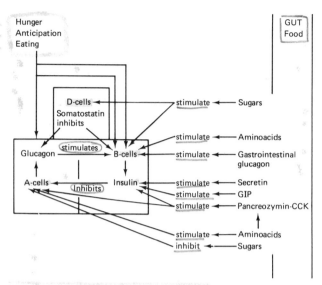

**Figure 24.5** Pancreatic hormone interactions.

glucose so that blood glucose concentrations may actually be reduced. Insulin does not affect the uptake of glucose by intestinal cells.

## Effects of absorption of food

Absorption of foodstuffs moves the metabolism into the storage phase. This is dominated by insulin, although glucagon may play a subsidiary role in protein absorption.

Absorption of glucose into the bloodstream raises the blood glucose concentration from 4.5 to 10 mmol/l and possibly higher. If the amount of glucose is excessive, or if the mechanisms for lowering blood glucose levels are ineffective, the blood concentration exceeds the renal tubular maximum rate of reabsorption and glucose appears in the urine. The glucose tolerance test consists of the administration of a known dose of glucose and a monitoring of blood glucose levels over the following two hours.

Once glucose has been absorbed into the bloodstream it stimulates insulin production by the B cells of the pancreas. The glucose will be taken up by many tissues because the concentration gradient into the cells is increased, but uptake will be greatest in those cells with membrane receptor sites for insulin – particularly muscle and adipose tissue. The liver takes up about half of the glucose absorbed at any one time and converts it into glycogen and triglyceride fats. Muscle and fat between them take up about 15%. In muscle the glucose is converted into glycogen, and protein synthesis is enhanced. In fat cells there is inhibition of lipolysis and uptake of the fats synthesised in the liver. If the meal consists solely of carbohydrate, secretion of glucagon is inhibited and its actions, which raise blood glucose concentrations, are prevented. The liver not only stores glucose in this immediately post-absorptive stage, but also ceases to export glucose.

Ingestion of a protein meal presents different problems. The gastro-intestinal hormones stimulate release of both insulin and glucagon. The levels of circulating glucagon, in particular, are increased, since the lack of stimulation from carbohydrate favours glucagon rather than insulin secretion. The aminoacids absorbed pass to the liver to be retained as part of the aminoacid pool stored there. Aminoacids with branched chains – valine, leucine and isoleucine – are taken up less readily and remain in the circulation so that they are higher in concentration in the mixture of aminoacids arriving at muscle cells. Insulin promotes protein synthesis in muscle: the consequent turnover of muscle protein results in the release of alanine and glutamine from the muscle cells. Alanine further stimulates glucagon release, and the glucagon promotes the transformation of aminoacids into glucose – most readily from alanine. The action of alanine on the pancreatic cells is promoted by cortisol. The production of new glucose is essential to prevent the insulin from lowering blood glucose concentrations at the same time as it promotes protein synthesis to allow storage of the ingested aminoacids (Fig. 24.5).

## Fasting

In the intervals between food intakes, metabolism is maintained by the stored nutrients. In the short term, a series of hormonal changes mobilise the glycogen and fat stores to maintain the levels of circulating nutrients. Later, gluconeogenesis is used to maintain the blood glucose levels and to supply glucose to those tissues which cannot use fatty acids.

As the tissues metabolise glucose, its concentration in the blood falls, and the major store of liver glycogen is mobilised. The liver has been called a glucose-secreting gland for this reason. The mobilisation of glucose takes place mainly in response to glucagon. Low blood concentrations of glucose do not stimulate insulin release so that any inhibition of glucagon secretion by insulin is removed. Glucagon can therefore promote glycogenolysis. Concentrations of glucagon in the blood rise for about two hours after a meal; in the later phases it stimulates gluconeogenesis also. After the overnight fast some 15% of energy requirements comes from gluconeogenesis. Since there is no protein intake as well as no carbohydrate intake during the fasting state, the turnover of muscle protein serves to maintain aminoacid levels in the blood. Alanine is converted into glucose, and glutamine reaches the kidney where it can be utilised in ammonia production. Fasting and starvation result in a metabolic acidosis, and so the ammonia assists in the excretion of acid. Glucagon also inhibits fat synthesis; normal fat cell metabolism therefore results in a net mobilisation of fatty acids.

## Starvation

Continued fasting results in the syndrome known as starvation. If there is a continued lack of uptake of nutrients the short-term adaptations are insufficient to satisfy the body's metabolic needs. In the longer term, the metabolism of the cells will be altered by hormonal activity and by the availability of substrates. The high level of glucagon and the low level of insulin in the blood remain almost constant over long periods of starvation but other changes occur because if the body continues to use protein as an energy source muscles will be weakened. In the carnivore, muscles are essential in food capture and so use of protein as an energy source is postponed until the terminal stages of starvation.

Low concentrations of glucose in the blood (hypoglycaemia) stimulate the adrenal medulla to produce adrenaline, which, like glucagon, stimulates glycogenolysis in both liver and muscle. Adrenaline also stimulates lipolysis, mobilising fatty acids from the fat stores: these can then act as alternative energy substrates. Indeed, when blood glucose concentrations are low, the fatty acids help to protect the glucose levels and reserve the glucose for tissues which must have it. The production of adrenal corticoids is also stimulated. These normally act in co-operation with glucagon, but in prolonged starvation their production increases markedly and they then produce a number of their own metabolic effects. They increase gluconeogenesis. This may result from, or be incidental to, the increase in protein catabolism stimulated by these hormones, the resultant increase in uptake of aminoacids by the liver and an increase in deamination and transamination. Finally the corticosteroids decrease the usage of glucose by peripheral tissues. In starvation the production of growth hormone is also stimulated. This promotes lipolysis to mobilise free fatty acids and reduces the entry of glucose into cells by antagonising the action of insulin. The high concentrations of fatty acids reaching the liver cause it to switch over to producing ketone bodies which serve as fuel for muscle and, most importantly, brain cells. Further, they inhibit protein breakdown in the muscle cells. However, the production of ketone bodies and ketoacids results in an acidosis.

Thus the substrates used by the different cells of the body in starvation differ from those used in times of relative plenty. Muscle cells use fatty acids and ketone bodies. Cardiac muscle cells normally use a high proportion of fatty acids and continue to do so in starvation. Brain cells switch from using glucose only to using increasing proportions of ketone bodies to conserve the remaining glucose.

The hormonal adaptations occurring during continued exercise and during post-trauma stress are related to those occurring during starvation – in all these situations the demands of the tissues exceed the ready availability of substrates.

## Hypoglycaemia and hyperglycaemia

Although the pathological condition of diabetes mellitus, resulting from insulin deficiency, is outside the scope of this book the effects of very low and very high concentrations of glucose in the blood will be described briefly.

Concentrations of glucose below 3 mmol/l are considered as constituting hypoglycaemia. In acute hypoglycaemia, such as occurs after an overdose of insulin, there is stimulation of the sympathetic nervous system resulting in sweating, tachycardia and anxiety. The relative lack of glucose causes muscular weakness and sensations of hunger. If the onset of hypoglycaemia is gradual, as in starvation, symptoms due to the decreasing supply of glucose to the brain may predominate: a lack of muscular co-ordination, manifested in such finely-controlled movements as speech, which may become slurred, followed by a slowing of mental processes and then convulsions resulting from disturbances of motor cortex activity. Coma finally results and may end in death.

Hyperglycaemia, such as that from insulin lack, causes glucose to appear in the urine (glycosuria), and this raises the osmotic activity of the renal tubular fluid to cause increased water loss – polyuria. This stimulates increased water intake but the associated salt loss leads to a reduction in extracellular fluid volume and dehydration often follows. If the hyperglycaemia continues this eventually results in coma and death. In insulin lack, hyperglycaemia is the result of the reduced ability of glucose to enter cells: the cells have, therefore, to use fats and proteins to a greater extent. The use of fats causes the production of an excess of ketone bodies and ketoacids and a consequent ketoacidosis.

# The control of calcium metabolism

Calcium is a most important ion in the functioning of the body. It is an extracellular ion, but is also found in cell membranes and cell organelles. In its ionised form it is essential for synaptic and neuromuscular transmission, for the contraction of all forms of muscle, for the majority of secretory processes which involve exocytosis, for the action of many hormones on target cells, and for blood clotting. The membrane potentials of excitable cells are stabilised by calcium ions, and if extracellular calcium levels are low, the membranes become more permeable to sodium ions and liable to spontaneous depolarisation and action potential generation. Calcium salts confer rigidity on the skeleton, making possible muscular activity and the maintenance of posture, and provide the hard structures of enamel and dentine which help to break up food for ingestion. Calcium metabolism is usually considered in terms of plasma concentrations of calcium, although these are only an index of a complex system of calcium balance. As with any homeostatic system, balance may be expressed simply by considering input and output. *Table 24.2* shows the daily balance in a normal adult male.

The tendency of calcium to form insoluble salts in the gut makes its absorption difficult, particularly in the higher pH of the small intestine. Even with active transport across the membranes, only about a half is actually absorbed. The main site is in the duodenum. Since calcium is also secreted in the digestive secretions, the net absorption may be only one tenth of the actual oral intake. Some of the faecal calcium comes from the desquamated cells of the epithelial lining of the gut. The calcium absorbed passes through the interstitial fluid to the blood plasma (Fig. 24.6). The process of absorption is controlled by an active metabolite of vitamin D, 1,25-dihydroxycholecalciferol. The estimation of dietary calcium requirements is difficult because the ability of the gut to absorb calcium depends both upon need and upon oral intake. Subjects with a low calcium intake are able to absorb calcium more efficiently so that a greater proportion of the intake is absorbed. This may be due to the ease with which the active transport mechanism can be saturated by excess calcium.

The concentration of calcium in the blood plasma is maintained very closely between 2.25 and 2.75 mmol/l. Of this, some 1.18 mmol/l are ionised and a further 0.16 mmol/l are present as inorganic complexes with hydrogen carbonate, citrate and other chelating agents. The remainder, bound to protein, forms a non-diffusible store in the blood. The bulk of this, 0.92 mmol/l, is bound to serum albumin and the rest, 0.24 mmol/l, to the globulins. The proportion of calcium ionised depends upon the concentration of protein, and also of electrolytes and the pH of plasma: as pH rises, binding increases – and so hyperventilation, by blowing off carbon dioxide, which is equivalent to losing acid, lowers the effective calcium ion concentration and can cause a condition of tetany. In tetany neuromuscular transmission is impaired but skeletal muscle undergoes spasm because both the muscle and nerve cells become hyper-excitable when the concentrations of ionic calcium are low. This can be demonstrated by tapping the facial nerve at the angle of the jaw to cause contraction of the facial muscles on that side, or by observing the characteristic spasmodic flexion of the wrist and thumb in more acute cases.

From the blood plasma, ionised calcium is able to exchange with interstitial fluid and with the ex-

Table 24.2 The balance sheet of calcium intake and loss

| Oral intake | 25.0 mmol | | |
|---|---|---|---|
| less non-absorbed | 10.0 mmol | Faecal loss | 10.0 mmol |
| Absorption | 15.0 mmol | | |
| less secretion into gut | 12.5 mmol | Faecal loss | 12.5 mmol |
| less urinary loss | 2.5 mmol | Urinary loss | 2.5 mmol |
| | 0.0 | | 25.0 mmol |

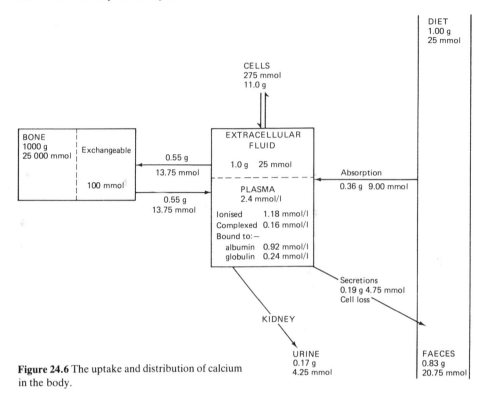

**Figure 24.6** The uptake and distribution of calcium in the body.

changeable calcium of bone – about 100 mmol in a total bone calcium content of 25 000 mmol (Fig. 24.6). Bone forms the principal store of body calcium and as it is well vascularised, there is a ready exchange of dissolved calcium ions in the tissue fluid with the surfaces of the apatite crystals. This is entirely a passive process. Even in the adult, however, bone is constantly being formed and resorbed, resulting in a normal active exchange of about 12 mmol/day, in contrast to the estimated 500 mmol/day of passive exchange. In the course of one year, all the calcium in the bones of an infant could be replaced; about 18% of the bone in an adult. This is important in considering the uptake of radioactive strontium, which behaves as a replacement ion for calcium. Fluoride also substitutes for hydroxyl ions in forming bone: as fluorapatite is more resistant to resorption, it accumulates. A dose of fluoride in a child with growing bones will be retained in bone to a much greater extent than a similar dose in an adult. The forming bone of the fetus and the infant impose a demand for calcium on the mother during pregnancy and lactation. The recommended daily intake of calcium during pregnancy is 1.5 g/day (37.5 mmol), and during lactation 2.0 g/day (50 mmol), in comparison with the requirement of 0.5–0.8 g/day (12.5–20 mmol/day) of an adult male. Calcium is lost from the plasma into the urine although most of the filtered load is reabsorbed in the kidney and only some 2.5 mmol/day appears in the urine. The fine control over urinary loss is exerted by the action of parathyroid hormone on the renal distal tubule.

Control of blood calcium concentrations is exerted by parathyroid hormone and calcitonin (Fig. 24.7). Increase in plasma ionised calcium concentrations stimulates calcitonin secretion and inhibits that of parathormone. Calcitonin inhibits bone resorption and reduces the number of osteoclast cells engaged in bone resorption. It also increases calcium excretion in urine. However, the importance of calcitonin is now being questioned. It may be important in young children forming bone, and possibly in pregnancy, where it may protect the maternal bone. It may also be necessary to control sudden intakes of calcium, but its action is transitory. The higher levels of calcitonin in the blood of male adults may account for the greater susceptibility of women to osteoporosis as they grow older. In osteoporosis calcium is lost from the bones and they become more liable to fracture. The opposing relationship between calcitonin and gastrin secretion suggests that calcitonin may have other functions in the body, as yet undiscovered.

A decrease in plasma calcium concentrations stimulates parathormone secretion, causing increased reabsorption of calcium in the distal tubule of the kidney and inhibiting renal phosphate

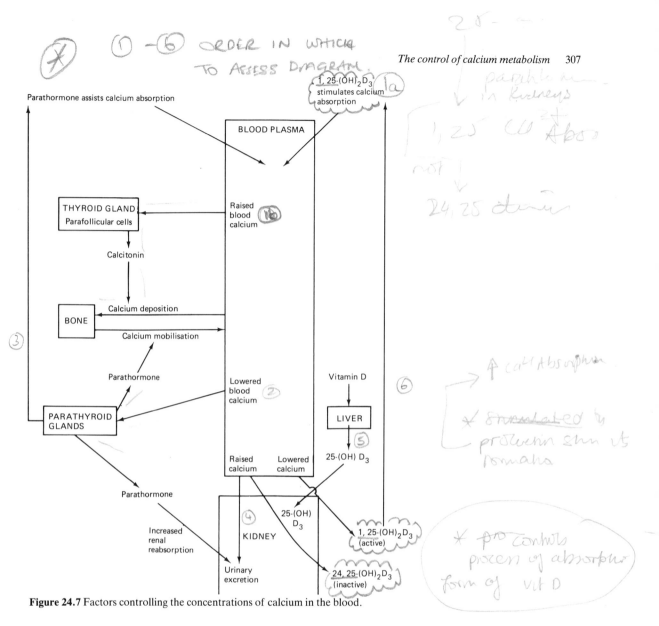

**Figure 24.7** Factors controlling the concentrations of calcium in the blood.

reabsorption. The second action of parathormone is to cause the kidneys to convert 25-hydroxycholecalciferol (the hydroxylated metabolite of vitamin $D_3$ formed in the liver) into 1,25-dihydroxycholecalciferol instead of a less active form, 24,25-dihydroxycholecalciferol. This circulates to the intestine where it increases both the passive absorption across the luminal cell membranes by affecting membrane permeability, and the active transport of calcium across the basal membranes by $Ca^{2+}$-ATPase activity. Both these effects depend upon the synthesis of calcium-binding protein. Prolactin also increases calcium absorption by stimulating formation of 1,25-dihydroxycholecalciferol. Parathormone is known to increase resorption of bone by increasing both the numbers and the activity of osteoclasts: but this action is possibly pharmacological or pathological rather than physiological – it occurs only if plasma ionised calcium levels are severely reduced. In physiological amounts parathormone appears to act in conjunction with the active metabolites of vitamin D to stimulate apposition of bone. Thus blood calcium levels are raised by ingestion and absorption of calcium under the control of both hormones, and maintained high by the renal reabsorption of calcium as a result of parathormone activity: whilst blood calcium levels are high bone apposition can occur. Major departures from the normal concentration of ionic calcium in the blood will cause parathormone to mobilise calcium from bone.

Other hormones affect bone development (Chap. 23): they may also affect plasma calcium levels.

Phosphate metabolism is not as closely controlled

as calcium metabolism, but is often connected with it – calcium phosphate is poorly soluble and forms, as apatite, the main salt of bone. Total body phosphorus is between 500 and 800 g (16–25 mol), of which about 90% is in bone. Plasma concentrations are around 4 mmol/l with about 1.3 mmol/l present as inorganic phosphates. Absorption occurs by active transport mainly in the jejunum and may be enhanced by calcium uptake. Excretion in the urine takes place after about 90% of the filtered load has been actively reabsorbed in the proximal convoluted tubules, a process inhibited by parathormone.

## Further reading

Newsholme, E.A. and Start, C. *Regulation in Metabolism,* John Wiley & Sons, London, 1974

Simkiss, K. *Bone and Biomineralisation,* Arnold, London, 1975

Tepperman, J. *Metabolic and Endocrine Physiology,* 5th edn, Year Book Medical Publishers, Chicago, 1987

Vaughan, J.M. *The Physiology of Bone,* Oxford University Press, Oxford, 1981

White, D.A., Middleton, B., and Baxter, M. *Hormones and Metabolic Control,* Edward Arnold, London, 1984

# 25

# The control of body temperature

## Normal temperature

Body temperature is generally believed to be constant at about 36.9°C: but is in fact subject to variation at different times of day, from one day to another, and from one individual to another. Before considering human responses to heat and cold it is necessary to define what is meant by normal temperature and what variations may be caused by the methods of measurement and the physiological circumstances.

Mammalian enzymes operate efficiently within a narrow range of temperature. All chemical reactions proceed more rapidly at higher temperatures, roughly doubling in rate with each increase of 10°C. At approximately 45°C, however, denaturation of most proteins, including enzymes, takes place.

The temperature in the core of tissue around the centre of the human body and in the arterial blood near the core is between 36°C and 37.5°C. This is the deep body, or core, temperature. It remains fairly constant in an adult male, at rest, in a post-absorptive state, and if measured at the same time of day. In the adult female body temperature is approximately 0.5°C higher in the post-ovulation phase of the menstrual cycle than in the pre-ovulatory phase. From the core, there is a progressive fall in temperature towards the skin surface and the air layers on the surface until ambient air temperature is reached. If the external temperature exceeds core temperature the gradient will be in the opposite direction. The site of measurement of body temperature is therefore important. Oesophageal temperature is probably the nearest to the temperature of the core that can be obtained by non-invasive methods, and the next is rectal temperature – although this may be increased by the generation of heat by the leg muscles, or even lowered by cool blood returning from the exposed limbs. Tympanic membrane temperature is the best approximation to brain temperature. Oral temperature, obtained by leaving a thermometer under the tongue in a closed mouth, is about 0.3°C lower than deep body temperature, provided the subject has not just been eating ice-cream or drinking hot coffee. A thermometer placed in the axilla, with the arm tightly against the side, gives a more variable reading, usually about 0.8°C lower than core temperature. Skin temperatures are widely variable since they depend also on the rate of loss or gain of heat to the environment.

Deep body temperature exhibits a circadian rhythm, being highest in the evening, and lowest early in the morning. The variation between these daily extremes may be as little as 0.6°C or as much as 1.5°C.

The range of deep body temperatures compatible with life is from 25°C to 42°C. Heat stroke develops at 43°C, and long exposure to 41°C will cause damage to brain cells. Below 28°C the subject cannot spontaneously return to normal temperatures. A decrease in deep body temperature is termed hypothermia, and an increase, hyperthermia or pyrexia – the last term usually being reserved for pathological rises in temperature. In hypothermia, all metabolic processes are slowed – respiration and heart rate decrease, blood pressure falls, and brain activity is damped. Death from hypothermia is not uncommon among the aged. Hypothermia may be deliberately induced in heart and brain surgery in order to slow the metabolism when blood flow, and hence oxygen supply, must be stopped for more than a few minutes.

## Heat gain

The body is constantly producing heat because the oxidation of foodstuffs yields energy which appears as heat. The maintenance of a constant body temperature implies, therefore, a continuing loss of heat (Fig. 25.1) which takes place normally because external temperatures are usually below deep body temperature. The daily production of heat amounts

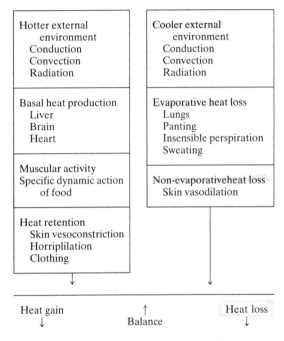

**Figure 25.1** The balance between heat gain and heat loss.

to some 7.0 MJ in basal conditions, but may rise as high as 25 MJ in heavy muscular work. An approximate figure for the average urban male would be around 12 MJ, and for the female, around 10 MJ. The basal metabolic rate indicates the resting metabolism of the tissues: of this the brain contributes about 1.31 MJ, the liver 1.45 MJ, and skeletal muscle about 1.4 MJ. If there were no heat loss this amount of heat would raise body temperature by about 1°C/hour. In addition to metabolic heat, there may be heat gained from radiated solar heat. Such gain is proportional to the exposed area of skin: it amounts to about 80% of the radiant heat in black- or brown-skinned subjects, and about 60% in light-skinned subjects. The main protection afforded by a dark skin is not against heat absorption but in its ability to absorb potentially damaging ultraviolet radiation. In hot air there can also be heat gain by direct conduction or convection.

## Heat loss

Heat loss can occur by conduction of heat from the skin to the layer of still air around the body, convection of heat to the free air layers, radiation from the skin, and evaporation of water (either diffused through the skin surface or actively secreted by the sweat glands). Conduction and convection are obviously linked. In a nude standing subject the air warmed by conduction rises up around the body to form a convective current extending 1.5 m over the head. Heat loss by this route is increased in the prone or supine subject because the rising air does not insulate the upper part of the body. If the surrounding air is moving, heat loss increases. Heat loss in still air from the nude subject is mainly by radiation. About 70% of heat loss is generally taken to be accounted for by convection and radiation; of this as much as 60% may be by radiation. In the clothed subject the situation is completely different.

Heat loss by any route may be expressed in the general equation

$$H = kA(T_s - T_a).$$

where $H$ is the loss in J/s, $k$ is the cooling constant, $A$ is the surface area, and $T_s - T_a$ is the difference between skin temperature and air temperature. This equation describes heat loss from the skin, and must be modified if heat loss from the core is intended. When heat loss is expressed as J/s/unit area, the $A$ term disappears from the equation and $k$ is then termed the body thermal conductance. The reciprocal of $k$ is the thermal insulation factor. Heat passes from the core to the skin to the still air layer to the free air – each of these transfers can be expressed in a similar equation. If the still air layer is increased, or the free air velocity decreased, heat loss is reduced. Clothing produces both of these conditions as well as preventing radiant heat loss.

Heat is lost by evaporation of water reaching the surface by three different routes: diffusion through the epidermal layer of the body surface (some 86–170 ml/day), the slow normal secretion from the sweat glands (insensible perspiration – 33–500 ml/day) as well as active, or sensible, perspiration, and water loss through the surfaces of the lungs and the respiratory tract, amounting to 600–700 ml/day. The rate of evaporative heat loss can be expressed in a similar fashion to that of non-evaporative heat loss:

$$E = c(P_s - P_a).$$

where c is the evaporative cooling constant and $P_s - P_a$ is the difference in vapour pressure of water in the atmosphere and at the skin surface. The latent heat of evaporation of water is 2.5 kJ/ml, so that if 600–800 ml/day are lost normally this will amount to 1.5–2.0 MJ/day, or 21–28% of total heat loss. Evaporative heat loss depends upon the vapour pressure of water in the air, so that in high humidities, temperatures above 32.5°C become intolerable, whereas 45°C can be suffered at low humidities. The maximum tolerable wet bulb temperature is just over 30°C in basal conditions. Subjects immersed in hot water, incidentally, may actually experience a net gain of water because water absorbed through the skin exceeds that lost by sweating.

## Temperature control

### Sensors

Temperature sensors may be classed as peripheral and central (Fig. 25.2). The peripheral are the skin hot and cold receptors, monitoring skin temperature. Their inputs pass to the thalamus and reticular activating system to reach the hypothalamus as well as the sensory cortex. The hypothalamic areas involved are in the pre-optic nuclei and the anterior area. The central receptors are also situated in these areas: they respond to changes in the blood temperature in the head. There may be other deep body temperature receptors.

### Controlling centre and set point

The hypothalamus contains the two controlling centres: one in the anterior hypothalamus and pre-optic area which controls mechanisms for heat loss, and a second reciprocally-linked centre in the posterior hypothalamus which controls mechanisms of heat retention or generation. The set point corresponds with a deep body temperature in the range 36–37.5°C. The normal set point may be raised in exercise or by information from peripheral cold receptors, and it is lowered during sleep. Bacterial toxins and inflammatory products termed pyrexins increase the production of prostaglandin $PGE_1$ which raises the set point. Aspirin and other antipyretic drugs, by inhibiting formation of the prostaglandin, restore the original set point. The transmitter substances in the hypothalamic centres are 5-hydroxytryptamine in the anterior, and noradrenaline in the posterior. The anterior centre maintains a tonic inhibitory influence over the posterior centre, which is not itself temperature-sensitive. In addition to the hypothalamic centres there is a lower centre situated in the medulla oblongata, which can operate if the brainstem is damaged above the level of the medulla.

### Output pathways and effector organs

Changes in body temperature may be achieved by varying heat gain or heat loss. A response to cooling may involve an increase in heat production and/or a

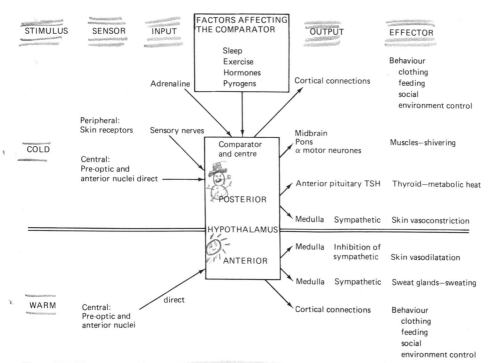

**Figure 25.2** The control of heat gain and heat loss – the sensors, input pathways, controlling centre, output pathways, and effectors.

reduction in heat transfer from the core to the surroundings. The response to heating may be a reduction in heat production, an increase in heat transfer to the surroundings, an increase in evaporative heat loss, or any combination of these. Where a discrepancy exists between information from skin and from core receptors, the core receptors are dominant in determining the physiological response. The core receptors are misalignment detectors, the peripheral disturbance detectors. Signals from the peripheral sensors pass directly to conscious levels and are associated with sensations of comfort or discomfort, so that they can also act as triggers for behavioural responses to environmental temperature.

## The response to cooling

The first response to cooling is a change in skin blood flow, so that transfer of heat from the core to the periphery is reduced. It is followed at the lower critical temperature by shivering – an increase in heat production resulting from the alternating activity of muscle fibres in opposing groups of muscles, producing no useful mechanical work. Later there may be an increase in metabolism, first in the liver and then in adipose tissue. The lower critical temperature is that at which metabolic compensation becomes necessary: for a resting man wearing no clothes the lower critical environmental temperature is around 27°C.

The blood vessels to the limbs are arranged with the main arteries and veins in close proximity. They form a counter-current loop which maintains a gradient of temperature from the distal cool end of the limb to the proximal warm attachment. Subsidiary vessels and collateral veins normally carry only a small proportion of the blood flow: flow through them is further restricted in cold conditions, increasing the efficiency of the counter-current exchanger. The vessels supplying the skin of the face, the hands and the feet are arranged in a complex interlocking series of arterio-venous anastomoses and capillary loops. Sympathetic nerves to these vessels and their glomus bodies adjust the flow to reduce the amount of warm blood flowing through, and cooled by, the cold skin. Sudden application of cold to the skin causes local vasoconstriction – possibly also by a direct stimulus to the vascular smooth muscle. This is usually painful. After a few moments vasodilatation occurs, due either to accumulation of metabolites or to a local axon reflex – rather than to central control. A cycle of alternate constriction and dilatation follows which conserves heat while providing nutrients and removing metabolites. The vasoconstrictor response is seen when the skin temperature falls below 33°C, but only if core temperature is below 37°C. This necessary combination of signals from the peripheral and central receptors is due to the action of the peripheral receptors as disturbance detectors, operating in this instance by altering the set point for the hypothalamic receptor/comparator system.

The human, with little body hair, is unable to increase the thickness of the layer of warm still air next to the skin by horripilation, the erection of the skin hairs by the pilomotor muscles under the control of sympathetic nerves. Goose pimples (U.S. goose bumps) are a vestige of this reflex.

In adults, the metabolic rate increases when core temperature falls below 36.5°C and skin temperature is below 33.5°C. A combination of temperatures must be specified because of the influence of the peripheral receptor signals on the set point of the central receptor/comparator. The increase in metabolic rate is almost entirely due to shivering. The shivering centre in the posterior hypothalamus is activated by impulses from peripheral cold sensors although there is over-riding inhibition from the anterior hypothalamus unless deep body temperature has also fallen. Descending tracts run through the midbrain tegmentum and the pons to reach the α motor neurones. Cold shivering begins, like the shivering of emotion, in the proximal parts of the limbs, but if cooling continues, is transferred to more central body regions. Shivering can increase the resting oxygen consumption between two and five times. Below the lower critical temperature, heat production increases linearly in proportion to the temperature discrepancy.

Non-shivering thermogenesis, an increase in basal metabolic rate, plays a very minor part in the correction of lowered deep body temperatures in adult humans. An increase in sympathetic activity, both nervous and via the adrenal medulla, may increase liver metabolism. Thyroid hormones are necessary as well as noradrenaline, but thyroid activity does not increase even over several hours of exposure to cold. The control centre is in the posterior hypothalamus, and resembles the shivering centre in its operation. Fat metabolism may increase in cold conditions. In infants the shivering reflex is not yet developed as a temperature response and so in them the main response to cold is non-shivering thermogenesis using brown adipose tissue fat as a substrate (see p. 228). Sympathetic nerves release noradrenaline which activates β receptors in the brown adipose tissue cells. Adrenaline has less effect than does noradrenaline.

The evidence for human acclimatisation to cold conditions is poor. Individual human beings seem unable to adapt physically, although circulating thyroid hormone levels are said to be higher in races living in the Arctic and their diet would support a higher level of metabolism. Further, they adapt by building up fat stores which function as insulating layers. The main human response to cold is a behavioural adaptation – clothing.

## The response to warming

The body does not respond to an increase in core temperature by reducing metabolic heat production. Indeed, above a certain critical temperature there is an increase in metabolic heat production due probably to the work done in sweating.

Transfer of heat from core to periphery, and from periphery to external environment, is enhanced by increasing blood flow to the skin by both vasodilatation and the transfer of blood to the arterio-venous shunts in the skin of the face, hands and feet rather than to the skin capillary beds. These arterio-venous shunts lie near the skin surface and enable a rapid flow of warm blood to reach the cool surface and transfer its heat. Above a certain core temperature of about 36.5°C, skin blood flow is proportional to skin temperature. This control is achieved by inhibition of the posterior hypothalamic sympathetic pathway by impulses from the anterior hypothalamus. The actual heat transfer to the air is by conduction and convection from the warm surface, and by radiation from the flushed skin.

When the core temperature exceeds 37°C, sweating is initiated provided the skin temperature is not below 34°C. Once again, there is a critical environmental temperature above which the physiological cooling response is observed in the resting unclothed human being. This upper critical environmental temperature is around 33°C. The temperature band between the lower and upper critical temperatures is termed the comfort zone.

The amount of sweating, or sensible perspiration, is directly proportional to the temperature difference from the normal resting temperature. The reflexes invoking vasodilatation and sweating are different and can be evoked independently. The amount of sweating and the skin temperature are not related since both vasodilatation and sweating change skin temperature fairly rapidly before there is any significant effect on core temperature. Despite the apparently slow process of cooling by evaporation from the skin, a rise in deep body temperature of 1°C can be dissipated in about 30 min by this means. Maximum rates of sweating can amount to 1.7 litres/h over short periods, or as much as 12 litres/day over longer periods. At this rate of perspiration 36 g of salt would be lost in 24 h – roughly 600 mmol each of sodium and chloride. The water loss would be more than a quarter of total body water – equivalent to twice the volume of the blood. Patients undergoing surgical treatment after being engaged in heavy manual labour under tropical conditions may well need stabilisation of their water balance before any surgery is attempted. A sweating rate of half of this maximum would result in a heat loss of 14 MJ/day.

The sweat glands are scattered over the whole of the body surface. Of the total evaporative heat loss about 50% is provided by the trunk, 25% by the legs, and about 25% by the head and arms. The sweat glands are eccrine glands secreting a fluid of low osmotic activity at slow flow rates because of ductal reabsorption of sodium and chloride, but at high flow rates the concentrations of electrolytes in the fluid approach those of extracellular fluid (see Tables 2.4 and 2.6). The fluid also contains lactic acid, from anaerobic metabolism, and urea, but very little glucose, because this has been metabolised to provide energy for secretion. The freshly secreted fluid is odourless and almost clear; bacterial action rapidly produces odours from the dried residue. Deodorants mask the developing odours; antiperspirants precipitate the salts in the ducts and block them. Sweat glands in general respond to sympathetic stimulation, the sympathetic postganglionic nerves releasing acetylcholine at the neuro-effector junctions. Some of the eccrine glands – particularly on the forehead, upper lip, palms and soles – respond to emotional stress, the others only to changes in deep body temperature. Such glands may be activated in fear and during attacks of vomiting or fainting. Presumably the stimuli are transmitted from the limbic brain via the hypothalamus and sympathetic nervous system. The apocrine sweat glands, confined to the axillary, pubic and nipple regions, are not stimulated by sympathetic nerves but by circulating adrenaline. They are analogous to glands with sexual functions in other animals and they develop at puberty under hormonal control. Their secretion is different from that of the eccrine sweat glands and has no thermoregulatory function.

Sweating is one of the responses to the heat gain of exercise: it is proportional to the rise in core temperature which is, in turn, proportional to the work performed. As such, it is independent of ambient temperature except in that the critical core temperature for initiating sweating is raised by a cold environment. Sweating during exercise is, however, often related directly to skin temperature. During exercise the set point of the hypothalamic control centre is raised, but it returns to its previous level within about one hour after exercise.

Total heat loss in a nude male at high ambient temperatures in dry air can reach about 3.4 MJ/h – about eight times the basal heat production.

Human acclimation to hot climates, like that to cold climates, is mainly by behavioural changes such as reducing the insulation provided by clothing. Perspiration increases with acclimation and the loss of sodium chloride in the secretion is reduced by more efficient reabsorption in the ducts, possibly caused by secretion of aldosterone from the adrenal cortex. Metabolic heat production is reduced in females but not in males. Human beings tolerate warm climates much better than cold, and their physiological responses to warm climates are more effective.

## Behavioural and other responses

The behavioural responses to temperature changes are very important in man both in the short and the long term. Clothing can be varied to suit the external temperatures, and heating used to change the environmental temperature. Housing is partly a response to climate. Heat may be gained by conscious increases in activity – stamping the feet, waving the arms, or even useful exercise – or by changing posture to reduce the surface available for heat loss. Huddling together can be a form of heat retention. The daily pattern of activity is often modified to suit the climate – as in the siesta.

Heat and cold also have psychological effects: heat produces listlessness whilst cold may enhance alertness, if not too extreme. Individuals with high or low basal metabolic rates due to increased or decreased rates of thyroid hormone secretion may have appropriately different temperature comfort zones.

There are also effects on hypothalamic centres in proximity to the temperature centres. Thirst may be induced by high temperatures and sweating; this helps to replace the lost water. Salt appetite may be stimulated. The rate of sweating, however, is not affected by a reduction in total body water or extracellular fluid volume. The satiety centre is more active in warm conditions, and the hunger centre in cold: how far these appetitive responses relate to metabolic needs, or to psychological effects, is not clear.

## Fever or pyrexia

The presence in the bloodstream of pyrexins from bacteria or damaged cells raises the hypothalamic set point for temperature.

The elevation of the set point causes body temperature to rise and heat-producing and heat-conserving reactions are initiated. Shivering increases heat gain and vasoconstriction causes the skin to appear pale. Higher body temperature may enhance the effectiveness of the body's defences by increasing the rates of enzyme reactions. When the fever reaches crisis and resolves, the set point falls again and heat loss is accelerated to restore the original body temperature. The signs, therefore, are sweating and flushing, due to vasodilatation. Antipyretic drugs such as aspirin which lower abnormally raised hypothalamic temperature set points, have similar effects. It is not unusual in many fevers for alternating phases of high temperature and resolution to occur, with appropriate physical signs.

## Further reading

Burton, A.C. and Edholm, O.G. *Man in a Cold Environment,* Edward Arnold, London 1965. (reprint Hafner, New York, 1969)

Garland, H.O. Altered Temperature. In *Variations in Human Physiology* (ed. Case, R.M.). Manchester University Press., Manchester, 1985

Hardy, R. *Temperature and Animal Life,* Arnold, London, 1972

Kerslake, D.McK. *The Stress of Hot Environments.* Cambridge University Press, Cambridge, 1972

# 26
# Exercise

Since physical exercise causes changes in the usage and demand for nutrients, and also produces a greater quantity of metabolites, it has effects on the cardiovascular system, the respiratory system and the organs storing energy substrates, as well as the tissues actually involved in the exercise.

The resting subject has already been characterised in various ways. A typical male subject resting in a post-absorptive state has a metabolic rate of around 4 kJ/min, or an equivalent oxygen consumption of about 10 mmol/min. Leisurely walking raises these values to about 5.8 kJ/min and 14 mmol/min. At rest the heart rate is about 72 beats/min, the stroke volume about 70 mls, and the cardiac output about 5 litres/min. The breathing rate is around 12/min and the tidal volume about 450 ml – a total ventilation of about 5.5 litres/min. The ratio of carbon dioxide output to oxygen usage, the respiratory quotient, is probably a little under 1.0. Deep body temperature is around 37°C and skin temperature about 30°C.

## Beginning exercise

Before any actual muscle effort begins, there is an increase in heart rate and in rate of ventilation (Fig. 26.1). These are not related to the concentrations of oxygen or carbon dioxide in the blood, which at this point would be unchanged – although they will now change as a result of the increased ventilation. This phase is therefore considered to be a phase of psychic stimulation. The cardiac effects are mediated first through a decrease in parasympathetic (vagal) activity and then through an increase in sympathetic activity. Cardiac output is increased mainly by an increase in heart rate whilst stroke volume changes relatively little. Blood pressure rises, diastolic to a lesser extent than systolic. Ventilation increases, mainly by an increase in tidal volume rather than rate of breathing. This results in a greater blowing-off of carbon dioxide from the lungs and the alveolar carbon dioxide concentration falls. The blood flow to the muscles increases, possibly because of sympathetic vasodilatation mediated by cholinergic or β adrenergic receptors. General sympathetic vasocontrol is unaffected.

The changes occurring immediately before exercise may be due in part to impulses from joint receptors and other proprioceptors but their role is probably more significant as exercise actually begins.

As a result of the psychic stimulation, sympathetic activity via the adrenal medulla is increased and more adrenaline begins to circulate. This initiates a redistribution of blood flow as a result of its differential action upon tissues, causing an increase in blood flow to the muscles and maintaining blood flow to the brain and heart.

## Early phase of exercise

In the early phase of exercise the psychic stimulus is amplified by the information reaching the brain from the muscle and joint receptors, which help to generate an immediate rapid increase in both cardiac output and ventilation. The increase in the first few seconds can be as much as half of the total increase seen during exercise. Although the changes are much greater than the anticipatory changes, the mechanisms are the same. Alveolar carbon dioxide concentrations fall to a minimum level.

Over the next 1–2 min a slower adjustment occurs as the muscular activity lowers the partial pressure of oxygen in the blood, raises that of carbon dioxide, and increases the tissue fluid concentrations of lactate, potassium ions, adenosine and adenosine monophosphate. The muscle produces heat, causing

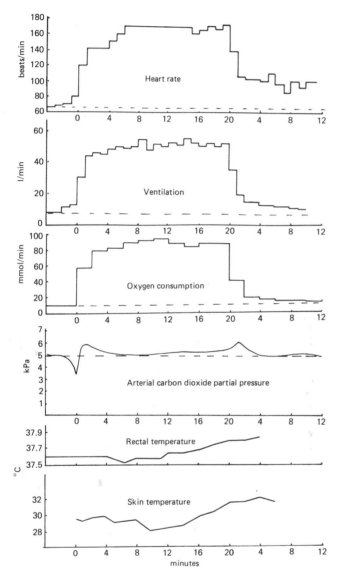

**Figure 26.1** Data from an experiment on exercise. Changes in heart rate, ventilation, oxygen usage, carbon dioxide elimination, rectal and skin temperatures before, during, and after a 20 min spell of exercise on a bicycle ergometer at a load of 150 watts (joules/s). The time calibrations are given as 0 for the start of the exercise, 20 for the end of the exercise period and the beginning of recovery, and then up to 12 min for the recovery. The dotted line across each trace represents the mean resting level.

the local temperature to rise. The blood returning to the lungs contains more carbon dioxide and less oxygen: the carbon dioxide acts on the medullary, and, to a lesser extent, the peripheral, chemoreceptors to stimulate both ventilation and cardiac output. The local stimuli within the muscle cause dilatation of the arterioles, metarterioles, and precapillary sphincters, increasing the number of open capillaries some 10–100 times. This incidentally decreases the diffusion distance for the oxygen from the blood to the muscle cells. In some muscles there is a capillary for every muscle fibre. The velocity of flow decreases because of the increased cross-sectional area of the total vascular bed and this, together with the increased amount of 2,3-diphosphoglycerate in the red blood cells, and the effect of the lower pH and higher temperature on the haemoglobin-oxygen dissociation curve, results in a much greater removal of oxygen from the blood. The gradient favouring oxygen diffusion into the muscle cells is increased because of the their greater oxygen usage. The venous blood from the muscles during exercise will have lost nearly three times as much oxygen as the venous blood in resting conditions (*Table 26.1*). Since blood flow increases over 30 times, the oxygen usage by the muscle can increase over 100-fold. However, in the early stages the oxygen supply cannot keep pace with the

**Table 26.1 Changes in oxygen carriage during exercise**

| Oxygen in venous blood | Resting | During exercise |
| --- | --- | --- |
| Partial pressure | 5.3 kpa | 2.0 kpa |
| % saturation of haemoglobin | 70% | 16% |
| Oxygen content | 6.7 mmol/l | 1.3 mmol/l |
| Oxygen extracted | 2.3 mmol/l | 7.7 mmol/l |

increased metabolic load and the muscle switches to a mixture of anaerobic and aerobic metabolism until the circulatory and respiratory adjustments are complete. In these situations, with moderate to severe workloads, the white 'fast' skeletal muscle fibres, with their greater potential for anaerobic work, are mobilised. Adrenaline stimulates glycogenolysis in these muscle cells. In anaerobic contraction ATP is regenerated from ADP by using energy from creatine phosphate or from the breakdown of glycogen to pyruvic and lactic acids. Oxygen is needed to oxidise pyruvic acid through the Kreb's cycle. In the red muscle fibres, which are used for continuing activity, or for exercise of low to moderate intensity, myoglobin functions as an intermediary oxygen store to maintain oxidative metabolism. During such exercise the main energy source is fat. The metabolic changes in the muscle cells are described on pp. 297–300.

Circulating carbon dioxide stimulates respiration, increasing the ventilation up to six or seven times by increasing both the rate and the depth. Oxygen usage may increase from around 10 mmol/min up to 80 or 90 mmol/min. As the tissue respiration is partially anaerobic an 'oxygen debt' is accumulated – further oxidation of the metabolic products will be needed later. The concentration of circulating lactic acid rises from 1 mmol/l to 10–15 mmol/l. This acid must be buffered: so carbon dioxide is lost via the lungs, and the respiratory quotient rises to 1.5 or even 2.0. The normal expiration of around 9 mmol/min of carbon dioxide rises towards 200 mmol/min. Despite this blood pH falls to around 7.3. The alveolar carbon dioxide concentration rises rapidly from the low level reached just before exercise to a new steady level higher than usual, as production of carbon dioxide first exceeds, and then is matched by, the capacity for excretion by ventilation. If exercise continues the partial pressure of carbon dioxide falls again as hydrogen ion accumulates in the blood.

The effects of carbon dioxide on the medullary cardiac control centres, together with increased sympathetic neural and hormonal activity, increase the cardiac output up towards 35 litres/min. The main increase is in the heart rate, which can reach 200/min in children, and progressively less in older subjects. The redistribution of blood flow continues.

As work increases above a threshold of about 30% of maximum, renal and splanchnic blood flow are progressively reduced until they are between 20 and 30% of normal – releasing about 2 litres of blood for other tissues. Blood flow to the muscles increases. Blood flow to the heart itself increases as the cardiac vessels dilate, mainly in response to the greater oxygen usage, but also in response to the production of metabolites. The blood flow to the brain appears to be unchanged. Very early in exercise the increased activity of the sympathetic nervous system causes vasoconstriction in the skin and the splanchnic circulations, but in the skin this rapidly gives way to vasodilatation as body temperature begins to rise. Although some heat is lost because of the increase in ventilation and the hypothalamic temperature control centre is reset to a higher set temperature, heat loss through the skin, both by evaporative and non-evaporative routes, is essential. During exercise both rectal and skin temperatures rise, the rectal by about 1°C in heavy exercise, the skin by about 2°C. Provided the humidity is low, heat loss by evaporation seems to be independent of the external temperature.

Both the redistribution of blood flow by vasoconstriction and the increased activity of the muscle pump increase the venous return and, therefore, at first, the stroke volume of the heart. Sympathetic activity, both nervous and hormonal, increases cardiac output both by increasing heart rate and stroke volume. However, as heart rate increases beyond a certain level, the time for filling is reduced, and the stroke volume reaches a plateau and then decreases again. The arterial side of the circulation will contain a greater than normal proportion of the blood volume (up to 30%, instead of only 20%) as venoconstriction reduces the volume of the venous capacitance vessels. The blood pressure increases, but not as much as the increase in cardiac output might suggest, since the vasodilatation in muscle, and to a lesser degree, skin, causes the peripheral resistance to fall markedly. Diastolic pressure, therefore, may show very little increase or even fall, although systolic may reach 24 kPa. In isometric exercise diastolic pressure usually rises.

The higher blood pressure in the muscle capillaries gives a steeper gradient of pressure to form the interstitial fluid; the increased volume of tissue fluid formed is removed by a more active pumping of fluid through the lymphatics by the exercising muscle. Contraction of the muscle fibres, however, compresses both the lymphatic and the blood capillaries so that flow through them may be prevented during powerful sustained contractions; this reduction in blood supply may terminate the activity of the muscle. Intermittent contractions permit flow to occur between the contractions and, indeed, assist in providing a pumping action along both veins and lymphatics. In isometric exercise flow from the muscle may be completely prevented. Painful swelling of the muscle results if accumulation of tissue fluid exceeds the ability of the lymphatics to remove it. The local metabolites raise the osmotic activity of the extravascular fluids increasing the movement of fluid out of the vessels; additionally many of these metabolites stimulate a sensation of pain.

## Steady exercise

During steady exercise cardiac output is linearly related to the usage of oxygen mainly as a result of variations in heart rate. Blood oxygen and carbon dioxide concentrations stay constant during moderate exercise because the ventilation is matched to the workload. The mechanisms by which these equilibria of oxygen consumption, cardiac output and blood gas concentrations are maintained are difficult to define because of the complexity of the interactions involved, but two main theories have been put forward. The 'central mechanism' theory envisages that the output from the motor cortex not only activates the muscles themselves, but indirectly, via the hypothalamus, causes vasodilatation in the muscles and increases cardiac output and ventilation either directly or through the medullary centres. The 'peripheral mechanism' theory is that chemoreceptors in the exercising muscle are stimulated by the increased concentrations of ions and ADP in the extracellular fluid and that signals from these receptors reach the brain to continue a co-ordinated response of the cardio-respiratory unit. The sensitivity of the blood gas and pH receptors may be increased during exercise (possibly due to the increased body temperature) but this seems unlikely since the venous blood from exercising muscles has no appreciable influence on the respiratory and cardiovascular responses. The baroreceptor reflex seems to be over-ruled, although the respiratory chemoreceptor reflexes appear to be reset to maintain respiration at the new higher level.

As exercise becomes more severe the pH of the blood falls, and hyperventilation ensues, driving off carbon dioxide and increasing the blood oxygen concentration. The respiratory quotient exceeds unity.

The redistribution of the blood flow, the opening up of the blood capillaries in the muscle, and the changes in lymph flow, result in changes in the number and proportions of circulating cells. The red blood cell count increases by about 10%, probably because the plasma volume decreases as fluid moves into the interstitial spaces of the muscles. In short bursts of strenuous exercise the number of lymphocytes increases by about 50%, and in prolonged exercise the neutrophil count may increase by up to 80%. These effects are partly due to the redistribution of water in the fluid compartments, and the opening up of smaller blood vessels where cells have been trapped. Greater flow through the lymphatic system releases more lymphocytes. Neutrophil counts may also be increased by the accumulation of metabolites in the muscles and by substances released from muscle cells maintaining contraction at their limits.

The principal hormonal change during exercise is an increase in circulating adrenaline concentrations. In addition to its effects on blood vessels and the heart, adrenaline increases the concentration of glucose in the plasma, up to 13 mmol/l being reported in severe exercise. Plasma free fatty acid concentrations also rise, because adrenaline favours lipolysis in adipose tissue. The muscles can use free fatty acids as an energy source, but only up to about 25 kJ/min, beyond which stored and circulating glucose must be used for the rest of their needs.

## Recovery from exercise

As exercise ceases there is an immediate drop in cardiac output and ventilation to about midway between the exercising and the previous resting levels. Blood pressure may drop abruptly because although cardiac output is less, the peripheral resistance remains low due to the continuing vasodilatation in muscle and skin. This effect is fairly rapidly remedied. The immediate changes on ceasing exercise are not due to changes in the blood but to a removal of the psychic and motor cortical drives. Proprioceptive inputs from the exercising muscles and moving joints also cease.

After severe exercise, the heart rate may remain elevated for some considerable time, although after moderate exercise most of the increase should have disappeared in the first 2 min. Ventilation falls slowly over about 90 min. The stimulus for these changes is lactic acid which has accumulated during the exercise, and represents the greater part of the 'oxygen debt'. The creatine phosphate debt is repaid rapidly, but the oxidation of lactic acid by the heart, the muscles and, most importantly, the liver, takes a considerable time. During this period oxygen consumption considerably exceeds the carbon dioxide output, and so the respiratory quotient falls – possibly even below 0.5.

Body temperature remains high because, even though the hypothalamic set point has now been lowered, the dispersal of heat from the exercise still takes time. For this reason skin temperature falls relatively slowly.

The pH of the blood returns to its resting level over about one hour after severe exercise.

The hormonal adaptations are rapidly reversed. The distribution of the blood returns to its more usual pattern, and synthesis of glycogen and triglyceride stores resumes.

## Exhaustion

The first sign of distress in exercise is usually a feeling of breathlessness. This is produced by afferent impulses from the muscle spindles of the respiratory muscles and the joint receptors of the thoracic cage; it is related to the amount of work necessary to maintain the appropriate ventilatory rate.

Exhaustion of a muscle or a muscle group is not normally due to a fatiguing of neural transmission, or neuromuscular transmission, or even actual muscle contraction. Indeed, of these three, only neuromuscular transmission is likely to be affected. Depletion of acetylcholine in the nerve terminal as a result of repetitive stimulation would stop muscle activity, but such depletion is unlikely to occur in physiological situations. Exhaustion is much more a subjective sensation, depending upon the discomfort related either to respiration or muscle contraction. This discomfort arises from the build-up of acid metabolites, particularly lactic acid. The capacity for continued exercise may depend upon the ability to maintain anaerobic metabolism within the limits of the ability to metabolise the lactic acid formed. If muscle tension is sufficiently great to prevent blood reaching the muscle, or if alternating contractions are not sufficiently spaced to allow blood flow during the relaxations, muscle fatigue occurs rapidly. Alternate work and rest periods of 2 min and 1 min give the highest overall working rate and Scout's pace (alternate walking and running) is efficient for the same reason. Although in prolonged exercise exhaustion and the depletion of carbohydrate stores run in parallel, the two are not directly related. However some of the symptoms of exhaustion may still be due to hypoglycaemia.

Various symptoms of discomfort may be produced in normal activities. The symptom of 'stitch' has never been satisfactorily explained but it may be due to the reduction in blood supply to the liver. Thus it is more frequently experienced soon after a meal, when the liver blood supply is particularly important. Walking or gentle running for long periods, when neither respiratory nor circulatory systems appear to be unduly loaded, is perhaps associated with slight damage to joints or muscles which can cause the discomfort described as fatigue and also the subsequent stiffness. On a different scale, route marching over metalled roads can result in the appearance of myoglobin in the urine, due to damage to muscle cells.

## Severe exercise and limiting factors

In strenuous exercise the amount of stored glycogen in the muscle may be a critical factor. In strenuous, short term anaerobic exercise, the glycogen store in a muscle lasts only some 20 s, and this terminates the activity. More usually, the muscle can employ a mixture of aerobic and anaerobic glycolysis to give a time limit of 3–4 min. However, if liver glycogen can also be utilised this will provide capacity for around 15 min exercise. Indeed, in sustained submaximal exercise the liver store can last for about an hour – but this is the limiting factor for maintaining the level of exercise. A further problem with anaerobic glycolysis is that it produces lactate, causing a fall in pH, which in turn inhibits phosphofructokinase and restrains further glycolysis. Hydrogen ions also compete with calcium ions for sites on troponin and inhibit the actual process of contraction.

As exercise continues, the fall in circulating glucose and fatty acids causes inhibition of insulin secretion and stimulation of the secretion of glucagon, cortisol, and growth hormone in addition to adrenaline. There may be sensors of metabolic activity in the muscles which can augment the stimuli to the endocrine glands. As a result of these endocrine activities, free fatty acids are mobilised from adipose tissue, and glucose is produced in the liver by glycogenolysis and gluconeogenesis. These hormonal changes and mechanisms resemble those seen in starvation (and in stress due to trauma).

In severe exercise, renal blood flow decreases markedly as the blood is redirected to more immediately vital organs. Antidiuretic hormone secretion increases, and renin, angiotensin and aldosterone levels in the blood rise. Sodium and water are thereby conserved although there is an increasing loss due to sweating.

## Training

Repetition of exercise, provided the exercise is in excess of that which the subject would normally undertake, leads to increasing efficiency, both within the muscle groups directly concerned and in the cardiovascular adjustments. Muscles not used in the exercise are not affected. With high intensity exercise the muscle cells hypertrophy: both the protein content and the number of nuclei in the cells increase and so the strength of contraction increases. This change develops over weeks or months. With less strenuous but continued exercise the muscles rapidly develop an increased capacity for aerobic metabolism. Finally, the capacity of the muscles to take up oxygen is increased by an increase in the content of myoglobin and an increase in the number of blood capillaries in unit volume by up to 50%.

Training has little effect on the rise in blood pressure during exercise but it reduces the peripheral resistance, particularly in muscles themselves. It therefore increases blood flow to muscle during exercise, most noticeably in severe exercise.

The resting heart rate and the heart rate during moderate exercise are lowered by training. This is due to hypertrophy of the cardiac muscle resulting in a greater stroke volume so that at rest the same cardiac output can be produced at a lower heart rate. Isometric training tends to increase selectively the thickness of the ventricular walls, whilst endurance training gives a more general hypertrophy. Vagal inhibition of the sinoatrial node is also increased as a result of training.

In training for severe exercise the changes are more complex and are related specifically to the muscle groups trained. If the leg muscles have been trained the maximum heart rate during severe exercise is reduced only when the exercise involves the legs: there is no reduction during arm activity unless the arm muscles also have been trained. This may be because the trained muscles are able to contract aerobically more readily, so that the concentrations of lactate and hydrogen ions do not increase so dramatically, resulting in less feedback of signals to the motor cortex as it exerts control over the respiratory and cardiac responses.

The maximum rate of ventilation increases with training.

When training ceases, and the exercise is no longer continued, the adaptations are lost. Bed rest, or lack of use of a limb, will lead to muscle wastage with loss of some of the cell protein.

## Fitness tests

Most fitness tests measure the ability to perform a specific exercise, or the rate of recovery from a specific exercise. Their relevance to anything other than that exercise is questionable: however, they may be meaningful at the extremes, or in assessing the degree of impairment of activity after disease or injury. Most include measurement of the pulse rate or blood pressure; some measure ventilation or oxygen usage. The tests are usually variants of a step test, in which the subject steps up and down a box of known height at a set rate, or of exercise on a treadmill or bicycle, on which rates of working can be set.

In the Flack test, once used for selection of aircrews, the subject was asked to maintain a pressure of 5.3 kPa by blowing for a period of at least 10 s. The high intrathoracic pressure inhibits the return of venous blood, and after a very brief rise the blood pressure falls some 3–4 kPa. In unfit subjects the fall in blood pressure is greater, whilst in trained athletes it is less or may even be converted into a rise.

## Energy usage and efficiency in exercise

The amount of energy consumed during various activities (*Table 26.2*) and the amount of actual work performed can be measured and used to calculate efficiency by expressing the the ratio of energy consumed to work done as a percentage. By definition, work is done only when movement occurs: isometric contraction therefore does not result in work and has zero efficiency in these terms. It is also impossible to take into account such factors as resistance to movements by, for example, water when swimming.

Eight hours sleep at 4.62 kJ/min consumes 2.22 MJ; the eight hours spent at home awake uses about twice this – 4.44 MJ; and the eight working hours vary between using 3.4 MJ by an office worker to 10.9 MJ by a manual labourer. A lumberjack may use 16.5 MJ/day. In one study housewives were found to use between 3.7 and 18 MJ/day. Although household chores are less demanding than sprinting, they are performed over a much longer timescale.

Shovelling coal or wet snow is a very inefficient pastime at about 3% efficiency; it is often rated as the most strenuous non-sporting exercise.

Sprinting is achieved by utilising to the full the anaerobic energy pathway, and then discharging the

**Table 26.2 Energy consumption during various activities**

| Activity | Cost (kJ/min) | Activity | Cost (kJ/min) |
|---|---|---|---|
| Lying down | 4.8 | Shovelling dry, wet snow | 6.0, 16.0 |
| Sitting | 5.0 | Walking, 3–6.5 km/h | 12.0–21.0 |
| Standing | 5.2 | Swimming, 0.3–1.8 m/s | 21.0–500 |
| Ironing | 9.0 | Skiing, 4–14 km/h | 34.0–126 |
| 'Tidying up' | 10.0 | Running on level 10–25 km/h | 41–126 |
| Cycling (6 days) | 11.0 | | |
| Dusting, mopping | 12.0 | | |
| Making beds | 18.0 | | |
| Scrubbing floors | 20.0 | | |
| Rowing | 24.0 | | |
| Hewing coal | 29.0 | | |
| Cycling, 45 km/h | 100.0 | | |

oxygen debt afterwards. Thus a sprinter running 100 km in 10.8 s actually performs 30 kJ of work while using 110 kJ of energy – an efficiency of 27%. The need for oxygen exceeds 13 mol/min, but the actual amount used during the sprint is only 0.022 mol in 10.8 s, a debt of 2.15 mol of oxygen being accumulated. Since the maximum oxygen supply that can be produced is only 0.26 mol/min, the limit on performance is the capacity for anaerobic oxidation. Running longer distances gives figures for efficiency around 20%.

A racing cyclist at 45 km/h consumes about 100 kJ/min at an efficiency of around 22%.

Swimmers are relatively inefficient: front crawl using arms and legs at 1.5 m/s costs 40 kJ/min with an efficiency of only 4%. Breast stroke at 1.2 m/s is roughly equivalent. Using fins increases the efficiency by about half to 6%.

A footballer has an average energy expenditure of 5 kJ/min.

Generally speaking, the maximum work output – as opposed to energy usage – works out at about 90 kJ/min over very short time periods and about 7.5 kJ/min over periods of several hours. These outputs correspond to an efficiency of around 18%.

Although regular exercise may help to maintain weight at a reasonable level, diet is equally important. The energy available from 500 g of body fat is about 4.65 MJ – the equivalent of the energy spent in sawing wood continuously for eight hours! Similarly, a typical British ice-cream cone or a pint of bitter provide about 0.7 MJ and would require over an hour's wood-sawing to eliminate the energy gain!

## Further reading

Brooks, G.A. and Fahey, T.D. *Exercise Physiology,* John Wiley & Sons, New York, 1984

Durnin, J.V.G.A. and Passmore, R. *Energy, Work and Leisure,* Heinemann, London, 1967

Evans, D.E. *Exercise,* In *Variations in Human Physiology,* (ed. Case, R.M.), Manchester University Press, Manchester, 1985

# 27

# Emergency situations

## Haemostasis

The term haemostasis signifies the stopping of blood flow from a wound. In the body it does not consist solely of the process of blood clotting with assistance from the blood platelets, but is made up of a complex series of interactions of tissues, cells and enzyme reactions. Blood clotting itself has already been described (pp. 48–51).

The process of haemostasis varies according to the size of the vessels which have been damaged. In small capillaries the loss of blood may be very small because the endothelial cells of the wall come into contact with each other and stick together. In larger capillaries the platelets adhere to the damaged area, forming a temporary plug, over which the endothelial cells slide to renew the wall.

A more complex sequence of events proceeds fairly rapidly in small vessels such as arterioles or venules up to about 150 µm in diameter. Its time course as observed in the thin tissue of a rabbit's ear is outlined below. The process in human tissues under physiological conditions is slower, probably by a factor of 5–10 times.

### The first few seconds

Injury to the wall of a blood vessel produces an almost instantaneous reaction of vasoconstriction. This is probably due to an axon reflex, although in other circumstances axon reflexes are always vasodilator. However, if sympathetic nerves are damaged (as they may be in the vessel wall) they may become active. Another possibility is that the smooth muscle cells contract in response to stimulation by the injury just as they contract in response to stretching (p. 94).

Damage to the vessel wall has other effects. The local endothelium ceases to produce prostacyclin ($PGI_2$) which inhibits platelet aggregation. The pattern of blood flow is changed by the injury itself and by the vasoconstriction.

The permeability of the red cell membranes is changed either by the changed stresses due to the alteration in the flow pattern of the blood, or by contact of the cells with the vessel wall as the axial stream of cells breaks up at a slower flow rate: ATP and ADP leak out of the cells. Adenosine diphosphate causes the platelets to become sufficiently adhesive to stick to any exposed tissue surfaces. This happens within a few milliseconds (as assessed from the rate of flow of platelets past the site of damage) but occurs only if red cells are present. Another possible activation mechanism for platelet adhesion is the formation of thromboxane $A_2$ as a result of distortion of platelet membranes in the slight turbulence at the injury site. The inhibitory influence of the prostacyclin from the normal endothelium is lost after damage to the endothelium. The von Willebrand factor (part of the factor VIII complex – see Table 3.3) is essential for platelet adhesion to take place.

The damage to the vessel wall allows the blood plasma to come into contact with the collagen fibres and the smooth muscle of the vessel wall instead of the endothelial cell lining. Contact with extravascular tissues and the contents of damaged cells may also occur. These processes activate the intrinsic clotting mechanism through factor XII, but also activate factor VII of the extrinsic system. The combination of factors XII and XI (the Contact Product) also helps to activate factor VII in these circumstances. Endothelial cells contain a thromboplastin which is released as they are damaged. They also release a plasminogen activator; and activated factor XII activates a plasminogen pro-activator.

## Between 3 and 15 seconds

In the next few seconds a number of processes occur simultaneously. There may be some red cell haemolysis, releasing more ATP and ADP, and also a factor which acts as a partial thromboplastin. Central release of adrenaline causes some general peripheral vasoconstriction. The intrinsic clotting system proceeds slowly through the cascade of reactions towards thromboplastin formation (Fig. 3.4). In the extrinsic system the interaction between the lipoprotein tissue factor, factor VII, and calcium ions, activates factor X. Activated factor X forms a complex with phospholipid, factor V and prothrombin; this releases thrombin. The thrombin increases the conversion of ATP to ADP; this raises the concentration of ADP sufficiently to maintain the local vasoconstriction and accelerate the process of platelet adhesion so that a primary platelet aggregation rapidly closes the defect in the vessel wall. The primary platelet aggregation is not stable unless supported by later fibrin formation. Platelets begin to lose their granules, but not yet to any significant extent.

## From 15 to 30 seconds

By the end of this period the primary platelet aggregation has closed the defect in these small vessels and fibrin filaments are being laid down on its surface by the thrombin. Thrombin also activates factor XIII to stabilise the forming strands of fibrin. Some platelet factors are released, but it is not until sufficient thrombin has been activated that appreciable changes can be observed in the platelets.

## After 30 seconds

During the second half minute a fibrin network develops over the surface of the primary platelet plug. The platelets, under the influence of thrombin, begin to lose their granules and disintegrate, releasing ATP and ADP, platelet fibrin stabilising factor, antiheparin, fibrinogen, and serotonin. This takes over the role of maintaining vasoconstriction. Formation and release of thromboxane $A_2$ acting together with serotonin and ADP, further increases platelet adhesion and aggregation. Contact of the platelets with collagen and thrombin stimulates thromboxane and prostaglandin endoperoxide ($PGG_2$ and $PGH_2$) formation. These, like thrombin itself, cause shape changes, aggregation and degranulation of the platelets. Platelet factor III, a phospholipoprotein, appears on the surface of the aggregation, and accelerates the extrinsic and, more particularly, the intrinsic pathways of thrombokinase production. The thrombin formed locally on the aggregation is soon present in sufficient concentration to overcome the counter-coagulation agents in the blood and so fibrin formation now proceeds rapidly. Platelets and red cells become trapped in the fibrin network, and this secondary aggregation produces a firm seal to the defect. Over 24 h the platelet-rich plug is covered and stabilised by a mass of fibrin. There is some evidence that the intrinsic clotting mechanism is more important than the extrinsic in this process.

Some time later clot retraction occurs, possibly by contraction of a contractile protein from the platelets, thrombosthenin, or by a contraction of a complex formed from the actin and myosin that have been identified in the platelets. Microtubules appear to be important in the platelet shape changes in the early stages, but play no role in clot retraction.

Fibroblasts move into the clot when the healing process begins, endothelial cells slide across the gap to re-establish the lining of the vessel, and finally the vessel wall is renewed.

The role of thromboxane $A_2$ and the prostaglandin endoperoxidases is not entirely clear since aspirin does not always inhibit haemostasis as might be expected if they were critically important.

In larger vessels the platelet plug may be insufficient to seal the gap and blood passes out into the wound, where platelets and fibrin together may form a seal over the vessel. In the tooth socket, where a number of vessels are damaged, the clot forms over the damaged vessels. Bleeding from a large vessel, however, cannot be stopped by haemostatic processes alone, but requires pressure on the vessel to slow the loss of blood and allow clotting to take place.

## Haemostasis as observed in the patient

The process of haemostasis is complex, involving many interactions and mutually accelerating processes. Plasmin and other factors operate against clot formation. Whilst tests of blood clotting may be performed in the test tube and deficiencies of specific factors identified, the whole process of haemostasis can be examined only in the patient. The crude test of bleeding time, which depends mainly on the vascular reactions to a calibrated stab, usually in the earlobe, tries to allow for factors other than blood clotting alone. The bleeding time measured in this way is much shorter (2–7 min) than the clotting time of 5–11 min in a glass capillary. Formation of the primary platelet plug is probably a key factor in determining bleeding time, since patients who lack factor VIII, or only the von Willebrand factor, have extended bleeding times. Patients in whom bleeding continues for apparently abnormally long periods after dental extractions do not necessarily suffer from deficiencies of clotting factors. Continual disturbance of a forming clot in the stage of the primary platelet plug, by repeated rinsing, or simply an inquisitive tongue, may result in bleeding beyond the normal time interval. The effect of a suture across the socket, or a gelatine pad, may simply be that of a defence against interference with a forming clot. Temperature is another factor which affects haemostasis: cold increases vasoconstriction, favouring the vascular response, whilst warmth increases the rate of reactions within the clotting sequence. Rinsing with warm saline after tooth extractions may help in stopping blood flow.

## The fibrinolytic system

The process of haemostasis involves the formation of fibrin: other mechanisms exist to break it down. The degradation and removal of excess or inappropriately formed fibrin is termed fibrinolysis. The fibrinolytic enzyme is plasmin, which is formed from plasminogen (p. 51) by the actions of various plasminogen activators. The most important of these is synthesised and secreted by the endothelium of blood vessels. It is released in response to circulating adrenaline, venous occlusion, and possibly also the action of thrombin. The mechanism of activation appears to be by the binding of the activator and the plasminogen to adjacent sites on the fibrin molecules. This sets in motion the process of fibrinolysis at those sites. It is possible that factor XII, which itself has been shown to be a weak stimulator of activator production, may also activate plasminogen via the production of kallikrein. This could provide a means of controlling fibrin produced when the intrinsic blood clotting mechanism only is activated by minor trauma to vessel walls. The blood cells, with the possible exception of the macrophages, are not capable of activating plasminogen.

Both plasma and platelets contain antiplasmins and antiplasmin activators. These restrict the action of plasmin to the sites where it is actually attached to the fibrin molecules.

## Hormonal effects on the process of haemostasis

Adrenaline, released in stress, not only causes vasoconstriction, but also increases platelet numbers by stimulating their release from the spleen, and, in the presence of fibrinogen and ADP, causes platelet aggregation. It increases the concentrations in the blood of factor VIII. All these effects help to promote clot formation when damage to tissues or blood vessels occurs. In addition, adrenaline stimulates release of plasminogen activator to assist in the breakdown of fibrin if necessary.

In high concentrations the corticosteroids, also released in stress, promote the release of factor VIII and enhance the effect of adrenaline on the plasma concentration of this factor.

The third hormone released in stress which can affect haemostasis is antidiuretic hormone. This increases the amount of factor VII in the blood and, like adrenaline, increases the amount of plasminogen activator. The hormones of stress, then, help provide for enhanced activity of the extrinsic clotting system and also of the fibrinolytic system, both protective measures.

An increased incidence of thrombus formation (clotting in the absence of damage to vessels, p. 51) has been observed in subjects taking contraceptive formulations over long periods of time. The effect appears to be due to the oestrogen content of the tablets. Oestrogens increase the activity of factors VII and X in the extrinsic clotting system, and factors XII and X in the intrinsic. Antithrombin activity is reduced, although plasminogen and plasmin are increased in concentration.

In pregnancy the main effect on the haemostatic system is an inhibition of urokinase activity causing decreased activation of plasminogen.

## Shock and the recovery from haemorrhage

In medical and surgical contexts the term shock is generally defined as a state of inadequate tissue perfusion. It is almost always associated with a major increase in activity of the sympathetic nervous system and the release of adrenaline. The definition is more often expressed in terms of blood pressure. Shock is an acute or chronic state of lowered blood pressure which results in inadequate tissue perfusion. This second definition gives a logical classification of the shock syndromes. Blood pressure falls (Chap. 24) because the blood volume falls, or because the peripheral resistance falls, or because the pump itself is failing. There are, therefore, three types of shock:

(i) *Hypovolaemic shock* in which blood volume is decreased. Such shock may be due to blood loss, to trauma, to surgery, or to burns. It is also called 'cold shock' because the patient's skin is cold, and feels clammy. The other symptoms are a low blood pressure, a rapid thready pulse, a rapid rate of ventilation, thirst, and either restlessness or lethargy.

(ii) *Low resistance shock* in which the peripheral resistance is decreased by vasodilatation generally or in the larger organs only (e.g. muscle or skin). This is termed 'warm shock' because the vasodilatation renders the skin warm to the touch. The condition may be due to emotional stress, to toxins produced by bacteria within the body, or to histamine released in severe allergic reactions (as in anaphylactic shock). The dental surgeon may be called on to treat two types of low resistance shock in patients – fainting, or syncope, and the anaphylactic shock of an allergic patient given penicillin or other drugs for a second time after a first sensitising dose.

(iii) *Cardiogenic shock* is the result of a fall in blood pressure due to heart disease. It is similar to hypovolaemic shock but is associated additionally with congestion, mainly in the lungs, but also in the viscera.

In addition to these general categories there are a few other syndromes termed shock – such as electrical shock, and spinal shock (the physical state seen for two to three weeks after transection of the spinal cord). Both of these are effects on the nervous system rather than directly on the cardiovascular system.

## Hypovolaemic shock

### Haemorrhage

In haemorrhage there is a loss of all the constituents of the blood – water, salts, proteins, and cells. The simpler effects of volume loss are complicated by these other losses. Acute haemorrhage results in a fall in blood pressure which normally returns to normal levels within a few hours unless the haemorrhage is severe. A loss of 20% of the blood volume (about 1 litre) can usually be compensated, but losses of between 25 and 30% are less readily tolerated, and over 30% the prognosis is poor. Although blood pressure is normally restored within a few hours, it may fall again subsequently, possibly to result in irreversible shock. Immediately after a moderate haemorrhage (350–1000 ml), the patient shows the typical symptoms of hypovolaemic shock. In more severe haemorrhage blood pressure falls, but in less severe cases arterial blood pressure remains near normal despite a reduction in the pulse pressure. The skin is cool, pale or greyish, possibly with signs of cyanosis. The respiratory rate is increased. The patient complains of thirst. The inadequate tissue perfusion results in an increase in anaerobic glycolysis, with increased production of lactic acid. There is usually, therefore, an acidosis due to the increased amounts of circulating lactic acid – as much as 9 mmol/l. The shock of haemorrhage is resolved by compensatory mechanisms which may occur immediately, or over a longer period.

### Immediate compensatory mechanisms

The initial symptoms of haemorrhagic shock are caused by the immediate compensatory mechanisms (Fig. 27.1). A reduction in blood pressure is sensed by the baroreceptors and the baroreceptor reflex inhibits vagal tone while enhancing the sympathetic

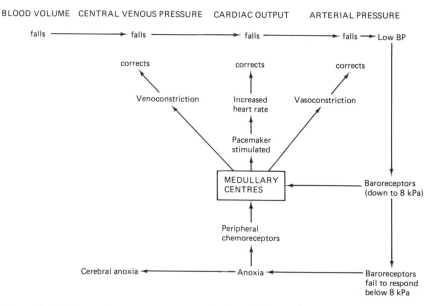

**Figure 27.1** The immediate compensatory mechanisms in haemorrhage. Note that there is a lower limit beyond which the baroreceptor response cannot produce any further compensatory reactions, and at this level there is a possibility of further reactions stimulated by hypoxia, although hypoxia of the medulla and higher centres will depress their activity.

outflow to the heart. As a result the heart rate increases to give tachycardia. Sympathetic impulses have an inotropic as well as a chronotropic effect. Sympathetic activity also causes venoconstriction in the venous blood reservoirs of the lungs, the liver, and the skin, and arteriolar vasoconstriction in the skin, skeletal muscle, the kidney, and the viscera. This may be accompanied by contraction of the smooth muscle in the spleen, although in man the spleen is relatively unimportant as a blood reservoir. The heart and the brain are the only organs spared in this general vasoconstriction. Indeed, accumulation of metabolites in the coronary circulation results in vasodilatation. The paleness and coldness of the skin is due to the vasoconstriction, and the signs of cyanosis to the stasis of blood in the skin capillaries. Patients suffering from hypovolaemic shock should not be kept too warm, since any cutaneous vasodilatation will exacerbate the condition, or given alcohol, which is a cutaneous vasodilator. There is severe visceral vasoconstriction. The ischaemia (lack of supply of blood) is particularly marked in the kidneys. Constriction of both the afferent and the efferent glomerular arterioles is observed although it is greater in the efferent. As a result the glomerular filtration rate may fall to a lesser extent than the severely reduced renal plasma flow. Thus the filtration fraction may actually increase. Little urine is formed and so urea and sodium ions are retained. If the period of hypotension is at all prolonged there is invariably some damage to the cells of the kidney tubules. All these effects are seen with relatively small falls in general arterial blood pressure.

The general vasoconstriction raises the peripheral resistance and maintains the diastolic blood pressure. Systolic pressure, however, falls, and so the pulse pressure is also reduced.

Constriction of the arterioles together with the fall in venous pressure as a result of the loss of blood volume results in lower than normal hydrostatic pressures in the capillaries. As a result fluid moves into the capillaries from the interstitial fluid. As much as a litre of blood volume may be restored in this manner in an hour. The haematocrit value falls as plasma volume is increased with the total number of cells unchanged. Loss of a litre of blood may cause the haematocrit value to fall to 35% over the next hour. Transfer of interstitial fluid in turn causes the removal of water from body cells. Although plasma osmolality may be unchanged, and so will not stimulate the hypothalamic osmoreceptors, the fall in blood pressure causes renin release and angiotensin II formation – resulting in thirst.

If systolic blood pressure continues to fall below about 8 kPa the baroreceptors become progressively less important since they are no longer stimulated. However, blood flow to the carotid and aortic bodies is decreased, and so there is an effective hypoxic stimulus. In addition the smaller number of red blood cells constitutes an anaemia, and the general reduction in blood flow results in stagnant hypoxia. Both hypoxia and acidosis stimulate the chemoreceptors. These then increase the ventilatory drive, causing rapid respiration, and also increase the sympathetic outflow from the medullary vasomotor centres. When blood pressures are too low to stimulate the baroreceptors, the chemoreceptors provide the only stimulus for further vasoconstriction.

Below about 5.3 kPa, the blood pressure compensatory mechanisms can no longer maintain an adequate blood flow to the brain. The cerebral ischaemia at first increases the sympathetic discharge to maximum levels, but later also stimulates vagal activity. The dull listlessness or torpor of patients in this condition reflects the falling oxygen supply to the brain. The adrenal medulla increases its output of catecholamines to emergency concentrations – increasing the output of adrenaline up to 50 times, and that of noradrenaline up to 10 times resting levels. This, acting via the reticular activating system, may overcome the cerebral damping so that the patient becomes restless and apprehensive, increasing muscular activity and respiration and thus increasing the rate of return of the blood to the heart.

### Longer term compensatory mechanisms

Longer term compensation results from more indirect effects (Fig. 27.2). The oncotic activity of the plasma is maintained at first by the mobilisation of stored plasma albumin from the liver. Nonetheless, in severe haemorrhage some dilution occurs. The liver begins an increased rate of synthesis of new plasma protein, but the loss of the proteins from a litre of blood cannot be restored in less than 3 or 4 days.

Renin secretion and consequent angiotensin II production stimulate aldosterone secretion: this supplements and continues the retention of sodium ions by the kidney, and maintains the osmolality of the blood by reducing salt loss (Fig. 27.2). Vasopressin (antidiuretic hormone) is released as the hypothalamus receives information about the low blood pressure from the baroreceptors and the volume receptors of the left atrium. This increases water retention – again supplementing and continuing the retention begun by the reduction in blood flow to the kidneys.

The fall in concentration of circulating red blood cells decreases the amount of oxygen available to the tissues. In the kidney this causes release of renal erythropoietic factor and the formation of erythropoietin, which stimulates red cell production in the bone marrow, bringing the reticulocyte count up to a peak in about 10 days. The total red cell count,

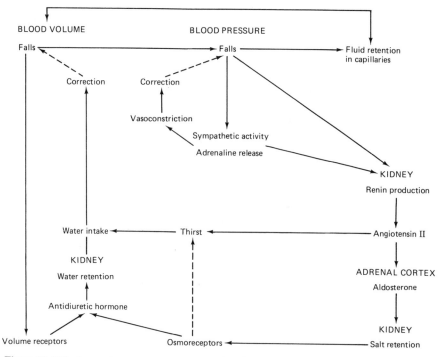

**Figure 27.2** The longer term compensatory reactions in haemorrhage. Stimulation of fluid volume receptors will also cause secretion of ACTH and affect aldosterone secretion.

however, takes 4–8 weeks to reach its former level. The red blood cells have increased concentrations of 2,3-diphosphoglycerate and, as a result, release oxygen more readily in the tissues. The red cell count must fall to half its normal level before increased respiration will result from the lack of oxygen, and so no respiratory symptoms of anaemia are observed. The white blood cells normally have a much more rapid turnover rate than do the red blood cells: their numbers are restored in about a week after a moderate haemorrhage as shown in Fig. 27.3).

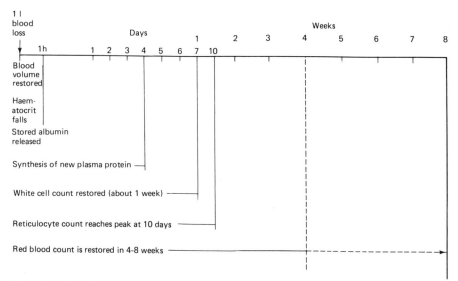

**Figure 27.3** The timescale of restoration of blood volume after a moderate haemorrhage.

### Irreversible shock

Loss of blood, depending upon the volume actually lost, ends either in recovery or death. Sometimes a partial recovery seems to occur, but the state of shock persists, and finally peripheral resistance falls as vasoconstriction can no longer be maintained. The heart is unable to compensate, slows, and the patient dies. Several factors are involved. Cerebral ischaemia results in a progressive loss of function in the medullary centres. Coronary ischaemia impairs the functioning of the heart. Hypoxia in the tissues results in vasodilatation, usually with stagnation because of the low blood pressure – and a vicious circle is set up.

### Other examples of hypovolaemic shock

Traumatic shock is caused by injury to muscle and bone. Internal bleeding results in hypovolaemia. In traumatic shock and in all the other forms of shock mentioned below, there is an increased secretion of thyroid hormones, and a resultant increase in basal metabolic rate. Cortisol secretion is also increased. These hormonal effects are part of the stress experienced after trauma and are dealt with later.

Surgical shock and wound shock are combinations of external and internal haemorrhage, often combined with dehydration. There may, however, be some differential loss of blood plasma due to the release of substances from damaged cells, which can increase the permeability of the capillaries with the effects described below.

In burn shock the burned surfaces lose plasma rather than blood. Further, the damaged cells of the burned tissues release proteins and peptides into the tissue fluid, reducing the oncotic gradient across the capillary walls and increasing the loss of fluid from the vascular compartment. Haemoconcentration often occurs: the packed cell volume value is raised. The increased concentration of red cells, and the increased viscosity due to this, results in greater resistance to flow through small vessels and actual plugging of small capillaries. The degree of shock depends upon the depth of the tissue damage and the area of the body surface involved. Deep burns over 75% of the body surface are invariably fatal.

## Low resistance shock

Shock due to vasodilatation in extensive capillary beds occurs in emotional disturbances, in certain bacterial infections, and in anaphylactic reactions.

It is most commonly seen as syncope, or fainting, a transient incident of shock. Loss of consciousness occurs as a result of cerebral ischaemia; the patient falls, and in the lying position blood then returns more easily to the heart and from there to the brain. Placing the dental chair in a horizontal position, or putting the patient's head between the knees may prevent loss of consciousness.

Although the term fainting is usually used of the collapse resulting from cerebral ischaemia, the sequence of events in syncope begins before this occurs. In emotional fainting there is a massive increase in sympathetic activity and release of adrenaline. This causes an increase in heart rate (tachycardia), an increase in stroke volume, and the two factors increase cardiac output. The circulating adrenaline causes cutaneous vasoconstriction, which appears as skin pallor, and a cold sweat. Emotional stimulation of the hypothalamus also results in increased production of antidiuretic hormone which enhances the vasoconstrictor effect. It decreases urine flow, but the effect of the vascular changes and possibly the hormone itself, gives a feeling of nausea and a desire to defaecate. The reasons for the train of events which follows are still not entirely clear but the present view is that the sudden increase in sympathetic activity causes the heart to attempt to increase cardiac output before the venous return is adequate to sustain it. This stimulates ventricular receptors as a result of inadequate filling of the ventricles and these switch off the sympathetic stimulus to replace it with a powerful vagal activation. This causes bradycardia – even, sometimes, cardiac arrest. The sympathetic nerves to blood vessels in skeletal muscle are activated, probably directly through the hypothalamus, to cause vasodilatation via the cholinergic endings. This causes blood to pool in muscle, and the size of the total muscular vascular bed is such that peripheral resistance and blood pressure drop catastrophically. This, together with the bradycardia, reduces blood flow to the brain, and results in a central inhibition of muscle tone, affecting particularly the muscles of posture. The standing patient falls. Since adrenaline has been circulating throughout the sequence of events, skin pallor and sweating persist.

The combination of slow heart rate due to the vagus, and the drop in peripheral resistance due to vasodilatation, have led to fainting being called a vasovagal attack.

Fainting may also occur after a blood loss of around 20% of the total blood volume, or after standing motionless for some time, as the blood pools in the limbs unless small muscular movements can be made. In these circumstances emotional factors may be absent. In hypovolaemic fainting the fall in blood pressure is detected by the baroreceptors; this results in tachycardia which cannot be sustained because of the reduction in blood volume and the consequently decreased venous return. A similar sequence of events to that described in emotional fainting follows. In postural syncope blood pressure falls because of the pooling of blood in the legs and the reduced central venous pressure: the baroreceptor reflex initiates tachycardia and this leads into the same train of events. In both these types of syncope the manifestations of sympathetic activity associated with emotional fainting may be absent.

## Cardiogenic shock

Cardiogenic shock results from cardiovascular disease. The heart in such conditions cannot pump as much blood as is returning from the tissues and so blood accumulates on the venous side of the circulation in the viscera, and also in the lungs – a state termed congestion. The situation arising has already been described in Chap. 16. Initially, the greater volume of blood on the venous side of the heart improves the efficiency of pumping by stretching the cardiac muscle fibres (the Starling effect) but eventually the stretching exceeds the compensatory effect. The dominant factor then becomes the Law of Laplace, which says that the force necessary to reduce the circumference of the chamber is proportional to the diameter – the more the cardiac chambers are stretched, the greater will be the force necessary to counteract this. The heart can improve its performance by muscle hypertrophy in response to the increased load, but eventually it ceases to be able to pump blood from the enlarged chambers.

# Stress

The term stress is one which everyone understands and no-one can define. It usually refers to chronic emotional or nervous reactions to barely tolerable situations, to overwork, or to conditions at the limits of physical or mental endurance. Such reactions are hormonal, involving mainly the adrenal glands, but also some hormones of the hypothalamus and the pituitary gland. They can be explained as the reactions normally invoked in the short-term for fight or defence. Over longer periods they become difficult to sustain and the hormonal adjustments which are appropriate for acute reactions are not easily consistent with the maintenance of normal homeostasis. As a result, there is a compromise which impairs both the ability to maintain a normal metabolic pattern and the ability to respond to new demands for defensive or attacking behaviour. Although the main interest in these conditions from a medical point of view is the attenuation in body defences which seems to be associated with them, it is worth pointing out that the human being functions more efficiently if subjected to a certain amount of stress.

## General stress reactions

Stress reactions can be invoked by stimulation of the limbic brain or of some parts of the hypothalamus. They are the normal responses of the sympathetic system to potentially dangerous situations. Acute stress – pain or emotion – is associated with changes in the cardiovascular system, the respiratory system, and the gastro-intestinal system (Fig. 27.4). The heart rate and pulse pressure increase, the normal baroreceptor reflex being inhibited by hypothalamic activity. Muscle blood flow increases because of sympathetic vasodilator impulses. Respiration increases in rate and depth – a reaction which might, by lowering the partial pressure of carbon dioxide in the lungs, help to anticipate muscular activity. Bronchioles are dilated. In the gastro-intestinal tract emotion produces varying responses. Thus some observations suggest that fear inhibits gastric motility and acid production, whilst anxiety and anger may increase acid secretion. In general, stress is associated with a reduction in motility throughout the gastro-intestinal tract.

In almost all stress situations adrenaline is released from the adrenal medulla. This enhances the sympathetic responses induced by activity in the sympathetic nervous system: the increase in heart rate, blood pressure, and peripheral vasoconstriction except in the blood supply to muscles. Bronchodilatation is also an effect of sympathetic activity, as possibly are the gastro-intestinal reactions. Adrenaline appears to enhance muscular activity when the muscles begin to be fatigued, possibly by causing some vasodilatation by release of kinins or of potassium ions in the fatiguing muscles. Liver glycogen is mobilised, causing hyperglycaemia. Blood clotting occurs more rapidly when circulating adrenaline levels are high: the mechanism behind this observation has not yet been completely elucidated, although some of the possible explanations have been given earlier in this chapter. Measurements of both adrenaline and noradrenaline levels in dental patients have been used to evaluate the degree of emotional stress during various procedures.

The hypothalamus also orchestrates other reactions in stress. Corticotrophin releasing hormone passes to the anterior pituitary where it increases the secretion of adrenocorticotrophic hormone which stimulates the adrenal cortex to produce cortisol. The physiological role of cortisol in stress is difficult to appreciate. It may collaborate with adrenaline in making available an adequate supply of fuel for muscular activity (see p. 206). A similar result is also attained by the relative increase in glucagon secretion and a decrease in insulin secretion. Equally the cortisol may be important in regulating the distribution of body water and salts and it is possible that it is involved in maintaining blood

330  *Emergency situations*

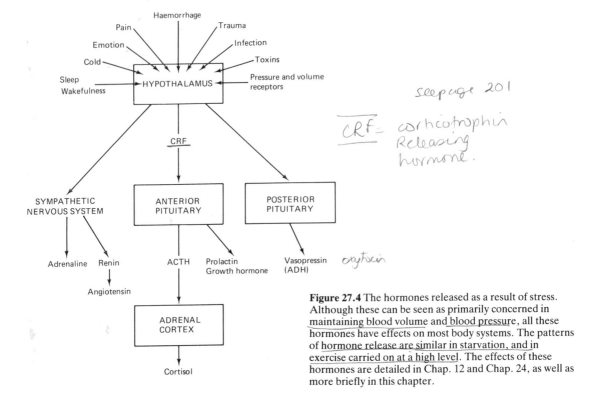

Figure 27.4 The hormones released as a result of stress. Although these can be seen as primarily concerned in maintaining blood volume and blood pressure, all these hormones have effects on most body systems. The patterns of hormone release are similar in starvation, and in exercise carried on at a high level. The effects of these hormones are detailed in Chap. 12 and Chap. 24, as well as more briefly in this chapter.

controlling kidney function. Vasopressin (antidiuretic hormone) secretion is also increased during stress – and again its exact role is unclear. The rate of urine production is decreased, but this is only marginally a defensive reaction. One integrative suggestion is that in stress the body endeavours to protect blood pressure and blood volume. Growth hormone is also released in stress as part of the metabolic adaptation.

## Chronic stress

In continued stress there is hypertrophy of the adrenal cortex and the increased levels of cortisol which are maintained cause some atrophy of lymphoid tissue and a reduction in the white blood cell count. Prolactin levels in blood are frequently raised. The cause and significance of this is unclear, although prolactin has an action in maintaining salt and water balance.

Stress appears to affect sexual activity by inhibiting gonadotrophin production in the anterior pituitary, presumably via suppression of LHRH secretion. Thus the regularity of the menstrual cycle can be disturbed in stressful situations. In the male there may be temporary impotence. Increased concentrations of prolactin in the blood have been found in both males and females suffering from stress – although lactation often seems to be inhibited by stress. In stress without physical trauma the secretion of thyroid hormones is reduced, in contrast to the situation in shock.

The principal oral effects which have been attributed to continued stress are an increased incidence of acute ulcerative gingivitis and the development of aphthous ulcers. The reasons for these conditions being associated with stress are difficult to explain. It may be that the continued hypersecretion of adrenaline can cause a chronic vasoconstriction in the gingivae and the resultant poor nutrition of the tissues renders them more susceptible to infection. Cortisol reduces the number of circulating white cells and may deplete the body defences in this way. Another possibility is behavioural: subjects living stressful lives may spend less time and effort on their oral hygiene and become more susceptible to infection as a result.

## Stress due to injury

Stress due to injury manifests itself in three phases. Acute stress is associated with a rapid increase in circulating glucose concentrations invoked first by an inhibition of insulin secretion and then, after 2–3 h, an increase in glucagon secretion. These effects may result from a release of adrenaline, which also causes vasoconstriction. The net effect of the hormonal changes is a mobilisation of fuel for any response to attack – be it fight or flight. Glucose is mobilised from the liver and within muscles from glycogen, and the sympathetic nerves to adipose tissue stimulate lipolysis to send free fatty acids and glycerol into the bloodstream.

Within 1–2 days of the acute response, an ebb phase occurs. Ontogenically, injury impairs the animal's ability to seek food, making a period of retrenchment necessary. In the ebb phase hyperglycaemia persists, but it is now due to a switching away from glucose utilisation as a source of energy. Adrenaline now becomes a major suppressing influence on insulin activity and growth hormone and the glucocorticoids all co-operate to reduce tissue uptake of glucose. Observations in man during this phase suggest that blood insulin levels fall, although the opposite is seen in some animals. The free fatty acids and ketone bodies resulting from lipolysis also switch cells away from glucose utilisation. As a result of the decreased oxidative metabolism in the cells in the ebb phase, body temperature may fall.

After some 3–6 days, provided metabolic failure does not ensue (Chap. 24), the metabolism enters a recovery, or flow, phase. This may last for some considerable time while healing continues. Metabolic rate rises, heat production increases, and both heart rate and respiration are increased. Blood glucose levels stay high, partly because muscle protein is now being used for gluconeogenesis under the influence of cortisol and the continuing high levels of glucagon in relation to insulin, and partly because fat continues to be used as the main energy source. Recovery from injury is almost invariably associated with loss of weight, in which the loss from the protein of skeletal muscle is some three times as great as that from adipose tissue.

In addition to the metabolic effects, stress due to injury or surgery may include disturbance of electrolyte concentrations in body fluids. Treatment must take these into consideration. In crush injuries there is loss of cell contents, raising concentrations of potassium in extracellular fluids. Loss of blood due to injury or surgery results in lowered levels of sodium as the body fluids equilibrate to restore blood volume. Later, sodium retention and potassium loss through the kidneys are increased in the compensatory reactions to hypovolaemia. Acidosis is commonly observed, but some surgical operations may result in a respiratory alkalosis.

## Inhibition of possibly beneficial stress reactions by steroid therapy

Corticotrophic compounds have been used in dentistry as anti-inflammatory agents in treating inflamed tooth pulps and aphthous ulcers. Although atrophy of the adrenal cortex due to long term administration of corticosteroids is unlikely with the amounts used dentally, it is significant when patients are undergoing systemic corticosteroid therapy: their reaction to stress, such as that involved in surgery, may then be inadequate and they may need further doses of steroids to enable them to recover normally.

## Further reading

Little, R.A. *Injury (shock)* In *Variations in Human Physiology* (ed. Case, R.M.), Manchester University Press, Manchester, 1985

Ogston, D. *The Physiology of Haemostasis,* Croom Helm, London, 1983

Porter, R. and Knight, J.(eds) *Energy Metabolism in Trauma,* CIBA, Churchill, London, 1970

Speirs, R.L. (1983) Some reflections on the mechanism of the common faint. *Dental Update* **10**: 644-650

# Appendix

**Table 1 Conversion of kPa to mmHG**

| kPa | mmHg |
|---|---|
| 0.25 | 1.9 |
| 0.4 | 3.0 |
| 0.8 | 6.0 |
| 1.0 | 7.5 |
| 1.3 | 10.0 |
| 1.5 | 11.5 |
| 2.0 | 15.0 |
| 2.2 | 16.5 |
| 2.4 | 18.0 |
| 2.5 | 19.0 |
| 2.7 | 20.0 |
| 3.0 | 22.5 |
| 3.2 | 24.0 |
| 3.3 | 25.0 |
| 3.4 | 25.5 |
| 3.5 | 26.5 |
| 3.7 | 28.0 |
| 4.0 | 30.0 |
| 4.3 | 32.5 |
| 4.4 | 33.0 |
| 4.8 | 36.0 |
| 5.0 | 37.5 |
| 5.2 | 39.0 |
| 5.3 | 40.0 |
| 6.0 | 45.0 |
| 6.2 | 46.5 |
| 6.3 | 47.5 |
| 7.0 | 52.5 |
| 8.0 | 60.0 |
| 9.0 | 67.5 |
| 9.5 | 71.5 |
| 10.0 | 75.0 |
| 10.6 | 80.0 |
| 11.0 | 82.5 |
| 12.0 | 90.0 |
| 12.8 | 96.0 |
| 13.0 | 97.5 |
| 13.3 | 100.0 |
| 15.0 | 112.5 |
| 16.0 | 120.0 |
| 20.0 | 150.0 |
| 21.3 | 160.0 |
| 24.0 | 180.0 |
| 25.0 | 188.0 |
| 101.1 | 760.0 |
| 250.0 | 1880.0 |
| 700.0 | 5250.0 |

1 kPa = 7.5 mmHg; 1 mmHg = 0.133 kPa

**Table 2 Conversion of mmol/l of gas to ml/dl**

| mmol/l | ml/dl |
|---|---|
| 0.1 | 0.224 |
| 0.2 | 0.448 |
| 0.3 | 0.672 |
| 0.4 | 0.896 |
| 0.5 | 1.120 |
| 0.6 | 1.344 |
| 0.7 | 1.568 |
| 0.8 | 1.792 |
| 0.9 | 2.016 |
| 1.0 | 2.240 |
| 2.0 | 4.48 |
| 3.0 | 6.72 |
| 4.0 | 8.96 |
| 5.0 | 11.20 |
| 6.0 | 13.44 |
| 7.0 | 15.68 |
| 8.0 | 17.92 |
| 9.0 | 20.16 |
| 10.0 | 22.40 |

1 mmol/l = 2.24 ml/dl; 1 ml/dl = 0.446 mmol/l

# Index

Abdominal expiration, 128, 242
Absorption in gut, 10, 40–41, 187–188, 302–303
 absorptive capacity, 188
 area available in small intestine, 40
 in colon, 188
 in duodenum, 188
 in ileum, 188
 in jejunum, 188
 in mouth, 188
 in small intestine, 40–41, 184, 187–188
 in stomach, 188
Acceleration, perception of, 153
Accessory gland saliva, 178, 179
Accessory salivary glands, 179, 180, 181, 182, 275
Acclimatisation to hot and cold, 312, 313
Accommodation of eye, 146, 147, 149
Acetoacetic acid, 299
Acetone, 296
Acetylcholine, 79–80, 81, 99, 139–140, 143–144,
 action in salivary glands, 274
 effect on pulp blood pressure, 235
 effect on sino-atrial node and cardiac muscle, 80, 99, 118
 effect on skeletal muscle, 80
 effect on smooth muscle, 95
 receptors, see Cholinergic receptors
 released by Renshaw cells, 81, 228
Acetyl cholinesterase, 79, 143–144
Acetyl CoA, 143, 298–299, 301
Acid
 as taste stimulus, 160, 272
 ingestion, effect on blood pH, 251–252
 production in dental plaque, 190
 secretion in stomach, 32
Acidaemia, 251, 253

Acidophils, 65, 66, 67
Acidosis, 240, 251, 255, 296, 304, 305, 325–326, 331
Acinarglands, 30–31
 see also Salivary glands, Sweat glands, Pancreatic glands
Acinus, 30, 31, 32, 34–36, 180, 183
 blood supply to, 180
 diffusion across, 183
 primary fluid, 31, 36
 secretion, 36
 pancreatic, 31
 salivary, 30–31, 34–35
 sweat gland, 30–31
Actin, 71, 85–88, 89, 94, 96, 323
Action potential, 75–78, 80, 83, 86, 217
 in nerve, 75–78
  composite, in mixed nerve, 78
  conduction, 77
  duration, 75–76
  frequency in relation to stimulus, 83–85
  generation in motorneurone, 81
  heat production, 76
  ionic basis, 75–76
 in cardiac muscle, 118
  conduction, 97
  duration, 86, 98
  ionic basis, 86, 98
  plateau, 97
 in smooth muscle, 94–95
  conduction velocity, 95
  ionic basis, 86, 95
  plateau, 95
 in skeletal muscle, 90–91
  conduction velocity, 95
  ionic basis, 86, 91
Active transport, 4, 12–14
 calcium, 9, 307
 chloride, 107, 188
 glucose, 14, 39, 188

Active transport (cont.)
 hydrogen carbonate, 184, 188
 in renal proximal convoluted tubule, 39, 172–173
 in small intestine, 41, 188
 iodine, 203
 of ions in salivary ducts, 36, 184
 sodium, see Sodium/potassium ATPase
 organic acids in kidney, 39, 173, 194
 transport maximum type, 14, 39
Actomyosin, 86
Acupuncture, 164
AcylCoA, 301, 302
Adaptation of sensory receptors, 84, 85, 157, 158, 162
Addison's disease, 203
Adenohypophysis, see Pituitary gland, anterior
Adenosine diphosphate (ADP), 71–72, 236, 318, 322–323, 324
Adenosine triphosphate (ATP), 8–9, 13, 38, 85, 92, 93, 195, 297–298, 299, 300, 322–323
Adenylate cyclase, 195–196
ADH, see Antidiuretic hormone
Adipocytes, 287, 296, 300, 301
Adipose tissue, 145, 287–288, 297, 298
 action of adrenaline, 205
 adrenergic receptors, 145
 aminoacid metabolism, 299
 blood flow, 112
 brown, 287, 288, 312
 effect of glucagon, 302
 glucose metabolism, 298–299
 glucose uptake, 303
 growth and development, 287, 291, 293
 in males and females, 15–16, 293
 lipid metabolism, 47, 299, 300–302, 318–319

335

336  *Index*

Adipose tissue (*cont.*)
  size of fat store, 300
  water content, 15–16
Adrenal androgens, 194, 287, 289, 291–292, 293
Adrenal cortex, 201, 205–207, 327, 330
  atrophy, 206, 331
  fetal intermediate zone, 203, 206, 289
  hypertrophy, 330
  zona fasciculata, 205
  zona glomerulosa, 205, 206
  zona reticulata, 205
Adrenal (suprarenal) glands, 199, 205–207
Adrenal medulla, 38, 142, 145, 205, 315
  innervation, 141–142
Adrenaline, 79, 81, 140, 144, 192, 193, 200, 204, 205, 238, 246, 296, 301, 302, 304, 312
  and blood clotting, 72, 323–324
  concentrations in plasma, 205
  control of apocrine sweat glands, 313
  differential action on blood vessels, 231
  effect on brown adipose tissue, 288, 312
  effect on cardiac rate and contractility, 99, 121, 232, 317–319
  effect on kidney, 210, 234
  effect on pulp blood pressure, 235
  effect on smooth muscle, 95
  excretion, 205
  in stress, 326–331
  metabolism, 205
  receptors, *see* Adrenergic receptors
  vasomotor effects, 232–235, 315–317, 326–331
Adrenocortical hormones, 194, 201, 205–206, 278, 284, 302, 304
Adrenocorticotrophic hormone (ACTH), 192, 197, 201, 203, 205, 206, 216, 239, 283, 291, 327, 329, 330
Aero-odontalgia, 167
Aerodynamic theory of sound production, 267
Aerophagy, 178
After-discharge, 220
Age, measures of
  bone, 286
  chronological, 285, 291
  dental, 286
  fetal, 285, 286
  mental, 290
  physical, 290
Ageing
  effect on blood pressure, 216
  effect on hearing, 151
  effect on pulse wave, 124
  effect on saliva composition, 276

Ageing (*cont.*)
  effect on taste perception, 159, 161
  of red blood cells, 13, 48, 55, 56, 63
  of white cells, 65–66
  loss of tissue elasticity, 130, 216
  osteoporosis, 306
Agglutination of red blood cells, 59
Agglutinins, 52, 59
Agonist muscles, 222
Agranulocytes, 65, 66, 67–70
Agranulocytosis, 65
Air
  atmospheric, composition, 107
  atmospheric, partial pressure of oxygen, 108, 253
Airways, 9, 29, 30, 126, 129
  innervation, 141
  irritation, 130, 243
  epithelium, 101, 234
  mucosa, 130
  obstruction, 246, 251
  protection of, 101, 243–244
  resistance, 104, 130, 131, 243
  velocity of air flow, 101
Alanine, 172, 296, 301, 303, 304
Albumin, 18, 28, 36, 43–47, 306
  carriage of ions, 46, 306
  daily production in adult, 44
  hormone binding, 203, 207
  in glomerular filtrate, 171
  in saliva, 27, 182
  store in liver, 326, 327
Alcohol, 81
  absorption in stomach, 187
  effect on hormone secretion, 200, 208
Aldosterone, 191, 194, 195, 201, 205–206, 209, 239, 240, 283, 313, 319, 326, 327
  effect on salivary glands, 206, 272, 276
  effect on intestine, 206
  metabolism and excretion, 206
Algesics, 165
Algogogues, 165, 167
Alkalaemia, 251, 253
Alkali
  buffering of, 252
  ingestion, effect on blood pH, 250, 251, 253
  production in dental plaque, 190
Alkaline tide, 33
Alkalosis, 240, 251
All or none rule, 76, 77, 80, 83, 91–92
Allergic reactions, 52, 53, 67, 206, 324
Alveolar air
  composition
    in non-perfused alveoli, 131
    stability, 112, 129
  partial pressures of gases, 11, 108
  partial pressure of oxygen, 11, 108, 109, 244
  partial pressure of carbon dioxide, 11, 108, 111, 315–317
Alveolar epithelium, 126, 130

Alveolar macrophages, 70
Alveolar process of jaw, 287
Alveolar ventilation, 129, 131, 248
Alveoli, 126, 127
  surface area, 126
Amacrine cells, 147–148
Amidopyrine, 65, 72
Amine precursor uptake decarboxylase system (APUD), 191
Aminoacids, 21
  absorption in gut, 41, 184, 187
  active transport, 14, 39, 41, 187
  free, concentration in plasma, 43, 47
  in saliva, 36, 181, 183
  metabolism, 6–7, 296, 299
  pool, 44, 47, 64, 296, 303
  reabsorption in renal proximal convoluted tubule, 39, 172
  uptake in liver, 302, 304
  uptake in muscle cells, 301
$\gamma$-Aminobutyric acid, 79, 81, 139–140, 148, 183
Aminopeptidase, 184–185, 185–186
Ammonia, 48
  daily excretion, 255, 296
  in dental plaque, 190
  in saliva, 183, 189
  production in distal tubule, 39, 174, 255, 304
Amylase,
  salivary, 35, 182, 185, 188, 209
    calcium binding, 183
    effect of flow rate, 182
    inactivation in stomach, 185
    optimal pH, 185
  pancreatic, 184, 185, 186
Anabolic hormones, 206–207, 287, 288, 296, 301
Anaemia, 55–58, 59, 62–63, 64, 65, 110, 246, 247, 326–327
Anaesthesia
  gaseous, 55, 112–113, 130, 243
  general, 77, 137, 219, 248
  in relation to stomach emptying, 178
  local, 77, 78, 144, 200, 219
    effect on taste, 161
  solubility, uptake and distribution of gases, 112–113
  use of muscle relaxants, 80, 128
Anal sphincters, 175, 176, 280
Analgesia, 165, 166
Anaphylatoxin, 67
Anaphylactic reactions, 328
Androgenic hormones, 110, 194, 202, 203, 205, 206–207, 208, 210, 282, 289–293
Anger, 135, 231, 329
Angina pectoris, 165
Angioneurotic oedema, 67
Angiotensin, 95, 192, 197, 200, 206, 209–210, 236, 238, 239, 319, 326, 330

Angiotensinogen, 209–210
Annulospiral endings, 82, 85, 155–156
Anorexia, 295
Anoxaemia, 246
Anoxia, 246
  *see also* Hypoxia
Antagonist muscles, 219, 222, 228
Anterior horn cells, 139, 219
Anterior pituitary gland, 122
  *see also* Pituitary gland, anterior
Antibodies, 45, 52, 59–60, 67, 70
  blood group, 59–60
  in saliva, 182, 188
  corticosteroid inhibition of, 206
Antidiuretic hormone (ADH), 38, 136, 174, 192, 200, 235, 237–238, 319, 324, 328, 330
Antidromic conduction, 77, 165, 218, 235
Antigen–antibody reactions, 22, 67, 69–70, 72, 211
  *see also* Immune response
Antigen-reactive cells, 68
Antigens, 52
  blood group, 59–60, 182
Antiheparin (platelet factor V), 71, 323
Antiplasmins, 71, 324
Antipyretic drugs, 311, 314
Antithrombins, 51, 324
Antitoxins, 52
Anxiety, 304, 329
Aorta, 120, 122
  arch of, 229–230, 240
  diastolic and systolic pressures, 123
  flow through, 101–102, 123
Aortic bodies, 231, 246–247, 326
Apatite, 8, 9, 189–190, 252, 253, 305, 308
Aphthous ulcers, 69, 330, 331
Apnoea, 244, 252, 253
Apocrine secretion, 29, 30
Apocrine sweat glands, 292, 313
Apoferritin, 41, 63, 64, 188
Apolipoproteins, 47
Appetite, 135–136, 283, 295, 314
Aqueous humour, 27, 146
Arachidonic acid, 194, 196, 197, 211
Arachnoid granulations, 25–27
Aragonite, 154
Arneth count, 66
Arterial pulse, 122, 124
Arteries, 122
  proportion of blood volume, 123
  smooth muscle cells, 93
  volume receptors in, 238
Arterioles, 122, 322
  innervation, 141, 145
  proportion of blood volume, 123
  proportion of smooth muscle, 123
  resistance to blood flow, 104, 123, 230, 232
Arteriovenous anastomoses, 122, 123, 232, 235, 312–313

Articulation of speech sounds, 268–269
Ascites, 22
Asphyxia, 246, 247, 248–249
Aspirin, 52, 72, 163, 187, 195, 311, 314, 323
Association cortex, 133, 135
Astigmatism, 146
Athletes
  nitrogen balance, 296, 297, 319
  training, 121, 320
ATP-ase, 9, 159
  *see also* Calcium ATPase, Sodium/potassium ATPase
Atria, 97, 117–118, 124–125
Atrial flutter, 121
Atrial natriuretic factor, 240, 239–240
Atrial pressure, 120–121
Atrial pressure receptors, 231, 238, 326
Atrial systole, 120, 125
Atrioventricular node, 97–98, 118
Atropine, 80, 143–144
Attitude, 223–225
Audioanalgesia, 165
Auditory apparatus, 150–152
  *see also* Hearing
Auditory cortex, primary, 133, 152
Auditory inputs, modulation, 152
Auditory pathways, 152
Auditory threshold, 151
Auerbach's plexus, 176
Autacoids, 193, 211
Auto-immune diseases, 69, 70
Autonomic nervous system, 132, 136, 140–145, 175, 232
  *see also* Parasympathetic nervous system, Sympathetic nervous system
Autophagosomes, 7
Autoregulation of blood pressure, 103, 233, 234
  in kidney, 239, 240
Axon, 73
  axoplasm, 16, 74
  hillock, 80–81
  reflex, 218, 235, 312, 322

B-lymphocytes, 69
  source in adult, 69
Babinski reflex, 220
Bacteria, 52, 65, 66–70
  in gut, 64
  in saliva, 179, 182, 189–190
Bacterial toxins, 311, 324
Ballisto-cardiograph, 121
Barbiturates, 40, 72, 137, 173
Baroreceptor reflex, 244, 318, 325, 329
Baroreceptors, 216, 229–231, 236, 238, 240, 325–326, 328
Basal ganglia, 47, 134, 140, 223, 225, 226–227
Basal metabolic rate (BMR), 295–296, 310, 312, 313, 314, 328
  relation to surface area, 295

Basement membrane, 11, 18–20, 21–22, 34, 36, 130, 159, 170
Basic electrical rhythms (BERs), 177
Basophils, 65, 66, 67, 179
  numbers, 66, 206
Bile, 33–34, 64, 187, 203, 205, 209, 278–280
  daily volume, 185
  excretion of hormones, 203, 205, 209
Bile pigments, 34, 64, 186
Bile salts, 33–34, 41, 186–187, 188, 194, 279–280
  reabsorption in ileum, 41, 186, 188
Bilirubin, 47, 64, 186
Biliverdin, 64, 186
Biological clock, *see* Circadian rhythms
Bipolar cells, 83, 146–148
Birth, 208, 217, 283, 284, 285, 288, 289–290, 291
  tissues complete at, 286–288
Biting force, 262
Bitter taste, 160–161
Bladder, 96, 168, 174, 240–241
  control, 240
  emptying reflex, 220
  innervation, 142–143
Bleeding, *see* Blood clotting, Haemorrhage, Haemostasis
Bleeding time, 72, 323
Blistering, 67
Blood, 42–72
  arterial
    hydrogen carbonate concentration, 111
    partial pressure of carbon dioxide, 108, 111
    partial pressure of oxygen, 108, 210
    control, 231, 247
    pH, 43, 111, 250, 317
    control, 246, 248
  buffering, 46, 246, 250, 252–253
  dissolved oxygen, 131
  distribution, 318
  glucose concentration, 39, 182, 201, 202, 204–206, 277, 276, 297–305, 318, 319, 330–331
    control of, 135, 199, 294–305
  mixed venous,
    hydrogen carbonate concentration, 43, 111
    oxygen partial pressure, 109, 244, 315, 316
    partial pressure of carbon dioxide, 108, 111, 318
    pH, 43, 108, 111, 250
  osmolality, 238, 326
    control, 216
  pooling in lower limbs, 124
  urea, 189
  viscosity, 56, 102, 103
    apparent, 104

338  Index

Blood clotting, 45, 48–51, 70–72, 102, 103, 212, 305, 322–323, 329
  defective, 50
  intravascular, 51
Blood clotting factors, 44, 48–51, 71, 322–324
  antibodies to, 51
  in saliva, 182, 189
  extrinsic pathway, 50, 322–324
  inhibition, 49, 51, 67
  intrinsic pathway, 50, 322, 323, 324
Blood clotting factor I, 49, 322
Blood clotting factor II, 49
Blood clotting factor III, 49, 71
Blood clotting factor IV, 49
Blood clotting factor V, 49, 50, 51, 71, 323
Blood clotting factor VII, 44, 49, 50, 51, 182, 322, 323, 324
Blood clotting factor VIII, 44, 49, 50, 51, 182, 323, 324
  factor VIII complex, 322
  factor VIII-like antigen, 51
Blood clotting factor IX, 44, 49, 50, 51
Blood clotting factor X, 49, 50, 51, 323, 324
Blood clotting factor XI, 49, 50, 71
Blood clotting factor XII, 49, 50, 51, 322, 324
Blood clotting factor XIII, 48, 49, 323
Blood flow
  control, 232–236
  coronary, 165, 234
  driving force, 100
  effect of hormones, 235
  effect of local chemical stimuli and autacoids, 211, 236, 318
  effect on apparent viscosity, 46
  effect on blood cell alignment, 66, 102
  measurement of, 121
  peak velocity, 102
  redistribution,
    in exercise, 315, 317, 318
    in shock, 326
    in stress, 329
  to adipose tissue, 113
  to brain, 112–113, 233, 246, 315, 317, 326
  to carotid and aortic bodies, 231, 247
  to dental pulp, 235
  to gastro-intestinal tract, 234
  to kidney, 169, 171, 234, 326
  to lungs, 126, 233
  to lung alveoli, 131
  to muscle, 93, 112, 165, 234, 315, 316, 317, 319, 328, 329
  to oral mucosa, 207, 208, 235
  to renal medulla, 173, 200
  to salivary glands, 36
  to skin, 112–113, 235, 313
    in exercise, 317
  to spleen, 234

Blood flow (cont.)
  to thyroid gland, 202
  to viscera, 112–113
    in exercise, 317
  velocity, 123
Blood group substances, 56, 59–60
  in saliva, 181, 182
Blood groups, 59–60
  ABO, 56, 59, 182, 290
  CDE, 60
  Duffy, 60
  inheritance of, 59
  Kell, 60
  Lewis, 60, 182
  Lutheran, 60
  MN, 60
  Rhesus, 60
Blood loss, 231, 234, 236, 272, 324–328, 331
  menstrual, 281
  recovery from, 324–328
  see also Haemorrhage
Blood plasma, see Plasma
Blood pressure, 123, 205, 209, 212, 229–232, 234, 238, 239, 240, 315, 317–320, 325–330
  arterial, 122–123, 125, 231, 280
  autoregulation in kidney, 169, 239
  capillary, 21, 123
  control, 216, 220, 229–232
  centres, 230, 247
  effect of hypothermia, 309
  effects of hormones, 235
  in shock, 324, 325
  in vascular tree, 123
  measurement, 125
  pulmonary circuit, 120
  receptors, see Baroreceptors
  venous, 123–125
Blood supply
  to bronchioles, 126
  to kidney, 169
  to limbs, 312
  to liver, 233
  to placenta, 208, 288
  to renal cortex, 169
  to renal medulla, 169
  to salivary glands, 180
  to slow muscle, 300
  to white muscle, 299
Blood vessels, 18, 122–124
  actions of autacoids, 211
  actions of prostaglandins, 212, 236
  counter-current exchange, 105–107, 173, 174, 235, 312
  cross-sectional area, 123
  damage to, 48, 49, 322, 324
  dilatation of, due to cholinergic stimuli, 230
  elastic tissue, 122
  flow rates, 123
  pressures in, 123
  ratio of wall thickness to diameter, 122

Blood volume, 324–328, 330, 331
  control, 206, 237, 238–240
  in lungs, 233
  proportions in different vessels, 123
  total, 121
Blood–brain barrier, 24–27
  permeability, 27, 246, 295, 301
Blushing, 235, 272
Body fluids, 14–29, 46, 237, 331
  buffering, 252–254
  equilibrium, 15, 237
  pH, 112, 240
  properties of marker substances, 15
Body temperature, 105, 295, 296, 309–314, 317, 318, 331
  control, 105, 216, 231, 235, 309–314
  methods of measurement, 309
  rise at ovulation, 203, 282, 309
  see also Fever
Body thermal conductance, 310
Body water
  control of, 238
  distribution, 46, 329
Body weight, 15, 16, 17, 18, 294, 296, 297
Bolus, 185, 189, 263–264
Bolus flow, 102–103
Bone, 287, 305–308
  action of calcitonin, 204
  buffer capacity, 252–254
  calcium content, 305
    and aging, 293
  development, 287, 306–307
  effect of parathormone, 204
  effect of vitamin D metabolites, 211, 306–307
  growth, 207, 287, 290
    at puberty, 291
    in males, 207, 292
  of jaws, deformation in chewing, 259
  osteoporosity, 293
  remodelling, 287, 306–307
  turnover in adults and children, 306
  water content, 15
Bone age, 286, 291
Bone marrow, 61–62, 63, 65, 69, 70, 71
  neoplasia, 65
  red, 61
  yellow, 61
Borborygmi, 178
Botulinus toxin, 80
Bowman's capsule, 11, 169–170
Bradycardia, 121, 252, 328
Bradykinin, 50, 182, 210, 236, 273, 276
Brain, 132–138
  activity, effect of hypothermia, 309
  blood flow, 26, 112, 229, 233, 246, 328
  capillaries, 19
  electrotonic conduction, 77
  energy usage, 134, 310
  fluid compartments, 25

Brain (cont.)
  glucose dependence, 297, 299
  higher centres, 219
  hypoxia, 248
  injury to, 220, 226, 248, 297
  insulin insensitivity, 204, 301
  ischaemia, 326, 328
  ketone body usage, 300, 302, 304
  neurotransmitters, 81, 140, 144
  peptides, 191, 200–201
  tissue fluid, 24, 246
  vasoconstriction, 252, 254
Breasts, development and secretion, 207–208, 283–284, 291–292
Breath holding, 244, 247, 251
Breathing, 124, 126–129, 216, 242–244
  capacity, 131
  control, 242–244, 263, 265–267
  rate, in adult, 128, 129, 315
  work of, 104, 130, 243
Bronchi, 101, 126, 130, 243
Bronchioles, 104, 126, 145
  resistance, 142
Bruising, 64
Brush border, 7
Bruxism, 135, 165
Buffering capacity, 250
  of cerebrospinal fluid, 27, 246
  of haemoglobin, 46, 61, 112
  of hydrogen carbonate, 250, 252
  of plasma proteins, 46
Buffering
  in blood, 46, 246, 250, 252
  in body fluids, 252
  in bone, 252–253
  in interstitial fluid, 46, 252
  in plasma, 46, 252
  in saliva, 277
Buffers
  haemoglobin, 46, 56, 112, 246
  hydrogen carbonate, 46, 250–251
  phosphate, 46
  plasma protein, 46, 246
Bulk flow, 14, 20, 100–104
Bulk transport, 12–13, 40, 45, 46–48, 100–113, 117
Bundle of His, 97, 118
Burn shock, 324, 328

Caeruloplasmin, 44, 47
Calcification, 8, 9, 182
  inhibitors, 185
  of teeth, 286
Calcitonin, 48, 192, 204, 209, 306–307
Calcium
  absorption in small intestine, 14, 41, 184, 188, 204, 211, 305–307
  active transport, 41, 188, 211, 307
  as second messenger, 145, 195, 197, 201, 274, 300
  ATPase, 307
  balance, 305
  channels, 36, 274

Calcium (cont.)
  daily requirement, 305–306
    in pregnancy and lactation, 306
  distribution in body, 306
  excretion, 306
    in urine, 305, 306
    in faeces, 187, 305, 306
  in bile, 34
  in blood clotting, 49–51, 323
  in bone, 305–306
    age effects, 293
    buffering, 252
    effect of acidaemia, 253
    exchangeable, 305, 306
    mobilisation, 307
  in cardiac muscle, 99
    action potential, 97–98
    contractility, 98
    role in contraction, 99, 121
  in complement activation, 52
  in CSF, 26
  in dentinal fluid, 23
  in exocytosis and protein secretion, 30, 274, 275, 305
  in interstitial fluid, 18
  in microbodies and mitochondria, 8, 9
  in platelet aggregation, 71
  in plasma, 18, 42, 204, 305, 306
    bound, 18, 46, 252, 305
    control, 305–308
      calcitonin, 204
      parotin, 182
    diffusible, 305
    effect of alkalaemia on ionic calcium levels, 252
    effect on vitamin D metabolism, 211
    ionised, 46, 191, 204, 252, 305, 307
  in presynaptic terminals, 79, 80, 305
  in saliva, 34, 183
    bound, 181, 183, 184
    ionised, 183
    precipitation of, 184
    role in bacterial aggregation, 190
  in skeletal muscle, 43, 85
    binding by troponin, 86
    in sarcoplasmic reticulum, 85
    role in contraction, 91, 319
  in smooth muscle, 94, 95
    action potential, 95
    membrane bound, 95
    permeability to, 95
  in sweat, 31, 34
  in synovial fluid, 28
  intake, 305, 306, 307
  intracellular ionic concentration, 6, 8, 9, 195, 196
  loss during pregnancy, 283
  metabolism, 305, 307
    control, 305
  reabsorption in kidney, 172, 204
    in distal tubules, 174, 306
    in loops of Henle, 174
  retention in pregnancy, 208
  uptake by basophils, 67

Calcium-binding protein, 41, 87, 188, 190, 198, 211, 307
Calmodulin, 87, 94, 195–196, 197, 275
cAMP, see Cyclic 3, 5-adenosine monophosphate
Capacitance vessels, 123, 233, 238, 317
Capillaries, 18–21, 55, 122–124, 322
  brain, 19
  continuous, 19, 21, 26
  density in slow muscles, 92
  diffusion of oxygen and carbon dioxide, 11
  discontinuous, 19, 20, 21, 22, 37
  exchange across, 17, 20, 21, 117, 122, 123, 238, 317, 328
  fenestrated, 19, 20, 21
  hydrostatic pressure in, 20–21
    in kidney, 123, 169, 171
  in endocrine glands, 20, 37
  in gut, 40–41
  permeability, 17, 20–22, 36, 43, 45, 53, 211, 328
  pores, 21, 46
    effective size, 21, 44
  proportion of blood volume, 123
  walls, 19
Carbamino compounds, 47, 56, 58, 110, 111, 112, 252
Carbohydrate
  absorption, 188
  as energy source, 295, 296
  digestion, 186
  in saliva, 181
  metabolism, 183, 251, 252, 296–304
  storage, 296–300, 319
    as glycogen, 298
    mobilisation, 297, 299
Carbon dioxide, 31, 173
  diffusion
    across cell membranes, 11, 21, 56, 111
    across lung capillaries, 11, 21, 56, 130
    rate of, 130
  effect on brain blood flow, 233
  effect on oxygen carriage, 110
  effect of excess, 248
  excretion by respiration, 251, 317
  in CSF, 26
  in saliva, 184, 190
  partial pressure in arterial blood, 108, 215, 242, 244, 245, 246, 250, 253, 254, 315, 316, 317
    control, 217, 242, 244–249
  partial pressure in lungs, 11, 108, 112, 329
  partial pressure in tissues, 108, 111
  partial pressure in venous blood, 11, 108
  production, 295, 315, 318
  solubility in plasma, 108, 111, 130, 250
  total daily production, 251
  transfer in lungs, 11, 131

Carbon dioxide (*cont.*)
  transport, 47, 107, 111–112
    in blood, 56, 111
    in plasma, 56, 111
    *see also* Carbamino compounds
Carbon monoxide poisoning, 246, 247
Carbonic acid, 32, 33, 56, 107, 111, 173, 184, 246, 250, 252, 254
Carbonic anhydrase, 32, 33, 56, 59, 111, 173, 184, 252, 253, 254
Carboxyhaemoglobin, 58
Carboxypeptidase, 184–186
Cardiac arrest, 328
Cardiac centres, 230
Cardiac cycle, 98, 118–121, 125
  electrical events, 118–119
  mechanical events, 119–121
  phases, 119–121
Cardiac glycosides, 99
Cardiac muscle, 86, 93, 96–99, 117–118, 320
  acetylcholine effect, 80
  action potentials, ionic basis, 86, 97–98
  blood flow, 234
  cell composition, 90
  contractility, 121
  contraction, 98–99
    role of calcium, 97–98, 121
  efficiency, 93
  electrical activity, 97
  energy substrate, 98, 300, 304
  filaments, 8
  gap and tight junctions, 6
  hypertrophy, 329
  inherent rhythmicity, 97, 98
  metabolism, 90, 99, 297, 300, 304
  oscillating potential, 97, 98
    ionic basis, 98
  pain, 165
  pre-potential, 98
  protein, 85–86
  refractory periods, 98
  resting potential, 97
  striations (A and I bands), 96
  sympathetic $\beta_2$ receptors, 145, 205
  volume tension curves, 98–99
Cardiac output, 104, 121–122, 170, 229, 230–232, 233, 244, 247, 315–320, 328
  control, 232
Cardiogenic shock, 325, 329
Cardiovascular centres, 231
Cardiovascular system, 117–125, 229–236, 324–331
  response to exercise, 315–320
Carlsen–Crittenden cup, 179
Carotid body chemoreceptors, 215, 231, 246–247, 253, 326
Carotid sinus, 216, 229–231, 240
Carriage of gases in the blood, 107–113
Cartilage, 28, 287, 292
  growth, 202, 287, 292
Cascade reactions, 48–50, 52–53

Catabolic hormones, 296, 301
Catalase, 7, 62
Catecholamine *ortho*-methyl transferase (COMT), 144, 205
Catecholamines, 38, 140, 144–145, 201, 205, 274, 296, 326
  receptors, *see* Receptors
  *see also* Adrenaline, Dopamine, Noradrenaline
CCK-Pz, *see* Cholecystokinin–pancreozymin
Cell damage
  in non-regenerative tissues, 286–287
  products of, 22, 49, 66, 163, 191, 206, 236, 314, 328
Cell division, 8
Cell junctions, 5, 6, 19, 20, 22
Cell loss in digestive tract, 175, 187
Cell membrane, 19, 52
  calcium, stabilising, 305
  calculated effective pore size, 9
  fluid model, 5
  movement across, 9–14
  permeability, 9, 10, 13, 14, 15, 36, 75, 88, 195, 198
    changes in action potential, 76
    effect of stretch in smooth muscle, 94
  phospholipids, 3–5, 52, 194, 196, 274, 296
  pores, 4, 9
  properties and function, 3–6
  proteins, 3–5, 195
  red blood cell, 55–56
    antigens, 56
  skeletal muscle, 85
Cell organelles, 6–9, 70
  calcium in, 8, 9, 305
Cell pH, 252, 255
Cell structure, 3–9
Cellular antibody response, 52
Cementum, 167, 288
  calcium content and ageing, 293
Central nervous system, 132–140
  synaptic transmission, 139
Central venous pressure, 124, 232, 329
Cephalic stimuli, *see* Gastrointestinal tract control
Cerebellum, 137, 140, 155–156, 224–227
Cerebral cortex, 132–135, 139, 225–227, 242
  and reflex activity, 218
  control of bladder sphincters, 241
  control of voluntary movement, 225–227
  electrical activity, 134–135
  feeding response, 273
  olfactory areas, 162
  role in stance, 223
  sensory area, 135, 164
  speech area, 265
  taste area, 160
  visual area, 146, 149

Cerebrospinal fluid (CSF), 17, 24–27, 233
  buffering, 27, 246
  composition, 26
  formation, 26
  pH, 215, 242, 248, 253, 246
  resorption, 27
  volume, 25
Cerebrum, *see* Brain
cGMP, *see* Cyclic guanosine monophosphate
Cheeks, role in mastication, 260, 262
Chemical transmission, 79–81, 139–140
Chemoreceptors, 82, 160, 163, 243
  pain, 163
  respiratory, 215, 231, 240, 242, 246, 316, 318, 326
  smell, 82
  taste, 82, 160
Chemotaxis, 23, 66, 69, 318
Chenodeoxycholic acids, 186
Chewing, 257–262
Chewing cycles, 258, 261–262
Children
  breathing rate, 128
  chewing, 262
  nitrogen balance, 296
  total body water, 15
  volume of swallow, 264
Chloride, 11, 21
  absorption in gut, 41, 184, 188
  active transport, 173, 188
  cell concentration, 3, 14, 16
  equilibrium potential, 74
  in aqueous humour, 28
  in CSF, 26
  in gastric secretion, 32
  in muscle cells, 90
  in plasma, 17, 18, 43
  in saliva, 31, 34, 183
  in sweat, 31, 34
  reabsorption in renal proximal convoluted tubule, 39, 171–172
  reabsorption in salivary striated duct, 36–37
  secretion, 31
  shift, 111
Chloride–hydrogen carbonate exchange, 33, 34, 37, 60, 111, 184
Chlorolabe, 149
Cholecalciferol (vitamin $D_3$), 211
Cholecystokinin–pancreozymin (CCK-Pz), 192, 209, 278, 279, 302
Choleglobin, 64
Cholesterol, 37, 47, 183, 188
  absorption, 188
  esters, 37, 183
  metabolism, 7, 186–187, 195, 201, 282
Cholic acid, 186
Cholinergic receptors, *see* Receptors
Cholinesterase, 59, 182, 274
  inhibitors, 80, 143

Chorda tympani nerve, 141, 160, 180, 272, 274
Chorionic (placental) gonadotrophin (hCG), 197, 282
Chorionic oestrogen, 283, 284
Choroid plexuses, 24, 26, 253
  blood supply, 26
Choroid layer of eye, 146
Chromatin, 6, 62, 198
Chronaxie, 76, 91
Chronotropism, 230, 326
Chylomicra, 22, 41, 47, 188, 299
Chyme, 187, 189, 279
Chymotrypsin, 184, 185, 186
Chymotrypsinogens, 186
Cigarette smoking, 184, 243
Cilia, 8, 9, 26, 87, 101, 152, 153, 161, 244
Ciliary body, 27, 146
  capillaries, 20
Ciliary muscle, 27, 146
Circadian rhythms, 65, 67, 191, 201, 276, 291, 295, 309
  aldosterone, 191
  cortisol, 205
Circulations, local, 233
Citrate, 43, 49, 251, 299, 305
  in saliva, 183
Citric acid cycle, see Tricarboxylic acid cycle
Clicks, 265, 269
Clonic theory of sound production, 267
Clot retraction, 72, 87, 323
Clothing, effect on heat loss, 310, 312, 313
Clotting time, 323
Co-transporter system, 31, 36, 274
Cochlea, 82, 150–152
Cochlea microphonic, 152
Coding of stimulus intensity, 76, 83, 84
  for sound intensity, 152
  for taste concentration, 160
Cold, 246, 314
  effect on pulp blood pressure, 235
  environments, 293, 311, 313
  receptors, 311
  response of skin vessels, 235
  shock, 324
Collagen, 19, 71, 252, 323
  fibres, 20, 293
Collagenase, 186
Colon, 176, 177, 178, 187, 188, 280
  absorption, 187–188
  aldosterone effect, 206
  chloride absorption, 184
  innervation, 143
  inter-organ control, 280
  motility, 280
  secretion, daily volume, 185
  secretion of mucus, 280
  sodium transport, 184, 206
  sphincter, 176
Colostrum, 289
Colour Index (CI), 58

Colour vision, 145, 148, 149
Coma, 304, 305
Comfort zones, 313, 314
Complement system, 44, 52, 59, 67, 70, 71
  inhibitors, 44, 53
Compliance
  of lung, 130, 131,
  of muscle, 88
Conditioned reflexes, 218, 221, 227, 272
Conduction of action potential, 77
  antidromic and orthodromic, 77
  saltatory, 77
  velocity of, 77
Conduction of heat, 310, 313
Cones, retinal, 83, 146, 147, 148, 149
Conjugation with glucuronide or sulphate in liver, 39, 64, 173, 194, 203, 205, 207, 208
Consciousness, loss of, 252
Consonants, 266, 269–270
Contact Product, 50, 322
Contraceptives, oral, 208, 324
Contractility, 98, 99, 121, 232
Control systems, 135, 215
  comparator, 216
  control centres, 216, 217, 230, 271, 311, 312
  controlled variable, 215, 216, 217, 245
  higher centres, 217, 220, 231, 242, 243, 247, 268
  input pathways, 216, 230, 247, 271
  output pathways, 216, 230, 248, 271
  sensors, 215, 216, 217, 247
  servo control, 215, 217, 228, 261, 268
  set level, 216, 217, 230, 311, 312, 313, 314, 317, 318
Control
  development, 290
  of blood calcium concentrations, 306
  of blood carbon dioxide partial pressure, 242, 244
  of blood distribution, 232
  of blood flow, 232
  of blood glucose concentrations, 294
  of blood oxygen partial pressure, 242, 244
  of blood pH, 250, 252, 253
    renal, 240, 254
  of blood pressure, 216, 220, 229, 235
  of blood volume, 237, 238, 240
  of body temperature, 105, 216, 220, 309
  of body water, 239
  of calcium metabolism, 305
  of cardiac output, 232
  of circulation, 220
    chemical, 234
    nervous, 234

Control (cont.)
  of digestion, 271
  of heart rate, 218, 220
  of internal environment, 220
  of kidney function, 240
  of large intestine, 280
  of mastication, 158
  of metabolism, 294, 297, 301, 302
  of movement, 134, 137, 155, 156, 225
  of oesophagus, 277
  of plasma composition, 220
  of plasma osmolality, 216, 220, 237
    factors affecting, 238
  of plasma potassium, 240
  of posture, 134, 155, 156, 258
  of rectum and anus, 280
  of respiration, 218, 220, 242, 243, 266
    conscious, 247
  of salivary secretion, 272
  of salt excretion, 239
  of small intestine, 278
  of sodium excretion, 256
  of speech, 265
  of stomach, 277
  of urinary bladder, 240
Convection of heat, 310, 313
Convergence, 81
  of eyes, 146
  of olfactory inputs, 163
  of pain and other sensory fibres, 164, 165
  of sensory inputs, 147, 166
  of visual inputs, 148, 149
Converting enzyme, 209, 210, 239
Convulsions, 304
Cooling constant, 310
Cooling, evaporative, 29
Copper, 46, 63
Core temperature, 105, 309–314, 315
  critical, 312, 313
  interaction with skin temperature, 312
Cornea, 146
  sensation, 163
Coronary blood flow, 229, 234, 317, 326
  effect of systole and diastole, 234
Coronary circulation, 145, 234
Coronary ischaemia, 165, 328
Corpus luteum, 194, 199, 203, 207, 208, 281, 282
Corticosteroids, 67, 69, 206, 238, 304, 324, 331
  absorption in mouth, 187, 206
Corticosterone, 194, 195, 205, 206,
Corticotrophin releasing hormone (CRH), 192, 200, 201, 216, 239, 329
Corticotrophin-like intermediate lobe peptide (CLIP), 192, 203, 207, 289
Cortisol, 48, 183, 194, 195, 201, 205, 206, 278, 283, 289, 290, 291, 302, 303, 319, 328, 329, 330, 331

Cortisol (*cont.*)
  blood levels, 201, 216
  circadian rhythm, 205
  metabolism and excretion, 205
Cortisone, 205
Coughing, 128, 138, 220, 243
Countercurrent exchange, 105–107, 173, 174, 235, 312
Countercurrent multiplication, 105, 173
Creatine phosphate, 92, 93, 317, 318
Creatinine, 254
  in saliva, 183
Crenation, of red blood cells, 55
Cricopharyngeal sphincter, 175, 178, 263, 278
Cristae ampullaris, 153, 154
Cristae of mitochondria, 9
Crossed extensor reflex, 220, 222, 228
Crown–rump length, 285
Crush injuries, 331
Curare, 80
Cutaneous sensations, 82, 157
Cyanocobalamine, 63, 188
Cyanolabe, 149
Cyanosis, 246, 248, 325, 326
Cyclic 3,5-adenosine monophosphate (cAMP), 99, 145, 194, 195, 196, 197, 200, 201, 204, 211, 238, 274, 296, 300, 301, 302
Cyclic guanosine monophosphate (cGMP), 145, 195, 196, 197, 274
Cyclopentano-perhydrophenanthrene ring, 194
Cytochromes, 9, 57, 62
Cytoplasm, 3, 288
Cytosol, 196
  calcium concentration, 195, 196

Dead space volume, 112, 129, 131, 248
  anatomical, 129
  physiological, 129
Deamination, 204, 255, 299, 304
Decamethonium, 80, 143
Decarboxylation of amines, 191
Decerebrate rigidity, 223
Decibel scale of loudness, 84, 151
Deep body temperature, *see* Core temperature
Defaecation, 177, 220, 241, 244, 247, 272, 280
Defence mechanisms, 48–53, 329, 330
  role of plasma proteins, 22–23, 45
  role of white blood cells, 65
Deflation reflex, 243
Deglutition, 262–264
Dehydration, 304, 328
Dehydroepiandrosterone (DHEA), 183, 194, 195, 206, 208, 283, 289, 292
Dendrites, 73, 81, 286
Dental abnormalities, 270
Dental age, 286, 291

Dental calculus formation, 183, 184, 185, 190
Dental caries, 15, 179, 182, 189, 283, 288, 295, 296
  and protein starvation, 296
  link with sucrose, 160
  protection against, 288
Dental enamel, 123, 167, 189, 283, 287, 288, 289, 305
  fluid, 22–23, 179
Dental pain
  hydrodynamic theory, 167
  mechanisms, 167
Dental plaque, 179, 189
  acid production, 190
  alkali production, 183, 190
  ammonia production, 183
  bacteria
    aggregation, 190
    polysaccharide synthesis, 190
  diffusion in, 190
  fluid, 190
  formation, 190
Dental pulp, 167
  blood pressure, 235
  pain receptors, 165
  referred pain, 165
  sensation, 163
Dental treatment, 144, 153, 260, 270
  adrenaline levels, 329
  effect on pulp blood pressure, 235
  haemostasis, 49, 61, 323
  impression taking, 161, 264
  in relation to neurotransmission, 81
  of bruxism, 135
  orthodontics, pubertal growth, 292, 293
  referred pain, 165
  tooth extraction, 260
  use of cortisol, 331
Dentinal fluid, 22–23, 167
Dentine, 8, 23, 167, 288, 305
  calcium content and aging, 293
  nerves, 167
  resorption, 288
Dentures, 262
  effect on taste sensation, 159, 260
  effects on speech, 270
  freeway space, 258
Deoxycholic acid, 186
Deoxyhaemoglobin, 109, 246
Depolarisation, 75, 76, 77, 305
Depressor muscles of mandible, 222, 258, 260, 261
Depressor nerves, 230
Desmosomes, 5, 6
Desoxyribonuclease, 7, 184, 186
Desoxyribonucleic acid (DNA), 6, 211
Deuterium oxide, 16
DHEA, *see* Dehydroepiandrosterone
Di-iodotyrosine, 203
Diabetes mellitus, 253, 254, 304
Diacylglycerol, 196, 197, 274
Diads, 96

Diaphragm, 100, 127, 241, 242, 244, 248, 280
Diarrhoea, 180, 185, 188, 189, 272, 280
Diastole, 98, 119, 121, 122, 124, 125
  blood flow in coronary vessels, 234
Diastolic blood pressure, 103, 123, 125, 230, 231, 244, 315, 317, 326
  pulmonary circulation, 120
  systemic circulation, 120
Diastolic interval, 121
Diastolic ventricular volume, 232
Dicrotic notch, 124
Dicrotic wave, 122, 124
Diet, 321
  calcium requirements, 305, 306
  carbohydrate intake, 294
  cholesterol, 47
  effect on saliva composition, 277
  high fat, low carbohydrate, 296
  high potassium, 256
  high protein, 277
  in cold environments, 312
  in relation to dental caries, 160
  intake, 43
  iodine content, 203
  iron content, 63
    sources of iron, 62
  low potassium, 240
  low protein, 22
  low sodium, 240
  normal, 240
  protein intake, 296
  salt content, 240
Diffusion, 9, 10, 11, 12, 14, 22, 36
  across cell membranes, 4, 14
  coefficient, 10
  distance, 10, 316
    for carbon dioxide in lungs, 21
    for oxygen in lungs, 21
  of gases, 108
  of sugar molecules, 40
  path, 130
    for oxygen in alveoli, length, 130
  passive, 9, 11
  rate, 10, 11, 21
    of carbon dioxide, 130
    of oxygen, 130
Digestion, 41, 177
  absorptive phase, 303
  control, 271–280
  of a meal, 204
  of carbohydrates, 186
  of fats, 186
  of nucleic acids, 186
  of proteins, 186
  of starch, 185, 186
  post-absorptive phase, 303
  role of, 40
  vasodilatation in gut, 232
Digestive enzymes, 185, 189
Digestive secretions, 29–37, 178–185
  water content, 188

Digestive tract, 175–190, 271–280, 286
  cell loss, 175
  functions, 190
  gases, 178
  innervation, 141, 142
  movements, 177, 178
Digitalis, 99
Digitonin, 55
Diglycerides, 183
Dihydrotestosterone (DHT), 207
1,25-Dihydroxycholecalciferol, $(1,25(OH)_2D_3)$, 41, 194, 198, 211, 305, 306
24,25-Dihydroxycholecalciferol, $(24,25(OH)_2D_3)$, 211, 307
Dipeptidases, 184, 185, 186
2,3-Diphosphoglycerate, (2,3-DPG), 56, 60, 109, 110, 111, 316, 327
  effect on haemoglobin–oxygen association curve, 109
Disaccharidases in small intestine, 186
Discrimination, 84
Distal nephron see Kidney, distal tubule
Disturbance detectors, 216, 238, 312
Disuse atrophy, 287
Diuretic, 212, 239
Divergence, 81
DNA, see Desoxyribonucleic acid
Donnan effect, see Gibbs–Donnan equilibrium
DOPA (dihydroxyphenylalanine), 144
Dopamine (dihydroxyphenylethylamine), 81, 140, 144, 145, 192, 201, 202, 246
Dorsal columns, 138, 156, 157, 158, 164
Dorsal horn, 138, 158, 163, 164, 165, 201, 227
  laminae, 158, 163
Dorsal root ganglia, 138, 219
Drinking, 216, 238, 262, 263
Drugs
  and taste perception, 161
  effect on white blood cells, 65
  lipid-soluble, absorption in gut, 40
  lipid-soluble, secretion in kidney, 39
Dry mouth, 238, 273
Duodenum, 175, 177, 178, 187, 208, 209, 278
  absorption in, 188
  chemosensors, 279
  inhibition of gastric activity, 278
  irritation, 278
  motility, 279
  mucosa, 209
  pH, 279
  ulcers, 277
Dynamic response, see Receptors
Dyspnoea, 129

Ear, 82, 150, 151, 153
  attenuation of sounds, 151
  inner, pain receptors, 165
Ebb phase of recovery, 331
Ectomorphs, 297
Edentulous subjects, 262, 270
Effector organs, 78, 82, 216, 218, 220, 230, 238, 248
Efficiency, 320, 321
  of muscle, 93
Ejaculation, 141, 292
Elastase, 184, 186
Elastic recoil
  in cardiac muscle, 117
  passive, of lungs, 104, 128
Elastic tissue in lung alveoli, 104, 126
Elastic–contractile model of muscle contraction, 88
Elasticity of lungs, 104, 130
Electrical shock, 325
Electroanalgesia, 165
Electrocardiogram (ECG), 98, 119
Electroencephalogram (EEG), 135
Electromagnetic wave receptors, 82
Electromyography, 91
Electron transfer chain, 9
Electrotonic spread, 75, 77, 79, 83, 95, 139
Elenin, 56, 59
Elephantiasis, 22
Elevator muscles of the mandible, 92, 222, 261, 258, 260, 261
Elliptocytes, 55
Embden–Meyerhof cycle, see Glycolysis
Emergency situations, 140, 169, 171, 232–234, 322–331
Emotion, 133, 135, 140, 162, 200, 201, 220, 235, 243, 247, 272, 273, 278, 280, 295, 329
  effect on hormone release, 330
  effect on pain sensation, 163, 164–165
  effect on salivary secretion, 272
  effect on taste sensation, 161
Emotional fainting, 328
Emotional stress, 238, 240, 251, 313, 324, 329
En passant innervation, 81, 180
Encephalins, 139, 164, 165, 166, 192, 202
End-diastolic volume, 121, 230, 232
End-systolic volume, 121, 230
Endocrine glands, 37–39, 135–136, 191, 199–212,
  capillaries, 19, 20, 37
Endolymph, 150, 151, 153–155
Endometrium, 207–208, 282
Endomorphs, 297
Endoplasmic reticulum, 6, 8, 29, 85, 93
  rough, 6, 7, 30, 20, 34, 37, 69
  smooth, 6, 37, 187
Endorphins, 139, 164, 165, 166, 192, 202

Endothelium, 17, 18–20, 21–22, 24, 26, 50, 61, 71, 122–123, 130, 170, 291, 322, 324
  prostaglandin production, 236, 323
Endplate potential, 79–80
Energy production, 198, 294, 295–296, 300
  in red blood cells, 56
  of brain, 134
Energy rich bonds, see Adenosine triphosphate, Creatine phosphate, Phosphate
Energy sources, 9, 111, 206, 295, 296, 297, 300, 304, 317, 318, 331
Energy stores, 9, 293, 297, 299–300
Energy usage, 9, 12, 294, 320–321
Enterogastrone, 192, 209, 278, 279
Enteroglucagon, 209, 278, 302
Enterohepatic circulation, 64, 186, 188, 279, 280
Enterokinase, 184, 186, 187
Enteroreceptors, 218
Enzymes
  activation by phosphorylation, 196
  of GI tract, 182, 185–186
  rate limiting, 195
Eosinophils, 65, 66, 67, 211
  numbers, 66, 206
Ependyma, 25, 26, 27
Epidermal growth factor, 287
Epiglottis, 262–263
  taste buds, 158
Epiphyses, 287
  closure, 293
Epithelial cell layers, 6, 24, 27, 29, 40, 286
Epithelial growth factor, 209, 278
Epsilon-aminocaproic acid, 51
Epulis, 283, 292,
Equilibrium potential, 74, 75, 80
Eructation, 178
Erythroblastosis fetalis, 60
Erythrocyte sedimentation rate (ESR), 54, 103
Erythrocytes, see Red blood cells
Erythrolabe, 149
Erythropoiesis, 61, 65
Erythropoietin, 63, 65, 71, 193, 210, 326
Eserine, 80, 143
Ether, 55, 112, 137
Ethylene diamine tetra acetic acid (EDTA), 49
Evan's Blue, 18
Evaporative water loss, 16, 189
  from lungs, 16, 189, 266
Evoked potential in cortex, 135
Excitability
  of cardiac muscle, 99
  of nerve, 76, 252, 305
Excitable tissues, 6, 73–99, 305
Excitatory post-synaptic potential (e.p.s.p.), 80, 81
Excretion, 168–174,187, 189, 280, 240–241
  of alkali, 255

Exercise, 22, 65, 109, 110, 121, 165, 200, 201, 202, 204, 205, 216, 251, 272, 302, 304, 315–321
  anaerobic, 319
  cardiovascular response, 121, 231, 232, 233, 234, 315–321
  effect on ventilation/perfusion ratio, 131
  heat production, 110, 311, 313–314, 316, 317, 320–321
  oxygen debt, 93, 317
  respiratory response, 126, 128, 129, 248, 315–321
Exocrine glands, 29–37
  capillaries, 20
  secretions, 29–37, 178–187
Exocrine secretion, 29, 30
Exocytosis, 13, 29–30, 34–35, 36, 37–39, 79, 188, 274–275
  role of calcium, 30, 36, 274, 305
Expanded tip receptors, 157
Expiration, 126, 128, 129, 130, 233, 242–244, 248, 251, 265–267
  forced, 130, 233, 244
Expiratory centre, 242–243
Expiratory reserve volume, 128–129
Expiratory volume, forced, in 1 second (FEV1), 129, 131
Extensor muscles, 219, 222–228
Extensor thrust, 220, 225
Exteroreceptors, 82, 152, 218, 227
Extinction of a reflex, 221
Extracellular fluid, 14, 16, 17–29, 36, 42–53, 150, 185, 252, 318
  composition, 9, 17, 18, 31, 36, 331
  osmotic activity, 216, 237–240
  pH, 250
Extracellular fluid volume, 16, 17, 18, 104, 174, 185, 189, 206, 209, 238–240, 304, 314
  adjustment, 20–22, 174, 238–241
  measurement, 17
Extrapyramidal tracts, 226
Extrinsic factor, 63
Eye, 82, 144, 145, 146, 153
  optical axis, 146
  reflexes, 149, 225

F-actin, 85
Facilitated diffusion, 14, 41, 60, 298
Facilitation, 81, 219
Faeces, 62, 64, 175, 187, 188, 205, 220, 280
  iodine loss, 203
  calcium loss, 305
Fainting, 123, 234, 244, 313, 325, 328
Fasting, 43, 202, 204, 297, 302, 303
Fat, 15, 287, 288, 296–297, 302, 331
  absorption in gut, 41, 63, 187–188
  accumulation in pregnancy, 208
  as energy source, 98, 111, 296, 303–305, 317
  brown, 287, 288, 312
  cells, 206, 296, 301, 303
  development, 297

Fat (cont.)
  content of the body, 296
  digestion, 186, 209
  droplets, 16, 37, 288, 296
  energy value, 295, 296
  in muscle fibres, 92
  inhibition of gastric emptying, 278
  metabolism, 204, 251, 252, 296, 299, 300, 312
    associated respiratory quotient, 111
    need for glucose, 296
  mobilisation, 296
  stores, 295, 296–297, 298, 303, 304, 312
  synthesis, 204, 298, 301, 304
  see also Lipid
Fatigue, 319, 329
  of muscle fibres types, 92
  of receptors, 85
  of reflexes, 219
  of synapses and neuromuscular junctions, 80
Fatty acids, 9, 41, 186–188, 209, 296–297, 298, 299–302, 304
  metabolism in liver, 304
  storage, 296–297
  synthesis, 301
Fatty acyl-CoA, 299
Fear, 135, 231, 247, 272, 273, 313, 329
Feedback control systems, 215–218, 291
  negative, 200, 201, 202, 203, 204, 207, 215–217, 218, 228, 230, 282, 291, 298, 300, 302
  positive, 48, 203, 215, 217, 243, 251, 261, 280, 282, 283, 291
Feeding centre, 273
  see also Hunger centre
Female, 207, 286
  abdominal breathing, 128
  acclimatisation to hot climates, 313
  adaptation for survival, 293
  adult, iron loss, 62
  bite force, 262
  body temperature, 309
  bone age, 286
  calcitonin levels, 204, 306
  chromosomes, 285
  development, 293
  effect of chronic stress, 330
  energy usage, 310
  erythrocyte sedimentation rates, 54
  extracellular fluid volume, 17
  fat content, 296
  haemoglobin concentrations, 58
  fetal growth, 289
  hair distribution, 291
  intracellular fluid volume, 16
  lung volumes, 129, 130
  metabolic rate, 207
  prepubertal growth rate, 289
  puberty, 290, 291
    changes in vocal cords, 267
    growth spurt, 291

Female (cont.)
  red blood cell count, 55
  sensitivity to odours, 162
  taste sensation, 161
  testosterone, 207
  total body water, 15
  voice, 265, 151
  volume of swallow, 264
Fenestrated capillaries, 19–20, 21, 170
Fenestrations, 19–20, 21, 26
Ferritin, 21, 41, 61, 62, 63, 64, 70
Fertilisation, 281–282
Feto-placental unit, 208
Fetus, 212, 283, 285–289
  adipose tissue, 287, 288
  adrenal glands, 203, 206, 208, 283, 289
  circulation, 208, 212, 288
  crown–rump length, 285, 288
  development, 283
    taste buds, 159
  growth, 203, 208, 288–290
    in female, 289
    in length, 288
    in male, 289
    in weight, 288
    rate, 289
  haemoglobin, 58, 110, 290
  iron requirement, 62
  nutrition, 288
  pancreas, 204
  pituitary gland, 289
  red blood cell production, 61–62
  thymus gland, 68
  thyroid gland, 288
Fever (or pyrexia), 65, 309, 314
Fibrillation, 121
Fibrin, 45, 48, 50, 51, 67, 71, 323, 324
  formation see Blood clotting
  stabilising factor (factor XIII), 48, 323
Fibrinogen, 42, 43, 45, 48, 49, 51, 54, 71–72, 103, 323–324
Fibrinolysis, 50, 51, 324
Fibrinopeptides, 45, 48
Fibroblast growth factor, 287
Fibroblasts, 206, 323
Fibrosis in lungs, 130
Fick's equation, 10, 40
Fick's principle, 104, 105, 122
Fight, 205, 329
Filtered load, 174
Filtration, 11, 20, 104, 170
  see also Renal glomerular capillaries
Filtration fraction, 105, 239, 326
Final common pathway, 81, 84, 222
Fitness tests, 320
Flack test, 320
Flavour, 158, 161
Flexor muscles, 219, 223, 224, 226, 228
Flexor responses, 220, 222, 227,
Flight, 205

Flow
  laminar and turbulent, 100–103
  measurement, 104
Flowerspray endings, 82, 85, 155, 156
Fluent aphasia, 266
Fluid
  compartments, 14–29, 45
    exchange between, 16, 239, 318
  intake, 175, 239
  loss, 328
  retention, 239
  secretion in gastro-intestinal tract, 185
Fluids
  of ear, 17, 24
  of eye, 17, 24, 27
Fluorapatite, 306
Fluoride, 15, 288, 306
  absorption in gastro-intestinal tract, 184, 187
  effect on osteoporosis, 293
  in saliva, 184
  reabsorption in kidney, 173
Flushing, 314
Folic acid, 63
Follicle stimulating hormone (FSH), 192, 194, 197, 201, 202, 203, 207, 208, 281, 282, 289, 291, 292
Folliculostatin, 282
Food intake, 175, 187, 204, 238, 257, 261, 262, 283, 294–295, 302
  effect on gastric secretion and motility, 277
  stimuli to salivation, 179, 272
Forced expiratory volume ($FEV_1$), 131
Formants, 268
Fovea, 147, 149
Frank–Starling principle, 121, 232, 329
Free fatty acids, 184, 205, 297, 299, 300, 302, 304, 318, 319, 331
  in muscle metabolism, 93, 98
  in plasma, 47, 187, 302
  monitoring by hypothalamus, 295
Freeway space, 257
Frequencies of speech sounds, 151, 267, 268
Friction, internal in muscle contraction, 90, 93
Frictional resistance to flow, 100, 104, 130
Fright, 205
Fructose, 39, 186, 187
Fructose-1,6-biphosphate, 301
Fructose-2,6-biphosphate, 301
Fructose-6-phosphate, 301
Fucose, 182
Functional residual capacity, 112, 129

G-actin, 85
Gagging, 264
Galactose, 14, 37, 39, 172, 182,
  absorption in gut, 41, 187

Gall bladder, 34, 186, 209, 278–279
  contraction, 271, 278
Ganglia, 132, 140, 142
Ganglion blocking agents, 143
Ganglion cells of retina, 84, 147, 148, 149
Gap junctions, 5, 93–95, 96–97, 118, 167
Gas exchange in lungs, 130
Gas solubility, 130
Gastric acid, 32, 33, 186, 187, 209, 251, 277, 278, 329
  secretion, 32, 208, 209, 211, 272, 277
Gastric contents
  gastrin stimulating, 277
  osmotic activity, 272
Gastric glands, 30, 32, 33, 208, 277, 280
  innervation, 141
Gastric inhibitory peptide (GIP), 201, 209, 302
Gastric lipase, 186
Gastric motility, 201, 208, 209, 271, 277, 278, 279, 329
Gastric mucosa, 186, 208, 209
Gastric mucus, 277
Gastric pH, 186
Gastric reflux, 178
Gastric secretion, 201, 209, 271, 277, 278, 279
  composition, 32–33
  daily volume, 185
Gastric ulcer, 277
Gastrin, 192, 201, 204, 208, 209, 277, 278, 279, 302, 306
Gastro-colic reflex, 280
Gastrointestinal tract, 10, 24, 175–190, 329
  absorption, 10, 40–41, 184, 187–188, 302–303
  bacteria, 175
  blood vessels, 176
  blood flow, 234
  control, 271–280
    cephalic phase, 271, 272, 277, 278
    conscious, 272
    hormonal, 272
    inter-organ, 272, 273, 278, 280
    nervous, 272
    psychic phase, 277
    within-organ, 271, 277
  food transit time, 178
  functions, 187
  hormones, 204, 272, 302, 303
  length, 175
  loss of epithelial cells, 175, 187
  loss of white cells, 65
  motility, 177–178, 271–272, 276–280, 302
  muscle of walls, 176
  nerve plexuses, 176–177, 277
  parasympathetic nerves, 176–177, 272

Gastrointestinal tract (cont.)
  secretions, 66, 184, 185, 271
    actions, 185, 187
    composition, 185
    volume, 185
  sites of secretion, 185
  sympathetic nerves, 176–177, 272
  water absorption, 188
Gastro-oesophageal (cardiac) sphincter, 175, 178, 209, 278
Gating system, in pain sensation, 164
Generator potentials, 83, 84, 148, 152, 158, 159
Genetic control of
  blood group, 59
  body build, 297
  complement inhibitor, 67
  haemophilia, 51
  hair growth, 293
  height, 290
  taste sensitivity, 161
Genitalia, external, 285, 288, 289
  innervation, 143
Gibbs–Donnan equilibrium, 6, 12, 18, 20, 26, 74, 171
Gingivae, 258, 262, 330
  blood flow, 235
  colour, 235
  inflammation, 69, 182, 283
  at puberty, 292
Gingival crevicular fluid, 22–23, 179, 181
Glands of von Ebner, 34, 180
Glial cells, 73, 148, 287
Glicentin, 209
  see also Enteroglucagon
Globulins, 36, 43, 44, 45, 46, 290, 305
  $\alpha$, 47, 63, 64, 205
  $\alpha_1$, 44
  $\alpha_2$, 44, 47, 48, 50, 53
  $\beta$, 44, 47, 48, 52, 182, 207, 239
  $\gamma$, 44, 45, 51, 52, 71
  see also Immunoglobulins
  in saliva, 182
Glomerulus, olfactory, 162
Glomerulus, renal see Renal glomerulus
Glottis, 101, 243–244, 265, 268, 280
Glucagon, 142, 145, 192, 197, 201, 204, 209, 296, 301–304, 319, 329–331
Glucagon-like immunoreactive factor (GLI), 209
  see also Enteroglucagon
Glucocorticoids, 201, 205–206, 296, 301–302, 331
Glucokinase, 298, 301
Gluconeogenesis, 205, 299, 300–304, 319, 331
Glucose, 9, 14, 36, 296, 297–300, 303–304, 319
  absorption in gut, 41, 188, 302
  in plasma, 43
  see also Blood glucose,
    Hyperglycaemia,
    Hypoglycaemia

346  Index

Glucose (*cont.*)
  intake, 299
  in aqueous humour, 27–28
  in cerebrospinal fluid, 26–27
  in glomerular filtrate, 171
  in saliva, 182
  in synovial fluid, 28
  in red cells, 60
  in urine, 172
  metabolism, 9, 56, 298
    in adipose tissue, 299
    in liver, 298, 301
    in muscle, 298
  reabsorption in renal proximal
    convoluted tubule, 39, 171, 172
  storage, 303
    as fat, 296, 297, 299
    as glycogen, 297
  tolerance test, 303
  transport in kidney, 39
  uptake, 303, 304, 331
    effect of insulin, 301
    in adrenal cortex, 201
    in red cells, 14
  utilisation, 205, 300, 301, 302, 304, 331
Glucose-1-phosphate, 41
Glucose-6-phosphatase, 298, 299, 301
Glucose-6-phosphate, 298, 299, 300
Glucose-dependent tissues, 297, 299, 300, 304
Glucosyltransferase, 298, 300, 301
Glucuronides, 34, 39, 64, 173, 203, 205, 206, 207, 208
Glutamate, 39, 79, 81, 140, 172, 182, 255, 296
Glutamine, 255, 296, 301, 303, 304
Glutathione, 56, 60
Glycerol, 41, 186, 296, 297, 299, 300–301, 302, 331
Glycine, 140, 172, 182, 183, 186
Glycocholate, 186
Glycogen, 204, 298–300, 301–303, 318, 319, 331
  in liver, 204, 296, 297, 299, 300
  in muscle, 92, 93, 94, 96, 297, 299, 317
  synthesis, 298, 301
  total store, 299
Glycogenolysis, 196, 205, 299–300, 301–302, 304, 317, 319
Glycolysis, 9, 33, 56, 57, 59, 66, 70, 298–299, 301
  aerobic, 297–298, 300
  *see also* Tricarboxylic acid cycle
  anaerobic, 56, 57, 59, 93, 98–99, 351, 297, 299, 301, 317, 319, 320–321, 325, 331
  *see also* Pentose monophosphate shunt
Glycoproteins, 19, 29, 33, 37, 190
  salivary pellicle, 178
  secretion, 35
  synthesis, 7
  *see also* Mucoproteins

Glycosuria, 172, 303, 304
Gnathodynamometer, 262
Goldman equation, 11, 74, 80
Golgi apparatus, 7, 8, 30, 37, 47, 274
Golgi tendon organs, 156, 228, 258
Gonadal hormone binding globulin (GBG), 48, 207
Gonadotrophic hormones, 191, 207, 281, 282, 283, 290, 330
  *see also* Follicle stimulating hormone, Luteinising hormone
Gonadotrophin releasing hormone (GnRH), 201, 202
Gradients
  of charge, 11, 12
  of concentration, 10, 11, 12
  of osmolality, 105, 106
    in kidney, 173
  of pressure, 11, 100
Granulocytes, 65–67, 290
Gravity, 262, 287
  effect on blood pressures, 124
  gravitational receptors, 153–154, 225
Growth and development, 61, 191, 202, 203, 206, 207, 285–293
  and hunger, 295
  of blood cells, 61, 290
  of bone, 287, 291–292
  of muscle, 287, 291, 292
  of the jaws, 290
  measures of, 285
  rate, 289
    in different body parts, 286–287, 293
  prepubertal, 288–289
  role of DHEA, 206–207
Growth factors, 182, 201, 202
  *see also* Somatomedin
Growth hormone (GH), 110, 193, 197, 201, 202, 211, 284, 287, 288–293, 296, 301–302, 304, 319, 330–331
Growth hormone release inhibiting hormone (GIH), *see* Somatostatin
Growth hormone releasing hormone (GRH), 192, 201, 202
Guanylate cyclase, 196
Gustatory pit, *see* Taste bud
Gut, *see* Gastrointestinal tract
Gut-associated lymphoid tissue, 69
Gynaecomastia, 293

H substance, 59, 182
Haem, 47, 57, 58, 64
Haematin, 58
Haematocrit value (Ht), 42, 54, 58, 102, 103, 104,
  after haemorrhage, 326–328
  in glomerular efferent arteriole, 171
Haematopoiesis, 61–62
Haematuria, 48
Haemocytoblasts, 61, 65, 69, 71

Haemoglobin, 48, 54, 55, 56, 57–59, 60, 63–64, 107, 108, 186, 252, 290
  $A_1$, 58
  $A_2$, 58
  affinity for carbon monoxide, 58
  breakdown products, 47, 64, 186
  buffering, 46, 56, 246, 251, 252
  carbamino compound, 47, 56
  concentration in blood, 55, 57, 58
  concentration in red blood cells, 57
  daily turnover, 63
  deoxyhaemoglobin, 109, 246
  estimation, 58
  fetal (F), 58, 110, 290
  Haldane standard, 58
  in urine, 48, 56
  in plasma, 48
  oxygen transport, 56, 60, 108–111
  S, sickle cell, 55, 58
  synthesis, 61, 62, 63
Haemoglobin-oxygen association/dissociation curve, 56, 58, 109–110, 244, 316
Haemolysis, 52, 55, 58, 59, 60, 63
Haemolytic disease, 64
  of the newborn, 60
Haemophilia, 51
Haemorrhage, 51, 65, 200
  recovery from, 324–328, 330
Haemosiderin, 62–63, 64, 70
Haemostasis, 48, 71–72, 236, 322–324
  *see also* Blood clotting
Hair cells, 151–152, 154–155
  vestibular, 153
Hair
  distribution in men and women, 290, 291, 292
  endorgans, 157–158
  growth at puberty, 291
  growth, genetic control, 293
Halothane, 112
Haptens, 52
Haptoglobins, 44, 48, 63
Haustral movements, 177
Hearing, 133, 134, 150–152
  *see also* Auditory
Heart, 96, 98, 100, 103, 117–122, 142, 220, 230
  blood supply *see* Coronary circulation
  force of contraction, 121
  innervation, 141–142
  intrinsic control, 121, 232
  muscle, *see* Cardiac muscle
  oxygen usage, 109
  size in males and females, 293
  sounds, 103
  spread of excitation, 118
  pain, 165
  *see also* Cardiac, Coronary
Heart failure, 21, 121, 324
Heart rate, 98, 99, 119–121, 137–138, 141, 142, 205, 230–232, 244, 247, 296, 304, 320, 326–331
  control, 218, 220

Heart rate (*cont.*)
  effect of hypothermia, 309
  in males and females, 293
  in relation to exercise, 315–320
  resting, 118
Heartburn, 178
Heat, 82, 318
  effect on pulp blood pressure, 235
  gain, 310, 311, 312–314
    from skin surface, 310
  loss, 310, 311, 312, 314, 317
    effect of clothing, 310, 313
    evaporative, 310, 311, 313, 317
      through lungs, 310
    non-evaporative, 105, 235, 310, 311, 313, 317
    through skin, 105, 235, 310
  metabolic,
    in adult, 295, 312
    in infant, 288
  production, 205, 310–314, 331
    due to adrenaline, 205
    due to progesterone, 208
    in action potential, 76
    in exercise, 315, 318
    in muscle, 90, 92–93
  retention, 235, 310, 311, 314
  stroke, 309
  *see also* Temperature
Height, 130, 285, 290, 292–293
  adult, 290, 293
  sex differences, 289, 292
Helmholtz resonators, 268
Henderson–Hasselbalch equation, 250, 251
Henry's Law, 108, 130
Heparin, 51, 67
Hering Breuer reflex, 243
Heterophagosomes, 7
Hexokinase, 298, 299
Hiccough, 244
High altitude, 56, 110, 251, 253–254
Histamine, 22, 53, 67, 72, 158, 163, 193, 206, 211, 235, 236, 277, 324
Histiocyte, 70
HL-A white cell antigens, 60
Holocrine secretion, 30
Homeostasis, 215, 217, 329
  *see also* Control, Internal environment
Hooke's Law, 89
Horizontal cells of retina, 148, 149
Hormones, 34, 37–39, 191–212
  and metabolism, 203–205, 301–302
  chemical nature, 193–195
  effects on blood pressure and flow, 235–236
  effects on haemostasis, 324
  glycoprotein, 192, 193, 201–203, 208, 210
  in plasma, 43, 193
  in saliva, 182, 193
  lipid soluble, 194–195
  mechanisms of action, 195–198, 305

Hormones (*cont.*)
  polypeptide, 48, 192, 193, 199, 200, 201, 202, 204, 207, 208, 209, 236, 239, 277, 283, 287
  receptors, 193, 196, 202, 204
    cell membrane, 195, 196, 201
    cytoplasmic, 198
    FSH and LH, 282
    mitochondrial, 196
    nuclear, 196
  releasing and release-inhibiting, 136, 192, 200–201
  secretion, 37–39
  storage, 37–39
  synthesis, 37, 194–195
  sites of production, 192, 193, 194, 199–212, 206
  trophic, 194, 201–203
  water-soluble, 193, 195–198
  *see also* Anabolic, Androgenic, Catabolic, Corticosteroid and Steroid hormones
Horripilation, 310, 312
Human chorionic somatomammotrophin (hCS), 193, 208
Human placental gonadotrophin (hCG), 193, 208, 283
Human placental lactogen (hPL), 208, 283
Human thyroid stimulatory immunoglobulin (HTSI), 203
Hunger, 135, 136, 220, 238, 294–295, 296, 304, 304
  centre, 136, 273, 283, 295, 314
  drive, 135, 297
Hyaluronate, 27, 28
Hydrocortisone, *see* Cortisol
Hydrogen carbonate, 3, 21, 56, 111, 186, 250–256
  buffering, 46, 250–253
  in saliva, 189
  exchange with chloride, *see* Chloride-hydrogen carbonate exchange
  in aqueous humour, 28
  in bile, 34, 186
  in blood, 43, 111, 251
    plasma, 43, 56, 111
  in CSF, 26
  in gastric mucus, 186
  in pancreatic secretion, 31, 186, 209
  in saliva, 31, 34, 183, 184
  in small intestine, 186
  in sweat, 31, 34
  reabsorption in kidney, 172, 251, 254–255
  secretion, 37, 184, 188
Hydrogen ions, 111–112, 160, 174, 246, 250–256
  effect on haemoglobin-oxygen dissociation curve, 109
  effect on peripheral chemoreceptors, 246
  receptors, 253

Hydrogen ions (*cont.*)
  secretion in gastric glands, 32
  secretion in kidney, 173, 174, 254
  excretion, 254
  *see also* pH
Hydrostatic pressure, 11, 20–22, 123, 124
  in pulmonary capillaries, 21
  in renal capillaries, 169, 170–171
Hydroxyapatite *see* Apatite
25-Hydroxycholecalciferol, 211, 306
5-Hydroxytryptamine, 95, 148, 163, 311
Hypercapnia, *see* Hypercarbia
Hypercarbia, 129, 246, 247, 248, 249
Hyperglycaemia, 304, 329, 331
Hyperkalaemic acidosis, 240
Hypermetropia, 146
Hyperphagia, 295, 297
Hyperplasia, 286
Hyperpnoea, 129, 189, 248
Hyperpolarisation, 75, 80, 83, 95, 99, 148
  in retina, 148
  of cardiac muscle, 118
Hypertension, 212, 230
Hyperthermia, 238, 309
Hyperthyroidism, 203
Hypertrophy, 286, 287, 319, 320, 330
Hyperventilation, 129, 233, 244, 247, 251, 266, 305, 318
Hypoglycaemia, 277, 278, 304, 319
Hypokalaemia, 188, 240
Hypothalamo–hypophyseal portal system, 122, 199
Hypothalamus, 38, 132, 135–136, 137, 140, 141, 164, 165, 191, 199–201, 202, 203, 205, 207, 210, 216, 220, 231, 234, 235, 237–240, 241, 247, 273, 281–284, 289, 291–292, 301, 311–314, 318, 326, 328, 329
  antidiuretic hormone release in shock, 236
  areas and nuclei, 135, 158, 200, 201, 237, 273, 295, 311, 312, 313
  control of
    blood osmolality, 237–240
    body temperature, 136, 311–314, 316, 317
    menstrual cycle, 281–282
    pubertal changes, 290–293
  hunger and satiety centres, 136, 295
  polypeptides, 38, 136, 139, 192, 199–201
  receptors for
    free fatty acids, 295
    glucose, 136, 199, 278, 294, 295
    oestrogens, 282, 291
    osmotic activity, 136, 189, 199, 216, 238, 326
    prostaglandins, 295
    testosterone, 292
    temperature, 136, 199, 216
  releasing hormones, 136, 192, 200–201

Hypothalamus (cont.)
  release-inhibiting hormones, 136, 192, 201
  sensory inputs, 135
  thirst centre, 238
  see also Antidiuretic hormone, Oxytocin
Hypothermia, 295, 309
Hypothyroidism, 296
  in infants, 288
Hypoventilation, 129, 251
Hypovolaemic shock, 324, 325, 326, 328
Hypoxaemia, 129, 246, 248
Hypoxia, 99, 129, 234, 235, 246, 247–248, 249, 252, 253–254, 325, 326, 328
  anaemic, 246, 326
  anoxic, 246
  cerebral, 248
  effect on pulmonary blood flow, 131, 233
  effect on red blood cell production, 55, 56, 58
  histotoxic, 246
  hypoxic, 129, 246
  stagnant, 246
  toxic, 246
Hysteria, 129, 251

I bands of muscle, 85, 87, 96
Idiopathic thrombocytic purpura, 70
Ileogastric reflex, 278
Ileum, 41, 63, 175, 186, 187, 188, 209, 278, 279, 302
  absorption in, 188
  absorption of bile salts, 189
  absorption of vitamin B12, 186
Immune response
  cellular, 70
  humoral, 52, 69
  primary, 52
  secondary, 52
  see also Antigen–antibody reactions
Immunity
  acquired, 52
  innate, 52
  racial, 52
Immunoblast cells, 69, 70
Immunoglobulins, 44, 45, 52, 60, 69, 203
  IgA, 44
  IgG, 44, 52, 69
  IgD, 44
  IgM, 44, 52, 69
  in saliva, 182
  synthesis, 69
Immunologically competent cells, 68
Incision, 257, 260
Infants
  blood cell counts, 290
  bone turnover, 306
  non-shivering thermogenesis, 288

Infants (cont.)
  response to cold, 312
  resting saliva flow rate, 273, 276
  taste sensation, 161
  total body water, 15
Inflammation, 22, 54, 66, 67, 102–103, 165, 167, 171, 195, 206, 211, 235, 236, 311
Inflammatory exudate, 23–23, 65
Inflation reflex, 243
Ingestion, 177, 257, 260–261
  of acid or alkali, 251, 252
Inherent rhythmicity of muscle, 93, 98, 176–177, 232
Inhibin, 192, 202, 207
Inhibitory postsynaptic potential (i.p.s.p.), 80, 81, 145
Inhibitory transmitters, 80, 81, 95, 140
Initial segment spike in motorneurone, 81
Injury, 201, 211
  acute response and compensatory reactions, 331
Innervation ratio, 92
Inositol-1,4,5-triphosphate, 196, 197, 274
Inotropism, 326
  positive inotropic effect, 230, 232
Inspiration, 126, 127, 129, 242–244, 247–248, 263, 266, 278
  centre, 242
  effect on thoracic vessels, 124, 233
  muscles of, 242
Inspiratory reserve volume, 128, 129
Insulin, 145, 191, 192, 195, 197, 201, 204, 209, 278, 287, 288, 290, 291, 295–298, 301–304, 319, 329–331
  action on liver, 204, 301–304
  deficiency, 296, 304
  sensitivity of tissues to, 204, 301
Insulinases, 204
Intercellular spaces, 5, 19, 20, 36, 43, 78, 96
Intercostal muscles, 127, 128, 242
Interferon, 70
Internal environment, 14, 82, 117, 215, 250
  control of constancy, 45–46, 174, 215, 217, 220, 237, 250
Internuncial neurones, 145, 158, 164, 166, 219, 228
Interoreceptors, 82, 132, 152, 155–157, 180
Interstitial cell stimulating hormone (ICSH), see Luteinising hormone
Interstitial fluid, 15, 17–18, 20–22, 26, 31–32, 33–34, 36, 37, 123, 305, 317–318, 326
  buffer capacity, 46, 252
  composition, 17–18, 31, 32
  volume, 17, 46
Intestine
  hormones, 201, 280
  insulin insensitivity, 301

Intestine (cont.)
  motility, 209, 211, 212, 278–280
  secretion, 184, 187
    daily volume, 185
    mucus, 278
  smooth muscle, 95, 176–177
  surface area, 40
Intra-abdominal pressure, 124, 128, 241, 278, 280
Intracellular fluid, 3, 9, 16, 17, 46, 252
  buffering, 252
  composition, 16, 74, 90
  of muscle cells, 16, 90
  of nerve cell, 16, 74
  osmotic activity, 16
  pH, 3, 16, 252, 255–256, 319
  volume, 16
Intracranial pressure, 233, 280
Intrapleural pressure, 127
  see also Intrathoracic pressure
Intrapulmonary pressure, 100, 130, 243
Intrathoracic pressure, 124, 130, 175, 233, 241, 244, 247, 320
  measurement, 130
Intravascular fluid see Plasma
Intrinsic factor, 63, 186
Inulin, 17, 104, 105
  clearance, 105
Iodide, 37, 38, 67, 181, 184
  in plasma, 203
  reabsorption in kidney, 203
Iodine, 38, 184, 202–203
Iodopsin, 149
Iris, 27, 146–147, 149
Iron II (ferrous), 57–58, 61, 62–64, 108
  absorption in gut, 41, 62, 63, 184, 188
  active transport in gut, 188
  balance, 62
  binding by apoferritin, 61, 62–64
  binding by transferrin, 47, 61, 64
  daily intake, 61, 62
  deficiency anaemia, 62, 63
  loss, 64
    in faeces, 62, 187
    in sweat, 62
    in urine, 62
  storage, 61, 62, 63
  total iron pool, 62
  transport in plasma, 47, 63
  uptake
    by basophils, 67
    by monocytes, 70
    by red blood cells, 56
Iron III, 58, 109
Irreversible shock, 325, 328
Ischaemia, 78, 165, 326, 328
Islets of Langerhans, 199, 204
Isomaltase, 186
Isometric
  contraction, 88, 93, 320
  of ventricles, 120

Isometric (*cont.*)
  exercise, 317, 320
  tension, 89, 92, 95
Isotonic contraction, 88, 89, 93
Isovolumetric contraction of
    ventricles, 120
Itching, 67, 157, 236


Jaundice, 64
Jaw
  depressor muscles, 222, 258, 261
  elevator muscles, 222, 258, 262
  growth, 290, 292
  jerk reflex, 219, 261
  opening reflex, 261
  rest position, 258
Jejunum, 41, 175, 187, 278, 308
Joints, 28–29
  receptors in, 137, 153, 156,
    224–225, 258, 260, 262, 268,
    315, 318, 319
Junction potentials, 81, 94–95
Junctions, en passant, 81
Juxtaglomerular apparatus, 170, 206,
    209–210, 212, 236, 238, 239,
    240, 327
  effect of adrenaline, 210


Kallidin, 210, 236, 276
Kallikrein, 46, 50, 182, 210, 211, 236,
    324
Kallikreininogen, 50, 210
Kell blood groups, 60
Ketoacids, 253, 299, 304, 305
Ketone bodies, 98, 296, 297, 298, 299,
    300, 301, 302, 304, 305, 331
  as substrate for brain cells, 304
Kidney, 39, 64, 103, 104, 106–107,
    141, 168–174, 237–241, 253–256,
    286, 331,
  action of parathormone, 204
  amino acid metabolism, 172, 304
  ammonia production, 304
  blood flow, 26, 103, 104–106, 169,
    170, 171, 173, 234, 326
    distribution in cortex and
      medulla, 240
    in exercise, 317–319
  circulation, 122, 168, 169
  collecting ducts, 169, 170, 172, 173,
    174, 200, 238, 255
  control of kidney function, 237–254
  cortex, 168–170
    gluconeogenesis, 302
  countercurrent exchange and
      multiplication, 106–107, 173–
      174
  distal convoluted tubule, 169–170,
    172, 174, 209, 254–255
  distal tubule, 37, 172, 173–174, 206,
    209, 238–240, 254–255, 306
    ammonia production, 174, 255
    function, 172, 174, 255

Kidney (*cont.*)
  distal tubule (*cont.*)
    reabsorption,
      of calcium, 172, 174, 204, 306,
        307
      of chloride, 172
      of fluoride, 174
      of hydrogen carbonate, 172
      of sodium, 172, 174
      of water, 172
    secretion of potassium, 172, 173,
      174
  enzymes, 209, 254–255
  excretion of acid, 304
  functions, 170–174
  hormone breakdown, 200, 202, 205
  hormone production, 210–211,
    *see also* Juxtaglomerular
      apparatus
  innervation, 141–143, 169, 171, 210
  insulin resistance, 204, 301
  ischaemia, 326
  loop of Henle, 106–107, 169–170,
    173–174
    counter current multiplication,
      106–107, 173–174
    reabsorption,
      of calcium, 172, 174, 204, 306,
        307
      of chloride, 172, 172
      of sodium, 172, 173
      of urea, 172, 173
      of water, 172
  medulla, 105–107, 168–170,
    173–174, 297
    blood flow, 105–106, 200, 235
    osmotic gradient, 105–106, 173–
      174, 238, 240
  oxygen usage, 109
  tubules, 39, 104, 122, 169–170
    tubular fluid, 171–174, 253–256,
      304
  proximal convoluted tubule, 39,
    169, 194, 240, 254, 255
    function, 171
    reabsorption,
      of aminoacids, 39, 172
      of calcium, 172
      of chloride, 39, 171–172
      of glucose, 39, 172
      of hydrogen carbonate, 172,
        173
      of iodide, 203
      of phosphate, 172, 173, 306,
        308
      of potassium, 172, 173
      of sodium, 39, 171, 172,
      of urea, 172
    secretion of organic acids and
      bases, 39, 173
  role in blood pH control, 254–256
  role in blood volume control
  role in sodium/potassium balance, 331
  tubular maximum for glucose
    reabsorption, 172, 303

Kidney (*cont.*)
  tubules, 39, 104, 122, 169–170
    tubular fluid, 171–174, 253–256,
      304
  vasa recta, countercurrent
      exchange, 105–106, 169, 173
  vasoconstriction, 326
  vitamin D metabolism, 194, 211,
    306
  *see also* Renal, Renin
Kinaesthesia, 152, 155, 228
Kinins, 53, 158, 165, 167, 192, 206,
  211, 236, 329
Korotkov sounds, 103, 125
Kreb's cycle, *see* Citric acid cycle
Kupfer cells, 70


Labour, 164, 217, 283
Labyrinthine receptors, 153–155, 224,
  227
Lactase, 185, 186
Lactation, 62, 200, 207, 281, 283–284,
  330
  calcium requirement, 306
Lacteals, 41
Lactic acid (lactate), 10, 31, 34, 41,
  43, 93, 251, 254, 297, 300, 301,
  313, 315, 317–319, 320, 325
Lactoperoxidase, 182, 184, 188
Laking of red blood cells, 58
Laminae of
  dorsal horn, 163
  medullary grey matter, 166
Laminar flow, 100, 101, 103, 104, 125
Language, 265, 266, 268
Lashley cannula or cup, 179
Latency of contraction, 91, 92
Law of Laplace, 121, 123, 126, 232,
  329
Learning, 133, 221, 226
Length tension curves of muscle,
  88–89, 95, 96, 98, 121, 232
Lens, 27, 145–148
Leucocytes, *see* White blood cells
Leucocytosis, 65
Leucopenia, 65
Leucopoietin, 65
Leucotaxins, 66
Leukaemia, 65
Leukotrienes, 211
Lewis blood groups, 60, 182
Ligament receptors, 156, 228
Light, perception of, 82, 83, 145, 148
Limbic brain, 133, 134, 135, 137, 141,
  145, 160, 161, 162, 165, 200, 216,
  231, 235, 240, 241, 247, 273, 278,
  294, 295, 313, 329
Limbic cortex, 132, 140, 201, 220
Liminal stimulus, 75, 76
Lingual mucosa, sensory receptors,
  259
Lipase, 41, 182, 185, 186, 187, 296,
  299, 300
  pancreatic, 41, 184, 186

Lipids, 187,
  in cell membrane, 3, 5
  in plasma, 43
  in saliva, 183
  inclusions, 69, 94
  of red cells, 60
  transport in plasma, 47, 298, 299
  uptake, 303
  see also Fats
Lipolysis, 41, 186–188, 211, 296, 299, 300–304, 318, 331
Lipoproteins, 21, 22, 44, 47, 71, 183, 188, 298, 299
  in blood clotting, 50, 323
  in plasma, 47
  lipase, 299, 301
β-Lipotrophin, 164, 192, 200, 201
Lisp, 270
Lithocholic acid, 186
Liver, 7, 33, 47, 64, 141, 186, 200, 205, 286, 301, 303, 319
  action of adrenaline, 231
  action of insulin, 204
  albumin store, 326
  aminoacid metabolism, 296, 299
  aminoacid uptake, 301, 302, 304
  ammonia formation, 296
  and jaundice, 64
  bile production, 186, 209, 279
  blood supply, 231, 233, 319
  capillaries, 20
  cell organelles, 7, 9,
  composition of lymph, 18
  effect of insulin, 204, 301
  energy usage, 310, 312
  fatty acid metabolism, 302, 304
  functions, 187
  gluconeogenesis, 302, 304
  glucose metabolism, 298, 301
  glucose uptake, 14, 204, 303
  glycogen, 204, 296, 297, 299, 300, 301, 302, 304, 329
    storage, 297, 319
    effect of glucagon, 302
  glycogenolysis, 205, 301
  haemosiderin storage, 63
  heparin synthesis, 51
  hormone metabolism, 194, 200, 202, 203, 204, 205, 206, 207, 208 211, 306
  hormone production, 199, 211, 287
  ketone body production, 300, 302
  lactic acid oxidation, 318
  lipid metabolism, 299
  lipid uptake, 299
  macrophages, 70
  plasma protein synthesis, 44–45, 51, 326
  protein metabolism, 47, 302
  red cell production, 61
  secretion, 184, 186, 187, 278–280
  storage of iron, 63
  vitamin D metabolism, 211, 306
Load velocity curves, 95
Locomotion, 222, 225

Long-acting thyroid stimulator (LATS), 203
Low resistance shock, 324, 328
Lower motor neurone, 222
Lungs, 126–131
  blood supply, 126, 233
    see also Pulmonary circulation
  capillaries, 11, 19, 21, 109, 129
  compliance, 130, 243
  converting enzyme, 209, 239
  elasticity, 104, 127, 130
  gas exchange in, 108–112, 129
  haemosiderin storage, 63
  macrophages, 70
  stretch receptors, 82, 243
  ventilation perfusion ratio, 131
  water loss from, 16, 189, 266
  see also Pulmonary
Lung function tests, 131
Lung volumes, 128–130
  in males and females, 293
Luteinising hormone (LH), 193, 194, 201, 202, 207, 208, 211, 281, 282, 284, 292
Luteinising hormone releasing hormone (LHRH), 192, 201, 202, 207, 282, 283, 291, 330
LH, see Luteinising hormone
Lutheran blood groups, 60
Lymph, 17, 18, 22, 24, 69
  capillaries, 20, 22, 40, 317–318
  circulation, 18, 20, 100
  composition, 18
  daily production, 18
  formation, 20, 21–23
Lymph nodes, 45, 52, 65, 68, 69, 70
Lymphoblasts, 69
Lymphocytes, 22, 28, 45, 65, 67–69, 70, 179, 318
  B-lymphocytes, 69
  formation, 69, 290
    in spleen, 234
  life span, 66, 68
  mobilisable or recirculating pool, 68
  numbers, 66, 206
  thymus-derived (T-lymphocytes), 52, 68
Lysins, 52
Lysosomes, 7, 38, 39, 67, 70
Lysozyme, 52, 67, 70, 182, 188

Macrocytes, 54
Macrophages, 7, 28, 68, 69, 70, 324
Macula (type of cell junction), 5–6
Macula densa of juxtaglomerular apparatus, 209
Maculas of the saccule and utricle, 154
Macula of retina, 147
Magnesium, 13, 21, 85
  effect of ions on synapses, 79
  in bile, 34
  in CSF, 26
  in plasma, 42
  in saliva, 34, 181
  in sweat, 31, 34

Magnet reaction, 225
Male, 315
  acclimatisation to hot climates, 313
  adult, extracellular fluid volume, 17
  adult, intracellular water, 16
  adult, iron loss, 62
  adult, total body water, 15
  bite force, 262
  bone age, 286
  calcitonin levels, 204
  chewing strokes, 262
  chromosomes, 285
  effect of chronic stress, 330
  energy usage, 310
  erythrocyte sedimentation rates, 54
  fat content, 296
  fetal growth, 288–289
  haemoglobin concentrations, 58
  hair distribution, 291
  lung volumes, 129, 130
  metabolic rate, 207, 315
  prepubertal growth rate, 289
  prolactin secretion, 202, 291
  puberty, 290, 292
    growth spurt, 291, 292
  red blood cell count, 55, 293
  sensitivity to odours, 162
  speech sounds, 151
  sperm production, 282
  taste sensation, 161
  voice, 265
    pubertal changes, 293
  volume of swallow, 264
Maltase, 184, 185, 186
Mammary glands, 19, 30, 200, 202, 208, 283, 284
  see also Breasts
Mandible
  growth, 287, 293
  protrusion, retrusion and rotation, 258, 261, 264
  rest position, 92, 258
Mannose, 37, 39, 182
Mass peristalsis, 177, 280
Masseter muscle, 165, 179, 217, 219, 258, 261
  silent period, 261
  tone, 258
Mast cells, 29, 67, 211
Mastication, 134, 138, 150, 217, 219, 257–262, 272
  closing stroke, 261
  control of, 134, 138, 158, 161
  forces, 259
  opening stroke, 261
  power stroke, 261
  muscles of, 258, 259
  reflexes, 257
  rhythm, 261
Matrix vesicles, 8
Meals
  carbohydrate, 43, 303
  fatty, 41, 47, 299
  high protein, 65, 303

Mean Corpuscular Haemoglobin
  Concentration (MCHC), 58, 63
Mean Corpuscular Haemoglobin
  Content (MCH), 58, 63
Mean corpuscular volume (MCV), 54
Mechanoreceptors, 82, 158, 163, 167,
  258, 259
Medulla oblongata, 137–138, 140,
  220, 311,
  chemoreceptors, 316
  control centres, 238
    cardiovascular, 216, 220,
      230–231, 235, 240, 317, 318
    defaecation, 280
    muscle movement, 226, 227
    respiration, 216, 220, 242–243,
      245, 247
    swallowing, 220, 263–264
    vomiting, 220, 278
  see also Reticular activating system
Megakaryoblast, 71
Megakaryocyte, 71
Megaloblasts, 63
Meissner's corpuscles, 157
Meissner's plexus, 176
Melanocyte stimulating hormones
  (MSH), 203
Membrane potential, 11, 12, 75–76,
  80–81, 90, 94–95, 98, 158, 305
  changes in action potential, 76, 95,
    98
  nerve, 74
  muscle, 86
    cardiac, 97
    skeletal, 91
    smooth, 94
  red cells, 60
  salivary gland cells, 36, 37
  resting, 36, 37, 75, 80, 91, 94, 97,
    119
Memory, 133
Menarche, 281, 285, 286, 292
Meniscocytes, 55
Menopause, 276, 281, 293
Menstrual cycle, 202–203, 207–208,
  276, 281–282, 284, 292, 330
  follicular phase, 207, 281
  luteal phase, 281–282, 309
  midluteal phase, 208
  proliferative phase, 281
  saliva changes, 276
  secretory phase, 281
Menstruation, 55, 62, 207, 208, 281,
  282, 291
Mercury, 47, 189
Merkel's discs, 157
Merocrine secretion, 29
Meromyosins, 86
Mesomorphs, 297
Messenger RNA, 6, 198
Metabolic acidosis, 304
Metabolic alkalaemia, 251
Metabolic failure, 331
Metabolism, 294–308, 329
  during food intake, 302

Metabolism (cont.)
  hormonal, 203, 204–206, 301
  daily production of acid, 252
  heat production, 92–93, 203, 310,
    312, 313, 331
  of cardiac muscle, 99
  of fats, 296, 299, 301
  of glycogen, 299–301
  of muscle, 93, 317
  of protein, 299, 296
  see also Basal metabolic rate
    control
Metabolites
  effect on circulation, 232–235, 317–
    318, 326
  of vitamin D, 306–307
Metarterioles, 122, 316
Methaemoglobin, 56, 58, 63, 109
Micelles, 41, 187, 188
Microbodies, 7
Microcytes, 54
Microsomes, 36, 64, 274
Microtubules and microfilaments, 8,
  9, 30, 35, 36, 70, 71, 275, 323
  proteins, 195
Micturition, 174, 241
Milk, 186, 289
  human, iron content, 63
  let-down reflex, 200, 263
  secretion, 202, 208, 284
Mineralocorticoids, 205–206
  see also Aldosterone,
    Corticosterone
Misalignment detectors, 216, 312
Mitochondria, 8–9, 144, 196, 288, 299
Mitosis, 8
MN blood groups, 60
Modulation of sensory inputs, 152,
  158, 165
Modulators (hormonal), 193, 194–195
Mono-iodotyrosine, 203
Monoamine oxidase (MAO), 144,
  205, 274
Monoblast, 69
Monocytes, 28, 29, 65, 66, 69, 70, 290
Monoglycerides, 41, 186, 187
Motilin, 192, 201, 209
Motor control, 133, 134, 137, 140,
  155–156, 225–228
  see also Posture
Motor cortex, 133–134, 135, 137, 138,
  140, 225–226, 265, 304, 318
Motor endplate, 79, 80, 85, 90, 264
Motor unit, 85, 92
Motorneurones
  A α, 81, 139, 140, 156, 217,
    219–220, 222, 227, 228, 252,
    312
  A γ, 217, 227, 228
Mouth, 175, 176, 187
  oral mucosa, 35
    absorption of corticosteroids, 206
    blood flow, 207, 208, 235
    dryness of, 238
    effects of stress, 330

Mouth (cont.)
    effects of puberty, 292
    sensory receptors, 84, 158, 258,
      259
    water intake, 216
    see also Taste
  oral resonators, 268
  temperature, 309
  pain, 165
  secretions, see Saliva
  volume, 260
Movement
  fine skilled, 92, 304
  finger, 225
  learned skilled, 227
  local control, 227
  skilled, 225
Mucins, see Glycoproteins,
  Mucoproteins
Mucoproteins, 28, 29, 30, 34, 35, 44,
  63, 175, 180, 181, 182, 188
Mucous cells, 30, 34, 35, 186
Mucous glands, 260
Mucous secretion, 176, 208, 272
Mucus, 29, 101, 186, 187
  production, 243
  in respiratory tract, 130
Muller's manoeuvre, 244
Mullerian duct inhibiting factor, 289
Muscarinic receptors, 144, 274
Muscle, 73, 85–99, 286, 288
  basal metabolism, 92
  blood flow, 112
  contractile-elastic model, 88, 90, 91
  filaments, 8
  glucose uptake, 14
  growth and development, 287
  heat production, 88, 310
  layers of gut, 176–178
  metabolism, 47, 92, 297
  proteins, 86
  pump, 20, 124
  relaxants, 80, 128
  spindles, 82, 85, 155–156, 157,
    217–218, 222, 228, 258,
    260–261, 268, 319
  tension, 87
  water content, 15, 17
  see also Cardiac muscle, Skeletal
    muscle, Smooth muscle
Muscles
  fast, 92, 299
  of inspiration and expiration, 127,
    242, 248, 266
  of mastication, 222, 257, 258, 261
  pain in, 259
  of vocalisation, 265, 266, 267
  red, 92
  slow, 91, 92
  white, 92
Muscularis mucosae, 176, 178, 209, 279
Myeloblast, 65, 71
Myelocyte, 65
Myeloid (white-cell-producing) cells,
  61, 65

Myenteric plexus, 176, 177, 264
Myo-elastic theory of sound
    production, 267
Myoepithelial cells, 35, 87, 180, 200,
    263, 274, 284
Myogenic activity, 93, 94, 95, 98, 177,
    234, 322
Myoglobin, 57, 62, 92, 98, 110, 317,
    319
Myopia, 146
Myosin, 8, 71, 85, 86, 87, 88, 89, 92,
    94, 96, 98, 323
Myosin ATP-ase, 92

Natriuretic hormone, 212, 239
Nausea, 254, 273, 278, 328
Negative supporting reaction, 223
Neonatal circulation, 212
Neonatal line, 289
Nephron, *see* Kidney tubule
Nernst equation, 11, 12, 74, 75, 158
Nernst potential, 11, 12, 75, 80, 91, 98
Nerve, 73–78
    membrane permeability, 74
    composition, 16, 74
    effect of anaesthetics, 77
    fibre types
        A, 78
        A α, 217
        A β, 158, 166
        A γ, 217
        A delta, 158, 163, 166
        B, 78
        C, 78, 158, 163, 166
    conduction velocity, 77
    sensory groups, 155–158
    free endings, 157, 158, 163, 258
    growth factor, 287
    metabolism, 297
    plexuses of gut, 176
    terminals, 77, 79, 80, 319
    in smooth muscle, 94
Nerves
    effect of pressure, 78
    growth and development, 285, 286
    hormone production, 193
    in dentine, 167
    mixed, 77
    to teeth, 77
    myelinated, 73, 74, 75, 77
Nervous system, 132–167
    parasympathetic, *see*
        Parasympathetic
    sympathetic, *see* Sympathetic
    fetal, effect of thyroid hormones,
        288
Neuraminic acid, 66
Neuraminidase, 182, 190
Neurilemma, 74
Neuro-effector junctions, 143–145
Neurohypophysis, *see* Pituitary gland,
    posterior
Neuromuscular junction, 79, 81, 94,
    319
    importance of calcium, 305

Neurone, *see* Nerve
Neuropeptides, 81
Neurophysins, 38, 200, 238
Neurotransmitters, 39, 79–82, 83,
    94–95, 143–145, 163, 166, 191,
    205, 209, 273–274
Neutrophils, 28, 65, 67
    count, 66, 318
Newton's Law, 102
Newtonian fluids, 46, 102, 103, 104
Nexus, 6
Nicotine, 200
Nicotinic receptors, 143
Nissl granules, 74
Nitrogen
    balance, 296
    in atmospheric air, 107
    in blood, 107
    retention in pregnancy, 208
    solubility in plasma, 108
Nitrous oxide, 55, 112, 129
Nociceptive reflex, 218
Nociceptors, 82, 163, 166, 220
Nodes of Ranvier, 73, 77
Noradrenaline, 79, 94, 140, 144, 145,
    192, 204, 205, 231, 273, 274, 296,
    311, 312, 326, 329
    effect on brown adipose tissue, 288
    effect on pulp blood pressure, 235
    effect on sino-atrial node, 118
    metabolism and excretion, 205
Normal saline, 55
Normetanephrine, 144
Normoblast, 61
Normocytes, 54
Nose, 161
    irritation of mucosa, 243
    resistance to airflow, 104
Nucleic acids, 6, 186
Nucleolus, 6
Nucleoproteins, 6, 7
Nucleosides, 186
Nucleus, 3, 6, 132, 196
    steroid hormone receptors, 196,
        198

Obesity, 15, 297
Occlusion of the teeth, 258
Occlusion of nerve pathways, 81, 220
Octapressin, 200
Odontoblast cells, 8, 23, 167, 288
Odours, 162
    chemical structure, 162
    lipid solubility, 162
Oedema, 22, 67, 124, 235, 240
    angioneurotic, 53
Oedema fluid, 18, 22, 130
Oesophagus, 175, 176, 177, 185, 263,
    264, 273
    control, 277
    innervation, 277
    irritation of, 273, 278
    measurement of intrathoracic
        pressure, 130
    smooth muscle, 277

Oesophagus (*cont.*)
    pressure, 130, 177, 278
    receptors, 264
    speech production, 178
    striated muscle, 277
    taste buds, 158
    temperature, 309
    temperature sensation, 158
    water intake sensors, 238
Oestriol, 194, 195, 206, 207, 208, 283
17β Oestradiol, 194, 195, 207, 208,
    281
Oestrogens, 183, 191, 194, 202, 203,
    205, 206, 207, 208, 210, 282, 283,
    284, 287, 289, 290, 291, 292, 324
    blood concentration, 291
Oestrone, 194, 195, 207, 208
Ohm's Law, 103
Olfaction, 134, 158, 161–162
Olfactory receptors, 82, 161
Oncotic activity (pressure), 20, 21, 22,
    45, 171, 326, 328
Ondine's syndrome, 242
Opsin, 148, 149
Opsonins, 52, 53
Optic disc, 146
Optic pathways, 145, 146, 147,
    148–149
Oral fluid, *see* Saliva
Organ of Corti, 150, 151, 152
Organic acids and bases, transport in
    kidney, 39, 173
Organiser molecules, 285, 288
Orthodromic conduction, 77
Osmolality, 43
    body fluids, 135
    definition, 237
    of extracellular fluid, 216
    of plasma, 200, 237
    of renal tubular fluid, 173
Osmolarity
    definition, 237
    plasma, 237
Osmoreceptors, in hypothalamus, 82,
    199, 200, 216, 237–238, 239, 327
Osmotic activity, 12, 20, 33, 55, 220,
    304
    intracellular, 16
    of extracellular fluid, 174, 237
    of gastric contents, 278
    of plasma, 56
    total, 43, 45
Osmotic fragility test, 55
Osmotic pressure, *see* Osmotic
    activity
Ossicular conduction, 151
Ossification centres, 287
Osteoclast cells, 70, 288, 306, 307
Osteoporosis, 306
Otolith organ, 154, 224
Ovarian follicles, 193, 202, 203, 207,
    281, 282
    granulosa and thecal cells, 282
    maturation, 203, 207
Ovary, 20, 37, 194, 199, 207, 208, 288

Ovulation, 203, 207–208, 211, 212, 281, 282, 291
Ovum, 207, 208
Oxygen
  binding by haemoglobin, 58, 108
  carriage in blood, 56, 107–111
  arteriovenous difference, 104, 244, 316
  in exercise, 317
  content, of expired air, 245
  debt, 93, 317, 318, 321
  diffusion
    gradient, 316
    in lung, 21, 56, 130
    rate, 130
  dissolved, 108, 247
  exchange in tissues, 11
  lack, 129, 248, 253, 327
    local, 78
  nascent or super, 67
  partial pressure
    at high altitudes, 253
    in air, 108
    in lung alveoli, 108, 109, 130, 244
    in blood, 215, 217, 231, 243, 245, 246, 247, 315
      arterial, 11
      mixed venous, 109
      venous, 130
    in blood, control, 242, 244
    in tissues, 11, 109
      effect on local circulation, 233
  solubility in plasma, 108
  supply, 93
    to the brain, 326
    to kidney, 63, 210
  uptake in lungs, 56, 109, 129, 130–131
  usage, 104, 203, 295, 315, 316, 317, 318, 320
    by the heart, 109
    in action potential, 76
    by the kidney, 109
    by the liver, 233
    by red blood cells, 13
    by skeletal muscle, 109
    whole body, 233, 312
Oxyhaemoglobin, 58, 109, 252
Oxytocin, 38, 95, 192, 197, 200, 201, 202, 207, 208, 217, 263, 283, 284

Pacemaker areas, 94, 97, 98, 99, 177
Pacinian corpuscles, 83, 85, 156, 158
Packed cell volume (PCV), see Haematocrit
Pain sensation, 78, 81, 82, 139, 157, 163–167, 247, 318, 329
  chemical stimuli, 82
  dental, mechanisms of, 167
  fibres, 78, 138, 158, 166, 180
  in salivary glands, 180
  inhibition, 167
  modulation, 164, 165, 202
  muscle and joint, 135, 165, 259
  oral, 165–167

Pain sensation (cont.)
  receptors, 82, 83, 161, 163, 165, 258
    in inner ear, 165
    in teeth, 165
    palatal, 260
    referred, 165
    in oral cavity, 165
    reflexes, 218
    surgical abolition, 164
    threshold, 165, 262
Pain-producing peptides, 83
Palate, 244, 258, 260, 268, 269
  cleft, 264, 268, 269
  hard, role in mastication, 257, 260
  soft, taste buds, 158
Pancreas, endocrine, 199, 204
  cells, 204, 302, 303
  fetal, 204
  hormones, see Glucagon, Insulin, Somatostatin interactions
Pancreas, exocrine, 7, 20, 30, 31, 33, 34, 175–176, 186, 187, 278–280, 290, 301–303
  amylase, 186
  enzymes, 279, 184–185
  innervation, 141
  lipase, 186
  secretion, 184–186, 187, 201, 209, 278–280
    daily volume, 185
Pancreozymin, see Cholecystokinin
Panting, 129, 189, 247, 310
Para–aminohippuric acid (PAH), 39, 105, 173
Paracrine secretion, 191
Paradoxical reflex, respiratory, 243, 253
Parasympathetic nervous system, 143
  control of
    bladder, 241
    gastrointestinal tract, 177, 272
    heart, 118
    muscles of gut, 178
    salivary glands, 180, 273
    vasodilatation, 232
Parathormone, see Parathyroid hormone
Parathyroid glands, 20, 199, 204
Parathyroid hormone (PTH), 173, 174, 191, 192, 197, 204, 211, 306, 307, 308
Parotid glands, 34–35, 141, 179–182
Parotid saliva, 178–183, 272, 275
Parotin, 182
Partial pressure, definitiion, 108
Parturition, 60, 208, 212
Pavlov, 221
Pellicle, salivary, 178, 190
Penicillin, 52, 325
Penis, 289, 292
  erection, 141, 232
Pentose monophosphate shunt, 56, 59, 66
Pepsin, 184, 185, 186, 187, 208, 278
Pepsinogen, 186, 187, 277, 279

Perceived intensity of sensation, 82, 84, 160, 165
Perception, 145
  of colour, 149
  of visual stimuli, 149
Pericardium, 24, 117
Pericytes, 20, 123
Perilymph, 150, 151
Periodic respiration, 253
Periodontal ligaments, 258, 259, 262
  mechanoreceptors. 158, 167, 258
Peripheral resistance, 229, 230, 231, 234, 235, 317, 318, 319, 324, 326, 328
Peristalsis, 177, 262, 280
  mass, 177, 280
Peristaltic waves, 174, 177, 178, 263
Permeability
  constant, 10–11
Permeases, 4
Peroxidase, 38, 62, 67, 69
Peroxisomes, 8
Perspiration, 310, 313
  see also Sweat
Peyer's patches, 69
pH, 319
  blood, 233, 246, 248, 316, 317
    arterial 43, 250
    after exercise, 318, 320
    control, 56, 173, 250–256
      by kidney, 254
      by respiration, 253
    effect on calcium binding in plasma, 46
    effect on cerebral blood flow, 233
    effect on haemoglobin–oxygen association curve, 109, 110
    mixed venous blood, 43, 250
  body fluids, 46, 112
    control, 240, 250–256
  cell fluid, 3, 16, 319
  cerebrospinal fluid, 26, 27, 242, 248
  dental plaque, 190
  duodenal contents, effect on gastrointestinal motility, 279
  extracellular fluid, 16, 250
  gradient in kidney, maximum, 254
  gastric secretion, 32, 186
  plasma, 42, 46, 174, 240, 305
  receptors, 318
  red blood cells, 60
  saliva, 183
    critical, 190
    effect on calcium binding, 183
  small intestine, 186
  stomach, 185, 186
  synovial fluid, 28
  of urine, 254
Phagocytes, 52
Phagocytin, 67
Phagocytosis, 7, 13, 38, 53, 66, 67, 70, 188
Phagosomes, 7
Pharynx, 175
  irritation of, 278

Pharynx (*cont.*)
  receptors for water intake, 216, 238
  resonating chambers, 268
  taste buds, 158
Phonation, 266, 267, 268
Phosphatases, 182, 190, 298
  acid, 7, 67, 70, 72, 159, 182
  alkaline, 67, 69
Phosphate, 3, 21, 23, 251, 300
  absorption in small intestine, 41, 308
  active transport, 308
  buffering, 46, 251, 252, 254
  excretion,
    effect of parathormone, 204
    in urine, 308
  high energy bonds, 8, 13, 86, 93, 297
  in CSF, 26
  in plasma, 43
  in saliva, 185
    inorganic, 185, 190
  metabolism, 307
  reabsorption in kidney, 172, 173, 306, 308
  retention in pregnancy, 208
  total body store, 308
Phosphofructokinase, 298, 299, 301, 319
Phosphoinositide-4,5 phosphate, 196, 197, 274
Phospholipid, 3, 39, 47, 186, 188, 323
  in blood clotting, 49–50, 71, 323
Phospholipid–lecithin, 126
Phosphorylase, 196, 197, 201, 204, 300, 302
Phosphorylase b kinase, 300
Photopic vision, 148
Photopigments, 146, 147, 148, 149
Photopsin, 149
Photoreceptors, 148
Physical fitness, 129
Physiological mesial drift, 259
Pinocytosis, 7, 13, 19, 21, 30, 39, 63, 173
Pipe resonators, 268
Pitch, 151, 265
Pituitary gland
  anterior, 164, 194, 199, 200, 201–203, 205, 207–208, 216, 239, 282–284, 292, 329–330
  intermediate lobe, 199, 203
  gonadotrophins, 282, 289, 291,
    *see also* Follicle stimulating hormone, Luteinising hormone
  posterior, 20, 199, 200, 201, 236, 237
Placenta, 60, 61, 199, 206, 207, 208, 212, 282–283, 288–289
  hormones, 194, 283, 284, 288, 289
Plasma, 15, 17, 18, 22, 24, 28, 33, 36, 42–53, 103, 117, 123, 130, 302, 324, 328
  adrenaline, 205

Plasma (*cont.*)
  aminoacid concentrations, 43, 47, 172, 202, 205, 204, 304
  blood group antibodies, 59
  calcium concentration, 42, 46, 204, 211, 305, 306
    ionised, 191, 306, 307
  carbon dioxide
    solubility, 108, 111
    transport, 56, 112
  cholesterol, 207
  copper transport, 46
  colloid pressure, *see* Oncotic activity
  composition, 17, 18, 42, 44
  cortisol concentration, 205, 210, 216
  fluoride, 184
  free fatty acid concentration, 187, 205, 302
  glucose concentration, 43, 204, 205
    *see also* Blood glucose
  growth hormone levels, 202
  hormones in, 43, 48, 135, 191, 193
  hydrogencarbonate concentration, 43, 172
  iodide content, 184, 203
  lipids, 43, 47, 298, 299
  magnesium concentration, 42
  nitrogen solubility, 108
  oestrogen concentrations, 207, 282, 291
  osmolality, 200, 237, 239, 326
  osmotic activity, 56, 238
    and thirst, 238
    control, 237
    total, 43, 45
  oxygen solubility in, 108
  pH, 46, 174, 240, 251, 305, 316, 318
  phosphate concentration, 43, 308
  potassium concentration, 42, 188, 206, 240
  progesterone concentrations, 208, 282
  protein content, 43 *see* Plasma proteins
  salt concentrations, 237
  sodium concentration, 42, 206, 240
  specific gravity, 54
  steroid hormones
    concentrations, 183
    transport, 194
  testosterone, 207, 291
  thyroid hormone concentrations, 202, 203, 204
  urea concentration, 43
  viscosity, 56
  vitamins, 43
  volume, 17, 42, 46, 238, 239, 318, 326
    measurement, 18
Plasma cells, 68, 69
Plasma membrane, *see* Cell membranes

Plasma proteins, 17, 18, 20–23, 26, 28, 36, 43–53, 63, 103, 171, 236, 239, 252
  binding for transport, 46
  calcium, 305
  hormones, 193, 205, 206, 207, 208
  iodine, 203
  ions, 46
  thyroid hormones, 203
  buffering, 46, 246, 252
  carbamino compounds, 58
  in glomerular filtrate, 171
  in infancy, 290
  in saliva, 181
  reabsorption in kidney, 173
  shape, 46
  synthesis after haemorrhage, 326, 327
  *see also* Albumin, Blood clotting factors, Complement, Fibrinogen, Globulins
Plasma skimming, 103
Plasmablast, 69
Plasmacytes, 45
Plasmin, 50, 51, 67, 323, 324
Plasminogen, 44, 50, 51, 324
Plasminogen activator, 322, 324
Plasticity of smooth muscle, 88, 95, 241
Platelets, 6, 8, 49, 70–72, 195, 322–324
  adhesion, 51, 71, 72, 322–323
  aggregation, 51, 71, 212, 322–324
  count, 66, 70, 324
  in infancy, 290
  degranulation, 71–72, 212, 323
  factors, 71, 182, 323
  lifespan, 66, 71
  primary plug, 71, 323
  prostaglandins, 212, 236, 322, 323
  release reaction, 71, 72, 212, 323
Pleasure centre, 165
Pleura, 24, 126–127
  adhesions, 127
  fluid, 17
Pneumocytes, 126
Pneumotaxic centre, 242
Pneumothorax, 127
Podocytes, 170
Poiseuille's Law, 103, 104
Polycythaemia, 56, 70
Polymorphonuclear neutrophils, 179
Polyuria, 304
Pons, 137–138, 141, 220, 223, 226, 227
  centres, 231, 242
Pores
  calculated effective size in cell membrane, 4
  in kidney distal tubule, 238
  in glomerular capillaries, 44, 170
  effective diameter in intestine, 40
  in capillaries, 46
    effective size, 21, 44

Pores (*cont.*)
  in cell membranes, 4, 9
  intestinal cell wall, 40
  nuclear, 6
Porphyrin ring, 57, 64
Portal circulations, 122, 180
  hypothalamic, 199, 200, 201, 216
  kidney, 169
Portal vein, 122, 233
Position sense, 152, 153, 155–156
Positive supporting action, 223, 225
Post-absorptive state, 43, 309, 315
Post-capillary sphincters, 123
Post-lysosomes, 7
Posterior pituitary gland, *see* Pituitary gland, posterior
Postsynaptic depolarisation, 79
Postsynaptic inhibition, 80
Postsynaptic membrane, 78, 79, 80, 81, 82
Postural syncope, 328
Posture, 92, 128, 131, 137, 222–225, 227, 231, 310, 328
  control, 134, 137, 155, 227, 258
  effect on circulation, 124, 232
  jaw position, 258
  maintenance of, 156, 218
  reflexes, 227
Potassium, 3, 11, 13, 14, 16, 21, 31, 36
  channels, 274
  equilibrium potential, 74
  excretion, 255, 331
    action of mineralocorticoids, 205
    hydrogen balance in urine, 256
  in aqueous humour, 28
  in bile, 34
  in CSF, 26
  in muscle, 90
  in nerve,
  in plasma, 42, 188, 206
    control, 240
  in saliva, 31, 34, 183
  in sweat, 31, 34
  in synovial fluid, 28
  in urine, 173
  reabsorption in kidney, 173, 240
  retention, 205
  secretion in kidney, 173, 240, 256
Power Law, 84
Precapillary sphincters, 122, 123, 232, 235, 316
Precipitins, 52
Pregnancy, 63, 128, 164, 178, 200, 207–208, 276, 281, 282–284, 288, 324
  calcium requirement, 306
  iron requirement, 62
Pregnenolone, 37, 201, 208
Presbyopia, 146
Pressure
  effect on nerves, 78
  of gases, 107
  gradients, 10, 103, 104
    in heart, 119
    in circulation, 124
    in mouth during speech, 269

Pressure (*cont.*)
  receptors, 82, 84, 85, 137, 153, 156, 164, 224, 330
  in mouth, 258, 259, 260
  relationship with flow, 104
  sensation, 138, 153, 157, 166, 222
Pressure-volume curve, 130
Presynaptic inhibition, 80, 164, 166, 228
Pro-erythroblasts, 61, 63, 210
Progesterone, 183, 194, 195, 202, 205, 206, 207–208, 281, 282, 283, 284
  plasma concentrations, 208
Prolactin, 193, 197, 201–202, 283–284, 290, 291, 292, 307, 330
Prolactin release inhibiting hormone (PIH), 192, 201
Prolactin releasing hormone (PRH), 192, 201, 202
Proline-rich peptides, 183
Prolymphocytes, 69
Promegakaryocyte, 71
Promyelocytes, 65
Proprioception, 137, 138, 152, 157, 223–224, 318
Proprioceptors, 82, 224, 225, 227, 258, 315
Prostacyclin (PGI2), 210, 211, 212, 236, 322
Prostaglandins, 71, 194, 207, 208, 210, 211, 212, 236, 295
  A (PGA), 212, 236
  E group (PGE), 211, 236
  $E_1$ (PGE$_1$), 212, 311
  $E_2$ (PGE$_1$), 212
  endoperoxides (PGG$_2$ and PGH$_2$), 194, 323
  $F_{2\alpha}$ (PGF$_{2\alpha}$), 212, 282, 283
  inhibition, 72, 195
  mechanism of action, 197
  synthesis, 192, 211
  *see also* Prostacyclin, Thromboxane
Prostate gland, 194, 207, 290, 292
Protagonist muscles, 222
Protein kinase, 195, 196, 201, 275, 300
Proteins
  digestion, 186, 209
  energy value, 295
  excretion in kidney, 39
  hormones, 193, 201, 202, 208
  in aqueous humour, 28
  in cells, as anion, 16
  in cell membrane, 3, 56, 59
  in cerebrospinal fluid, 26
  in synovial fluid, 28
  metabolism, 251, 252, 296, 299, 304, 305
  of muscle, 86, 94
  of saliva, 181
  secretion, 28–30
  starvation, 63, 296
  synthesis, 6, 65, 204, 211, 296, 301, 302, 305
    effect of steroid hormones, 198
  in fetus, 288
  *see also* Plasma proteins

Proteoglycans, 5, 211
Prothrombin, 44, 49, 51, 186, 323
Prothrombinase, 49
Protoporphyrin IX, 57, 62
Protrusor muscles, 222, 260, 261
Psychic stimuli, 271, 272, 302, 315, 318
PTH, *see* Parathyroid hormone
Puberty, 207, 283, 285, 286, 287, 289, 290–293, 313
  changes in vocal cords, 267
Pulmonary
  circulation, 21, 100, 117, 118, 126, 229
    systolic and diastolic pressures, 120, 122
    reservoir function of veins, 232, 233
  congestion, 325, 329
  fibrosis, 56
  oedema, 21
  vessels, effect of hypoxia, 233
  *see also* Lungs
Pulse, 124, 324
  pressure, 296, 325, 326, 329
  rate, *see* Heart rate
  wave, 122, 124
  velocity, 123
Pupil, 144, 146, 149
Purkinje fibres, 97, 98, 118
Pus, 23
Pyramidal tracts, 139, 226
Pyrexia, 309, 314
Pyrexins, 211, 311, 314
Pyrophosphatase, 182
Pyrophosphate in saliva, 185
Pyruvate, 10, 93, 296, 298, 299, 301, 317
Pyruvate dehydrogenase, 9

QRS complex, 119

Radiation of heat, 310, 313
Radiation, effect on lymphocytes, 69
Radius of tube, effect on flow characteristics, 101, 104, 123
Rage, 135, 272
Reabsorption
  in kidney, 39, 104, 171–172, 200, 251, 306
  in salivary ducts, 33, 37
Reactive hyperaemia, 235
Receptive field, 84, 147, 163, 166
Receptor potential, 83, 84, 85, 148, 152
Receptors, 82–85, 79–81
  neurotransmitter
    α adrenergic, 145, 205, 231, 232, 234, 235, 274
    $α_1$ adrenergic
    $α_2$ adrenergic, 145
    β adrenergic, 99, 145, 231, 232, 234, 288, 312, 315
    $β_1$ adrenergic, 145, 205, 274
    $β_2$ adrenergic, 145, 205
    cholinergic, 143–144

Receptors (*cont.*)
  neurotransmitter (*cont.*)
    muscarinic cholinergic, 144, 274
    nicotinic cholinergic 143
  hormonal, 195–198
  sensory, 82–85, 145, 146, 148, 159, 162, 167, 215, 218, 238
    rapidly and slowly adapting, 157
    static and dynamic, 155, 156
Recovery or flow, phase, 331
Recruitment, 220
Rectoflatus, 178
Rectum, 176, 177, 220, 280
  distension, 271
  innervation, 143
  temperature, 309, 316, 317
Red blood cell count, 54, 55, 56, 58, 63, 64, 66, 206, 318, 326
  after haemorrhage, 327
  in infants, 290
  in males and females, 293
Red blood cells, 6, 14, 22, 54–64, 102–103, 107–111, 130, 297, 323, 328
  abnormal forms, 55
  aging, 13, 48, 70
  buffering capacity, 46, 61, 252
  cell membrane, 4, 5, 56, 130, 322
  chloride/hydrogen carbonate exchange, 60
  composition, 56
  dimensions, 54, 55, 66, 112
  functions, 56
  haemolysis, 47, 48, 52, 56, 62, 64, 186, 234, 323
    after birth, 290
  hyperchromic, 59
  hypochromic, 59, 63
  insulin resistance, 204, 301
  life, 63, 66
  loss in haemorrhage, 325–326
  macrocytic, 63
  maturation, 63, 207
  membrane charge, 60, 103
  microcytic, 63
  normochromic, 59, 63
  osmotic swelling, 55
  pH, 60
  production, 56, 61, 62, 63, 65, 207, 210, 234, 326
    hormonal effects, 55
    after birth, 290
  role in carbon dioxide transport, 112
  water content, 57
Red flare, 235
Red muscle, 300, 302, 317
Red tache, 235
Referral of pain, theories, 165
Reflexes, 218–221, 222, 227, 271, 295
  conditioned, 218, 221, 227, 272
  development of, 290
  latency, 219
  masticatory, 257, 261
  milk letdown, 200

Reflexes (*cont.*)
  monosynaptic, 220, 262
  of cranial nerves, 138
  postural, 227
  protective, 243, 261
  reflex arc, 218, 219
  reinforcement, 221
  spinal, 138, 241
  stretch, 218, 219, 222
  visual, 149
Refractory period, 76, 77, 83, 91, 94
  absolute
    of cardiac muscle, 98
    of nerve, 76
    of skeletal muscle, 79, 91
  relative
    of cardiac muscle, 98
    of nerve, 77
    of skeletal muscle, 91
Regenerative tissues, 286–287
Relaxin, 192, 208, 283
Renal clearance, 104–105
Renal corpuscles, 169–170
Renal erythropoietic factor (REF), 63, 209, 210, 326
Renal filtration fraction, 171
Renal glomerular arterioles, 169, 170, 171, 209, 239, 326
Renal glomerular capillaries, 11, 20, 104, 169–171,
  effective pore size, 44, 46, 48
  filtration pressure, 11, 169, 170–171, 239, 240
Renal glomerular filtrate, 39, 171
  composition, 171
  pH, 254
Renal glomerular filtration, 11, 104
Renal glomerular filtration rate (GFR), 105, 171, 326
Renal glomerulus, 11, 168–169, 170–171
Renal plasma flow, effective (ERPF), 105, 171, 326
Renin, 46, 206, 209, 210, 212, 236, 238, 239, 240, 319, 326, 327, 330
  secretion, 145
Renin–angiotensin system, 239
Renshaw cell, 81, 140, 228
Residual volume, 129
Resistance
  peripheral, 46, 118, 121–123, 138, 229, 234, 328
  to flow, 100, 103–104, 130
  vessels, 123
Resonance, 265, 266
Resonators, 268, 269
Resorption of bone, 306
Respiration, 104, 126–131, 231, 250, 318, 326, 327, 331
  artificial, 128, 242, 244
  control, 27, 220, 242–251
  during swallowing, 263
  effect of exercise, 316, 320
  effect of hypercarbia, 249
  effect of hypothermia, 309

Respiration (*cont.*)
  effect of hypoxia, 249
  minute volume, 130
  voluntary, 216, 242
Respiratory
  acidaemia, 251
  alkalaemia, 251
  alkalosis, 331
  centres, 216, 242, 243, 246, 247–248, 251
  chemoreceptors, 246–247, 248, 251, 253
  control of pH, 253–254
  depth, 128, 129, 137, 138, 242, 248, 317, 329
  failure, 251
  muscles, 251, 319
  quotient, 111, 252, 295, 299, 315, 317, 318
  rate, 128, 129, 137, 138, 242, 248, 315, 317, 325, 326, 329
  system, 100, 101, 104, 126–131, 251, 315, 329
Retching, 264
Reticular activating system, 84, 85, 134, 137–138, 140–141, 152, 155, 164, 166, 226, 227, 264, 273, 326
Reticulo-endothelial system, 61–62, 63, 70
Reticulocyte, 62, 326, 327
Retina, 19, 82, 83, 145, 146, 147, 148
Retinene1, 148
Retrusors of mandible, 222, 261
Reynold's number, 101, 102, 104
Rheobase current, 76
Rhesus blood groups, 60
Rhodopsin, 147, 148
Rhythm generators, 242, 248, 261
Ribonuclease, 7, 67, 182, 186
Ribonucleic acid (RNA), 6, 211, 302
Ribosomes, 6, 20, 30, 47
Righting reflexes, 223, 224, 225
RNA, *see* Ribonucleic acid
Rods, 83, 146, 147, 148
Rotation of body, perception of, 153
Rouleaux, 54, 102
Ruffini's end organs, 156, 158, 258
Running, 319, 320
Russell bodies, 69

Saccades, 149
Saccule, 150–154, 225
Saliva, 23, 29–37, 65, 66, 178–185, 272–277, 278
  aminoacids, 183
  ammonia, 181, 183
  amylase, *see* Amylase
  antibacterial activity, 182, 184, 189–190
  blood clotting factors, 182
  blood group substances, 181
  buffering, 184, 277
  calcium content, 34, 181, 183
  carbon dioxide content, 184
  cells, 65, 66, 179

Saliva (*cont.*)
  chloride content, 34, 181, 183, 184
  citrate, 183
  composition, 31, 161, 180, 181, 183
    changes with age, 276
    effect of diet, 277
    effect of duration of stimulation, 275–276
    effect of flow rate, 34, 275
    effect of hormones, 276
    effect of nature of stimulus, 276
    effect of time of day, 276
    in relation to plasma, 276
  cortisol, 181, 205
  creatinine, 183
  daily volume, 179, 185
  enzymes, 181, 182
  flow rate, 275, 296
  fluoride content, 181, 184
  free sugars, 182
  functions, 189–190
  hormones in, 182, 183, 193, 205
  hydrogen carbonate content, 34, 181, 183, 184, 185
  hypotonicity, 37
  interactions with oral and dental tissues, 190
  lactate, 183
  leucocytes, 65, 66, 188
  lipid content, 181, 183
  lubricant action, 188
  mixed, 178, 275
    composition, effect of flow rate, 181, 275
    modification in striated ducts, 36
  mucoprotein secretion, 273
  pH, 183, 185
  pH and calcium binding, 183
  phosphate concentration, 181, 185
  potassium concentration, 34, 181, 183
  protective functions, 188
  protein content, 181, 182, 190
  protein secretion, 273
  pyrophosphate in, 185
  rate of flow, changes with age, 276
  rest transient, 183
  sodium concentration, 34, 181, 183
  thiocyanate, 181
  unstimulated flow, 179, 273
  urea content, 181, 183
  uric acid, 183
  vitamins, 183
Salivary glands, 30, 33, 34–37, 141, 179–180, 203, 295
  anterior lingual, 35
  blood supply, 180, 232, 273
  control, 138, 160, 272–275
  cortisol metabolism, 205
  distension, 180
  ducts, 34, 180, 183, 184
    aldosterone effect, 206
    innervation, 274
    reabsorption in, 313
  inhibition of secretion, 144

Salivary glands (*cont.*)
  innervation, 180, 273, 274
  lymphoid tissue, 182
  myoepithelial cells, innervation, 274
  serous cells, 35
  stimulation, 145, 160, 179, 221, 273, 276, 278
    cephalic phase, 272
    effect of local peptides, 276
    gustatory, 179, 272
    interorgan stimuli, 273
Salivary sediment, 179
Salivatory centres, 138, 160, 272
Salt taste, 160, 276
Salt balance, 202, 206, 240, 283, 330
  control, 239
Salt intake, 161, 240, 314
Salt loss, 33, 304, 326
  diarrhoea and vomiting, 185
  in haemorrhage, 325
  in sweating, 313
Salt retention, 202, 237, 239
Saltatory conduction, 77
Sarcomere, 85, 86, 87
Sarcoplasmic reticulum, 85, 91, 93, 95, 96, 98
Satiety centre, 283, 295, 314
Schwann cells, 73
Scotopic vision, 148
Scotopsin, 148
Scratch reflex, 222, 225
Second messengers, 30, 145, 193, 194, 195–198, 305
Secondary sexual characteristics, 207, 282, 291, 293
Secretin, 191, 192, 201, 204, 209, 278–280, 302
Secretion, 24–41, 274, 313
  endocrine, 37–41
  exocrine, 24, 29–37
  in the kidney, 39, 174
    hydrogen ion, 174, 254
    organic acids and bases, 173
    potassium, 174
  role of Golgi apparatus, 7
Secretions, 29, 60, 175, 182
  gastrointestinal, 185–187
    control, 278–280
    volume, 185
Secretomotor nerves, 141, 180, 273, 274, 277
Secretors, 182
Secretory granules, 7, 30, 35, 36, 274, 275
Secretory potentials, 36
Secretory vesicles, 30, 35, 36, 275
Seeding, 190
Segmentation, 177
Selective permeability, 10, 11, 12
Semicircular canals, 150–151, 153–154, 224–225
Seminal vesicles, 290, 292
Seminiferous tubules, 202, 282
Semipermeable membrane, 10, 11

Sensation, 134, 135, 145–167
Sensitivity, 84
Sensitivity, dynamic and static, 84
Sensory cortex, 84, 134–135, 145, 164, 311
Sensory nerves, 73, 82–84, 132, 158, 166, 271
  classification, 77
Sensory pathways, 139
Sensory receptors, 82–85
Sensory unit, 84, 147
Serotonin, 71, 72, 81, 95, 140, 166, 236, 323
Serum, 42
Sex-linked characteristics, 51
Sexual behaviour, 135, 220, 330
Sexual dimorphism, 285, 290, 293
Sham rage, 273
Shivering, 288, 311, 312, 314
Shock, 234, 236, 243, 246, 324–329
  compensatory mechanisms, 325
Sialic acid, 28, 182, 190
Sialin, 183
Sickle cell anaemia, 55, 64
Sighs, 243, 244
Singing, 248
Sinoatrial node, 97, 98, 118, 320
Sinusoids, 20, 60, 201, 205
Skeletal muscle, 85–93, 134, 155, 286, 288
  action potential, 86, 90–91
  aminoacid metabolism, 299, 301
  blood flow, 112, 232, 234, 326, 328, 329
    in exercise, 316–319
  composition, 16, 90
  contraction, 86, 87–90, 91–93, 317–318, 319
    effect on blood flow, 234, 317
    velocity and force, 89, 91
  effect of insulin, 301
  efficiency, 93
  energy sources, 92, 297, 299, 300
  excitation–contraction coupling, 79
  fibres, 85
  fibre types, 92
  filaments, 8
  glucose uptake, 303
  glycogen, 297, 299, 302
  grading of contraction, 220
  growth and development, 287, 291, 292, 293
  heat production, 88, 90, 92, 93, 310, 315
  lymph formation, 22, 318
  metabolism, 47, 93, 297, 298, 304
    basal, 92, 210
    lipid, 299
  movements, 90, 222, 225–228
  oxygen usage, 109
  pain, 165
  proprioceptors, 137, 153, 224–225, 315
  proteins, 85–86
    turnover, 301, 304, 331

Skeletal muscle (*cont.*)
  role of calcium, 90–91
  spasm, 165
  tension, 87, 88–89, 90, 91–92
  tone, 92, 223, 296
  vasodilatation, 205, 231, 232
  wasting, 287, 296
  water content, 17
  work, 88, 93
Skin, 82, 158, 206, 286, 324
  barriers, 52
  blood flow, 112, 235, 246, 312, 313
    in exercise, 317
    in relation to temperature, 231, 296, 311, 312
  capillaries, 19, 105–106, 110
  cells, 7
  circulation, 105–106, 218, 235
  colour, 235, 246, 314, 328
  heat gain from sun, 310
  heat loss, 296, 310
  sensation, 82, 84, 137, 157, 158
  temperature, 110, 309, 310, 311, 312, 313, 315, 317, 318
    effect of exercise, 316
    interaction with core temperature, 312
  temperature receptors, 158, 216, 311
  triple response, 218
  vapour pressure of water at surface, 310
  vitamin D synthesis, 211
Skull, growth, 292, 293
Sleep, 135, 138, 179, 201, 202, 238, 242, 244, 264, 273, 292, 297, 311, 320
Slow muscle, 300
Small intestine, 41, 47, 69, 175, 177, 178, 186, 187, 188, 206, 208, 209
  action of vitamin D metabolites on calcium absorption, 204, 211, 307
  aldosterone effect, 206
  area available for absorption, 40
  cell surface enzymes, 186
  control, 278
    cephalic phase, 278–279
  hormone production, 199, 201
  hydrogen carbonate, 186
  lipid absorption, 41
  motility, 209
  pH, 186
  secretion, 209
  sodium absorption, 206
Smell brain, 133, 162
Smell, 82, 133–134, 161–163
  in cephalic phase of digestive control, 271–273, 277
  interaction with taste, 161
  power law, 84
  theories of sensation, 162
Smoking, effect on taste, 161
Smooth muscle, 6, 8, 85, 86, 87, 93–96, 132, 141, 145, 195

Smooth muscle (*cont.*)
  action of hormones, 95, 200, 208
  action potential, 86, 94–95
  bladder, 96, 174, 241
  blood flow, 234
  composition, 16, 90, 94
  contraction, 95–96
  extracellular space, 93
  intrinsic rhythmicity, 96, 177, 232
  length tension curves, 95–96
  multi-unit, 93, 94
  oscillating resting potential, 94
  pain due to distension, 165
  proteins, 94
  plasticity, 94, 96
  role of calcium in action potential, 95
  single unit, 93, 94, 95
  uterus, 200, 208
  vascular, 93, 94, 95, 122, 123, 200, 232, 322
  visceral, 93, 94, 95, 96, 176–177
Sneezing, 138, 220, 243
Sniffing, 244
Snoring, 244
Sodium, 11, 13, 14, 16, 21, 31, 36, 95, 188
  absorption in gut, 41, 184
  binding by albumin, 46
  channels, voltage dependent, 75, 76, 77, 91
  equilibrium potential, 74
  excretion, 256
    action of mineralocorticoids, 205
  in aqueous humour, 28
  in bile, 34
  in CSF, 26
  in muscle, 90
  in nerve, 16
  in plasma, 42, 236, 240
  in saliva, 31, 34, 183
  in sweat, 31, 34
  in synovial fluid, 28
  loss from cells during buffering, 252
  loss in sweating, 319
  membrane permeability to, 75
  reabsorption,
    in kidney, 39, 171, 172, 173, 174, 206, 209, 240
    in salivary striated ducts, 37, 276
  retention, 240, 252, 312, 314, 315, 326, 490, 505, 516
  total body, 174
Sodium/potassium ATPase, 13, 16, 39, 63, 107, 176, 206, 295
Sodium/potassium pumps, 13, 14, 30, 31, 36, 39, 41, 74, 99, 148, 188, 196, 239, 240
  *see also* Sodium/potassium ATPase
Solubility, 130
  coefficient, 108
  of gases, 108, 112
    in blood, 107
  of carbon dioxide, 130, 250
  of oxygen, 130

Somatomedin C, 192, 197, 202, 211, 287
Somatomedins, 211, 287, 290
Somatosensory cortex, 133, 134, 135, 137
Somatostatin, 158, 192, 197, 201, 202, 204, 209, 302, 303
Somatotopic localisation *see* Topographical localisation
Somatotrophic hormone (SH) *see* Growth hormone
Sound, 265, 268
  conduction through bone, 150
  frequencies, 82, 151, 265
    tonotopic localisation, 152
  intensity, 151, 265, 267
    coding, 152
  perception, 150, 151
  power law, 84
  sensitivity, 151
  transmission in ear, 151–152
Specific gravity of blood cells and plasma, 54
Spectrin, 55, 56
Speech, 133, 134, 216, 225, 243, 248, 265–270, 304
  airflow, 265–269
    interruption of, 266, 269
  loudness, 151, 265
  production, 189, 267, 268
Spermatozoa, 202, 207, 282, 285, 292
Sphering of red cells, 55
Sphygmomanometer, 125
Spinal cord, 132, 138–140, 164, 219, 220, 222, 225, 227, 228
  injury, 220, 280
  transection, 325
Spinal man, 280
Spinal shock, 325
Spleen, 45, 52, 60, 61, 63, 64, 65, 69, 70, 71, 326
  blood supply, 234
  capillaries, 20
  lymphocyte production, 68–69, 234
  macrophages, 70
  platelet release, 324
  red blood cell production, 234
Stance, *see* Posture
Standard man, 15, 16, 17
Standing, 92, 223, 328
Starch, 182, 186, 188
  digestion, 185, 186
Starling effect, *see* Frank Starling principle
Starvation, 63, 201, 202, 293, 296, 297, 302, 304, 319
Static posture, *see* Posture
Statokinetic reflexes, 223, 224
Steatorrhea, 63
Step test, 320
Stercobilinogens, 64
Stercobilins, 64
Stereognosis, 157, 259, 260, 261
Sternal puncture, 61

Steroid hormones, 37, 193–195, 201, 205–208
   in saliva, 183
   mechanisms of action, 198
   metabolism, 39, 173, 194
      in fetus, 289
   synthesis, 7, 9, 37, 194, 195, 201, 206, 207, 208
   transport in plasma, 48, 194
Stimulus
   to action potential, 75–77, 82–83
      threshold (liminal), 76, 77, 79, 80, 83, 91
   to sensory receptor, 82–84
      intensity coding, 83–84
Stitch, 319
Stomach, 175, 177, 185, 189, 208
   antibacterial role, 186
   antrum, 277
   as storage organ, 297
   chief cells, 186
   control, 277
   distension, 294
   hormone production, 199
   hormones, 201, 208
   inter-organ control, 278
   irritation of, 273
   protective role, 101
   protein digestion, 186
   rate of emptying, 178
   volume, 178
   water intake sensors, 238
   within-organ control, 277
   *see also* Gastric
Storage of metabolic substrates, 296, 299, 302
   carbohydrate, 297–299
   fat, 245, 297, 299
   protein, 47, 61
Streamline flow, 100, 101, 104
Stress, 200, 201, 202, 205, 206, 216, 238, 302, 319, 324, 328, 329–331
Stretch receptors, 82, 85, 217, 222, 224, 228, 229, 238, 260
   in bladder, 241
   in JGA, 209
   muscle *see* Muscle spindles
   reflex, 218, 219, 222
   thoracic, 243
   *see also* Baroreceptors, Blood volume receptors
Striated ducts of salivary glands, 35, 36, 37, 180, 183, 276
Striated muscle, *see* Cardiac muscle, Skeletal muscle
Stroke volume of heart, 121, 205, 230, 232, 315, 317, 320, 328
Stromatin, 56, 59
Sublingual glands, 34, 179, 180
Sublingual saliva, 180, 181, 182
Submandibular glands, 34, 35, 179, 180
Submandibular saliva, 60, 179, 181, 182, 183, 275
Submucous plexus, 176, 177

Substance P, 148, 158, 163, 164, 166, 235, 236, 274, 276
Substantia gelatinosa, 138, 163, 164, 166
Succinyl choline, 80, 143
Sucking, 259, 260, 262, 263
Suckling, 200, 260, 262, 284
Sucrase, 184, 185, 186
Sucrose, 17, 160, 186
Sulphate, 21, 43, 173, 203, 207, 251
Summation, 76, 80, 81, 220
   spatial and temporal, 76
Surface tension, 126, 127
Surfactant, 126, 130
Surgical shock, 328
Swallow, 178, 262, 263
   duration, 263
   volume of, 264
Swallowing, 150, 177, 185, 188, 220, 244, 247, 257, 259, 262–264 277
   centre, 263
Sweat, 31, 33, 62, 313, 328
Sweat glands
   apocrine, 142,162, 313
   eccrine, 30, 33, 142, 144, 310, 313
Sweating, 220, 296, 304, 310–312, 313, 314, 319, 328
   loss of water and salts, 313
Sweet taste, 160
Swimming, 244, 248, 251, 320, 321
Sympathetic nerves
   cholinergic, 144, 313, 315, 328
   effect on cardiac contractility, 121
   effect on muscles of gut, 178
   development, 287
   postganglionic, 205
   preganglionic, to adrenal medulla, 205
   to blood vessels, 230, 322
   to brown fat, 288
   to gastrointestinal tract, 177, 272
   to heart, 118, 230
   to hepatic vessels, 233
   to kidney, 169, 171, 210, 234, 240
   to pulmonary veins, 233
   to salivary glands, 180, 273
   to skin blood vessels, 312
   vasodilator, 329
Sympathetic nervous system, 141–143, 144, 145, 205, 232, 304, 313, 324, 329
Synapses, 78, 81, 143–144, 159, 219
   electrical transmission in CNS, 139
   fatigue, 80
Syncope, 325, 328
Synergist muscles, 219, 222, 226
Synovial joints, 28, 257
   fluid, 17, 24, 28
Systemic circulation, 46, 100, 117–125, 126, 229
   systolic and diastolic pressures, 120, 244
Systole, 98, 119–120, 125, 231
   blood flow in coronary vessels, 234
   turbulent flow, 102

Systolic arrest, 99
Systolic blood pressure, 120, 121, 122, 125, 230, 231, 244, 293, 315, 317, 326
   in pulmonary circulation, 120

T wave of ECG, 119
T-lymphocytes, 68–70
T-system, *see* Transverse tubules
T-tubules, *see* Transverse tubules
$T_3$, *see* Tri-iodothyronine
$T_4$, *see* Thyroxine
Tachycardia, 121, 231, 304, 326, 328
Taenia coli, 177
Taste, 82, 134, 158–161, 271
   and chemical structure, 160, 161
   buds, 82, 158, 159, 160, 189, 260
   coding of concentration, 84, 160
   electrical, 160
   genetic variation, 161
   hot, 161
   preferences, 240
   primary, 160
   receptors, 82, 85, 158, 159, 240, 259
   sensation, 159
      effect of ageing, 161
      effect of drugs, 161
      effect of smoking, 161
      effect of temperature, 160
      in infants, 161
      need for saliva, 189
   stimuli
      effect on saliva flow, 272
      effect on gastric secretion, 277
   theories of sweet and bitter taste sensation, 160
   thresholds, 159, 160, 161, 240, 260
Taurocholate, 186
Teeth, 257, 263
   blood supply, 167
   deciduous, 286, 288, 289
   eruption, 286, 289
   grinding, 135, 165
   haemostasis after extractions, 323
   innervation, 77, 167
   occlusion, 258
   pain, mechanisms, 167
   permanent, 286, 287, 289
   protective action of saliva, 189
   pulps, 331
   sensation, 158, 167
Teeth apart swallow, 264
Temperature, 82, 216, 220, 314, 316, 324
   power law, 84
   and hunger, 295
   axillary, 309
   centres, 314
   critical, 312, 313
   effect on haemoglobin–oxygen association curve, 109, 110
   effect on haemostasis, 324
   effect on Nernst potential, 11
   effect on rate of reaction, 309
   effect on taste sensation, 160

360  Index

Temperature (cont.)
  environmental, 105, 238, 309, 310, 312, 314, 317
  receptors, 82, 84, 85, 158, 164
    central, 311
    cold, 158, 311
    hypothalamic, 199, 216
    in mouth, 259, 260
    skin, 158, 216, 311
    warm, 158
  regulation, 106, 212, 247, 288, 309–314
  sensation, 138, 157, 158
  and pain, 164, 166
  set level, 311–314, 317, 318
  see also Core temperature
Temporomandibular joint, 28, 257, 258, 262
  pain, 165, 135
Tendon jerk, 222
Tendon receptors, 156, 224, 228, 258
Tension velocity curve, 89
Tension
  in muscle, 90, 91, 98
  of a gas in solution, 11, 108
  of vocal cords, 267
Tension/volume curve, 98
Testes, 37, 194, 199, 202, 206, 207, 208, 288, 290, 291, 292, 297
  interstitial cells (of Leydig), 203, 207, 282, 288
Testosterone, 48, 183, 194, 195, 202, 203, 205, 206, 207, 276, 282, 287, 289, 290, 291, 292, 293
  blood concentration, 291
  metabolism and excretion, 207
  in fetus, 289
Testosterone-binding protein (TBG), 44
Tetanic contraction, 89, 91, 92, 98, 267
Tetanus, see tetanic contraction
Tetany, 252, 305
Thalamus, 84, 133, 134, 137, 138, 140, 145, 152, 156, 160, 164, 166, 227
Thermal insulation factor, 310
Thermogenesis, 295
  non-shivering, 288, 312
  non-shivering, in infants, 312
Thiocyanate, 17, 181–182, 184, 189
Thirst, 135–136, 189, 209, 216, 220, 238, 239, 314, 324, 325, 326, 327
Thirst centre, 136, 238, 239
Thorax, 127, 130
  compliance, 130, 243
  negative pressure in, 124
Threshold stimulus, 75, 76, 77, 79, 83, 91, 94, 98, 118
  skeletal muscle, 80
Thrombin, 48, 49, 50, 51, 71, 186, 323, 324
Thrombocythaemia, 70
Thrombocytopenia, 70, 72
Thrombokinase (thromboplastin), 49, 322, 323

Thrombopoietin, 71
Thrombosthenia, 72
Thrombosthenin, 71, 72, 87, 323
Thromboxane A2, 212, 236, 322, 323
Thromboxane endoperoxidase (PGG2), 323
Thrombus formation (thrombosis), 51, 324
Thymic-dependent cells, see T-lymphocytes
Thymus gland, 61, 68, 69
Thyroglobulin, 37, 38
Thyroid gland, 37, 184, 199, 201, 202, 203
  colloid, 202, 203
  fetal, 288
  follicles, 37, 202, 203
  parafollicular C cells (ultimobranchial tissue), 204, 307
Thyroid hormones, 110, 195, 196, 198, 200, 202, 203–204, 284, 287–288, 290, 293, 295–296, 301, 302, 307, 312, 314, 328, 330
  binding by plasma proteins, 203
  mechanism of action, 196–198
  synthesis, 38
  see also Thyroxine, Tri-iodothyronine
Thyroid stimulating hormone (TSH), 136, 193, 194, 197, 200, 201, 202, 203, 208, 211
Thyrotrophic hormone, see Thyroid stimulating hormone
Thyrotrophin releasing hormone, 136, 192, 197, 200, 201, 202, 203
Thyroxine, 38, 48, 192, 197, 196, 203, 276, 283, 295
Thyroxine binding globulin (TBG), 44, 48, 203
Thyroxine binding pre-albumin, 203
Tickle, 157
Tidal volume, 128, 243, 315
Tight junctions, 5, 20, 26, 34, 36, 79, 84, 93, 96, 146
Timed vital capacity (FEV1), 129, 131
Tissue fluid, 15, 17–18, 20–23, 24, 28, 124, 317, 328
  see also Interstitial fluid
  composition, 17, 315
Tissues
  macrophages, 7, 69
  partial pressures
    of carbon dioxide, 111
    of oxygen, 109
  perfusion, 324, 325
  pressure, 36, 171
  products of damage to, 22, 49, 65, 165, 167, 211, 235, 236, 286, 328
  regenerative and non-regenerative, 286
  water content, 15
Tone, 92, 93, 95, 175, 220, 223
Tongue, 84, 158, 179, 180, 257, 258
  effect on oral resonator, 268

Tongue (cont.)
  papillae, 158, 180
  role in mastication, 257, 259, 260
  taste buds, 158
  touch and temperature receptors, 158
Topographical localisation, 133, 134, 137, 149
  in sensory cortex, 133
  of colour, 149
  of pain sensation, 164
  of sound frequencies, 152
  trigeminal fibres, 166
Total body water, 15, 16, 17, 185, 189, 314
  measurement, 15
Total iron carrying capacity of the blood, 47
Total lung capacity, 129
Total peripheral resistance (TPR), 229
Touch sensation, 78, 82, 153, 157, 166
  discrimination, 84
  receptors, 82, 84, 137, 164, 222
    in mouth, 258, 259, 260, 271
  fine, 138
  stimuli, 222, 225
Trachea, 101, 126, 243
  resistance to airflow, 104
  turbulence, 101
Trained athletes, 121, 320
Training, 297, 319, 320
  of breathing, 266
  effect on nitrogen balance, 296
  of masticatory muscles, 262
Transamination, 172, 296, 304
Transcellular fluids, 16, 17, 24–29
Transcortin, 44, 48, 205
Transduction of nerve stimuli, 82, 145, 150, 274
  of dental pain, 167
  of smell, 162
  of sound, 150
Transfer RNA, 6
Transferrin, 44, 47, 61, 62, 63
Transmission cells for pain sensation, 164, 166
Transmitter substances, 79, 83, 94, 139, 143, 144, 164, 228, 236, 246
Transmural pressure, 123, 124, 229
Transplant rejection, 69
Transport maximum systems, 39, 172, 173
Transport in blood, 46–53, 56, 57, 65, 107–113
Transverse tubules, 85, 91, 96
Trauma, 70, 200, 272, 302, 324, 330
Traumatic shock, 304, 328
Treppe, 91
TRH, see Thyrotrophin releasing hormone
Tri-iodothyronine, 38, 192, 196, 198, 203, 295
  binding in plasma, 203
Triads, 85

Tricarboxylic acid cycle, 9, 298–299, 317
Trigeminal nerve, 138, 139, 141, 161, 165, 258
  neuralgia, 81
  nuclei, 165–166, 261
  spinal tract, 165–166
Triglyceride lipase, 301, 302
Triglyceride synthesis, 41, 301
Triglycerides, 41, 47, 92, 183, 186, 188, 288, 298, 299, 300, 301, 302, 303
1,24,25-Trihydroxycholecalciferol, 211
Triple response, 165, 235, 236
Tritiated water, 16
Tropomyosin, 85, 86, 87, 94
Troponin, 85, 86, 87, 94, 319
Trypsin, 48, 184, 185, 186
Trypsinogens, 186, 187
TSH, see Thyroid stimulating hormone
Tube radius, effect on resistance, 104
d-Tubocurarine, 80, 128
Tumour cells, 22
Tumours, effect on nerves, 78
Turbulent flow, 101, 102, 103, 104, 125, 267
Two-point discrimination, 84, 157
Tyrosine, 38, 144, 186, 202, 203, 205, 299

U wave, of electocardiogram 119
Ultimobranchial tissue, 199, 204
Unconsciousness, 248
Unit membrane, see Cell membranes
Universal donors and recipients, 59
Unloading reflex, 261
Upper respiratory tract, see Airways
Uraemia, 189
Urea, 21, 27, 31, 34, 36, 190, 296, 313, 326
  concentration in plasma, 43
  diffusion in distal tubule, 238
  in saliva, 183, 189
  in urine, 238
  reabsorption in kidney, 172, 173
Urease, 182
Uric acid
  metabolism, 7
  in saliva, 183, 189
Uricosomes, 8
Urination, 174, 240
Urine, 15, 16, 39, 65, 105, 174, 304, 319, 326, 330
  alkaline, 254, 256
  calcium excretion, 306
  concentration of, 107, 173–174, 236, 238
  production, 39, 173, 328
  glucose, 172
  haemoglobin loss, 56
  hypertonic, 107
  nitrogen, 296
  pH, 254

Urine (cont.)
  phosphate excretion, 308
  potassium, 173
  salt content, 238
  urea content, 238
  volume, 174, 200
    daily, 174
Urobilinogens, 64
Urobilins, 64
Urokinase, 324
Uterus, 207–208, 281–283, 290, 292
  contraction, 217, 218
  endometrium, 207–208, 281–282
  innervation, 141
  smooth muscle, 95, 200, 208, 212, 281, 283
Utilisation time, 76

Vacuoles see Vesicles
Vagina, 281, 282, 290, 292
Vagus nerve, 138–139, 141, 160, 165
  actions on cardiovascular system, 99, 118, 230, 320, 325, 326, 328
  actions on gastrointestinal system, 177, 204, 208, 264, 277, 302
  actions on respiratory system, 242–243, 247
Valsalva's manoeuvre, 244
Van't Hoff equation, 12
Varicosities, 81, 94
Vasa recta, 105, 106, 107, 169, 200, 238, 240
Vascular obstruction, 103
Vascular permeability, 67, 235
Vascular smooth muscle, 145, 212, 312
Vasoactive intestinal peptide (VIP), 201, 209, 232, 273, 274
Vasoconstriction, 46, 123, 138, 141, 145, 205, 230, 232, 239, 244, 246, 247, 314, 322, 323, 324, 328
  cerebral, 233
  due to
    adrenaline, 145, 205, 231, 328
    angiotensin II, 209, 236, 239
    hypothalamic hormones, 200
    kinins, 46
    noradrenaline, 231
    thromboxane, 212
    vasopressin, 200
  renal, 171
Vasodilatation, 46, 67, 145, 208, 211, 230, 232, 234, 314, 318, 324, 326, 328
  due to
    adrenaline, 231
    bradykinin, 50, 53, 236
    VIP, 209
  in salivary glands, 182, 273
  sympathetic, 205
Vasomotor centre, 141, 230, 240
Vasopressin see Antidiuretic hormone
Vasovagal attack, 328
Veins, 122, 145
  as blood reservoirs, 123, 124, 326

Veins (cont.)
  great see Venae cavae
  pressure in, 124, 326
    see also Central venous pressure
  proportion of blood volume, 123
  valves, 124
  wall thickness, 123
Venae cavae, 118, 119, 120, 122, 123, 238
Veno-arterial blood shunt, 131
Venoconstriction, 123, 211, 230, 233, 317, 325, 326
Venodilatation, 124
Venous pulse, 120, 124–125
Venous return, 98, 124, 230, 231, 232, 234, 244, 280, 317, 320, 326, 328
Ventilation, 128–130, 131, 242–249, 251–253, 315–320
  alveolar, 129, 131, 248
  control, 242–249
  depth, 242, 243, 317
  in exercise, 101, 315–320
  inadequate, 131
  maximum voluntary, 129–131
  minute volume, 130
  rate, 243, 319, 320, 324
  total, 129
  see also Respiration, Respiratory
Ventilation/perfusion ratio, 131
Ventricles of the brain, 24
Ventricles of the heart, 117–121, 124, 230
  contraction, 98, 118–119, 125
  filling of, 119, 328
  muscle, 97–99
  pressures, 118, 120, 121, 122, 124
  receptors, 231, 238
Venules, 18, 122, 123, 145, 322
  ratio of wall thickness to diameter, 123
Vesicles, 7, 19, 20, 21, 29, 30, 37, 38
  synaptic, 74, 79, 81, 94, 144, 180
Vestibular apparatus, 82, 137, 153, 155, 224, 225, 227
Vibration sense, 82, 157, 158
Villikinin, 209, 279
VIP, see Vasoactive intestinal peptide 273
Viscera
  blood flow, 112
  control, 141, 277
  pain, 165
Viscerorceptors, 82
Viscosity
  effect on flow, 101
  effect on resistance, 104
  effect on turbulence, 101
  of blood, 46, 54, 56, 102, 103, 104, 328
  of gastric contents, 278
Vision, 83–84, 133, 134, 145–150
  acuity, 148
  central and peripheral, 147
  colour, 148

Vision (*cont.*)
  defects due to lens or cornea, 146
  near, 149
  *see also* Accommodation
  night, 148
  perception, 149
  photopic, 148
  receptors, 82, 137, 146, 224, 227
  scotopic, 148
Visual cortex, 146, 149, 152, 224
Visual field, 146, 149
Visual purple, 148
Visual receptors, 82, 137, 146, 224
Vital capacity, 128, 129, 131
  timed ($FEV_1$), 131
Vitamin $A_1$, 148
Vitamin B in saliva, 183
Vitamin $B_{12}$, 63, 186, 188
  absorption in ileum, 63, 186, 188
Vitamin C (ascorbic acid)
  in saliva, 183
  in taste buds, 159
Vitamin D, 41, 48, 173, 194, 209, 211, 305, 307
  transport, 44
  metabolites, 198, 211, 305
Vitamin K, 51, 183
Vitamins
  absorption in gut, 184, 188
  in plasma, 43
  in saliva, 183
Vitreous humour, 27, 146, 148
Vocal cords, 101, 263, 265–268
  false, 265, 266, 267, 268
  length before and after puberty, 267, 293
  tension, 266
Vocalisation, 265–270

Volume receptors, 326, 327
  in circulation, 238
Voluntary motor activity, 220, 222, 225–228
Vomiting, 138, 178, 180, 185, 220, 244, 247, 251, 254, 264, 273, 278, 313
Von Willebrand factor, 51, 72, 322, 323
Vowels, 266, 268

Walking, 218, 225, 226, 315, 319
Warm receptors, 311
Warm shock, 324
Warming, response to, 312
Water
  absorption in gut, 24, 41, 184, 188, 206
  balance, 15, 189, 202, 206, 237, 283, 330
  content of tissues, 15, 57
  daily intake, 188
  diffusion through epidermis, 310
  intake, 15, 161, 216, 238, 297, 304
  sensors, 238
  loss, 15, 188, 238, 297, 304, 319
    diarrhoea and vomiting, 185
    evaporative, 16
    from lungs, 16, 266
    in haemorrhage, 325
    in sweating, 313
  reabsorption in kidney, 39, 171, 172, 200, 238
  retention, 202, 207, 236, 239, 319, 326
Water layer unstirred, 41, 187
Weal, 22
Weber–Fechner Law, 84

Weight, 294, 321
  gain, 240, 297
  loss, 240, 331
Whispering, 267
White blood cell count, 64, 65, 66, 206, 290, 330
White blood cells, 22, 23, 28, 29, 53, 54, 60, 64–70, 179, 182, 185, 206, 327
  antigens, 60
  axial flow, 102
  functions, 65
  granules, 66, 67
  loss in haemorrhage, 325
  maturation, 65
  turnover, 61, 65
White tache, 235
Whole saliva, 178–179, 180–183
  composition, 181
  proportions of gland secretions, 179
  *see also* Saliva
Wilson's disease, 47
Withdrawal reflex, 220
Work, 88, 295, 299, 313, 317, 320, 321
  of breathing, 104, 130, 243, 319
  muscular, 310, 317, 318
Wound shock, 328

X chromosomes, 51, 285
Xerostomia (dry mouth), 276

Y chromosome, 285, 288
Yawning, 150, 244

Z lines, 85, 87, 94, 96
Zonula occludens, 19
Zonule, 5,
  of lens, 146
Zymogen granules, 7

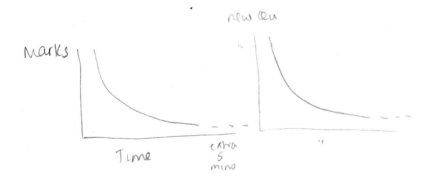

∴ don't exceed allotted time!!
make rough essay plans to look
at day before exam.